MW01193736

OXFORD STATISTICAL SCIENCE SERIES

OXFORD STATISTICAL SCIENCE SERIES

---

1. A.C. Atkinson: *Plots, transformations, and regression*
2. M. Stone: *Coordinate-free multivariable statistics*
3. W.J. Krzanowski: *Principles of multivariate analysis: a user's perspective*
4. M. Aitkin, D. Anderson, B. Francis, and J. Hinde: *Statistical modeling in GLIM*
5. Peter J. Diggle: *Time series: a biostatistical introduction*
6. Howell Tong: *Non-linear time series: a dynamical system approach*
7. V.P. Godambe: *Estimating functions*
8. A.C. Atkinson and A.N. Donev: *Optimum experimental designs*
9. U.N. Bhat and I.V. Basawa: *Queuing and related models*
10. J.K. Lindsey: *Models for repeated measurements*
11. N.T. Longford: *Random coefficient models*
12. P.J. Brown: *Measurement, regression, and calibration*
13. Peter J. Diggle, Kung-Yee Liang, and Scott L. Zeger: *Analysis of longitudinal data*
14. J.I. Ansell and M.J. Phillips: *Practical methods for reliability data analysis*
15. J.K. Lindsey: *Modelling frequency and count data*
16. J.L. Jensen: *Saddlepoint approximations*
17. Steffen L. Lauritzen: *Graphical models*
18. A.W. Bowman and A. Azzalini: *Applied smoothing methods for data analysis*
19. J.K. Lindsey: *Models for repeated measurements Second editon*
20. Michael Evans and Tim Swartz: *Approximating integrals via Monte Carlo and deterministic methods*
21. D.F. Andrews and J.E. Stafford: *Symbolic computation for statistical inference*
22. T.A. Severini: *Likelihood methods in statistics*

Likelihood Methods in Statistics

THOMAS A. SEVERINI

Department of Statistics
Northwestern University
Evanston, Illinois

OXFORD
UNIVERSITY PRESS

Great Clarendon Street, Oxford OX2 6DP

Oxford University Press is a department of the University of Oxford.
It furthers the University's objective of excellence in research, scholarship,
and education by publishing worldwide in

Oxford New York

Athens Auckland Bangkok Bogotá Buenos Aires Calcutta
Cape Town Chennai Dar es Salaam Delhi Florence Hong Kong Istanbul
Karachi Kuala Lumpur Madrid Melbourne Mexico City Mumbai
Nairobi Paris São Paulo Singapore Taipei Tokyo Toronto Warsaw

with associated companies in Berlin Ibadan

Published in the United States
by Oxford University Press Inc., New York

First published 2000

A catalogue record of this book is available from the British Library

Library of Congress Cataloging in Publication Data

ISBN 0 19 850650 3 (Hbk)

Typeset by Newgen Imaging Systems (P) Ltd., Chennai, India
Printed in Great Britain
on acid-free paper by

T.J. International Ltd, Padstow

To Karla

Preface

Many commonly used statistical methods are based on the likelihood function; these include maximum likelihood estimates and tests and confidence intervals based on the likelihood ratio statistic. A general theory of likelihood-based inference must rely on asymptotic approximations. The classical theory of likelihood inference is based on first-order asymptotic theory, such as that given by the central limit theorem, along with a higher-order asymptotic theory based on Edgeworth series expansions. More recently, an alternative approach has been developed that has several important features. One is the recognition that it is desirable to condition on relevant ancillary statistics. Another is that probability approximations are based on saddlepoint and closely related approximations that generally have very high accuracy. A third aspect is that, for models with nuisance parameters, inference is often based on marginal or conditional likelihoods, or approximations to these likelihoods. These methods have been shown to often yield substantial improvements over the more classical methods.

The goal of this book is to provide an introduction to this modern approach to likelihood inference, suitable for students and statisticians who are not specialists in either likelihood theory or higher-order asymptotics. The book also tries to provide an up-to-date account of recent results in the field, which has been undergoing rapid development. The book is designed for readers with at least some background in classical likelihood theory, although, for completeness, the classical theory is covered here as well. Adequate preparation is a knowledge of graduate-level statistical theory, such as that contained in the books Bickel and Doksum (1977) and Casella and Berger (1990).

Chapters 1 through 3 provide a summary of the concepts of theoretical statistics, asymptotic approximations and likelihood-based methods, respectively, that are used in the later chapters. Chapter 4 discusses the classical first-order theory of likelihood-based inference and Chapter 5 considers classical higher-order theory, such as Edgeworth expansions and Bartlett corrections. The more recent theory of likelihood-based inference is covered in Chapters 6 through 9. The asymptotic theory of conditional inference is presented in Chapter 6, including a discussion of conditional distribution of the maximum likelihood estimate and related topics, such as approximate ancillarity and the approximation of sample space derivatives. The modified signed likelihood ratio statistic is derived in Chapter 7, along with several approximations to that statistic. Conditional and marginal likelihood functions are presented in Chapter 8 and Chapter 9 considers the modified profile likelihood function along with several approximations.

The emphasis in the book is on the development of statistical methods and a description of the underlying theory rather than on the statement and proof of precise mathematical results. At the end of each chapter, the interested reader can find references to more rigorous results, if available. Each chapter also contains numerous exercises, designed to illustrate the results presented in the chapter. For instructors, or those engaged in independent study, solutions to all of the exercises are available from the author.

I am grateful to Jason Osborne and Shuguang Huang for providing many important comments on an earlier version of the manuscript. The support of the National Science Foundation is also gratefully acknowledged.

Evanston T. A. S.
January 2000 severini@northwestern.edu

Contents

1 Some basic concepts 1

1.1 Introduction 1
1.2 Exponential family models 4
1.3 Transformation models 7
1.4 Cumulants 11
1.5 Sufficiency 16
1.6 Ancillary statistics 19
1.7 Normal distribution theory 22
1.8 Discussion and references 23
1.9 Exercises 24

2 Large-sample approximations 27

2.1 Introduction 27
2.2 Central limit theorem 28
2.3 Edgeworth series approximations 31
2.4 Saddlepoint approximation of densities 37
2.5 Approximation of integrals and sums 40
2.6 Saddlepoint approximations for distribution functions 46
2.7 Saddlepoint approximations for lattice variables 49
2.8 Saddlepoint approximations for multivariate distributions 53
2.9 Stochastic asymptotic expansions 54
2.10 Approximation of conditional distributions 58
2.11 Laplace approximations 66
2.12 Discussion and references 68
2.13 Exercises 69

3 Likelihood 73

3.1 Introduction 73
3.2 Some properties of the likelihood function 75
3.3 The likelihood principle 77
3.4 Regular models 80
3.5 Log-likelihood derivatives 85
3.6 Information 88
3.7 Methods of inference 96
3.8 Discussion and references 100
3.9 Exercises 102

4	**First-order asymptotic theory**	105
4.1	Introduction	105
4.2	Maximum likelihood estimates	105
4.3	The likelihood ratio statistic	113
4.4	The score and Wald statistics	120
4.5	Confidence regions	123
4.6	The profile likelihood function	126
4.7	Nonregular models	129
4.8	Discussion and references	133
4.9	Exercises	134
5	**Higher-order asymptotic theory**	138
5.1	Introduction	138
5.2	Some preliminary results	138
5.3	Maximum likelihood estimates	141
5.4	Likelihood ratio statistic	154
5.5	Saddlepoint approximations	163
5.6	Discussion and references	171
5.7	Exercises	172
6	**Asymptotic theory and conditional inference**	175
6.1	Introduction	175
6.2	Log-likelihood derivatives	176
6.3	Conditional distribution of maximum likelihood estimates	183
6.4	Stable inference	196
6.5	Approximation of the conditional model	204
6.6	Approximate ancillarity	209
6.7	Approximation of sample space derivatives	218
6.8	Discussion and references	234
6.9	Exercises	235
7	**The signed likelihood ratio statistic**	238
7.1	Introduction	238
7.2	Normalizing transformations	239
7.3	One-parameter models	241
7.4	Scalar parameter of interest in the presence of a nuisance parameter	248
7.5	Approximations to R^*	261
7.6	Discussion and references	274
7.7	Exercises	275
8	**Likelihood functions for a parameter of interest**	278
8.1	Introduction	278
8.2	Conditional likelihood functions	279
8.3	Marginal likelihood functions	298
8.4	Integrated likelihood functions	306

8.5 Inference based on a pseudo-likelihood function 309
8.6 Discussion and references 319
8.7 Exercises 320

9 The modified profile likelihood function 323
9.1 Introduction 323
9.2 Profile likelihood 323
9.3 Modified profile likelihood 327
9.4 Calculation of the modified profile likelihood without an explicit nuisance paramter 337
9.5 Approximations to the modified profile likelihood 340
9.6 Discussion and references 352
9.7 Exercises 352

Appendix: Data sets used in the examples 354

References 362

Author index 374

Subject index 377

1 Some basic concepts

1.1 Introduction

The purpose of this chapter is to review some basic concepts of probability and statistics that will play a central role in the subsequent chapters, as well as to introduce some notation and terminology.

In statistical inference observational or experimental data are modeled as the observed values of certain random variables. Let Y denote a random variable with density function $p(y)$; here the term 'density function' will be used to mean a density function with respect to either Lebesgue measure or counting measure. For simplicity however quantities such as $\mathrm{E}(Y)$ will always be written as integrals; for example,

$$\mathrm{E}(Y) = \int p(y)\, dy.$$

In such cases the range of integration will be understood to be the support of the distribution, the set of y such that $p(y) > 0$; the same notation will be used when Y is a vector, $Y = (Y_1, \ldots, Y_n)$. In general, upper-case letters will be used to denote random variables, with the corresponding lower-case letter used to denote particular values of the random variable.

In a statistical analysis the function $p(\cdot)$ is typically unknown and the goal is to draw conclusions about $p(\cdot)$ on the basis of observing $Y = y$. The starting point for such an analysis is generally the specification of a *model* for the data. A model consists of a set of possible density functions $\mathcal{P}$ such that we are willing to proceed as if the true density function lies in that set. In parametric statistical inference, the set of possible density functions is assumed to be parameterized by a finite-dimensional parameter θ taking values in a set Θ so that

$$\mathcal{P} = \{p(\cdot; \theta) \colon \theta \in \Theta\}.$$

In this formulation, Θ is taken to be a subset of d-dimensional Euclidean space for some integer d. The function $p(y; \theta)$ is called the model function.

The symbol p will generally be used to represent a probability density function with the argument of the function indicating the random variable under consideration, at least when there is no danger of confusion. For instance, $p(y; \theta)$ will be used to denote the density of a random variable Y and $p(z; \theta)$ will be used to denote the density of a random variable Z.

In a parametric model with parameter θ, quantities such as probabilities and expectations will depend on the value of θ under consideration. To emphasize this we will often write probabilities as $\Pr(\cdot; \theta)$ and expectations as $\mathrm{E}(\cdot; \theta)$.

Alternatively, the data can be modeled nonparametrically. A nonparametric model is one which does not have a parametric representation. In this book we focus entirely on parametric models.

A statistical model is often developed by analyzing the process that gives rise to the data. Of course, any model-building process of this type involves some idealization.

Example 1.1 Measurement model
Consider the problem of trying to determine the value of a certain physical constant, for example, the weight of an object. Let μ denote the unknown value of that constant. Let Y denote an experimental determination of μ such that Y is equal to μ plus some experimental 'error'. Hence, we might write

$$Y = \mu + \epsilon$$

where ϵ is a random variable. Often it is assumed that the distribution of the error term is known except for a scale factor. In these cases we may write

$$Y = \mu + \sigma\epsilon$$

where $\sigma > 0$ is an unknown scale parameter and ϵ is a random variable with a known distribution, for example, a standard normal random variable. Then the density function for Y is given by $f((y - \mu)/\sigma)/\sigma$ where $f(\cdot)$ denotes the density function of ϵ. In the most general case μ can take any value in $\Re$ so that the model for Y may be written

$$\mathcal{P} = \{p(\cdot; \theta)\colon\ p(y; \theta) = f((y - \mu)/\sigma)/\sigma,\ \theta = (\mu, \sigma) \in \Re \times \Re^{+}\}.$$

Suppose now that instead of a single determination of Y we have determinations $Y_1, \ldots, Y_n$, such that

$$Y_j = \mu + \sigma\epsilon_j, \quad j = 1, \ldots, n$$

where $\epsilon_1, \ldots, \epsilon_n$ each have known density function $f(\cdot)$. In this case we may take $Y = (Y_1, \ldots, Y_n)$ and, under the further assumption that $\epsilon_1, \ldots, \epsilon_n$ are independent random variables, the density function of Y may be written

$$p(y; \theta) = \prod_{j=1}^{n} f((y_j - \mu)/\sigma)/\sigma.$$

A formal statement of the model $\mathcal{P}$ is now possible, although it is rarely necessary to carry out that step. ∎

Example 1.2 Poisson process
Suppose that certain events occur randomly through time such that the following properties hold. The stochastic properties of the events do not change with time,

the chance of two or more events occurring simultaneously is negligible, the chance of one event in an small interval is approximately proportional to the length of the interval, and the number of events occurring before a time t is statistically independent of the number events occurring in any time interval taking place after time t. More precisely, let $N(t)$ denote the number of events that occur up to and including time t. Assume that

(a) for any $0 < t_1 < \cdots < t_m$, the random variables $N(t_1) - N(0), \ldots, N(t_m) - N(t_{m-1})$ are independent
(b) $\lim_{h \to 0} \Pr\{N(t+h) - N(t) \geq 2\}/h = 0$ for all t
(c) $\lim_{h \to 0}[\Pr\{N(t+h) - N(t) = 1\} - \lambda h] = 0$ for some $\lambda > 0$, for all t.

Then the stochastic process $\{N(t)\colon\ t > 0\}$ is a Poisson process.

Suppose that the process is observed until n events have occurred and let $Y_1, \ldots, Y_n$ denote the inter-event times. Then $Y_1, \ldots, Y_n$ are independent random variables each having an exponential distribution with rate parameter λ. The density function of $Y = (Y_1, \ldots, Y_n)$ is given by

$$p(y; \lambda) = \lambda^n \exp\{-\lambda(y_1 + \cdots + y_n)\}, \quad y_j > 0, \ \ j = 1, \ldots, n. \qquad \blacksquare$$

Clearly the parameterization of a model is not unique. For instance, in the Poisson process example, the distribution of Y could also be parameterized by the mean of the Y_j, $\mu = 1/\lambda$; the density function of Y in terms of μ is given by

$$\tilde{p}(y; \mu) = \mu^{-n} \exp\{-(y_1 + \cdots + y_n)/\mu\}, \quad y_j > 0, \ \ j = 1, \ldots, n.$$

More generally, given a model with model function $p(y; \theta)$, the model may be parameterized by any one-to-one function of θ. One requirement of a parameterization is that it be *identifiable*; that is, there must be exactly one value of the parameter θ corresponding to a given element of $\mathcal{P}$.

Selection of a parameterization is arbitrary, although in certain cases a particular parameterization may be more useful if, for example, it simplifies the interpretation of the results. A desirable property of methods of inference is that the conclusions drawn about the distribution of Y do not depend on the parameterization used.

In many cases the random variables under consideration do not all have the same probability distribution. In particular, we may observe independent random variables $Y_1, \ldots, Y_n$ such that Y_j has density function $p_j(\cdot; \theta)$. Often there are constants $x_1, \ldots, x_n$, which may be vectors, such that $p_j(\cdot; \theta) = p(\cdot; \theta, x_j)$. In these cases, the Y_j are often called response variables and the x_j are called explanatory variables or covariates. Primary interest is often in the relationship between Y_j and x_j.

Example 1.3 Regression models

Let $Y_1, \ldots, Y_n$ denote independent random variables such that for each $j = 1, \ldots, n$, Y_j may be written

$$Y_j = \mu_j + \sigma\epsilon_j$$

where ϵ_j is distributed according to a known density f. Let $x_1, \ldots, x_n$ denote covariates such that $\mu_j = g(x_j; \beta)$ for some known function g, where β is an unknown parameter.

If $g(x_j; \beta)$ is a linear function of β, for instance $g(x_j; \beta) = x_j^T \beta$, then the model is said to be a *linear model*; otherwise it is a nonlinear model. An example of a nonlinear model is

$$g(x_j; \beta) = \frac{\beta_1 x_j}{\beta_2 + x_j},$$

which is used in enzyme kinetics; in this model x_j is a scalar.

Often f is taken to be the standard normal density leading to the normal-theory linear model when, in addition, $g(x_j, \beta)$ is linear in β. ∎

1.2 Exponential family models

A particularly useful general class of statistical models are those in the exponential family. Suppose that the density function of Y is of the form

$$p(y; \theta) = \exp\{s(y)^T c(\theta) - k(\theta) + D(y)\}, \quad y \in \mathcal{Y} \tag{1.1}$$

where $\mathcal{Y}$ is a set not depending on θ, s is a function defined on $\mathcal{Y}$ taking values in a subset of $\Re^m$, c is a function defined on Θ taking values in a subset of $\Re^m$ and D is a function defined on $\mathcal{Y}$. Then the model is said to be in the exponential family. Suppose that $\Theta \in \Re^d$ and that the set $\{c(\theta) : \theta \in \Theta\}$ is also d-dimensional. Then $\mathcal{P}$ is said to be a (m, d)-exponential family model; if $d = m$ then the model is said to be of full rank or, simply, *full*; if $d < m$ then the model is said to be *curved*.

Note that the representation (1.1) for a given density $p(y; \theta)$ is not unique, since, for example, we may replace $s(y)$ by $s(y)/2$ and $c(\theta)$ by $2c(\theta)$. This nonuniqueness however does not cause any problems for inference about θ. The dimension m of the vectors $s(y)$ and $c(\theta)$ may be reduced if either $s(y)$ or $c(\theta)$ satisfies a linear constraint; we will assume that the representation (1.1) is minimal in the sense that the dimension m is the smallest possible.

Example 1.4 Bernoulli random variables
Let $Y_1, \ldots, Y_n$ denote independent random variables such that for each $j = 1, \ldots, n$,

$$\Pr(Y_j = 1; \theta) = 1 - \Pr(Y_j = 0; \theta) = \theta.$$

The density function of Y_j may be written

$$p(y_j; \theta) = \theta^{y_j}(1 - \theta)^{1 - y_j}, \quad y_j = 0, 1;$$

the parameter θ takes values in the set $\Theta = (0, 1)$.

The density function of the vector $Y = (Y_1, \dots, Y_n)$ is given by

$$p(y; \theta) = \theta^{\sum y_j}(1-\theta)^{n-\sum y_j}, \quad y_j = 0, 1, \quad j = 1, \dots, n;$$

the function $p(y; \theta)$ may be written

$$p(y; \theta) = \exp\left\{\left(\sum y_j\right)\log(\theta/(1-\theta)) + n\log(1-\theta)\right\}, \quad y \in \{0, 1\}^n$$

so that this model is in the one-parameter exponential family with $c(\theta) = \log(\theta/(1-\theta))$, $s(y) = \sum y_j$, $k(\theta) = n\log(1-\theta)$, $D(y) = 0$, and $\mathcal{Y} = \{0, 1\}^n$. ∎

Example 1.5 Normal-theory measurement model
Consider the measurement model discussed in Example 1.1 with $\theta = (\mu, \sigma) \in \Re \times \Re^+$ and $f(\cdot) = \phi(\cdot)$, the standard normal density. Then the density of $Y = (Y_1, \dots, Y_n)$ is given by

$$\begin{aligned} p(y; \theta) &= \prod_{j=1}^{n} \phi((y_j - \mu)/\sigma)/\sigma \\ &= (2\pi\sigma)^{-n/2}\exp\left\{-\frac{1}{2\sigma^2}\sum(y_j-\mu)^2\right\}, \quad y \in \Re^n. \end{aligned}$$

This function can be rewritten as

$$\exp\left\{-\frac{1}{2\sigma^2}\sum y_j^2 + \frac{\mu}{\sigma^2}\sum y_j - \left(n\frac{\mu^2}{2\sigma^2} + \frac{n}{2}\log\sigma^2\right) - \frac{n}{2}\log(2\pi)\right\}.$$

This is clearly of the form (1.1) with $s(y) = (\sum y_j^2, \sum y_j)$, $c(\theta) = (-1/(2\sigma^2), \mu/\sigma^2)$, $k(\theta) = n\mu^2/2\sigma^2 + (n/2)\log\sigma^2$ and $D(y) = (n/2)\log(2\pi)$. This is a full two-parameter exponential family model. ∎

Example 1.6 Normal distributions with common mean
Let $Y_1, \dots, Y_n$ follow the normal-theory measurement model considered in Example 1.5, with mean μ and standard deviation σ_1 and let $X_1, \dots, X_m$ follow the same model but with mean μ and standard deviation σ_2; assume further that the Y_j and X_j are independent. Hence, the the two sets of observations have different variances but a common mean; the parameter for the model is $\theta = (\sigma_1, \sigma_2, \mu) \in \Theta = \Re^+ \times \Re^+ \times \Re$.

The density function for $Y = (Y_1, \dots, Y_n)$ may be written

$$\begin{aligned} p(y; \theta) = \exp\Bigg\{ &-\frac{1}{2\sigma_1^2}\sum y_j^2 + \frac{\mu}{\sigma_1^2}\sum y_j - n\left(\frac{\mu^2}{2\sigma_1^2} + \frac{1}{2}\log\sigma_1^2\right) \\ &- \frac{n}{2}\log(2\pi)\Bigg\} \end{aligned}$$

and the density function for $X = (X_1, \ldots, X_m)$ may be written

$$p(x; \theta) = \exp\left\{ -\frac{1}{2\sigma_2^2} \sum x_j^2 + \frac{\mu}{\sigma_2^2} \sum x_j - m\left(\frac{\mu^2}{2\sigma_2^2} + \frac{1}{2}\log\sigma_2^2\right) - \frac{m}{2}\log(2\pi)\right\}.$$

It follows that the joint density of (Y, X) is of the form

$$\exp\left\{ -\frac{1}{2\sigma_1^2}\sum y_j^2 - \frac{1}{2\sigma_2^2}\sum x_j^2 + \frac{\mu}{\sigma_1^2}\sum y_j + \frac{\mu}{\sigma_2^2}\sum x_j - \left(n\frac{\mu^2}{2\sigma_1^2} + \frac{n}{2}\log\sigma_1^2 + m\frac{\mu^2}{2\sigma_2^2} + \frac{m}{2}\log\sigma_2^2\right) - \frac{n}{2}\log(2\pi) - \frac{m}{2}\log(2\pi)\right\}.$$

This density is of the form (1.1) with $c(\theta) = (-1/(2\sigma_1^2), \mu/\sigma_1^2, -1/(2\sigma_2^2), \mu/\sigma_2^2)$.

Note that, although $c(\theta)$ has four components, θ is three-dimensional. It follows that this model is a (4, 3)-curved exponential family model. ∎

Note that, as illustrated by Examples 1.4 and 1.5, if $Y_1, \ldots, Y_n$ are independent identically distributed random variables and the distribution of Y_j is in the exponential family, then the distribution of $Y = (Y_1, \ldots, Y_n)$ is in the exponential family. This fact follows directly from the form (1.1) of the density function of Y_j.

Suppose that the model is a full exponential family model and consider the parameterization $\eta = c(\theta)$. The parameter η is said to be the *natural* or *canonical* parameter of the distribution. The density of Y in terms of η may be written

$$\exp\{s(y)^T\eta - k(\theta(\eta)) + D(y)\} = \exp\{s(y)^T\eta - k_0(\eta) + D(y)\} \tag{1.2}$$

where $k_0(\eta) = k(\theta(\eta))$. Note that the function k_0 is given by

$$k_0(\eta) = \log \int \exp\{s(y)^T\eta + D(y)\}\,dy$$

and the set of all η for which this integral is finite is called the natural parameter space of the model; the natural parameter space may be viewed as the largest set of η values for which (1.2) defines a probability density. The statistic $s(Y)$ is called the *canonical statistic* of the model. The canonical statistic may be used to define another convenient parameter, the *mean parameter*, given by

$$\mu \equiv \mu(\theta) = \mathrm{E}\{s(Y); \theta\}.$$

Example 1.7 Bernoulli random variables
Consider the Bernoulli random variables $Y_1, \ldots, Y_n$ defined in Example 1.4. The density function for Y may be written

$$\exp\{s(y)\eta - n\log(1+\exp\{\eta\})\}$$

where $\eta = \log(\theta/(1-\theta))$ and $s(y) = \sum y_j$. The natural parameter space for the model is $(-\infty, \infty)$. The mean parameter for this model is $n\theta$; equivalently, we may take the mean parameter to be simply θ. ∎

Example 1.8 Normal-theory measurement model
In the normal-theory measurement model, considered in Example 1.5, the natural parameter of the model is given by $\eta = (\eta_1, \eta_2)$ where $\eta_1 = -1/2\sigma^2$ and $\eta_2 = \mu/\sigma^2$. The canonical statistic of the model is given by $(\sum y_j^2, \sum y_j)$. The natural parameter space is $(-\infty, 0) \times \Re$ which corresponds to the parameter space $\Re \times \Re^+$ for (μ, σ). The mean parameter may be taken to be $(\mu^2 + \sigma^2, \mu)$. ∎

Models with covariates may also be in the exponential family. In this case if $Y_1, \ldots, Y_n$ are independent and the distribution of Y_j is a full-rank exponential family distribution, the distribution of $Y = (Y_1, \ldots, Y_n)$ may not be a full-rank exponential family distribution.

Example 1.9 Exponential regression model
Let $Y_1, \ldots, Y_n$ denote independent random variables such that Y_j has an exponential distribution with rate parameter λ_j so that

$$p(y_j; \lambda_j) = \lambda_j \exp\{-\lambda_j y_j\}, \quad y_j > 0$$

where $\lambda_j > 0$. Suppose that $\lambda_j = \theta x_j$, where $x_1, \ldots, x_n$ are fixed positive constants and $\theta > 0$. Then $Y = (Y_1, \ldots, Y_n)$ has a density of the form

$$\exp\Big\{-\theta \sum x_j y_j + n\log\theta + \sum \log x_j\Big\}$$

which is in the one-parameter exponential family.

Now suppose that $\log\lambda_j = \theta x_j$ where $x_1, \ldots, x_n$ are fixed constants and $\theta \in \Re$. Then Y has a density of the form

$$\exp\Big\{-\sum \exp\{\theta x_j\} y_j + \theta \sum x_j\Big\}.$$

This distribution is not a full-rank exponential family distribution. It may be viewed as an $(n, 1)$-exponential family distribution with function $c(\theta)$ given by

$$c(\theta) = (\exp\{\theta x_1\}, \ldots, \exp\{\theta x_n\}).$$

∎

1.3 Transformation models

Another important class of models is transformation models. Consider the measurement model discussed in Example 1.1. In this model, Y is of the form $Y = \mu + \sigma\epsilon$

where $\theta = (\mu, \sigma) \in \Re \times \Re^+$ and ϵ has a known distribution. Let $T = b + cY$ denote an affine transformation of Y where b and c are constants with $c > 0$. Then T follows the same measurement model as Y with μ replaced by $b + c\mu$ and σ replaced by $b\sigma$. Let $g(b, c)$ denote this transformation so that $T = g(b, c)Y$. The distribution of $g(b, c)Y$ is the same as the distribution of Y except with different parameter values.

The measurement model may be described by $Y = g(\mu, \sigma)\epsilon$ and, hence, the model may be described by the set of possible transformations, together with the distribution of ϵ. Let $\mathcal{G}$ denote the set of possible transformations which, in this case, is given by

$$\mathcal{G} = \{g(\mu, \sigma): -\infty < \mu < \infty,\ \sigma > 0\}.$$

The set $\mathcal{G}$ forms a group with respect to the binary operation $\circ$ given by

$$g(\mu_1, \sigma_1) \circ g(\mu_2, \sigma_2) = g(\mu_2 + \sigma_2\mu_1, \sigma_1\sigma_2).$$

The identity element of the group is $g(0, 1)$.

The same considerations apply to the model for n observations $Y_1, \ldots, Y_n$ from the measurement model. In that case

$$g(b, c)Y = \begin{pmatrix} b + cY_1 \\ \vdots \\ b + cY_n \end{pmatrix}. \tag{1.3}$$

The space $\mathcal{Y}$ of possible Y, in this case $\Re^n$, can be partitioned into a collection of *orbits* such that each point in $\mathcal{Y}$ is on exactly one orbit and any two points y, $\tilde{y}$ in $\mathcal{Y}$ are on the same orbit if and only if $\tilde{y} = gy$ for some $g \in \mathcal{G}$. It follows that each point in $\mathcal{Y}$ can be represented by its orbit together with its position on that orbit.

For instance, in the measurement model the orbit of a point y may be identified by $a = (a_1, \ldots, a_n), a_j = (y_j - \bar{y})/s$ where $\bar{y} = \sum y_j/n$ and $s^2 = \sum(y_j - \bar{y})^2/(n-1)$. The position on the orbit is given by $(\bar{y}, s)$. Hence, an observation y may equivalently be expressed by $(\bar{y}, s, a)$ since $(\bar{y}, s, a)$ is clearly a function of y and $y = g(\bar{y}, s)a$; here we are assuming that the transformation taking a to y is unique, in which case the action of $\mathcal{G}$ is said to be *free*.

By definition, any two points y, $\tilde{y}$ that are related by a transformation $g \in \mathcal{G}$ are on the same orbit and, hence, have the same value of a. The statistic a is therefore said to be *invariant* since its value does not depend on whether we observed y or gy, $g \in \mathcal{G}$. The statistic a is said be a *maximal invariant* if any other invariant statistic is a function of a; an equivalent condition is that

$$a(y) = a(\tilde{y}) \quad \text{if and only if } \tilde{y} = gy \ \text{ for some } g \in \mathcal{G}.$$

Since $Y = g(\mu, \sigma)\epsilon$ where $\epsilon = (\epsilon_1, \ldots, \epsilon_n)^T$, $a(Y) = a(\epsilon)$; it follows that $A = a(Y)$ has a known distribution, i.e., the distribution of $a(Y)$ does not depend on the value of θ.

Note that the group $\mathcal{G}$ may be viewed as operating both on the sample space $\mathcal{Y}$ by (1.3) and on the parameter space Θ by

$$g(b, c)(\mu, \sigma) = g(b, c) \circ g(\mu, \sigma).$$

In the measurement model any point in Θ may obtained by transforming the point corresponding to $\mu = 0$ and $\sigma = 1$, i.e.,

$$(\mu, \sigma) = g(\mu, \sigma)(0, 1).$$

In this case the action of $\mathcal{G}$ on Θ is said be *transitive* and the parameter space Θ is equivalent to the set of transformations $\mathcal{G}$.

Consider a statistic T, defined on $\mathcal{Y}$ and taking values in Θ; such a statistic is often, but not always, an estimator of θ. We say that T is *equivariant* if, for each $g \in \mathcal{G}$,

$$T(gy) = gT(y), \quad y \in \mathcal{Y}.$$

Note that in the left-hand side of this equation, g is acting on $\mathcal{Y}$, while in right-hand side of this equation, g is acting on Θ. For instance, in the measurement model, $(\bar{y}, s)$ is equivariant since

$$\frac{1}{n} \sum (b + cy_j) = b + c\bar{y}$$

and

$$\left[\frac{\sum (b + cy_j - (b + c\bar{y}))^2}{n - 1}\right]^{1/2} = b\left[\frac{\sum (y_j - \bar{y})^2}{n - 1}\right]^{1/2}.$$

An equivariant statistic may be used to form a maximal invariant statistic. Let T be equivariant and let $a \equiv a(y) = T(y)^{-1}y$. Then

$$T(gy)^{-1}gy = (gT(y))^{-1}gy = T(y)^{-1}g^{-1}gy = T(y)^{-1}y,$$

which shows that a is invariant. To show that a is a maximal invariant statistic, note $a(y) = a(\tilde{y})$ implies that $\tilde{y} = T(\tilde{y})T(y)^{-1}y$, yielding the result. The observation y may be expressed as $(T(y), T(y)^{-1}y)$ where $T(y)^{-1}y$ gives the orbit of y and $T(y)$ gives the position on that orbit. In typical cases, $T(y)$ may be viewed as an estimate of the unknown parameter and $T(y)^{-1}y$ may be viewed as a vector of 'residuals'. In the measurement model example, with $T(y) = (\bar{y}, s)$,

$$T(y)^{-1}y = \frac{y - \bar{y}}{s}.$$

A transformation model is simply a generalization of these ideas. Let Y denote a random variable taking values in a set $\mathcal{Y}$ and let $\mathcal{G}$ denote a group of transformations such that each $g \in \mathcal{G}$ maps $\mathcal{Y}$ back to $\mathcal{Y}$. Let $\{p(\cdot; \theta) \colon \theta \in \Theta\}$ denote the model for Y; we assume that Θ may be taken to be equivalent to $\mathcal{G}$. If the distribution of Y has parameter θ then the distribution of gY has parameter θ_1 which we will denote by $g\theta$. Let e denote the identity element of the group and let ϵ denote a random variable with density function $p(\cdot; e)$. Then Y has the same distribution as $\theta\epsilon$. An observation y may be represented by the value of a maximal invariant statistic a, which indicates the orbit of y, together with a statistic indicating the position on that orbit.

Example 1.10 Linear regression model
Let Y denote an $n \times 1$ vector random variable of the form

$$Y = x\beta + \sigma\epsilon$$

where x denotes a known $n \times k$ design matrix, $\sigma > 0$ is an unknown scalar parameter, β is an unknown $k \times 1$ parameter vector, and ϵ is an $n \times 1$ vector random variable with known distribution. Hence, the parameter of the model is $\theta = (\beta, \sigma)$ which takes values in the set $\Theta = \Re^k \times \Re^+$. The transformation $g(\beta, \sigma)$ corresponding to a parameter value $\theta = (\beta, \sigma)$ is given by

$$g(\beta, \sigma)Y = \sigma Y + x\beta$$

and

$$g(\beta_1, \sigma_1) \circ g(\beta_2, \sigma_2) = g(\sigma_1\beta_2 + \beta_1, \sigma_1\sigma_2).$$

The identity element is $e = (0, 1)$ and the distribution of ϵ is the same as the distribution of Y under this parameter value; here 0 denotes the zero vector.

A maximal invariant statistic is given by $a = (y - x\hat{\beta})/\hat{\sigma}$ where $\hat{\beta} = (x^T x)^{-1} x^T y$ is the least-squares estimator of β and $\hat{\sigma}^2 = (y - x\hat{\beta})^T (y - x\hat{\beta})/n$; note that other estimators may be used in this context. Hence, the data y is equivalent to a together with $\hat{\beta}, \hat{\sigma}$. ∎

Example 1.11 Poisson process
Let $Y_1, \ldots, Y_n$ denote the inter-event times of a Poisson process, as discussed in Example 1.2. The density function of $Y = (Y_1, \ldots, Y_n)$ is given by

$$p(y; \lambda) = \lambda^n \exp\{-\lambda(y_1 + \cdots + y_n)\}, \quad y_j > 0, \ j = 1, \ldots, n.$$

Consider the transformation $g(\lambda)y = \lambda y$ with $g(\lambda_1) \circ g(\lambda_2) = g(\lambda_1\lambda_2)$. Let $\bar{y} = n^{-1} \sum y_j$. A maximal invariant statistic is $(y_1/\bar{y}, \ldots, y_{n-1}/\bar{y})$ which, together with $\bar{y}$ is equivalent to $(y_1, \ldots, y_n)$. ∎

The transformation models discussed above are often called *pure* transformation models since the model for the data may be described entirely by group of transformations $\mathcal{G}$. Another useful type of model is a *composite* transformation model. A composite transformation model is parameterized by a parameter $\theta = (\psi, \lambda)$, with parameter space $\Theta = \Psi \times \Lambda$, such that for each fixed value of the index parameter ψ, the model given by $\{p(\cdot; \psi, \lambda) : \lambda \in \Lambda\}$ forms a transformation model with group $\mathcal{G}$. It is assumed that the group $\mathcal{G}$, along with other aspects of the group structure, are the same for each ψ.

Example 1.12 Gamma distribution
Let $Y_1, \ldots, Y_n$ denote independent random variables each distributed according to the density function

$$\frac{\lambda^\psi}{\Gamma(\psi)} y^{\psi-1} \exp\{-\lambda y\}, \quad y > 0$$

where $\psi > 0$ and $\lambda > 0$ are unknown parameters. For each fixed ψ this model is a transformation model with the same group structure as that considered in Example 1.11. Hence, it is a composite transformation model with index parameter ψ and group parameter λ. ∎

Example 1.13 Exponential regression model
Let $Y_1, \ldots, Y_n$ denote independent random variables such that Y_j has an exponential distribution with rate parameter λ_j where $\log \lambda_j = \alpha + \beta x_j$; here $x_1, \ldots, x_n$ are fixed scalar constants and α, β are unknown scalar parameters each taking values in $\Re$. This model is a composite transformation model with group parameter α and index β. To see this, note that, for fixed β, $(Y_1, \ldots, Y_n)$ is a vector of exponential random variables with rate parameters

$$\exp\{\alpha\} \, (\exp\{x_1\beta\}, \ldots, \exp\{x_n\beta\}).$$

Hence, as in Example 1.11, the distribution of $\exp\{-\alpha\}(Y_1, \ldots, Y_n)$ does not depend on α. A maximal invariant statistic is $(y_2/y_1, \ldots, y_n/y_1)$. ∎

1.4 Cumulants

Let Y denote a real-valued random variable and let $M_Y(t) = \mathrm{E}[\exp\{tY\}]$ denote the moment-generating function of Y. The cumulant-generating function of Y is given by $K_Y(t) = \log M_Y(t)$ which may be expanded

$$K_Y(t) = \kappa_1 t + \tfrac{1}{2}\kappa_2 t^2 + \tfrac{1}{6}\kappa_3 t^3 + \cdots .$$

The constants $\kappa_1, \kappa_2, \ldots,$ are called the cumulants of Y and may be obtained by differentiation of $K(t)$:

$$\kappa_j = \frac{d^j}{dt^j} K(t) \bigg|_{t=0} .$$

It is straightforward to show that $\kappa_1 = \mathrm{E}(Y)$, the mean of Y, and $\kappa_2 = \mathrm{Var}(Y)$, the variance of Y. In general, the jth cumulant of Y is a function of the first j moments of Y.

The relationship between cumulant and moments may be obtained by equating two expressions for the cumulant-generating function. Since the moment-generating function $M_Y(t)$ may be expanded

$$M_Y(t) = 1 + \sum_{j=1}^{\infty} m_j \frac{t^j}{j!},$$

where $m_1, \ldots,$ denote the moments of Y, it follows that

$$K_Y(t) = \sum_{j=1}^{\infty} \kappa_j \frac{t^j}{j!} = \log\left(1 + \sum_{j=1}^{\infty} m_j \frac{t^j}{j!}\right).$$

Using a Taylor's series expansion for the function $\log(1+x)$ around $x = 0$ yields

$$\log\left(1 + \sum_{j=1}^{\infty} m_j \frac{t^j}{j!}\right) = m_1 t + \frac{1}{2!}(m_2 - m_1^2)t^2$$
$$+ \frac{1}{3!}(m_3 - m_1 m_2 - 2(m_2 - m_1^2)m_1)t^3 + \cdots.$$

From this, it follows that

$$\kappa_1 = m_1, \qquad \kappa_2 = m_2 - m_1^2, \qquad \kappa_3 = m_3 - m_1 m_2 - 2(m_2 - m_1^2)m_1,$$

and so on. Similar identities may be obtained relating the cumulants to the central moments $\mu_1, \ldots,$ by starting with the moment-generating function for a mean-corrected version of Y, $Y - \mathrm{E}(Y)$.

Note that, like the moment-generating function, the cumulant-generating function may not exist. We will say that the cumulant-generating function exists whenever the corresponding moment-generating function $M(t)$ is finite for all t in a neighborhood of 0. Since the jth cumulant κ_j is a function of the first j moments of Y, κ_j may still be defined provided that these moments exist.

Cumulants have a number of useful properties. Let a, b denote constants and let $W = aY + b$. Then $K_W(t) = bt + K_Y(at)$ so that $\kappa_1(W) = b + a\kappa_1(Y)$ and for $j = 2, \ldots,$

$$\kappa_j(W) = a^j \kappa_j(Y).$$

Let $Y_1, \ldots, Y_n$ denote independent random variables each with the same distribution as Y and let $S = \sum Y_j$. Then $K_S(t) = nK_Y(t)$ so that $\kappa_j(S) = n\kappa_j(Y)$.

It is often convenient to work with standardized cumulants. Suppose a random variable Y has cumulants $\kappa_1, \kappa_2, \ldots$. Define the jth standardized cumulant of Y by $\rho_j = \kappa_j / \kappa_2^{j/2}$. Note that, except for ρ_1, the standardized cumulants are not affected by changes in location or scale, i.e.,

$$\rho_j(a + bY) = \rho_j(Y), \quad j = 2, \ldots.$$

If S is the sum of n independent random variables then, under weak conditions, $\kappa_j(S) = O(n)$ and $\rho_j(S) = O\big(n^{-(j-2)/2}\big)$.

Example 1.14 Laplace distribution
Let Y denote a random variable with a standard Laplace, or double exponential, distribution so that Y has density $\exp\{-|y|\}/2$, $-\infty < y < \infty$. The moment-generating function of Y is given by

$$M_Y(t) = \int_{-\infty}^{\infty} \exp\{ty\} \frac{1}{2} \exp\{-|y|\}\, dy = \frac{1}{1 - t^2}, \quad |t| < 1.$$

The cumulant-generating function is therefore given by $K_Y(t) = -\log(1-t^2)$. Since

$$K_Y(t) = t^2 + \frac{1}{2}t^4 + \cdots$$

Y has mean 0, variance 2, third cumulant 0, and fourth cumulant 12. ∎

Example 1.15 One-parameter exponential family distribution
Let Y denote a random variable with a distribution in the one-parameter exponential family so that Y has a density function of the form

$$\exp\{\eta S(y) - k(\eta) + D(y)\}.$$

Note that the function $k(\cdot)$ may be written

$$k(\eta) = \log \int \exp\{\eta S(y) + D(y)\}\, dy$$

so that for any η in the interior of the natural parameter space, $S = S(Y)$ has moment-generating function

$$\begin{aligned} M_S(t) &= \int \exp\{tS(y)\} \exp\{S(y)\eta - k(\eta) + D(y)\}\, dy \\ &= \exp\{k(\eta + t) - k(\eta)\}, \quad |t| < t_0 \end{aligned}$$

for some $t_0 > 0$. Hence, the cumulant-generating function of S is given by

$$K_S(t) = k(\eta + t) - k(\eta), \quad |t| < t_0$$

and the cumulants of S are given by $k'(\eta)$, $k''(\eta)$, and so on. The function k is often called the cumulant function of the exponential family. ∎

Example 1.16 Binomial distribution
Let Y denote a random variable with a binomial distribution with index n and success probability θ so that Y has density

$$p(y; \theta) = \binom{n}{y} \theta^y (1-\theta)^{n-y}, \quad y = 0, \ldots, n$$

where $0 < \theta < 1$. This density function may be written

$$\begin{aligned} &\exp\left\{y \log(\theta/(1-\theta)) + n \log(1-\theta) + \log \binom{n}{y}\right\} \\ &= \exp\left\{y\eta - n \log(1 + \exp\{\eta\}) + \log \binom{n}{y}\right\} \end{aligned}$$

where $\eta = \log(\theta/(1-\theta))$. It follows that Y has cumulant-generating function

$$K_Y(t) = n \log(1 + \exp\{\eta + t\}) - n \log(1 + \exp\{\eta\})$$

so that

$$\kappa_1 = K_Y'(0) = n\frac{\exp\{\eta\}}{1+\exp\{\eta\}}, \qquad \kappa_2 = K_Y''(0) = n\frac{\exp\{\eta\}}{[1+\exp\{\eta\}]^2},$$

$$\kappa_3 = K_Y'''(0) = n\frac{\exp\{\eta\}[1-\exp\{\eta\}]}{[1+\exp\{\eta\}]^3}.$$

In terms of the original parameter θ, we have

$$\kappa_1 = n\theta, \qquad \kappa_2 = n\theta(1-\theta), \qquad \kappa_3 = n\theta(1-\theta)(1-2\theta).$$ ∎

Now consider a vector random variable Y of dimension m. The moment-generating function of Y is given by

$$M(t) = \mathrm{E}[\exp\{t^T Y\}]$$

and the cumulant-generating function of Y is given by $K(t) = \log M(t)$; as in the univariate case, we will say that this cumulant-generating function exists provided that $M(t)$ is finite for all t is some neighborhood of 0. The (joint) cumulants of $Y = (Y_1, \ldots, Y_m)$ may be obtained by differentiating $K(t)$; specifically, the cumulant of order $(i_1, \ldots, i_m)$ is given by

$$\kappa_{i_1\cdots i_m} = \frac{\partial^{i_1+\cdots+i_m}}{\partial t_1^{i_1}\cdots\partial t_m^{i_m}} K(t)\Big|_{t=(0,\ldots,0)}.$$

Note that the cumulant of order $(i, 0, \ldots, 0)$ is simply the ith cumulant of Y_1. Joint cumulants are often written using the notation $cum(\cdot, \ldots, \cdot)$ so that

$$\mathrm{cum}(Y_1) = \kappa_{10\cdots0}, \qquad \mathrm{cum}(Y_1, Y_2) = \kappa_{110\cdots0},$$

$$\mathrm{cum}(Y_1, Y_1, Y_2, Y_3) = \kappa_{2110\cdots0}$$

and so on.

Suppose that $m = 2$ so that $Y = (Y_1, Y_2)$. Then it is straightforward to show that

$$\kappa_{10} = \mathrm{E}(Y_1), \qquad \kappa_{01} = \mathrm{E}(Y_2), \qquad \kappa_{20} = \mathrm{Var}(Y_1),$$

$$\kappa_{02} = \mathrm{Var}(Y_2), \quad \text{and} \quad \kappa_{11} = \mathrm{Cov}(Y_1, Y_2).$$

If Y_1 and Y_2 are independent, then

$$\begin{aligned} M_Y(t) &= \mathrm{E}(\exp\{t_1 Y_1 + t_2 Y_2\}) \\ &= \mathrm{E}(\exp\{t_1 Y_1\})\mathrm{E}(\exp\{t_2 Y_2\}) = M_{Y_1}(t_1) M_{Y_2}(t_2) \end{aligned}$$

so that

$$K_Y(t) = K_{Y_1}(t_1) + K_{Y_2}(t_2).$$

It follows that all joint cumulants of (Y_1, Y_2), involving both Y_1 and Y_2 are 0. If the cumulant-generating function of (Y_1, Y_2) exists, then the converse holds: if all joint cumulants of (Y_1, Y_2) are 0, then Y_1 and Y_2 are independent. The same result holds for more than 2 random variables.

Example 1.17 Linear functions of Poisson random variables
Let X_1, X_2 denote independent Poisson random variables with means λ_1, λ_2, respectively. Let $Y_1 = X_1 + X_2$ and $Y_2 = X_1 - X_2$. Note that X_j has cumulant-generating function

$$K_{X_j}(t) = \lambda_j(\exp\{t\} - 1)$$

which may be obtained by either direct computation or by noting that a Poisson distribution is in the one-parameter exponential family.

The joint cumulant-generating function of $Y = (Y_1, Y_2)$ is given by

$$\begin{aligned} K_Y(t) &= \log \mathrm{E}(\exp\{t_1 Y_1 + t_2 Y_2\}) \\ &= \log \mathrm{E}(\exp\{(t_1 + t_2)X_1 + (t_1 - t_2)X_2\}) \\ &= \lambda_1(\exp\{t_1 + t_2\} - 1) + \lambda_2(\exp\{t_1 - t_2\} - 1). \end{aligned}$$

For example,

$$\mathrm{cum}(Y_1, Y_2) = \lambda_1 - \lambda_2, \qquad \mathrm{cum}(Y_1, Y_2, Y_2) = \lambda_1 + \lambda_2,$$

and

$$\mathrm{cum}(Y_1, Y_1, Y_2, Y_2) = \lambda_1 + \lambda_2.$$

∎

Example 1.18 Exponential family distributions
Let Y have an exponential family distribution with density of the form

$$\exp\{s(y)^T \eta - k(\eta) + D(y)\}.$$

Then, as in Example 1.15, the joint cumulant-generating function of $S = S(Y)$ is given by

$$K_S(t) = k(\eta + t) - k(\eta).$$

∎

Example 1.19 Multinomial distribution
Let $Y = (Y_1, \ldots, Y_m)$ denote a random variable with a multinomial distribution with density of the form

$$p(y; \theta) = \binom{n}{y_1, \ldots, y_m, y_{m+1}} \theta_1^{y_1} \cdots \theta_m^{y_m} \theta_{m+1}^{y_{m+1}}$$

where $y_{m+1} = n - y_1 - \cdots - y_m$ and $\theta_{m+1} = 1 - \theta_1 - \cdots - \theta_m$; here $y_1, \ldots, y_m$ are nonnegative integers such that $\sum y_j \leq n$ and $\theta_1, \ldots, \theta_m$ are nonnegative parameters such that $\sum \theta_j < 1$. This density may be written in exponential family form as

$$\begin{aligned} \exp\Big\{ &\sum y_j \log\left(\theta_j/(1 - \sum \theta_j)\right) + n \log\left(1 - \sum \theta_j\right) \\ &+ \log \binom{n}{y_1, \ldots, y_m, y_{m+1}} \Big\}. \end{aligned} \tag{1.4}$$

Let $\eta_j = \log(\theta_j/(1 - \sum \theta_j))$. Then (1.4) may be written

$$\exp\left\{\sum y_j\eta_j - n\log\left(1 + \sum \exp\{\eta_j\}\right) + \log\binom{n}{y_1, \ldots, y_m, y_{m+1}}\right\}.$$

Hence, the cumulant-generating function of Y is given by

$$K_Y(t) = n\log\left(1 + \sum \exp\{\eta_j + t_j\}\right) - n\log\left(1 + \sum \exp\{\eta_j\}\right).$$

From this it follows easily that, for example,

$$\begin{aligned}
\text{Cov}(Y_i, Y_j) &= \frac{\partial K_Y}{\partial t_i \partial t_j}(t)\Big|_{t=(0,\ldots,0)} \\
&= \begin{cases} -n\exp\{\eta_i + \eta_j\}/[1 + \sum \exp\{\eta_j\}]^2 & \text{if } i \neq j \\ n\exp\{\eta_j\}/[1 + \sum \exp\{\eta_j\}] - n\exp\{2\eta_j\}/[1 + \sum \exp\{\eta_j\}]^2 & \text{if } i = j. \end{cases}
\end{aligned}$$

In terms of the original parameterization, $\text{Cov}(Y_i, Y_j) = -n\theta_i\theta_j$ if $i \neq j$ and $\text{Var}(Y_j) = n\theta_j(1 - \theta_j)$. ∎

In some cases, we are interested in cumulants of functions of $Y = (Y_1, \ldots, Y_m)$. For instance, we may be interested in $\text{cum}(Y_1Y_2, Y_3Y_4)$, the covariance of Y_1Y_2 and Y_3Y_4. Cumulants of this type are often called *generalized cumulants*. Generalized cumulants may be written in terms of standard cumulants by utilizing the relationship between cumulants and moments. Generalized cumulants often arise in the calculation of the cumulants of certain series expansions; these will be considered in Section 5.1.

1.5 Sufficiency

Consider a random variable Y with density function $p(\cdot; \theta)$, $\theta \in \Theta$. In statistical inference, the goal is to draw conclusions about the unknown parameter θ on the basis of observing $Y = y$. Such conclusions are typically based on a statistic, a function $T(Y)$. The use of a statistic involves some reduction of data and, therefore, simplifies the analysis. Too much reduction however may involve some loss of information about θ. It is important to know how much data reduction is possible without any information loss. This is the idea behind sufficiency.

Let $T(Y)$ denote a statistic and suppose the value of $T(Y)$ is known, say $T(Y) = t$. The statistic T is sufficient if knowing the exact value of Y gives no additional information about θ, in addition to that provided by the knowledge that $T(Y) = t$. More formally, T is sufficient if the conditional distribution Y given $T = t$ does not depend on the value of θ. Sufficient statistics are often determined using the *factorization theorem* which states that a statistic T is sufficient if and only if there exist functions g and h such that

$$p(y; \theta) = g(T(y), \theta)h(y)$$

for all θ, y.

It is clear from the factorization theorem that sufficient statistics are not unique. In general we want the sufficient statistic to be as 'small' as possible, i.e. we want the sufficient statistic to yield the coarsest partition of the sample space. A sufficient statistic is said to be *minimal* if it is a function of any other sufficient statistic. If T_1 and T_2 are both minimal sufficient statistics then they must be equivalent in the sense that

$$T_1(y) = T_1(\tilde{y}) \quad \text{if and only if} \quad T_2(y) = T_2(\tilde{y}).$$

We will often refer to *the* minimal sufficient statistic with the understanding that this actually refers to an equivalence class of statistics.

For a model consisting of a finite set of densities, $\mathcal{P} = \{p_1(y), \ldots, p_k(y)\}$, with common support, the minimal sufficient statistic is given by

$$T(Y) = \left(\frac{p_2(Y)}{p_1(Y)}, \ldots, \frac{p_k(Y)}{p_1(Y)}\right).$$

To see this, first note that, taking $\Theta = \{1, \ldots, k\}$ and $p(y; \theta) = p_\theta(y)$,

$$p(y; \theta) = \frac{p_\theta(y)}{p_1(y)} p_1(y)$$

so that $T(Y)$ is sufficient by the factorization theorem. Let $\tilde{T}(Y)$ denote another sufficient statistic. Again by the factorization theorem,

$$p(y; \theta) = g(\tilde{T}(y), \theta)h(y)$$

so that

$$\frac{p_\theta(y)}{p_1(y)} = \frac{g(\tilde{T}(y), \theta)}{g(\tilde{T}(y), 1)}.$$

It follows that $\tilde{T}(Y)$ is a function of $T(Y)$.

For a general model $\mathcal{P} = \{p(y; \theta) \colon \theta \in \Theta\}$, let $\{\theta_1, \ldots, \theta_k\}$ denote a subset of Θ such that

$$T(Y) = \left\{\frac{p(Y; \theta_2)}{p(Y; \theta_1)}, \ldots, \frac{p(Y; \theta_k)}{p(Y; \theta_1)}\right\}$$

is sufficient for $\mathcal{P}$. Let $\tilde{T}(Y)$ denote another sufficient statistic for $\mathcal{P}$. Then $\tilde{T}(Y)$ is also sufficient for the model $\{p(y; \theta) \colon \theta = \theta_1, \ldots, \theta_k\}$. It follows that $\tilde{T}(Y)$ is a function of $T(Y)$. Hence $T(Y)$ is minimal sufficient.

Example 1.20 Poisson distribution

Let $Y_1, \ldots, Y_n$ denote independent Poisson random variables each with mean θ so that each Y_j has density function

$$\theta^y \exp\{-\theta\}/y!, \quad y = 0, 1, \ldots$$

where $\theta > 0$. Then $Y_1, \ldots, Y_n$ has density function

$$p(y;\theta) = \frac{\theta^{y_1+\cdots+y_n}\exp\{-n\theta\}}{y_1!y_2!\cdots y_n!}.$$

By the factorization theorem, $Y_1 + \cdots + Y_n$ is a sufficient statistic. To show that this statistic is minimal sufficient, consider the parameter space consisting of two points, $\Theta_1 = \{1, 2\}$. Since

$$\frac{p(y;2)}{p(y;1)} = 2^{y_1+\cdots+y_n}\exp\{-n\}$$

$Y_1+\cdots+Y_n$ is minimal sufficient for the model with parameter space Θ_1 and, hence, minimal sufficient for the original model. ∎

The same basic approach can be used for any exponential family model.

Example 1.21 Exponential family distributions
Let $Y_1, \ldots, Y_n$ denote independent random variables each with a distribution in the exponential family and having a density of the form

$$\exp\{s(y)^T c(\theta) - k(\theta) + D(y)\}. \tag{1.5}$$

Then

$$p(y;\theta) = \exp\left\{\sum s(y_j)^T c(\theta) - nk(\theta) + \sum D(y_j)\right\}$$

so that, by the factorization theorem, $\sum S(Y_j)$ is sufficient. To show that this statistic is minimal sufficient, consider the submodel with parameter space $\Theta_1 = \{\theta_0, \theta_1, \ldots, \theta_m\}$. The minimal sufficient statistic for the model with parameter space Θ_1 has components

$$\sum s(Y_j)^T[c(\theta_1) - c(\theta_0)], \ldots, \sum s(Y_j)^T[c(\theta_m) - c(\theta_0)] = s(Y_j)^T C$$

where C denotes the matrix with jth row $c(\theta_j) - c(\theta_0)$, $j = 1, \ldots, m$. It follows that this sufficient statistic is equivalent to $\sum S(Y_j)^T$ provided that C is of full rank. If C is not of full rank, there exist constants $\alpha_1, \ldots, \alpha_m$ such that

$$\sum \alpha_j[c(\theta_j) - c(\theta_0)] = 0.$$

Hence, $\sum s(Y_j)$ is minimal sufficient provided that $\theta_0, \ldots, \theta_m$ can be chosen so that $c(\theta_1) - c(\theta_0), \ldots, c(\theta_m) - c(\theta_0)$ are linearly independent. This is generally possible provided that the exponential family representation (1.5) is in minimal form. ∎

Statistical inference is generally based on the minimal sufficient statistic; this is sometimes referred to as the *sufficiency principle*. One justification for the sufficiency principle is the fact that, roughly speaking, corresponding to any statistical procedure is a procedure based only on the sufficient statistic that is as good, or better, than the original one. Hence, one only needs to consider procedures based on the minimal sufficient statistic.

Example 1.22 Rao–Blackwell theorem
Let Y denote a random variable, possibly vector valued, distributed according to a density $p(y;\theta)$, $\theta \in \Theta \subset \Re$, and let S denote a one-dimensional sufficient statistic for the model. Consider the problem of estimating θ and suppose that we decide to measure the quality of an estimator X by the mean squared error, given by

$$R(\theta; X) = \mathrm{E}[(X-\theta)^2; \theta] = \mathrm{E}[(X - \mathrm{E}(X;\theta))^2; \theta] + \mathrm{Var}(X;\theta);$$

the smaller the value of $R(\theta; X)$ the better the estimator X is.

Let T denote an estimator with mean squared error $R(\theta; T)$. Then, if T is not a function of S, there exists an estimator $g(S)$ such that $R(\theta; g(S)) \leq R(\theta; T)$ for all $\theta \in \Theta$ with strict inequality for at least one value of θ. A proof of this result relies on the following result concerning conditional expectation. If X_1, X_2 are random variables, then $\mathrm{E}(X_1) = \mathrm{E}[\mathrm{E}(X_1|X_2)]$ where $\mathrm{E}(X_1|X_2)$ denotes the conditional expected value of X_1 given X_2. Applying this result to $\mathrm{E}(X_1^2)$ as well, yields the following result concerning conditional variances:

$$\mathrm{Var}(X_1) = \mathrm{Var}[\mathrm{E}(X_1|X_2)] + \mathrm{E}[\mathrm{Var}(X_1|X_2)].$$

Take $g(S) = \mathrm{E}(T|S)$; note that, by the sufficiency of S, this conditional expected value does not depend on θ. Then

$$\mathrm{E}[g(S);\theta] = \mathrm{E}[\mathrm{E}(T|S);\theta] = \mathrm{E}(T;\theta)$$

and

$$\mathrm{Var}(g(S);\theta) = \mathrm{Var}[\mathrm{E}(T|S);\theta] = \mathrm{Var}(T;\theta) - \mathrm{E}[\mathrm{Var}(T|S);\theta].$$

It follows that $R(\theta; g(S)) \leq R(\theta; T)$ for all θ, with equality only if $\mathrm{E}[\mathrm{Var}(T|S);\theta] = 0$ which would imply that T is a function of S.

A formal statement of this result is known as the *Rao–Blackwell theorem.* ∎

For independent identically distributed observations with support not depending on the parameter, a minimal sufficient statistic with fixed dimension, not depending on n, exists only for exponential family models. Hence, for a large class of models no reduction or very little reduction by sufficiency is possible.

For independent identically distributed observations $Y_1, \ldots, Y_n$, the order statistics $Y_{(1)}, \ldots, Y_{(n)}$, where $Y_{(1)} \leq \cdots \leq Y_{(n)}$, are sufficient.

1.6 Ancillary statistics

Consider a model for a random variable Y. A statistic A having a distribution not depending on the parameter θ is said to be *distribution constant.* If, in addition, A is a function of the minimal sufficient statistic then A will be said be an *ancillary statistic.* It should be noted that ancillary statistics are often defined to be simply statistics with a distribution not depending on θ, i.e. a distribution-constant statistic. If however A is not a function of the minimal sufficient statistic, then A would be eliminated from consideration by sufficiency so we include that additional requirement here. Clearly, by itself, an ancillary statistic provides no information regarding θ. Hence, one approach to statistical inference is to treat an ancillary statistic A as a fixed constant, rather than as a random variable; that is, instead of using the distribution of Y as the basis for inference, we use the conditional distribution of Y given $A = a$, where a denotes the observed value of a. The following well-known example illustrates the main point.

Example 1.23 Two measuring instruments
Consider the measurement of a physical constant, as in Example 1.1. Suppose however that there are two measuring instruments with different known precisions. Suppose that the observation Y follows the model

$$Y = \mu + \sigma\epsilon$$

where ϵ has a normal distribution. Assume that $\sigma = \sigma_1$ if the first instrument is used and $\sigma = \sigma_2$ if the second instrument is used, where σ_1, σ_2 are known and $\sigma_1 > \sigma_2$. Suppose that the instrument actually used is chosen randomly, with probability $1/2$ that either instrument is used; let $A = 1$ if instrument 1 is selected and $A = 2$ otherwise.

The data are therefore y, a and the model function is given by

$$p(y, a; \mu) = \frac{1}{2}\phi\left(\frac{y-\mu}{\sigma_1}\right)\frac{1}{\sigma_1}I_{\{1\}}(a) + \frac{1}{2}\phi\left(\frac{y-\mu}{\sigma_2}\right)\frac{1}{\sigma_2}I_{\{2\}}(a). \tag{1.6}$$

Here $I_{\{\cdot\}}(\cdot)$ denotes an indicator function.

Suppose that $A = 1$ is observed. What model function should be used for inference? The function (1.6) takes into account that, although measuring instrument 1 was used, we *might* have used instrument 2. Alternatively, we could use the conditional density of Y given $A = 1$:

$$p(y|a = 1; \mu) = \phi\left(\frac{y-\mu}{\sigma_1}\right)\frac{1}{\sigma_1}.$$

In this case, the random variable Y is simply treated as a normally distributed random variable with unknown mean μ and known variance. It seems clear that the conditional approach is the correct one in this example. ∎

This example gives support to the *conditionality principle* which states that in models with an ancillary statistic A, inference should be based on the conditional distribution of the data given $A = a$ where a denotes the observed value of A. The idea behind this principle is essentially the same as that in the previous example: since the distribution of A does not depend on θ there seems to be little reason for analyzing A as a random variable. Furthermore, conditioning on $A = a$ makes the analysis more relevant to the data actually observed.

Example 1.24 Uniform distribution

Let $Y_1, \dots, Y_n$ denote independent observations, each distributed according to a uniform distribution on the interval $(\theta, \theta+1)$ where $\theta > 0$ is an unknown parameter. Then

$$p(y; \theta) = 1, \quad \theta \leq \min\{y_1, \dots, y_n\} \leq \max\{y_1, \dots, y_n\} \leq \theta + 1.$$

Hence, the minimal sufficient statistic for this model is $(y_{(1)}, y_{(n)})$ where $y_{(1)}$ denotes the minimum of the sample and $y_{(n)}$ denotes the maximum of the sample. Inference about θ may be based on the distribution of $(Y_{(1)}, Y_{(n)})$ which has density function

$$n(n-1)[y_{(n)} - y_{(1)}]^{n-2}, \quad \theta \leq y_{(1)} \leq y_{(n)} \leq \theta + 1.$$

Here $A = Y_{(n)} - Y_{(1)}$ is an ancillary statistic, with density given by

$$n(n-1)a^{n-2}, \quad 0 \leq a \leq 1.$$

Inference about θ may then be based on the conditional distribution of $(Y_{(1)}, Y_{(n)})$ given $Y_{(n)} - Y_{(1)} = a$, where a denotes the observed value of A. That is, since the distribution of the range of the data does not depend on θ, in probability calculations we restrict attention to those values of $Y_{(1)}, Y_{(n)}$ that have the same range as that actually observed. ∎

For a transformation model, we may use a maximal invariant statistic as an ancillary. As discussed in Section 1.3, if Y follows a transformation model with parameter θ and $T(Y)$ is an estimator of θ, then $T(y)^{-1}y$ is a maximal invariant and the corresponding statistic is distribution constant. Provided that it is also a component of the minimal sufficient statistic, we may take $a = T(y)^{-1}y$.

There are difficulties however in extending the idea of conditioning to a general principle. The main difficulty is that ancillary statistics are not unique and, in contrast to sufficiency, there is no simple criterion for choosing between alternative ancillary statistics. The following example, also well known, illustrates this point.

Example 1.25 Special multinomial distribution

Consider a multinomial distribution with four cells with cell probabilities

$$\frac{1-\theta}{6}, \quad \frac{1+\theta}{6}, \quad \frac{2-\theta}{6}, \quad \frac{2+\theta}{6}$$

where $|\theta| < 1$. Consider n independent trials and let $Y_1, \dots, Y_4$ denote the cell frequencies. It is straightforward to show that $(Y_1, \dots, Y_4)$ is the minimal sufficient statistic.

Let $A_1 = Y_1 + Y_2$. Then A_1 has a binomial distribution with parameters n and $1/3$; hence A_1 is an ancillary statistic. Let $A_2 = Y_2 + Y_3$. Then A_2 has a binomial distribution with parameters n and $1/2$ so that A_2 is also ancillary. Note that neither one of A_1, A_2 is a function of the other and that (A_1, A_2) is not ancillary. Hence, it is not possible to carry out the conditionality principle without some basis for choosing between A_1 and A_2. ∎

In some cases, nontrivial ancillary statistics do not exist. In particular, if the minimal sufficient statistic S is *complete* then if T is a statistic that is distribution constant, it follows that T and S are independent. A statistic S is said to be complete with respect to a given model if

$$\mathrm{E}[h(S); \theta] = 0 \quad \text{for all } \theta \in \Theta$$

implies that

$$\Pr[h(S) = 0; \theta] = 1 \quad \text{for all } \theta \in \Theta.$$

Suppose that S is complete and that T is distribution constant. Then for any set B, $p_B = \Pr(T \in B)$ is a constant, not depending on θ. Let $h(S) = \Pr(T \in B|S) - p_B$; note that, by sufficiency of S, $h(S)$ does not depend on θ and that

$$\mathrm{E}[h(S); \theta] = \mathrm{E}[\Pr(T \in B|S) - p_B] = \mathrm{E}[\Pr(T \in B|S)] - \Pr(T \in B) = 0.$$

It follows from completeness that

$$\Pr[\Pr(T \in B|S) = \Pr(T \in B); \theta] = 0 \quad \text{for all } \theta.$$

That is, the conditional distribution of T given S is the same as the unconditional distribution of T. It follows that T and S are independent.

Now suppose that A is an ancillary statistic so that A is a function of S and is distribution constant. Then any two functions of A must be independent. It follows that A must be a constant.

The most common case in which a complete sufficient statistic exists is in a full exponential family model, in which case the canonical statistic is a complete sufficient statistic. Hence, in those models, inference may proceed without consideration of any possible ancillary statistics.

1.7 Normal distribution theory

The normal distribution plays a central role in statistics and hence we now briefly review some useful distribution theory based on the normal distribution.

Let $Y = (Y_1, \ldots, Y_k)^T$ denote a random vector with a multivariate normal distribution. The multivariate normal distribution is parameterized by two parameters, a mean vector μ and a covariance matrix Σ. The matrix Σ is assumed to be nonnegative definite so that $b^T \Sigma b \geq 0$ for any vector b. The matrix Σ consists of the covariances of all pairs (Y_i, Y_j); the covariance of Y_i, Y_j is given by Σ_{ij}; if Σ is a diagonal matrix then the elements of Y are independent.

The multivariate normal distribution may be characterized by the fact that for any vector b, $b^T Y$ has a (univariate) normal distribution with mean $b^T \mu$ and variance $b^T \Sigma b$; if $b^T \Sigma b = 0$ this normal distribution is taken to be degenerate, with all of its mass at $b^T \mu$. If Σ is positive definite, then the density function of Y is given by

$$(2\pi)^{-k/2} |\Sigma|^{-1/2} \exp\left\{ -\frac{1}{2}(y - \mu)^T \Sigma^{-1} (y - \mu) \right\}.$$

Let B denote a $m \times k$ matrix of full rank, $m \leq k$ and let $X = BY$. Then X also has a multivariate normal distribution with mean $B\mu$ and covariance matrix $B \Sigma B^T$.

In particular, if $\Sigma = D_k$, the identity matrix of rank k, and B is a $k \times k$ orthogonal matrix so that $BB^T = D_k$, then X has also covariance matrix D_k. Note that these results imply that the marginal distribution of any subset of the random variables in Y is also multivariate normal.

Let B_1 denote a $m \times k$ matrix and let B_2 denote a $(k - m) \times k$ matrix such that the $k \times k$ matrix formed by combining B_1 and B_2 is nonsingular. Let $X_1 = B_1 Y$ and $X_2 = B_2 Y$. Then the conditional distribution of X_2 given X_1 is multivariate normal with mean

$$B_2\mu + B_2\Sigma B_1^T (B_1 \Sigma B_1^T)^{-1}(X_1 - B_1\mu)$$

and covariance matrix

$$B_2\Sigma B_2^T - B_2\Sigma B_1^T (B_1\Sigma B_1^T)^{-1} B_1 \Sigma B_2^T.$$

Let Y denote a multivariate normal random vector with mean 0, the zero vector, and covariance matrix D_k. Then $Y^T Y$ has a chi-square distribution with k degrees of freedom. More generally, if B is a $k \times k$ matrix, then $Y^T BY$ has a chi-square distribution if and only if B is idempotent so that $BB = B$; in this case the degrees of freedom of the chi-square distribution is equal to the rank of B, which, since B is idempotent, is equal to the trace of B. Let B_1 and B_2 denote two $k \times k$ idempotent matrices. Then $Y^T B_1 Y$ and $Y^T B_2 Y$ are independent if and only if $B_1 B_2 = 0$.

1.8 Discussion and references

There are many excellent references describing in detail the basic concepts of statistics; see, for example, Bickel and Doksum (1977), Cassella and Berger (1990), Cox and Hinkley (1974), Lindsey (1996), and Pace and Salvan (1997).

Exponential family models play a central role in statistical theory and methodology. The mathematical properties of exponential families are given in Barndorff-Nielsen (1978) and Brown (1988). The book by McCullagh and Nelder (1989) on generalized linear models illustrates the wide range of application of models based on the exponential family. Curved exponential families are considered in Efron (1975, 1978), Amari (1985) and Kass and Vos (1998).

Transformation models are a generalization of location-scale and regression models. General treatments of transformation models are given by Fraser (1968, 1979), Barndorff-Nielsen, Blæsild, and Erikson (1989), Berger (1985, Chapter 6), and Eaton (1988).

The properties of cumulants and generalized cumulants are discussed in McCullagh (1987) and Kendall and Stuart (1977).

It is generally accepted that statistical inference based on a given model should depend on the data only through the sufficient statistic; this is sometimes referred to as the *sufficiency principle*. One justification for this principle is the result by Bahadur (1954) that states that given any statistical procedure there is an equally good procedure that depends on the data only through the sufficient statistic; this result generalizes the Rao–Blackwell theorem described in Example 1.22. Many properties of sufficiency are discussed in Lehmann (1983, Chapter 1). The relationship between sufficient statistics of fixed dimension and exponential family distributions is considered by Brown (1964), Barndorff-Nielsen and Pedersen (1968), and Andersen (1970). The Rao–Blackwell theorem is due to Rao (1945) and Blackwell (1947).

The idea that statistical inference should be based on the conditional distribution given an ancillary statistic has been advanced by many writers; see, for example, Cox (1958), Efron and Hinkley (1978), Hinkley (1982) and McCullagh (1987, Chapter 8). As noted in Section 1.6, however, there are difficulties in a general 'conditionality principle'; these difficulties are discussed in Barnard and Sprott (1971), Basu (1964), Cox (1971), Kalbfleisch (1975), Lehmann (1986, Chapter 10), Lloyd (1985, 1992), and Reid (1995). It was suggested by Welch (1939) that conditional procedures are in some cases less efficient than unconditional ones; however see Barnard (1976, 1982) and Severini (1995). The properties of ancillary statistics are summarized by Buehler (1982). Example 1.23 is due to Cox (1958); Example 1.25 is due to Basu (1964). The result outlined at the end of Section 1.6 stating that ancillary statistics and complete sufficient statistics are independent is known as Basu's theorem (Basu 1955, 1958); see also Lehmann (1981).

Normal distribution theory is considered in Johnson and Kotz (1970a,b; 1972) and Rao (1973, Chapter 3).

1.9 Exercises

1.1 Let $X_1, \ldots, X_n$ denote independent identically distributed random variables each with density function

$$p(x) = \sqrt{\theta}(2\pi x^3)^{-3/2} \exp\Big\{-\tfrac{1}{2}\theta(x + x^{-1} - 2)\Big\}, \quad x > 0$$

where $\theta > 0$. Is this distribution in the one-parameter exponential family?

1.2 Suppose Y is a real-valued random variable with density function

$$p_Y(y; \theta) = \exp\{\theta y + k(\theta) + S(y)\} I_A(y)$$

and X is a real-valued random variable with density function

$$p_X(x; \theta) = \exp\{\theta x + \tilde{k}(\theta) + \tilde{S}(y)\} I_{\tilde{A}}(y)$$

where $\theta \in \Theta \subset \Re$. Show that:

(a) if $k = \tilde{k}$, then Y and X have the same distribution

(b) if $E(Y; \theta) = E(X; \theta)$ for all $\theta \in \Theta$ then Y and X have the same distribution

(c) if $\mathrm{Var}(Y; \theta) = \mathrm{Var}(X; \theta)$ for all $\theta \in \Theta$ then Y and X do not necessarily have the same distribution.

1.3 Show that the natural parameter space of an exponential family model is convex.

1.4 Let Y denote a nonnegative continuous random variable with density p. Suppose that Y is observed only if $Y \leq y_o$ where y_o is a known constant.

(a) Find the density function of Y given that Y is observed; denote this density by p_o.

(b) Suppose that the density p is in the one-parameter exponential family. Under what conditions, if any, is p_o also in the one-parameter exponential family?

1.5 Consider a one-parameter exponential family model with natural parameter η and canonical statistic y. Suppose that y has bounded support. Show that the natural parameter space of the model is $\Re$.

1.6 A family of distributions that is closely related to the exponential family is the family of exponential dispersion models. Suppose that a scalar random variable X has a density of the form

$$p(x; \eta, \sigma^2) = \exp\{[\eta x + k_0(\eta)]/\sigma^2 + S(x, \sigma^2)\} I_A(x), \qquad \eta \in H$$

where for known σ^2 the density p satisfies the conditions of a one-parameter exponential family distribution and H is an open set. The set of density functions $\{p(\cdot; \eta, \sigma^2): \eta \in H, \ \sigma^2 > 0\}$ is said to be an exponential dispersion model.

(a) Find the cumulant-generating function of X.

(b) Suppose that a random variable Y has the density function $p(\cdot; \eta, 1)$, that is, it has the distribution as X except that σ^2 is known to be 1. Find the cumulants of X in terms of the cumulant of Y.

(c) Let $W = \sigma^2 Y$. Is the distribution of W an exponential dispersion model?

1.7 Let $Y_1, \ldots, Y_n$ denote independent identically distributed random variables, each uniformly distributed on the interval (θ_1, θ_2), $\theta_1 < \theta_2$.

(a) Show that this is a transformation model and identify the group of transformations. Show the correspondence between the parameter space and the transformations.

(b) Find a maximal invariant statistic.

1.8 Let X denote a random variable with frequency function

$$p(x) = c\theta^{x-1}, \qquad x = 1, 2, \ldots$$

where $0 < \theta < 1$ and c is a constant.
Find the cumulant-generating function of X and the first three cumulants.

1.9 Let X denote a random variable with density function

$$p(x) = cx^{\alpha-1} \exp\{-\beta x\}, \quad x > 0$$

where $\alpha > 0$, $\beta > 0$, and c is a constant.
Find the cumulant-generating function of X and the first three cumulants.

1.10 Let Y be a scalar random variable with density function $p(y)$ that is symmetric about 0. Assume that the cumulant-generating function of Y, $K_Y(t)$, exists for all $|t| \leq t_0$ for some $t_0 > 0$. Show that all the odd cumulants of Y are 0.

1.11 Consider n independent random variables $Y_1, \ldots, Y_n$ such that all the cumulants of Y_j exist and are bounded for all j. Let $S = \sum Y_j$, $\bar{Y} = S/n$, and $Y = \sqrt{n}[\bar{Y} - \mathrm{E}(\bar{Y})]$. Show that $\kappa_j(S) = O(n)$, $\kappa_j(Y) = O(n^{-(j-2)/2})$ and $\rho_j(S) = O(n^{-(j-2)/2})$.

1.12 Let Y denote a random vector of dimension m and let

$$X = a + BY$$

where a is a constant vector of dimension m and B is an $m \times m$ matrix. Give an expression for the cumulants of X in terms of the cumulants of Y.

1.13 Consider a sequence of random variables $X_1, X_2, \ldots, X_n$ which each take the values 0 and 1. Assume that

$$\Pr(X_j = 1) = 1 - \Pr(X_j = 0) = \phi, \quad j = 1, \ldots, n$$

where $0 < \phi < 1$ and that

$$\Pr(X_j = 1|X_{j-1} = 1) = \lambda, \quad j = 2, \ldots, n.$$

(a) Find $\Pr(X_j = 0|X_{j-1} = 1)$, $\Pr(X_j = 1|X_{j-1} = 0)$, $\Pr(X_j = 0|X_{j-1} = 0)$.

(b) Find the requirements on λ so that this describes a valid probability distribution for $X_1, \ldots, X_n$.

(c) Suppose that ϕ and λ are unknown parameters. Find a three-dimensional sufficient statistic for the model.

1.14 Let $Y_1, \ldots, Y_n$ denote independent normally distributed random variables each with mean θ and standard deviation $c\theta$ where $\theta > 0$ and c is a known positive constant. Show that

$$A = \frac{\sum Y_j}{[\sum (Y_j - \bar{Y})^2]^{1/2}},$$

where $\bar{Y} = n^{-1} \sum Y_j$, is an ancillary statistic.

1.15 Let $Y_1, \ldots, Y_n$ denote independent identically distributed normal random variables each with mean 0 and variance 1. Let $\bar{Y} = n^{-1} \sum Y_j$ and let $T = (T_1, \ldots, T_n)$ where $T_j = (Y_j - \bar{Y})/S$ where $S = \sum (Y_j - \bar{Y})^2$. Show that T and $\bar{Y}$ are independent.

2
Large-sample approximations

2.1 Introduction

In many cases it is difficult or impossible to find the exact sampling distribution of a statistic. In these cases, we may use a large-sample approximation. Let $Y_1, \ldots, Y_n$ denote a sequence of real-valued random variables and let F_n denote the distribution function of Y_n. Consider the problem of approximating F_n. In typical applications, n denotes the underlying sample size so that, for example, Y_n may be the sample mean based on n independent observations.

The basic tool for approximating distributions is convergence in distribution. Let Y denote a real-valued random variable with distribution function F. If, as $n \to \infty$, $F_n(t)$ converges to $F(t)$ for each t at which F is continuous, then we say that Y_n *converges in distribution* to Y, written $Y_n \xrightarrow{\mathcal{D}} Y$. Alternatively, we might say that the Y_n is approximately distributed according to F or that F is the asymptotic distribution of Y_n. To approximate probabilities regarding Y_n we use the distribution function F.

Example 2.1 Minimum of uniform random variables
Let $X_1, \ldots, X_n$ denote independent random variables each uniformly distributed on the interval $(0, 1)$ and let $Y_n = n \min(X_1, \ldots, X_n)$. Then

$$\begin{aligned} F_n(t) = \Pr(Y_n \leq t) &= 1 - \Pr(X_1 > t/n, \ldots, X_n > t/n) \\ &= 1 - \left(1 - \frac{t}{n}\right)^n, \quad 0 \leq t \leq n. \end{aligned}$$

It is straightforward to show that

$$\lim_{n\to\infty} \log[1 - F_n(t)] = -t, \quad t \geq 0$$

so that

$$\lim_{n\to\infty} F_n(t) = 1 - \exp\{-t\}, \quad t \geq 0.$$

Let Y denote a random variable with density function $\exp\{-y\}$, $y > 0$. Then $Y_n \xrightarrow{\mathcal{D}} Y$ as $n \to \infty$. Note that this example is exceptional in that the exact distribution of Y_n is known. ∎

A concept closely related to convergence in distribution is *convergence in probability*; a sequence of random variables $Y_1, Y_2, \ldots$ is said to converge in probability to 0, written $Y_n \xrightarrow{p} 0$, if for any $\epsilon > 0$

$$\lim_{n\to\infty} \Pr(|Y_n| > \epsilon) = 0.$$

It is straightforward to show that this is equivalent to $Y_n \xrightarrow{\mathcal{D}} 0$, where here 0 denotes the random variable identically equal to 0. In general, $Y_n \xrightarrow{p} Y$ if $Y_n - Y \xrightarrow{p} 0$; note that this is a stronger condition than $Y_n \xrightarrow{\mathcal{D}} Y$. By Chebychev's inequality $Y_n - \mathrm{E}(Y_n) \xrightarrow{p} 0$ provided that $\mathrm{Var}(Y_n) \to 0$ as $n \to \infty$.

One way to show that Y_n converges in distribution to Y is to use cumulant-generating functions. Suppose that Y_n has cumulant-generating function $K_n(t)$ which defined for all $|t| < t_0$, $t_0 > 0$ and Y has cumulant-generating function $K(t)$ also defined for all $|t| < t_0$. Then Y_n converges in distribution to Y provided that $K_n(t)$ converges to $K(t)$ for all $|t| < t_0$.

2.2 Central limit theorem

Convergence in distribution is often established using some version of the *central limit theorem*. In its simplest form the central limit theorem states that if $X_1, \ldots, X_n$ are independent identically distributed random variables each with mean μ and standard deviation σ then

$$Y_n = \frac{X_1 + \cdots + X_n - n\mu}{\sqrt{n}\sigma}$$

converges in distribution to Z, a standard normal random variable.

A simple proof of the central limit theorem is available whenever the X_j have a cumulant-generating function $K(t)$. Then the cumulant-generating function of Y_n is given by

$$K_n(t) = -\frac{\sqrt{n}t\mu}{\sigma} + nK\left(\frac{t}{\sqrt{n}\sigma}\right).$$

Using a Taylor's series expansion, it can be shown that

$$K_n(t) = \frac{1}{2}t^2 + \frac{1}{6}\frac{K'''(0)}{\sigma^3}\frac{1}{\sqrt{n}} + \cdots$$

so that, as $n \to \infty$, $K_n(t)$ converges to $t^2/2$, the cumulant-generating function of the standard normal distribution.

The central limit theorem is available in much more general settings than the one considered here. A relatively general form of the central limit theorem applies to *triangular arrays*. A triangular array of random variables is a two-dimensional array of the form $\{X_{nj}, j = 1, \ldots, n; n = 1, \ldots, \}$ such that, for each fixed n, $X_{n1}, \ldots, X_{nn}$ are independent. Let $\mu_{nj} = \mathrm{E}(X_{nj})$, $\sigma_{nj}^2 = \mathrm{Var}(X_{nj})$, and let

$$Y_n = \frac{\sum_{j=1}^n (X_{nj} - \mu_{nj})}{\left[\sum_{j=1}^n \sigma_{nj}^2\right]^{1/2}}.$$

Then Y_n converges in distribution to a standard normal random variable provided that there exists a $\delta > 0$ such that

$$\lim_{n\to\infty} \sum_{j=1}^{n} \frac{1}{\left[\sum_{j=1}^n \sigma_{jn}^2\right]^{1+\delta/2}} \mathrm{E}\left[|X_{nj}|^{2+\delta}\right] = 0;$$

this is known as *Lyapounov's condition.*

The conclusion of these central limit theorems is essentially always the same: $[\bar{X} - \mathrm{E}(\bar{X})]/[\mathrm{Var}(\bar{X})]^{1/2}$ is approximately normally distributed.

Example 2.2 Sample mean of Laplace random variables
Let $Y_1, \ldots, Y_n$ denote independent random variables each with a standard Laplace distribution, as discussed in Example 1.14. Since the standard Laplace distribution has mean 0 and variance 2,

$$\frac{\sqrt{n}\bar{Y}}{\sqrt{2}} \xrightarrow{\mathcal{D}} N(0, 1) \quad \text{as } n \to \infty$$

where $N(0, 1)$ denotes a random variable with a standard normal distribution. ■

Example 2.3 Sample mean of Poisson random variables with differing means
Let $Z_1, \ldots, Z_n$ denote independent Poisson random variables such that Z_j has mean j/n. Consider the asymptotic distribution of $\bar{Z}$.

First note that, since the definition of Z_j depends on the value of n (through its mean, j/n), the random variables in this example actually form a triangular array. Let X_{nj}, $j = 1, \ldots, n$, denote independent Poisson random variables such that X_{nj} has mean $\mu_{nj} = j/n$. Then the random variable $\bar{Z}$ described above has the same distribution as $\bar{X}_n = \sum X_{nj}/n$.

Since the variance of a Poisson random variable is equal to its mean, $\mathrm{Var}(X_{nj}) = \sigma_{nj}^2 = j/n$ and hence

$$\sum_{j=1}^{n} \sigma_{nj}^2 = \sum_{j=1}^{n} \mu_{nj} = \frac{n+1}{2}.$$

Let

$$Y_n = \frac{n\bar{X}_n - (n+1)/2}{[(n+1)/2]^{1/2}}.$$

It is straightforward to show that

$$\mathrm{E}[X_{nj}^3] = 1 + 3\left(\frac{j}{n}\right)^2 + \left(\frac{j}{n}\right)^3$$

so that, taking $\delta = 1$,

$$\sum_{j=1}^{n} \frac{1}{\left[\sum_{j=1}^{n} \sigma_{jn}^2\right]^{1+\delta/2}} \mathrm{E}[|X_{nj}|^{2+\delta}] = \left(\frac{n+1}{2}\right)^{-3/2} \left[\frac{9}{4}n + 2 + \frac{3}{4}\frac{1}{n}\right];$$

it follows that Lyapounov's condition holds and hence Y_n is asymptotically distributed according to a standard normal distribution. That is, $\bar{X}_n$ is approximately normally distributed with mean $(n+1)/(2n)$ and variance $(n+1)/(2n^2)$. ∎

The central limit theorem is also applicable to vector-valued random variables. For instance, let $X_1, \ldots, X_n$ denote independent identically distributed random vectors each with mean vector μ and covariance matrix Σ, which is assumed to be positive definite. Then $\sqrt{n}\Sigma^{-1/2}(\bar{X} - \mu)$ converges in distribution to a multivariate normal distribution with mean vector 0 and covariance matrix equal to the identity. That is, $\bar{X}$ is approximately distributed according to a multivariate normal distribution with mean vector μ and covariance matrix $n^{-1}\Sigma$.

Example 2.4 Multinomial distribution

Consider n independent identically distributed random variables taking values in the set $\{1, \ldots, m+1\}$ such that the value j has probability θ_j. Let Y_j denote the number of times that the value j is observed. Then $Y = (Y_1, \ldots, Y_m)$ has a multinomial distribution with density of the form

$$p(y; \theta) = \binom{n}{y_1, \ldots, y_m, y_{m+1}} \theta_1^{y_1} \cdots \theta_m^{y_m} \theta_{m+1}^{y_{m+1}}$$

where $y_{m+1} = n - y_1 - \cdots - y_m$ and $\theta_1 + \cdots + \theta_{m+1} = 1$; this distribution was discussed in Example 1.19.

Let e_j denote the vector of length m with all elements equal to 0 except the jth, which is equal to 1. For $j = 1, \ldots, n$, let X_j be equal to e_k if the jth observation is equal to k. Hence, $X_1, \ldots, X_n$ are independent identically distributed random vectors each with mean vector $(\theta_1, \ldots, \theta_m)^T$. Note that $Y = \sum X_j$; from Example 1.19 it follows the X_j have covariance matrix

$$\Sigma = \begin{pmatrix} \theta_1(1-\theta_1) & -\theta_1\theta_2 & \cdots & -\theta_1\theta_m \\ -\theta_2\theta_1 & \theta_2(1-\theta_2) & \cdots & -\theta_2\theta_m \\ \vdots & \vdots & & \vdots \\ -\theta_m\theta_1 & -\theta_m\theta_2 & \cdots & \theta_m(1-\theta_m) \end{pmatrix}.$$

It follows from the central limit theorem that Y/n is approximately normally distributed with mean vector $(\theta_1, \ldots, \theta_m)$ and covariance matrix $n^{-1}\Sigma$. ∎

2.3 Edgeworth series approximations

Let $X_1, \ldots, X_n$ denote independent identically distributed real-valued random variables with mean μ, standard deviation σ, and cumulant-generating function $K(t)$. Let $Y_n = \sqrt{n}(\bar{X} - \mu)/\sigma$. Then $K_n(t)$, the cumulant-generating function of Y_n, may be expanded

$$K_n(t) = \frac{1}{2}t^2 + \frac{1}{6}\frac{K'''(0)}{\sigma^3}\frac{t^3}{\sqrt{n}} + \cdots . \tag{2.1}$$

A first-order approximation to the cumulant-generating function $K_n(t)$ is given by $t^2/2$, the cumulant-generating function of the normal distribution. Hence, to first-order, the distribution of Y_n may be approximated by the standard normal distribution.

A more accurate approximation to the distribution of Y_n may be obtained by retaining more terms in the approximation to $K_n(t)$. For instance, including the next term in the expansion (2.1) yields the approximation

$$\frac{1}{2}t^2 + \frac{1}{6}\frac{K'''(0)}{\sigma^3}\frac{t^3}{\sqrt{n}} = \frac{1}{2}t^2 + \frac{1}{6}\frac{\kappa_3}{\sigma^3}\frac{t^3}{\sqrt{n}}.$$

We may obtain an approximation to the density function of Y_n by inverting this cumulant-generating function; that is, by finding the density function that corresponds to this cumulant-generating function. To do this, first note that the corresponding moment-generating function is given by

$$\exp\left\{\frac{1}{2}t^2 + \frac{1}{6}\frac{\kappa_3}{\sigma^3}\frac{t^3}{\sqrt{n}}\right\} = \exp\{t^2/2\}\left[1 + \frac{1}{6}\frac{\kappa_3}{\sigma^3}\frac{t^3}{\sqrt{n}} + \cdots\right].$$

Hence, to invert this expansion, we need to find a density $p_n(y)$ such that

$$\int_{-\infty}^{\infty} \exp\{ty\}\, p_n(y)\, dy = \exp\{t^2/2\}\left[1 + \frac{1}{6}\frac{\kappa_3}{\sigma^3}\frac{t^3}{\sqrt{n}} + \cdots\right].$$

Note that

$$\int_{-\infty}^{\infty} \exp\{ty\}\,(-1)^r \frac{d^r\phi(y)}{dy^r}\, dy = t^r \exp\{t^2/2\}, \quad r = 0, 1, \ldots;$$

this may be shown by induction, using integration by parts. Hence, an approximation to the density function of Y_n based on retaining the $1/\sqrt{n}$ term in the expansion for the moment-generating function is given by

$$\phi(y)\left[1 + \frac{\kappa_3}{6\sigma^3}(y^3 - 3y)\frac{1}{\sqrt{n}}\right].$$

In general, an approximation to the density function of Y_n is given by

$$\phi(y)\left[1 + \frac{\kappa_3}{6\sigma^3}H_3(y)\frac{1}{\sqrt{n}} + \left(\frac{\kappa_4}{24\sigma^4}H_4(y) + \frac{\kappa_3^2}{72\sigma^6}H_6(y)\right)\frac{1}{n} + \cdots\right]. \tag{2.2}$$

Here the functions $H_j(y)$ are the Hermite polynomials, defined by

$$H_r(y)\phi(y) = (-1)^r \frac{d^r \phi(y)}{dy^r};$$

in particular,

$$H_2(y) = y^2 - 1, \qquad H_3(y) = y^3 - 3y, \qquad H_4(y) = y^4 - 6y^2 + 3,$$
$$H_5(y) = y^5 - 10y^3 + 15y.$$

An approximation to the distribution function of Y_n may then be obtained by integrating the approximation to the density. This procedure is simplified by noting that

$$\int_{-\infty}^{y} H_r(t)\phi(t)\,dt = (-1)^r \int_{-\infty}^{y} \frac{d^r \phi(t)}{dt^r}\,dt = (-1)^r \frac{d^{r-1}\phi(t)}{dt^{r-1}}\Big|_{-\infty}^{y}$$
$$= -H_{r-1}(y)\phi(y).$$

Hence, an approximation to the distribution function F_n of Y_n is given by

$$\Phi(y) - \phi(y)\left[\frac{\kappa_3}{6\sigma^3}H_2(y)\frac{1}{\sqrt{n}} + \left(\frac{\kappa_4}{24\sigma^4}H_3(y) + \frac{\kappa_3^2}{72\sigma^6}H_5(y)\right)\frac{1}{n} + \cdots\right]. \quad (2.3)$$

Typically, either the $n^{-1/2}$ term or the $n^{-1/2}$ term together with the n^{-1} term are used when approximating the distribution function of Y_n. The error of the approximation is one power of $n^{-1/2}$ greater than the last included term. For instance, if the approximation includes only the $n^{-1/2}$ term, then the error is of order n^{-1}.

The Edgeworth series approximation to the distribution of Y_n shows that the main contribution to the error in the normal approximation to this distribution is $\rho_3 = \kappa_3/\sigma^3$, the standardized skewness. When the underlying distribution is symmetric, then the normal approximation is accurate to order $O(n^{-1})$. Also, when the density function of Y_n evaluated at $y = 0$ is required or, equivalently, when the density function of $\bar{X}$ evaluated at $x = \mu$ is required, then the usual normal approximation is accurate to order $O(n^{-1})$, instead of $O(n^{-1/2})$.

The Edgeworth expansion given in (2.2) or (2.3) is for the sample mean standardized to have mean 0 and unit variance. A similar expansion for the unstandardized sample mean may be obtained by transforming (2.2) and is given by

$$\frac{\sqrt{n}}{\sigma}\phi(\sqrt{n}(\bar{x} - \mu)/\sigma)\left[1 + \frac{\kappa_3}{6\sigma^3}H_3(\sqrt{n}(\bar{x} - \mu)/\sigma)\frac{1}{\sqrt{n}} + \cdots\right].$$

In some cases, the mean and variance of Y_n are not exactly equal to 0 and 1, respectively; this may happen, for instance, when the mean and variance of X_j must be approximated. Suppose that Y_n has mean $\mu/\sqrt{n} + O(n^{-3/2})$ and variance $1 + \tau/n + O(n^{-2})$; then the cumulant-generating function of Y_n may be expanded

$$\left(\frac{\mu}{\sqrt{n}} + O(n^{-3/2})\right)t + \frac{1}{2}\left(1 + \frac{\tau}{n} + O(n^{-2})\right)t^2 + \frac{1}{6}\frac{K'''(0)}{\sigma^3}\frac{t^3}{\sqrt{n}} + \cdots.$$

The argument leading to (2.3) may be repeated to yield an expansion for the distribution function of Y_n of the form

$$\Phi(y) - \phi(y)\left\{\frac{\mu}{\sqrt{n}} + \frac{\kappa_3}{6}H_2(y)\frac{1}{\sqrt{n}} + \frac{1}{2}\left(\frac{\mu^2}{n} + \frac{\tau}{n}\right)y\right.$$
$$\left.+\left[\frac{1}{24}\kappa_4 H_3(y) + \frac{1}{72}\kappa_3^2 H_5(y) + \frac{1}{6}\kappa_3\mu H_3(y)\right]\frac{1}{n} + O\left(n^{-3/2}\right)\right\}.$$

An important property of Edgeworth series approximations is that they apply to fixed values of y, the argument at which the density or distribution function of Y_n is to be evaluated. As noted above, an Edgeworth expansion for Y_n may be transformed to obtain an approximation for the density or distribution function of

$$\bar{X} = \mu + \sigma Y_n/\sqrt{n}.$$

The resulting approximation however only applies when evaluated at an argument x of the form

$$x = \mu + \sigma y/\sqrt{n}$$

where y is a fixed value. This is sometimes referred to as the *moderate deviation* range of values of x. It is generally preferable to have an approximation for the density or distribution function of $\bar{X}$ that applies for fixed values of the argument x; this is referred to as the *large deviation* range. An approximation that has high accuracy in the large deviation range can be expected to perform well in the tails of the distribution of $\bar{X}$. In terms of the random variable Y_n, the moderate deviation range of y is $y = O(1)$ while the large deviation range is $y = O(\sqrt{n})$.

Note that an Edgeworth series approximation is not limited to sample means. The same type of approach can be applied to any statistic that has a cumulant-generating function of the same general form as that of a sample mean.

The Edgeworth approximation as described here applies only to continuous random variables. We now consider the case in which the X_j take values in the set of nonnegative integers; similar results are available for any lattice random variables. As before, $K_n(t)$, the cumulant-generating function of $Y_n = \sqrt{n}(\bar{X} - \mu)/\sigma$ may be expanded

$$K_n(t) = \frac{1}{2}t^2 + \frac{1}{6}\frac{K'''(0)}{\sigma^3}\frac{t^3}{\sqrt{n}} + \cdots.$$

This expansion for the cumulant-generating function may be converted to an expansion for the density of Y_n by finding a function p, depending on n, such that

$$\sum_{j=0}^{\infty} \exp\{t(j/\sqrt{n} - \sqrt{n}\mu)/\sigma\} p[j/(\sigma\sqrt{n}) - \sqrt{n}\mu/\sigma]$$

$$= \exp\left\{\frac{1}{2}t^2 + \frac{1}{6}\frac{\kappa_3}{\sigma^3}\frac{t^3}{\sqrt{n}}\right\} = \exp\{t^2/2\}\left[1 + \frac{1}{6}\frac{\kappa_3}{\sigma^3}\frac{t^3}{\sqrt{n}} + \cdots\right]. \tag{2.4}$$

The leading terms in the expansion of p may then be used to approximate the density of Y_n; note that Y_n takes values in the set

$$\left\{0, \frac{1}{\sigma\sqrt{n}}, \frac{2}{\sigma\sqrt{n}}, \ldots\right\}.$$

To find such a function p, we first approximate the sum in (2.4) by an integral. Let f denote a sufficiently smooth function defined on the real line such that f and its derivatives vanish at ∞. The Euler–MacLaurin formula states that for any $m = 0, 1, \ldots,$

$$\sum_{j=0}^{\infty} f(j) = \int_0^{\infty} f(x)\,dx + f(0)/2 - \sum_{i=1}^{m} B_{2i} f^{(2i-1)}(0) + R_m$$

where B_{2i} are the Bernoulli numbers and the remainder R_m is given by

$$R_m = -\frac{1}{(2m+2)!}\int_0^{\infty} B_{2m+2}(x) f^{(2m+2)}(x)\,dx;$$

here $B_r(x)$ denotes the rth Bernoulli polynomial.

Consider a function of the form

$$f(x) = \exp\{t(x/\sqrt{n} - \sqrt{n}\mu)/\sigma\}\phi^{(r)}\left(\frac{x}{\sigma\sqrt{n}} - \sqrt{n}\frac{\mu}{\sigma}\right).$$

Let $o(\epsilon_n)$ denote a generic term that goes to 0 faster than any power of n^{-1}. Then

$$f^{(k)}(0) = \sum_{i=0}^{k} \binom{k}{i} t^{k-i} \exp\{-\sqrt{n}\mu/\sigma\}\phi^{(r+i)}(-\sqrt{n}\mu/\sigma)(\sigma\sqrt{n})^{-k}$$

$$= o(\epsilon_n), \quad k = 0, \ldots.$$

Hence,

$$\sum_{i=1}^{m} B_{2i} f^{(2i-1)}(0) = o(\epsilon_n), \quad m = 1, \ldots.$$

It is straightforward to show that for $m = 1, \ldots,$ $R_m = O(n^{-(m+1/2)})$. Hence,

$$\sum_{j=0}^{\infty} \exp\{t(j/\sqrt{n} - \sqrt{n}\mu)/\sigma\}\,\phi^{(r)}\left(\frac{j}{\sigma\sqrt{n}} - \sqrt{n}\frac{\mu}{\sigma}\right)$$

$$= \int_0^\infty \exp\{t(x/\sqrt{n} - \sqrt{n}\mu)/\sigma\}\,\phi^{(r)}\left(\frac{x}{\sigma\sqrt{n}} - \sqrt{n}\frac{\mu}{\sigma}\right)dx + o(\epsilon_n)$$

$$= \sigma\sqrt{n}\int_{-\infty}^{\infty} \exp\{t\mu\}\,\phi^{(r)}(u)\,du + o(\epsilon_n)$$

$$= \sigma\sqrt{n}t^r \exp\{t^2/2\} + o(\epsilon_n).$$

It follows that we may take

$$p(y) = \frac{1}{\sigma\sqrt{n}}\phi(y)\left[1 + \frac{\kappa_3}{6\sigma^3}H_3(y)\frac{1}{\sqrt{n}} + \left(\frac{\kappa_4}{24\sigma^4}H_4(y) + \frac{\kappa_3^2}{72\sigma^6}H_6(y)\right)\frac{1}{n} + \cdots\right];$$

in this expression, y is a value in the support of Y_n. That is, the Edgeworth expansion for $\sigma\sqrt{n}\,\Pr(Y_n = y)$ in the lattice case is the same as the Edgeworth expansion for the density of Y_n in the continuous case.

For the distribution function, however, the situation is somewhat different. Since the X_j are discrete taking integer values, Y_n is discrete with jumps of order $O(n^{-1/2})$. Since the Edgeworth approximation to the distribution function is a continuous distribution, the order of the error of the approximation cannot be less than $O(n^{-1/2})$, no matter how many terms are included. There are however versions of the Edgeworth approximation that do have error of order $O(n^{-3/2})$.

Example 2.5 Sample mean of Laplace random variables
Let $Y_1, \ldots, Y_n$ denote independent random variables each with a standard Laplace distribution, as discussed in Example 2.2. Since the first four cumulants of the standard Laplace distribution are given by 0, 2, 0, 12, respectively, using a Edgeworth approximation

$$\Pr\{\sqrt{n}\bar{Y}/\sqrt{2} \le t\} = \Phi(t) - \frac{1}{32}(y^3 - 3y)\phi(y)\frac{1}{n} + O(n^{-3/2}).$$ ∎

Example 2.6 Chi-square distribution
Let $Z_1, \ldots, Z_n$ denote independent standard normal random variables and let $X_n = \sum_{j=1}^n Z_j^2$. Then X_n has a chi-square distribution with n degrees of freedom. We will consider approximations to the chi-square distribution function based on an Edgeworth expansion.

It is straightforward to show that the first four cumulants of Z_j^2 are 1, 2, 8, and 48, respectively. Let

$$Y_n = \sqrt{n}\left(\frac{X_n/n - 1}{\sqrt{2}}\right).$$

Then $\Pr(Y_n \leq y)$ may be approximated by

$$\Phi(y) - \phi(y)\left[\frac{2}{3\sqrt{2}}H_2(y)\frac{1}{\sqrt{n}} + \left(\frac{1}{2}H_3(y) + \frac{1}{9}H_5(y)\right)\frac{1}{n}\right].$$

For instance, consider approximation of $\Pr(X_n \leq 1.145)$ for the case $n = 5$; the exact probability is approximately 0.05; note that $\Pr(X_n \leq 1.145) = \Pr(Y_n \leq -1.219)$. The first-order approximation, based on the normal distribution, is 0.111; the second-order approximation which includes the $1/\sqrt{n}$ term in the Edgeworth expansion is 0.092, while the third-order approximation is 0.069. ∎

Example 2.7 Sum of Poisson random variables
Let $X_1, \ldots, X_n$ denote independent Poisson random variables each with mean μ. Recall that all cumulants of X_j are equal to μ. Hence, an Edgeworth expansion for $\sqrt{(n\mu)}\Pr(Y_n = y)$ is given by

$$\phi(y)\left[1 + \frac{1}{6}\frac{1}{\sqrt{(n\mu)}}H_3(y) + \left(\frac{1}{24}H_4(y) + \frac{1}{72}H_6(y)\right)\frac{1}{n\mu}\right].$$

Let $S = \sum X_j$. Then $\Pr(Y_n = y)$ is equivalent to $\Pr[S = n\mu + y\sqrt{(n\mu)}]$.

Consider the case $n\mu = 4$ and suppose we are interested in approximating $\Pr(S = 6)$. This is equivalent to $\Pr(Y_n = 1)$. The Edgeworth expansion approximation including only a skewness correction is 0.1008; if both correction terms are included, the approximation is 0.1050. The exact probability is 0.1042. ∎

Edgeworth series approximations also apply to vector-valued random variables. The derivation is essentially the same as in the univariate case. The cumulant-generating function of the normalized sample mean can be expanded up to a specified power of $n^{-1/2}$. The truncated expansion may then be inverted to yield an approximation to the density function of the standardized sample mean. The leading term in the approximation is the multivariate normal density and the later terms are based on higher-order cumulant arrays.

More specifically, let $X_1, \ldots, X_n$ denote independent identically distributed random vectors each of dimension m. Let $K(t)$ denote the cumulant-generating function of X_1 and let κ_i denote the mean of X_{1i}, κ_{ij} denote the covariance of X_{1i}, X_{1j}, κ_{ijk} denote $\text{cum}(X_{1i}, X_{1j}, X_{1k})$, and so on. We will write Σ for the covariance matrix with (i, j)th element κ_{ij}. An Edgeworth expansion for the density of

$Y_n = (\sum X_j - \mathrm{E}(X_j))/\sqrt{n}$ evaluated at y is of the form

$$\phi_m(y; \Sigma)\left[1 + \frac{1}{6\sqrt{n}} \sum_{i,j,k} \kappa_{ijk} H_{ijk}(y; \Sigma) + \frac{1}{24n} \sum_{i,j,k,l} \kappa_{ijkl} H_{ijkl}(y; \Sigma) \right.$$

$$\left. + \frac{1}{72n} \sum_{i,j,k,l,q,r} \kappa_{ijk}\kappa_{lqr} H_{ijklqr}(y; \Sigma) + \cdots \right];$$

here $\phi_m(y; \Sigma)$ denotes the m-dimensional multivariate normal density with mean vector 0 and covariance matrix Σ and $H_{ijk}(y; \Sigma)$, $H_{ijkl}(y; \Sigma)$ and $H_{ijklqr}(y; \Sigma)$ are the multivariate forms of the Hermite polynomials. These multivariate Hermite polynomials may be obtained by differentiating the density $\phi_m(y; \Sigma)$, similar to the way in which the univariate Hermite polynomials were defined. For instance,

$$H_{ijk}(y; \Sigma) = z_i z_j z_k - \lambda_{jk} z_i - \lambda_{ik} z_j - \lambda_{ij} z_k$$

where $z = \Sigma^{-1} y$ and the matrix with elements λ_{ij} is Σ^{-1}.

2.4 Saddlepoint approximation of densities

Let $X_1, \ldots, X_n$ denote independent identically distributed random variables each with cumulant-generating function $K(t)$, $|t| < t_0$, and density function $p(x)$. Assume that the X_j have a continuous distribution; the lattice case will be considered in Section 2.7. For $|\lambda| < t_0$ define

$$p(x; \lambda) = \exp\{x\lambda - K(\lambda)\}\, p(x).$$

Note that

$$\int p(x; \lambda)\, dx = \int \exp\{x\lambda\}\, p(x)\, dx \exp\{-K(\lambda)\} = 1$$

so that, for each λ, $p(x; \lambda)$ defines a density function. The cumulant-generating function of the density $p(x; \lambda)$ is $K(\lambda + s) - K(\lambda)$ and, hence, the cumulants are given by $K'(\lambda)$, $K''(\lambda)$, and so on.

Let $S_n = X_1 + \cdots + X_n$ and let $p_n(s; \lambda)$ denote the density function of S_n under the density $p(x; \lambda)$ for the X_j. Note that the actual density of S_n is given by $p_n(s) \equiv p_n(s; 0)$.

Since

$$p_n(s; \lambda) = \int p(x_1; \lambda) \cdots p(x_n; \lambda)\, dx_1 \cdots dx_n,$$

where the integral is over all $(x_1, \ldots, x_n)$ such that $\sum x_j = s$, it is straightforward to show that

$$p_n(s; \lambda) = \exp\{s\lambda - nK(\lambda)\}\, p_n(s);$$

this holds for any λ, $|\lambda| < t_0$. Let $\hat{p}_n(s;\lambda)$ denote an approximation to $p_n(s;\lambda)$. Then an approximation to $p_n(s)$ is given by

$$\exp\{nK(\lambda) - s\lambda\}\, \hat{p}_n(s;\lambda).$$

The idea behind the saddlepoint approximation is to choose λ so that $\hat{p}_n(s;\lambda)$ is an accurate approximation to $p_n(s;\lambda)$; the value of λ chosen will depend on s. Since we are not directly approximating the density of interest, $p_n(s;0)$, the saddlepoint approximation is often referred to as an *indirect* approximation, in contrast to the Edgeworth approximation which is referred to as a *direct* approximation.

We have seen that the normal approximation to a density function is very accurate when it is evaluated at the mean. Hence, the value of λ is chosen so that the point s at which the density of S is to be evaluated corresponds to the mean of $p(\cdot;\lambda)$. That is, given s, choose $\lambda = \hat{\lambda}_s$ so that

$$s = \mathrm{E}(S_n;\lambda) = n\mathrm{E}(Y_1;\lambda) = nK'(\lambda).$$

It follows that $\hat{\lambda}_s$ satisfies $nK'(\hat{\lambda}_s) = s$. Note that, since $K''(\lambda)$ is the variance of X_j under $p(x;\lambda)$, $K''(\lambda) > 0$ for all λ so that $K(\lambda)$ is a convex function which implies that the equation $nK'(\hat{\lambda}_s) = s$ has at most one solution. We assume that s is such that a solution exists.

For the approximation $\hat{p}_n(s;\lambda)$ we can use an Edgeworth approximation incorporating one correction term; since the evaluation of the density is at the mean of S_n and the variance of the X_j under $p(x;\lambda)$ is $K''(\lambda)$, the approximation is given by

$$[2\pi nK''(\hat{\lambda}_s)]^{-1/2}.$$

It follows that an approximation to $p_n(s)$ is given by

$$\hat{p}_n(s) = \exp\{nK(\hat{\lambda}_s) - s\hat{\lambda}_s\}\, [2\pi nK''(\hat{\lambda}_s)]^{-1/2}.$$

Let $\bar{X} = S_n/n$. Then an approximation to the density of $\bar{X}$ is given by

$$\exp\{n[K(\hat{\lambda}_x) - x\hat{\lambda}_x]\}\, [2\pi K''(\hat{\lambda}_x)/n]^{-1/2}$$

where $\hat{\lambda}_x$ satisfies $K'(\hat{\lambda}_x) = x$. The error of this approximation is of order $O(n^{-1})$, the order of the error of the Edgeworth series approximation used.

An important property of the saddlepoint approximation is that, because the Edgeworth expansion is always evaluated at 0, the error is $O(n^{-1})$ for any value of the argument x; hence, the error is $O(n^{-1})$ in the large deviation range of x.

The saddlepoint approximation does not necessarily integrate to 1 so that the accuracy of the approximation may be improved by renormalization. The renormalized approximation for the density of $\bar{X}$ is given by

$$\bar{c}\exp\{n[K(\hat{\lambda}_x) - x\hat{\lambda}_x]\}\, [2\pi K''(\hat{\lambda}_x)/n]^{-1/2}$$

where $\bar{c}$ is chosen so that the approximation integrates to 1. In typical cases, $\bar{c}$ must be determined numerically; we will refer to this as the renormalized saddlepoint approximation.

In general, renormalization reduces the error in the approximation from $O(n^{-1})$ to $O(n^{-3/2})$, but only for x in the moderate deviation range, $x = \mathrm{E}(\bar{X}) + O(n^{-1/2})$. To see this, suppose that a two-term Edgeworth series approximation is used in forming the saddlepoint approximation. In this case, the approximation is given by

$$\exp\{n[K(\hat{\lambda}_x) - x\hat{\lambda}_x]\} \times [2\pi K''(\hat{\lambda}_x)/n]^{-1/2}\left[1 + \frac{1}{8}\frac{K^{(4)}(\hat{\lambda}_x)}{K''(\hat{\lambda}_x)^2}\frac{1}{n} - \frac{5}{24}\frac{K'''(\hat{\lambda}_x)^2}{K''(\hat{\lambda}_x)^3}\frac{1}{n}\right];$$

the error in this approximation is of order $O(n^{-3/2})$. For x in the moderate deviation range, we may write

$$1 + \frac{1}{8}\frac{K^{(4)}(\hat{\lambda}_x)}{K''(\hat{\lambda}_x)^2}\frac{1}{n} - \frac{5}{24}\frac{K'''(\hat{\lambda}_x)^2}{K''(\hat{\lambda}_x)^3}\frac{1}{n} = \bar{c} + O(n^{-3/2})$$

for some constant $\bar{c}$ of the form $1 + O(n^{-1})$. Hence, the density of $\bar{X}$ evaluated at x may be written

$$\bar{c}\exp\{n[K(\hat{\lambda}_x) - x\hat{\lambda}_x]\}\,[2\pi K''(\hat{\lambda}_x)/n]^{-1/2}[1 + O(n^{-3/2})],$$

for x of the form $\mathrm{E}(\bar{X}) + O(n^{-1/2})$, establishing the result.

Example 2.8 Sample mean of Laplace random variables
Let $Y_1, \ldots, Y_n$ denote independent random variables each with a standard Laplace distribution, as discussed in Example 2.5. Since the cumulant-generating function is given by $K(t) = -\log(1 - t^2)$, $|t| < 1$,

$$K'(t) = \frac{2t}{1 - t^2}$$

and the equation $K'(\lambda) = y$ may be reduced to a quadratic equation. This quadratic has two solutions, but only one in the interval $(-1, 1)$,

$$\hat{\lambda}_y = \frac{(1 + y^2)^{1/2} - 1}{y}.$$

The saddlepoint approximation to the density of $\bar{Y}$ is therefore given by

$$\frac{\sqrt{n}\exp\{n\}}{2^n(2\pi)^{1/2}}\frac{|y|^{2n-1}}{[(1 + y^2)^{1/2} - 1]^{n-1/2}}\frac{\exp\{-n(1 + y^2)^{1/2}\}}{(1 + y^2)^{1/4}}.$$

The accuracy of this approximation may be improved by renormalization. The required constant $\bar{c}$ may be determined by numerical integration; for instance, for $n = 5$, $\bar{c} \doteq 1.056$. ∎

2.5 Approximation of integrals and sums

One drawback of the saddlepoint method is that, unlike Edgeworth expansions, saddlepoint approximations cannot generally be integrated analytically. Hence, to approximate tail probabilities, further approximation is needed. In this section, we consider a general approach to approximating integrals and sums; in the following section, these methods are used to construct a saddlepoint approximation to a distribution function of a random variable.

Let $h_n(x)$ denote a sequence of functions of a real variable x such that h_n and its derivatives are of order $O(1)$ as $n \to \infty$; for notational simplicity we will drop the dependence of h on n. Consider approximation of the integral

$$\int_z^\infty h(x)\sqrt{n}\phi(\sqrt{n}x)\,dx \tag{2.5}$$

where $\phi(\cdot)$ denotes the standard normal density function. Let $g(x) = [h(x)-h(0)]/x$. Then

$$\begin{aligned}\int_z^\infty h(x)\sqrt{n}\phi(\sqrt{n}x)\,dx &= \int_z^\infty h(0)\sqrt{n}\phi(\sqrt{n}x)\,dx \\ &\quad + \int_z^\infty xg(x)\sqrt{n}\phi(\sqrt{n}x)\,dx \\ &= h(0)[1-\Phi(\sqrt{n}z)] + \int_z^\infty xg(x)\sqrt{n}\phi(\sqrt{n}x)\,dx\end{aligned}$$

where $\Phi(\cdot)$ denotes the standard normal distribution function.

Using integration by parts, we may write

$$\begin{aligned}\int_z^\infty xg(x)\sqrt{n}\phi(\sqrt{n}x)\,dx &= -\frac{1}{n}g(x)\sqrt{n}\phi(\sqrt{n}x)\,\Big|_z^\infty \\ &\quad + \frac{1}{n}\int_z^\infty g'(x)\sqrt{n}\phi(\sqrt{n}x)\,dx.\end{aligned}$$

Assuming that $g(x)$ is such that

$$g(x)\phi(x) \to 0 \quad \text{as } |x| \to \infty,$$

it follows that

$$\begin{aligned}\int_z^\infty xg(x)\sqrt{n}\phi(\sqrt{n}x)\,dx &= \frac{1}{n}g(z)\sqrt{n}\phi(\sqrt{n}z) \\ &\quad + \frac{1}{n}\int_z^\infty g'(x)\sqrt{n}\phi(\sqrt{n}x)\,dx\end{aligned}$$

so that (2.5) may be written

$$h(0)[1-\Phi(\sqrt{n}z)] + \frac{1}{n}g(z)\sqrt{n}\phi(\sqrt{n}z) + \frac{1}{n}\int_z^\infty g'(x)\sqrt{n}\phi(\sqrt{n}x)\,dx.$$

Taking $z = -\infty$ in this expression yields the result

$$\int_{-\infty}^{\infty} h(x)\sqrt{n}\phi(\sqrt{n}x)\,dx = h(0) + \frac{1}{n}\int_{-\infty}^{\infty} g'(x)\sqrt{n}\phi(\sqrt{n}x)\,dx$$

so that

$$h(0) = \int_{-\infty}^{\infty} h(x)\sqrt{n}\phi(\sqrt{n}x)\,dx - \frac{1}{n}\int_{-\infty}^{\infty} g'(x)\sqrt{n}\phi(\sqrt{n}x)\,dx$$

and hence

$$\begin{aligned}\int_{z}^{\infty} & h(x)\sqrt{n}\phi(\sqrt{n}x)\,dx \\ &= [1 - \Phi(\sqrt{n}z)]\int_{-\infty}^{\infty} h(x)\sqrt{n}\phi(\sqrt{n}x)\,dx + \frac{1}{n}g(z)\sqrt{n}\phi(\sqrt{n}z) \\ &\quad + \frac{1}{n}\left\{\int_{z}^{\infty} g'(x)\sqrt{n}\phi(\sqrt{n}x)\,dx - \int_{-\infty}^{\infty} g'(x)\sqrt{n}\phi(\sqrt{n}x)\,dx[1 - \Phi(\sqrt{n}z)]\right\}.\end{aligned} \tag{2.6}$$

The final term in this expression can be approximated further. Note that the integral

$$\int_{z}^{\infty} g'(x)\sqrt{n}\phi(\sqrt{n}x)\,dx$$

is of the same form as (2.5) with $g'(x)$ replacing $h(x)$. Let $Q_z = 1 - \Phi(\sqrt{n}z)$ and let

$$R_0 = \int_{z}^{\infty} g'(x)\sqrt{n}\phi(\sqrt{n}x)\,dx - \int_{-\infty}^{\infty} g'(x)\sqrt{n}\phi(\sqrt{n}x)\,dx\ Q_z.$$

Using (2.6) we have that

$$\begin{aligned}R_0 &= \frac{1}{n}g_1(z)\sqrt{n}\phi(\sqrt{n}z) \\ &\quad + \frac{1}{n}\left\{\int_{z}^{\infty} g_1'(x)\sqrt{n}\phi(\sqrt{n}x)\,dx - \int_{-\infty}^{\infty} g_1'(x)\sqrt{n}\phi(\sqrt{n}x)\,dx\ Q_z\right\}\end{aligned}$$

where the function g_1 is given by $g_1(x) = [g'(x) - g'(0)]/x$.

This process can be continued indefinitely. Let

$$R_1 = \int_{z}^{\infty} g_1'(x)\sqrt{n}\phi(\sqrt{n}x)\,dx - \int_{-\infty}^{\infty} g_1'(x)\sqrt{n}\phi(\sqrt{n}x)\,dx\ Q_z.$$

Again, using (2.6), it follows that

$$R_1 = \frac{1}{n} g_2(z) \sqrt{n} \phi(\sqrt{n} z) + \frac{1}{n} \left\{ \int_z^\infty g_2'(x) \sqrt{n} \phi(\sqrt{n} x)\, dx - \int_{-\infty}^\infty g_2'(x) \sqrt{n} \phi(\sqrt{n} x)\, dx\, Q_z \right\}$$

where $g_2(x) = [g_1'(x) - g_1'(0)]/x$.

Let $g_{j+1}(x) = [g_j'(x) - g_j'(0)]/x$. Continuing in this way shows that

$$\int_z^\infty h(x) \sqrt{n} \phi(\sqrt{n} x)\, dx = [1 - \Phi(\sqrt{n} z)] \int_{-\infty}^\infty h(x) \sqrt{n} \phi(\sqrt{n} x)\, dx + \frac{1}{n} g(z) \sqrt{n} \phi(\sqrt{n} z) + \frac{1}{n} \left\{ \left[\frac{1}{n} g_1(z) + \frac{1}{n^2} g_2(z) + \cdots \right] \sqrt{n} \phi(\sqrt{n} z) \right\}.$$

Hence,

$$\int_z^\infty h(x) \sqrt{n} \phi(\sqrt{n} x)\, dx = [1 - \Phi(\sqrt{n} z)] \int_{-\infty}^\infty h(x) \sqrt{n} \phi(\sqrt{n} x)\, dx + \frac{1}{n} \sqrt{n} \phi(\sqrt{n} z) [g(z) + R(z)]$$

where

$$R(z) = \frac{1}{n} g_1(z) + \frac{1}{n^2} g_2(z) + \cdots.$$

Note that the functions $g_j(x)$ are, in general, well behaved provided that $h(x)$ is a well-behaved function. For instance, suppose that h is of the form

$$h(x) = \sum_0^\infty h_j x^j / j!.$$

Then

$$g(x) = \sum_0^\infty \frac{h_{j+1}}{j+1} x^j / j!, \qquad g_1(x) = \sum_0^\infty \frac{h_{j+3}}{(j+3)(j+1)} x^j / j!$$

and so on.

Let $\bar{c}$ denote a constant such that

$$\bar{c} \int_{-\infty}^\infty h(x) \sqrt{n} \phi(\sqrt{n} x)\, dx = 1;$$

assume that $\bar{c} = 1 + O(n^{-1})$. Then, using $\bar{c}h(\cdot)$ in place of $h(\cdot)$, we have that

$$\int_z^\infty \bar{c}h(x)\sqrt{n}\phi(\sqrt{n}x)\,dx$$
$$= [1 - \Phi(\sqrt{n}z)] + \frac{1}{n}\sqrt{n}\phi(\sqrt{n}z)\left[\bar{c}\frac{h(z) - h(0)}{z} + O(n^{-1})\right].$$

Since

$$\bar{c}[h(z) - h(0)] = h(z) - h(0) + O(n^{-1})[h(z) - h(0)]$$

it follows that

$$\int_z^\infty \bar{c}h(x)\sqrt{n}\phi(\sqrt{n}x)\,dx$$
$$= [1 - \Phi(\sqrt{n}z)] + \frac{1}{n}\sqrt{n}\phi(\sqrt{n}z)\left[\frac{h(z) - h(0)}{z} + O(n^{-1})\right].$$

Hence,

$$\int_z^\infty \bar{c}h(x)\sqrt{n}\phi(\sqrt{n}x)\,dx \tag{2.7}$$

may be approximated by

$$[1 - \Phi(\sqrt{n}z)] + \frac{1}{n}\sqrt{n}\phi(\sqrt{n}z)\frac{h(z) - h(0)}{z}. \tag{2.8}$$

Note that knowledge of $\bar{c}$ is not needed to compute this approximation.

The error of (2.8) as an approximation to (2.7) is given by

$$\frac{1}{n}\sqrt{n}\phi(\sqrt{n}z)O(n^{-1}) = \phi(\sqrt{n}z)O(n^{-3/2})$$

and the relative error of the approximation may be written

$$\frac{\phi(\sqrt{n}z)O(n^{-3/2})}{[1 - \Phi(\sqrt{n}z)] + (1/n)\sqrt{n}\phi(\sqrt{n}z)[(h(z) - h(0))/z + O(n^{-1})]}$$
$$= \left(\frac{1 - \Phi(\sqrt{n}z)}{\phi(\sqrt{n}z)/(\sqrt{n}z)}\frac{1}{z} + \frac{h(z) - h(0)}{z}\right)^{-1} O(n^{-1}).$$

Consider the case in which z is fixed. Note that

$$\frac{1 - \Phi(t)}{\phi(t)/t} \to \begin{cases} -\infty & \text{as } t \to -\infty \\ 0 & \text{as } t \to 0 \\ 1 & \text{as } t \to \infty. \end{cases} \tag{2.9}$$

Hence,

$$\left(\frac{1 - \Phi(\sqrt{n}z)}{\phi(\sqrt{n}z)/(\sqrt{n}z)}\frac{1}{z} + \frac{h(z) - h(0)}{z}\right)^{-1} = O(1)$$

as $n \to \infty$. It follows that, for fixed z, the relative error of the approximation is $O(n^{-1})$.

Now consider the case in which $z = O(n^{-1/2})$. Note that, using (2.9),

$$\frac{1 - \Phi(\sqrt{n}z)}{\phi(\sqrt{n}z)/(\sqrt{n}z)}$$

and $[h(z) - h(0)]/z$ are both of order $O(1)$ so that

$$\frac{1 - \Phi(\sqrt{n}z)}{\phi(\sqrt{n}z)/(\sqrt{n}z)} \frac{1}{z} + \frac{h(z) - h(0)}{z} = O(\sqrt{n}).$$

It follows that the relative error in this case is $O(n^{-3/2})$.

To summarize

$$\int_z^\infty \bar{c}h(x)\sqrt{n}\phi(\sqrt{n}x)\, dx \doteq [1 - \Phi(\sqrt{n}z)] + \frac{1}{n}\sqrt{n}\phi(\sqrt{n}z)\frac{h(z) - h(0)}{z}.$$

The relative error of the approximation is $O(n^{-1})$ for fixed z and $O(n^{-3/2})$ for $z = O(n^{-1/2})$. A first-order approximation, given by $1 - \Phi(\sqrt{n}z)$, may also be used; the relative error of this approximation is $O(n^{-1/2})$ for $z = O(n^{-1/2})$.

For discrete distributions, approximation of a distribution function involves calculation of a sum. Hence, we now consider the approximation of a sum of the form

$$\frac{1}{n}\sum_{j=k}^{\infty} \sqrt{n}\phi(\sqrt{n}f(j/n))h(j/n) \tag{2.10}$$

where h and f are smooth $O(1)$ functions and f is invertible; for simplicity, assume that $f'(x) > 0$ for all x. Here $k = O(n)$ so that in (2.10) the points of evaluation of f and h are $O(1)$. The approach we take is to first approximate the sum by an integral and then approximate the integral using the methods discussed earlier in this section.

First consider the integral

$$\int_{j/n}^{(j+1)/n} h(x)\sqrt{n}\phi(\sqrt{n}f(x))\, dx. \tag{2.11}$$

Let $u = n(x - j/n)$ so that (2.11) may be written

$$\int_0^1 \frac{1}{\sqrt{n}} h(j/n + u/n)\phi(\sqrt{n}f(j/n + u/n))\, du;$$

expanding the functions h and f, this integral may be written

$$\frac{1}{\sqrt{n}}\int_0^1 \phi(\sqrt{n}(f(j/n) + f'(j/n)u/n + \cdots))[h(j/n) + h'(j/n)u/n + \cdots]\, du$$

$$= \frac{1}{\sqrt{n}}\phi(\sqrt{n}f_j)h_j \int_0^1 \exp\{-f_j' f_j u\}\, du[1 + O(n^{-1})]$$

$$= \frac{1}{n}\frac{\sqrt{n}\phi(\sqrt{n}f_j)h_j}{f_j f_j'}[1 - \exp\{-f_j f_j'\}][1 + O(n^{-1})]$$

where $f_j = f(j/n)$, $h_j = h(j/n)$, $f_j' = f'(j/n)$.

Hence,

$$\sqrt{n}\phi(\sqrt{n}f_j)h_j$$
$$= \frac{nf_jf_j'}{1-\exp\{-f_jf_j'\}}\int_{j/n}^{(j+1)/n}\sqrt{n}\phi(\sqrt{n}f(x))h(x)\,dx[1+O(n^{-1})]$$
$$= n\int_{j/n}^{(j+1)/n}\sqrt{n}\phi(\sqrt{n}f(x))h(x)\frac{f'(x)f(x)}{1-\exp\{-f'(x)f(x)\}}\,dx[1+O(n^{-1})].$$

It follows that

$$\frac{1}{n}\sum_{k=j}^{\infty}\sqrt{n}\phi(\sqrt{n}f(k/n))h(k/n)$$
$$= \int_{j/n}^{\infty}\sqrt{n}\phi(\sqrt{n}f(x))\,h(x)\frac{f'(x)f(x)}{1-\exp\{-f'(x)f(x)\}}\,dx[1+O(n^{-1})]. \quad (2.12)$$

To approximate this integral, we need to write it in the form (2.7). Consider the change of variable $y = f(x)$ and let g denote the inverse function so that $x = g(y)$. Then the integral in (2.12) may be written

$$\int_{f_j}^{\infty}\sqrt{n}\phi(\sqrt{n}y)h(g(y))\frac{y}{1-\exp\{-f'(g(y))y\}}\,dy.$$

This integral is of the form (2.7), with the function h in (2.7) taken to be

$$h(g(y))\frac{y}{1-\exp\{-f'(g(y))y\}}.$$

The approximation (2.8) is therefore given by

$$1-\Phi(\sqrt{n}f_j)+\frac{1}{\sqrt{n}}\phi(\sqrt{n}f_j)\left[\frac{h_j}{1-\exp\{-f'(g(f_j))f_j\}}-\frac{h(g(0))}{f_jf'(g(0))}\right]; \quad (2.13)$$

note that the lower limit of the integral is f_j. Hence, (2.13) provides an approximation to the sum (2.10); the error of this approximation is $O(n^{-1})$.

This approximation is based on using the integral (2.11) as the basis of an approximation for the jth term in the sum. Another approach is to use the integral

$$\int_{(j-1/2)/n}^{(j+1/2)/n}h(x)\sqrt{n}\phi(\sqrt{n}f(x))\,dx \quad (2.14)$$

in place of (2.11). The rest of the derivation is essentially the same and, hence, only a brief sketch is given.

Using the change of variable $u = n(x - j/n)$ and expanding the functions h and f, (2.14) may be written

$$\frac{1}{\sqrt{n}}\phi(\sqrt{n}f_j)h_j \int_{-1/2}^{1/2} \exp\{-f_j' f_j u\}\, du$$
$$= \frac{1}{n}\frac{\sqrt{n}\phi(\sqrt{n}f_j)h_j}{f_j f_j'} 2\sinh(f_j' f_j/2)\,[1 + O(n^{-1})].$$

It follows that

$$\frac{1}{n}\sum_{k=j}^{\infty} \sqrt{n}\phi(\sqrt{n}f(k/n))h(k/n)$$
$$= \int_{(j-1/2)/n}^{\infty} \sqrt{n}\phi(\sqrt{n}f(x))h(x)\frac{f'(x)f(x)}{2\sinh[f'(x)f(x)/2]}\, dx[1 + O(n^{-1})]$$

and this integral may be approximated by

$$1 - \Phi(\sqrt{n}f_j) + \frac{1}{\sqrt{n}}\phi(\sqrt{n}f_j)\left[\frac{h_j}{2\sinh\{f_j' f_j/2\}} - \frac{h(g(0))}{f_j f'(g(0))}\right] \tag{2.15}$$

where f_j, f_j', and h_j are all calculated using $j - 1/2$. Hence, (2.15) also provides an approximation to the sum (2.10) with error $O(n^{-1})$. The approximation (2.15) is called the *continuity-corrected* approximation.

2.6 Saddlepoint approximations for distribution functions

Consider the problem of approximating $\Pr(\bar{X} \geq t)$. Using the saddlepoint approximation for the density of $\bar{X}$, this probability can be approximated by

$$\bar{c}\int_t^{\infty} \exp\{n[K(\hat{\lambda}_x) - x\hat{\lambda}_x]\}\,[2\pi K''(\hat{\lambda}_x)/n]^{-1/2}\, dx. \tag{2.16}$$

To approximate this integral using the approximation derived in Section 2.5, we first need to write (2.16) in the form (2.7).

Let

$$r(x) = \text{sgn}(\hat{\lambda}_x)\,\{2[x\hat{\lambda}_x - K(\hat{\lambda}_x)]\}^{1/2}.$$

Note that, for fixed x, the function $x\lambda - K(\lambda)$ is uniquely maximized at $\lambda = \hat{\lambda}_x$ and since this function is 0 at $\lambda = 0$, it follows that

$$x\hat{\lambda}_x - K(\hat{\lambda}_x) \geq 0.$$

It follows from the fact that $r(x)^2 = 2[x\hat{\lambda}_x - K(\hat{\lambda}_x)]$ that

$$r'(x)r(x) = \hat{\lambda}_x + [x - K'(\hat{\lambda}_x)]\,\frac{d\hat{\lambda}_x}{dx} = \hat{\lambda}_x.$$

Hence, $r(x)$ is a strictly increasing function of x. Note that, since $K(0) = 0$, this implies that $r(x) = 0$ if and only if $\hat{\lambda}_x = 0$.

The integral in (2.16) may be written

$$\bar{c}\int_t^\infty \left(\frac{n}{2\pi}\right)^{1/2} \exp\left\{-\frac{n}{2}r(x)^2\right\} K''(\hat{\lambda}_x)^{-1/2}\,dx. \tag{2.17}$$

Let $z = r(x)$; note that $dx/dz = z/\hat{\lambda}_x$. Then (2.17) may be written

$$\bar{c}\int_{r(t)}^\infty \left(\frac{n}{2\pi}\right)^{1/2} \exp\left\{-\frac{n}{2}z^2\right\} K''(\hat{\lambda}_x)^{-1/2}\frac{z}{\hat{\lambda}_x}\,dz$$

where $x = r^{-1}(z)$. This is of the form (2.7) with

$$h(z) = \frac{z}{\hat{\lambda}_x[K''(\hat{\lambda}_x)]^{1/2}}, \qquad x = r^{-1}(z).$$

Note that

$$h(0) = \frac{1}{K''(0)^{1/2}}\lim_{z\to 0}\frac{z}{\hat{\lambda}_x} = \frac{1}{K''(0)^{1/2}}\lim_{z\to 0}\frac{1}{d\hat{\lambda}_x/dz}$$

where

$$\frac{d\hat{\lambda}_x}{dz} = \frac{d\hat{\lambda}_x}{dx}\frac{dx}{dz} = \frac{1}{K''(\hat{\lambda}_x)}\frac{z}{\hat{\lambda}_x} \tag{2.18}$$

since $K'(\hat{\lambda}_x) = x$ implies that $d\hat{\lambda}_x/dx = 1/K''(\hat{\lambda}_x)$. It follows from the last part of (2.18) that

$$h(0) = K''(0)^{1/2}\lim_{z\to 0}\frac{\hat{\lambda}_x}{z}$$

so that $h(0) = 1/h(0)$. Hence, $h(0) = \pm 1$; z and $\hat{\lambda}_x$ however have the same sign so that $h(0) > 0$. It follows that $h(0) = 1$. Also,

$$\begin{aligned} h(r(t)) &= \frac{r(t)}{\hat{\lambda}_x K''(\hat{\lambda}_x)^{1/2}}, \qquad x = r^{-1}(r(t)) = t \\ &= \frac{r(t)}{\hat{\lambda}_t K''(\hat{\lambda}_t)^{1/2}}. \end{aligned}$$

An approximation to the integral (2.17) is given by (2.8) with

$$\frac{h(r(t)) - h(0)}{r(t)} = \frac{1}{\hat{\lambda}_t K''(\hat{\lambda}_t)^{1/2}} - \frac{1}{r(t)}.$$

Hence, $\Pr(\bar{X} \geq t)$ may be approximated by

$$1 - \Phi(\sqrt{n}r) + \frac{1}{n}\left[\frac{1}{\hat{\lambda}_t K''(\hat{\lambda}_t)^{1/2}} - \frac{1}{r}\right]\sqrt{n}\phi(\sqrt{n}r), \qquad r = r(t).$$

This approximation has relative error $O(n^{-3/2})$ for fixed r, corresponding to t of the form $t = \mathrm{E}(\bar{X}) + O(n^{-1/2})$ and relative error $O(n^{-1})$ for $r = O(\sqrt{n})$, corresponding to fixed values of t.

A similar approximation may be derived for $X = n\bar{X}$. Since $\Pr(X \geq x) = \Pr(\bar{X} \geq x/n)$, an approximation to $\Pr(X \geq x)$ is given by

$$[1 - \Phi(\sqrt{n}r)] + \frac{1}{n}\left[\frac{1}{\hat{\lambda}_x K''(\hat{\lambda}_x)^{1/2}} - \frac{1}{r}\right]\phi(r) \tag{2.19}$$

where

$$r \equiv r(x/n) = \text{sgn}(\hat{\lambda}_x)\,\{2[x\hat{\lambda}_x/n - K(\hat{\lambda}_x)]\}^{1/2}$$

and $\hat{\lambda}_x$ satisfies $K'(\hat{\lambda}_x) = x/n$.

Let $K_n(t)$ denote the cumulant-generating function of X; note that $K_n(t) = nK(t)$. Then (2.19) may be written

$$[1 - \Phi(r_n)] + \left[\frac{1}{\hat{\lambda}_x K_n''(\hat{\lambda}_x)^{1/2}} - \frac{1}{r_n}\right]\phi(r_n)$$

where

$$r_n = \text{sgn}(\hat{\lambda}_x)\,\{2[x\hat{\lambda}_x - K_n(\hat{\lambda}_x)]\}^{1/2}$$

and $\hat{\lambda}_x$ satisfies $K_n'(\hat{\lambda}_x) = x$. The error of this approximation is of order $O(n^{-1})$ for fixed x and is of order $O(n^{-3/2})$ for x of the form $\text{E}(X) + O(\sqrt{n})$.

Example 2.9 Sample mean of Laplace random variables
Consider the sample mean of n independent Laplace random variables, as in Example 2.8, and the approximation of the tail probability $\Pr(\bar{X} \geq t)$. It is straightforward to show that here

$$r = \text{sgn}(t)\left\{2\left[(1+t^2)^{1/2} - 1 + \log\left(\frac{2[(1+t^2)^{1/2} - 1]}{t^2}\right)\right]\right\}^{1/2}$$

and

$$\hat{\lambda}_t K''(\hat{\lambda}_t)^{1/2} = (1+t^2)^{1/2}[(1+t^2)^{1/2} - 1]^{1/2}.$$

For instance, consider the case $n = 8$ and $t = 1$; note that for $n = 8$, the standard deviation of $\bar{X}$ is 1/2. Then $r = 0.672$ and $\hat{\lambda}_t K''(\hat{\lambda}_t)^{1/2} = 0.910$; the approximation to the tail probability $\Pr(\bar{X} \geq 1)$ is therefore approximately 0.020. ∎

Example 2.10 Chi-square distribution
Let Y_n denote a chi-square random variable with n degrees of freedom and consider approximation of $\Pr(Y_n \leq y)$; an Edgeworth series approximation to this probability was given in Example 2.6. The cumulant-generating function of the chi-square distribution with n degrees of freedom is given by

$$K_n(t) = -\frac{n}{2}\log(1 - 2t), \quad |t| \leq \frac{1}{2}.$$

The solution to $K_n'(t) = y$ is given by $\hat{\lambda}_y = -(n-y)/(2y)$ so that

$$r_n = \text{sgn}(y-n)\{y - n + n\log(n/y)\}^{1/2}$$

and $K_n''(\hat{\lambda}_y) = 2y^2/n$. It follows that $\Pr(Y_n \geq y)$ may be approximated by

$$[1 - \Phi(r_n)] + \left[\frac{\sqrt{2n}}{n-y} - \frac{1}{r_n}\right]\phi(r_n)$$

and $\Pr(Y_n \leq y)$ may be approximated by

$$\Phi(r_n) - \left[\frac{\sqrt{2n}}{y-n} - \frac{1}{r_n}\right]\phi(r_n). \tag{2.20}$$

Consider the case $n = 5$ and $y = 1.145$; as noted in Example 2.6, the exact probability is 0.05. Then $r_n = -1.875$ so that (2.20) becomes

$$\Phi(-1.875) - [-0.820 + 0.533]\phi(-1.875) \doteq 0.0501. \qquad \blacksquare$$

2.7 Saddlepoint approximations for lattice variables

The saddlepoint approximations discussed thus far have been limited to means and sums of continuous random variables. We now consider the case in which $X_1, \ldots, X_n$ are independent identically distributed lattice random variables. For simplicity, we assume that the X_j take values in the set of nonnegative integers.

The basic approach used to derive the saddlepoint approximation for the density of $S = \sum X_j$ given in the continuous case considered in Section 2.4 holds for the lattice case as well. The density function of S may be approximated by

$$\exp\{nK(\lambda) - s\lambda\}\,\hat{p}_n(s;\lambda)$$

where $\hat{p}(s;\lambda)$ is an approximation to the density of S under the distribution with parameter λ, and λ may be chosen to make that approximation as accurate as possible.

Let

$$Y_n = \frac{S - nK'(\lambda)}{\sqrt{n}K''(\lambda)^{1/2}}.$$

Then $\Pr(S = s;\lambda) = \Pr(Y_n = y;\lambda)$ where

$$y = \frac{s - nK'(\lambda)}{\sqrt{n}K''(\lambda)^{1/2}}.$$

Using an Edgeworth expansion to approximate $\Pr(Y_n = y;\lambda)$ we have that

$$\Pr(S = s;\lambda) = \frac{1}{\sqrt{n}K''(\lambda)^{1/2}}\phi(y)\Big[1 + \frac{1}{6}\frac{K'''(\lambda)}{K''(\lambda)^{3/2}}H_3(y)\frac{1}{\sqrt{n}} + O(n^{-1})\Big].$$

If λ is chosen so that $nK'(\hat{\lambda}_s) = s$, then $x = 0$ and

$$\Pr(S = s; \hat{\lambda}_s) = \frac{1}{\sqrt{n}K''(\hat{\lambda}_s)^{1/2}}\phi(0)[1 + O(n^{-1})].$$

It follows that the saddlepoint approximation to the density of S is given by

$$\exp\{nK(\hat{\lambda}_s) - s\hat{\lambda}_s\}\,[2\pi nK''(\hat{\lambda}_s)]^{-1/2}$$

where $\hat{\lambda}_s$ satisfies $nK'(\hat{\lambda}_s) = s$.

This result can be used to obtain the saddlepoint approximation to the density of $\bar{X}$; note that in the lattice case, no Jacobian term is needed for this transformation. Hence, the saddlepoint approximation for the density of $\bar{X}$ in the lattice case is given by

$$\exp\{n[K(\hat{\lambda}_x) - x\hat{\lambda}_x]\}\,[2\pi nK''(\hat{\lambda}_x)]^{-1/2}$$

where $\hat{\lambda}_x$ satisfies $K'(\hat{\lambda}_x) = x$. The error of these approximations is $O(n^{-1})$.

Example 2.11 Sum of Poisson random variables

Let $X_1, \ldots, X_n$ denote independent Poisson random variables each with mean μ. Consider the problem of approximating the density function of $S = \sum X_j$; of course, it is well known that S has a Poisson distribution with mean $n\mu$. The cumulant-generating function of the Poisson distribution is given by $K(t) = \mu(\exp\{t\} - 1)$ so that $\hat{\lambda}_s$ is given by $\hat{\lambda}_s = \log(s/(n\mu))$. The saddlepoint approximation to the density of S is therefore

$$\frac{\exp\{(s - n\mu) - s\log(s/(n\mu))\}}{(2\pi s)^{1/2}} = \frac{(n\mu)^s\exp\{-n\mu\}}{(2\pi)^{1/2}s^{s+1/2}\exp\{-s\}}, \quad s = 0, 1, \ldots.$$

The exact density is

$$\frac{(n\mu)^s\exp\{-n\mu\}}{s!}, \quad s = 0, 1, \ldots;$$

the difference between the approximation and the exact density is that the approximation replaces the $s!$ in the denominator with Stirling's approximation.

For instance, for the case in which $n\mu = 4$ and $s = 6$, the saddlepoint approximation is 0.1057 while the exact value is 0.1042. ∎

Now consider the problem of approximating the tail probability $\Pr(\bar{X} \geq t)$. Note that $\bar{X}$ takes values $0, 1/n, 2/n, \ldots$. Assume that t is a lattice point and, hence, may be written $t = j/n$ for some integer j. Using the saddlepoint approximation for the density of $\bar{X}$, the tail probability under consideration may be approximated by

$$\sum_{k=j}^{\infty}\exp\{nK(\hat{\lambda}_k) - k\hat{\lambda}_k\}\,[2\pi nK''(\hat{\lambda}_k)]^{-1/2}$$

where $\hat{\lambda}_k$ satisfies $K'(\hat{\lambda}_k) = k/n$.

This sum is of the form (2.10) with

$$f(k/n) = \operatorname{sgn}(\hat{\lambda}_k)\left\{2\left[\frac{k}{n}\hat{\lambda}_k - K(\hat{\lambda}_k)\right]\right\}^{1/2}$$

and $h(k/n) = K''(\hat{\lambda}_k)^{-1/2}$. To compute the approximation (2.13) note that, as in the continuous case, $f_j' f_j = \hat{\lambda}_j$ and $f(k/n) = 0$ if and only if $\hat{\lambda}_k = 0$. It follows that $g(0)$ is the value of k such that $\hat{\lambda}_k = 0$, i.e., $g(0) = K'(0)$. Hence,

$$f'(g(0)) = \lim_{\hat{\lambda}_k \to 0} \frac{\hat{\lambda}_k}{\{2[(k/n)\hat{\lambda}_k - K(\hat{\lambda}_k)]\}^{1/2}};$$

by L'Hospital's rule this limit is equal to

$$K''(0)^{-1} \lim_{\hat{\lambda}_k \to 0} \frac{\{2[(k/n)\hat{\lambda}_k - K(\hat{\lambda}_k)]\}^{1/2}}{\hat{\lambda}_k} \equiv K''(0)^{1/2} \frac{1}{f'(g(0))}.$$

Hence, $f'(g(0)) = K''(0)^{-1/2}$. Since $h(g(0)) = K''(0)^{-1/2}$, it follows that the approximation (2.13) is given by

$$[1 - \Phi(\sqrt{n} f_j)] + \frac{1}{n}\left[\frac{1}{[1-\exp\{-\hat{\lambda}_j\}]K''(\hat{\lambda}_j)^{1/2}} - \frac{1}{f_j}\right]\sqrt{n}\phi(\sqrt{n} f_j) \tag{2.21}$$

where $f_j = f(j/n)$; the error of this approximation is $O(n^{-1})$.

To approximate the lower tail probability, $\Pr(\bar{X} \le j/n)$, we use the fact that

$$\Pr(\bar{X} \le j/n) = 1 - \Pr(\bar{X} \ge j/n) + \Pr(\bar{X} = j/n).$$

From (2.21), we have that $1 - \Pr(\bar{X} \ge j/n)$ may be approximated by

$$\Phi(\sqrt{n} f_j) - \frac{1}{n}\left[\frac{1}{[1-\exp\{-\hat{\lambda}_j\}]K''(\hat{\lambda}_j)^{1/2}} - \frac{1}{f_j}\right]\sqrt{n}\phi(\sqrt{n} f_j)$$

and since

$$\begin{aligned}\Pr(\bar{X} = j/n) &= \exp\{nK(\hat{\lambda}_j) - j\hat{\lambda}_j\}\,[2\pi n K''(\hat{\lambda}_j)]^{-1/2}[1 + O(n^{-1})] \\ &= \frac{\phi(\sqrt{n} f_j)}{\sqrt{n} K''(\hat{\lambda}_j)^{1/2}}[1 + O(n^{-1})],\end{aligned}$$

it follows that $\Pr(\bar{X} \le j/n)$ may be approximated by

$$\begin{aligned}&\Phi(\sqrt{n} f_j) - \frac{1}{n}\left[\frac{1}{[1-\exp\{-\hat{\lambda}_j\}]K''(\hat{\lambda}_j)^{1/2}} - \frac{1}{f_j}\right]\sqrt{n}\phi(\sqrt{n} f_j) + \frac{\phi(\sqrt{n} f_j)}{\sqrt{n} K''(\hat{\lambda}_j)^{1/2}} \\ &= \Phi(\sqrt{n} f_j) - \frac{1}{n}\left\{\left[\frac{1}{1-\exp\{-\hat{\lambda}_j\}} - 1\right]\frac{1}{K''(\hat{\lambda}_j)} - \frac{1}{f_j}\right\}\sqrt{n}\phi(\sqrt{n} f_j) \\ &= \Phi(\sqrt{n} f_j) + \frac{1}{n}\left[\frac{1}{[1-\exp\{\hat{\lambda}_j\}]K''(\hat{\lambda}_j)^{1/2}} + \frac{1}{f_j}\right]\sqrt{n}\phi(\sqrt{n} f_j).\end{aligned} \tag{2.22}$$

This last expression uses the fact that

$$\frac{1}{1-\exp(-x)} - 1 = -\frac{1}{1-\exp(x)}.$$

The continuity-corrected approximation (2.15) may be used in place of (2.13) to approximate $\Pr(\bar{X} \geq j/n)$. The form of the resulting approximation is identical to (2.21) except that the term $1-\exp(-\hat{\lambda}_j)$ is replaced by $2\sinh(\hat{\lambda}_j/2)$, leading to the approximation

$$[1-\Phi(\sqrt{n}f_j)] + \frac{1}{n}\left[\frac{1}{2\sinh(\hat{\lambda}_j/2)\,K''(\hat{\lambda}_j)^{1/2}} - \frac{1}{f_j}\right]\sqrt{n}\phi(\sqrt{n}f_j); \tag{2.23}$$

recall that in (2.23) f_j and $\hat{\lambda}_j$ are calculated using $j-1/2$ in place of j.

To calculate the lower tail probability we could, in principle, use the same approach used to derive (2.22): that is, we could use approximations to $\Pr(\bar{X} \geq j/n)$ and $\Pr(\bar{X} = j/n)$. When constructing a continuity-corrected approximation however it is more natural to have an an approximation to $\Pr(\bar{X} \leq j/n)$ that is based on f_j calculated using $j+1/2$. Hence, we approximate $\Pr(\bar{X} \leq j/n)$ by an approximation to $1-\Pr(\bar{X} \geq (j+1)/n)$:

$$\Phi(\sqrt{n}f_j) + \frac{1}{n}\left[\frac{1}{2\sinh(-\hat{\lambda}_j/2)\,K''(\hat{\lambda}_j)^{1/2}} + \frac{1}{f_j}\right]\sqrt{n}\phi(\sqrt{n}f_j). \tag{2.24}$$

In this expression, f_j and $\hat{\lambda}_j$ are calculated using $(j+1)-1/2 = j+1/2$ in place of j. Note that the relationship between the lower tail probability approximations (2.22) and (2.24) is the same as that between the upper tail probability approximations (2.21) and (2.23).

Example 2.12 Sum of Poisson random variables

Let $X_1, \ldots, X_n$ denote independent Poisson random variables each with mean μ and let $S = \sum X_j$; the saddlepoint approximation to the density of S was considered in Example 2.11. Here we consider approximation of the tail probability $\Pr(S \geq j) = \Pr(\bar{X} \geq j/n)$. Using the fact that $\hat{\lambda}_j = \log(j/(n\mu))$, it is straightforward to show that

$$f(j/n) = \operatorname{sgn}(j-n\mu)\left\{2\left[\frac{j}{n}\log\left(\frac{j}{n\mu}\right) - \left(\frac{j}{n}-\mu\right)\right]\right\}^{1/2}$$

and $K''(\hat{\lambda}_j) = j/n$. Let

$$z = \operatorname{sgn}(j-n\mu)\left\{2\left[j\log\left(\frac{j}{n\mu}\right) - (j-n\mu)\right]\right\}^{1/2}.$$

Then, using (2.21), $\Pr(S \geq j)$ may be approximated by

$$1-\Phi(z) + \left[\frac{\sqrt{j}}{j-n\mu} - \frac{1}{z}\right]\phi(z).$$

Using (2.23), the continuity-corrected approximation is given by

$$1 - \Phi(z) + \Big[\frac{\sqrt{n\mu}}{j - 1/2 - n\mu} - \frac{1}{z}\Big]\phi(z)$$

where z is calculated using $j - 1/2$.

For instance, consider the case $n = 5$, $\mu = 1$, and $j = 10$. The exact tail probability based on the fact that S has a Poisson distribution with mean 5, is 0.03183. Here $z = 1.9654$ so that the approximation based on (2.21) is 0.03183. For the continuity-corrected approximation, $z = 1.7875$ and the probability approximation is 0.03189. Hence, for this example, the continuity-corrected approximation is slightly less accurate, although the difference would rarely be important in practice. ∎

If the minimal lattice of the X_j is of the form $a + bk$, $k = 0, 1, 2, \ldots$, the approximations must be modified slightly. Consider the approximation (2.21). Let $Z_j = (X_j - a)/b$, $j = 1, \ldots, n$; then the minimal lattice of the X_j is $0, 1, \ldots$, and, hence, (2.21) is a valid approximation to the tail probability $\Pr(\bar{Z} \geq j/n)$. Let $K_Z(t)$ denote the cumulant-generating function of the Z_j and let $K_X(t)$ denote the cumulant-generating function of the X_j; then $K_Z(t) = K_X(t/b) - at/b$. It follows that $\hat{\lambda}_z$, the solution to $K'_Z(\hat{\lambda}_z) = j/n$ and $\hat{\lambda}_x$ the solution to $K'_X(\hat{\lambda}_x) = a + bj/n$ are related by $\hat{\lambda}_z = b\hat{\lambda}_x$. Hence, the value of f_j in (2.21) does not depend on whether it is calculated using the X_j or the Z_j. However,

$$[1 - \exp\{-\hat{\lambda}_z\}]K''_Z(\hat{\lambda}_z)^{1/2} = [1 - \exp\{-b\hat{\lambda}_x\}]K''_X(\hat{\lambda}_x)^{1/2}/b.$$

It follows that $\Pr(\bar{X} \geq a + bj/n)$ may be approximated by

$$[1 - \Phi(\sqrt{n f_j})] + \frac{1}{n}\Big[\frac{b}{[1 - \exp\{-b\hat{\lambda}_j\}]K''(\hat{\lambda}_j)^{1/2}} - \frac{1}{f_j}\Big]\sqrt{n}\phi(\sqrt{n f_j}).$$

It is worth noting that, as $b \to 0$

$$\frac{1 - \exp\{-b\hat{\lambda}_j\}}{b} \to \hat{\lambda}_j$$

so that the lattice approximation approaches the one given for continuous distributions.

2.8 Saddlepoint approximations for multivariate distributions

We now consider saddlepoint approximations for the density of $\bar{X}$ where $X_1, \ldots, X_n$ are independent identically distributed random vectors of dimension m. Let $K(t)$ denote the cumulant-generating function of X_1. The derivation of the saddlepoint approximation closely parallels that given in the univariate case and, hence, only a brief sketch is given.

Let $p(x)$ denote the density of X_1 and for an m-dimensional vector λ, define

$$p(x; \lambda) = \exp\{\lambda^T x - K(\lambda)\}\, p(x).$$

Let $S = X_1 + \cdots + X_n$ and let $p_n(s; \lambda)$ denote the density function of S under the density $p(x; \lambda)$ for the X_j. Then

$$p_n(s; \lambda) = \exp\{\lambda^T x - nK(\lambda)\}\, p_n(s)$$

where $p_n(s)$ denotes the actual density of S. For a given value of λ, we can approximate $p_n(s)$ by

$$\exp\{nK(\lambda) - \lambda^T x\}\, \hat{p}_n(s; \lambda)$$

where $\hat{p}_n(s; \lambda)$ is an approximation to $p_n(s; \lambda)$.

Let $\hat{\lambda}_s$ denote the value of λ that solves the set of equations $nK'(\lambda) = s$; note that, since λ is a vector, here $K'(\lambda)$ is a vector as well; then $s = \mathrm{E}(S; \hat{\lambda}_s)$. To approximate $\hat{p}_n(s; \hat{\lambda}_s)$ we can use an Edgeworth series approximation; since the evaluation of the density is at the mean of S, the approximation is given by

$$(2\pi)^{-m/2} |nK''(\hat{\lambda}_s)|^{-1/2}$$

where $K''(\hat{\lambda}_s)$ is the $m \times m$ matrix of second derivatives of $K(\cdot)$ evaluated at $\hat{\lambda}_s$. The corresponding approximation to the density of $p_n(s)$ is given by

$$\frac{\exp\{nK(\hat{\lambda}_s) - \hat{\lambda}_s^T s\}}{(2\pi)^{m/2} |nK''(\hat{\lambda}_s)|^{1/2}}.$$

The relative error of this approximation is $O(n^{-1})$ for fixed values of s; as in the univariate case, the accuracy of the approximation may be improved by renormalization.

2.9 Stochastic asymptotic expansions

Let $Y_1, Y_2, \ldots, Y_n$ denote a sequence of random variables. Thus far, when considering approximations to Y_n we have focused on approximations for the distribution function or density function of Y_n. In many cases however it is convenient to approximate the random variable Y_n directly by other random variables, the properties of which are well understood.

Example 2.13 Log odds ratio
Let nX_n denote a binomial random variable with index n and success probability θ, $0 < \theta < 1$, and let $Y_n = \log(X_n/(1 - X_n))$. Hence, X_n may be viewed as the proportion of successes in the 'sample' and Y_n denotes the corresponding log odds ratio. Using a Taylor's series expansion of $\log(X_n/(1 - X_n))$ around $X_n = \theta$, the mean of X_n, yields the following stochastic asymptotic expansion for Y_n

$$Y_n = \log\left(\frac{\theta}{1-\theta}\right) + \frac{1}{\theta(1-\theta)}(X_n - \theta) + \frac{1}{2}\frac{2\theta - 1}{\theta^2(1-\theta)^2}(X_n - \theta)^2 + \cdots. \tag{2.25}$$

Hence, instead of analyzing Y_n directly, we can analyze the first few terms in this expansion. ∎

For such an expansion to be useful we need to be able to truncate it at some point and establish that the ignored terms are negligible in some sense. In carrying this out, the following notation is often useful. Let $V_1, V_2, \ldots$ denote a sequence of random variables and let c_n denote a sequence of real numbers. We will write $V_n = o_p(c_n)$ if V_n/c_n converges in probability to 0. We will write $V_n = O_p(c_n)$ if V_n/c_n is bounded in probability in the sense that given $\varepsilon > 0$ there exists constants C and N, possibly depending on ε, such that

$$\Pr\Big(\Big|\frac{V_n}{c_n}\Big| \geq C\Big) \leq \varepsilon \quad \text{for all } n \geq N.$$

Hence, V_n/c_n is not growing arbitrarily large in magnitude. It is straightforward to show that if V_n converges in distribution to some random variable V, then $V_n = O_p(1)$. If V_n converges in probability to a constant c we may write either $V_n = O_p(1)$ or, more informatively, $V_n = c + o_p(1)$.

There are relatively simple rules for manipulating these symbols that make working with them particularly convenient. For instance,

$$O_p(c_n)O_p(d_n) = O_p(c_n d_n), \qquad O_p(c_n)o_p(d_n) = o_p(c_n d_n),$$

and

$$O_p(c_n) + O_p(d_n) = O_p(\max(c_n, d_n)).$$

Also, if $Y_n \xrightarrow{\mathcal{D}} Y$ and $V_n = o_p(1)$ then

$$Y_n + V_n \xrightarrow{\mathcal{D}} Y$$

and if $V_n = c + o_p(1)$ then

$$\frac{Y_n}{V_n} \xrightarrow{\mathcal{D}} \frac{Y}{c}.$$

These results are often referred to as *Slutsky's theorem.*

Example 2.14 Log odds ratio
Note that the expansion (2.25) may be written

$$Y_n = \log\left(\frac{\theta}{1-\theta}\right) + \frac{1}{\theta(1-\theta)}(X_n - \theta) + R_n(\theta).$$

The remainder term R_n is given by

$$R_n(\theta) = \frac{1}{2}\frac{2\theta_n - 1}{\theta_n^2(1-\theta_n)^2}(X_n - \theta)^2$$

where θ_n lies on the line connecting X_n and θ; it follows that $\theta_n = \theta + o_p(1)$ and, hence, that

$$\frac{1}{2}\frac{2\theta_n - 1}{\theta_n^2(1-\theta_n)^2} = O_p(1).$$

Since, by the central limit theorem, $\sqrt{n}(X_n - \theta)$ converges in distribution to a normally distributed random variable, $n(X_n - \theta)^2 = O_p(1)$, or, equivalently, $(X_n - \theta)^2 = O_p(n^{-1})$. It follows that $R_n(\theta) = O_p(n^{-1})$ so that we may write

$$Y_n = \log\left(\frac{\theta}{1-\theta}\right) + \frac{1}{\theta(1-\theta)}(X_n - \theta) + O_p(n^{-1}).$$ ∎

In some cases, the random variable under consideration may depend on the parameter θ. Suppose that, under the distribution with parameter θ, $X_n(\theta) = O_p(n^r)$. Can we conclude that the derivative $X_n'(\theta)$ is also of order $O_p(n^r)$? Since

$$X_n'(\theta) = \lim_{\delta\to 0} \frac{X_n(\theta+\delta) - X_n(\theta)}{\delta},$$

the order of $X_n'(\theta)$ depends on the behavior of $X_n(\theta+\delta)$ for δ near 0. If $X_n(\theta+\delta) = O_p(n^r)$ for all sufficiently small δ, then $X_n'(\theta) = O_p(n^r)$. It may however happen that $X_n(\theta+\delta)$ is of larger order than $X_n(\theta)$.

For instance, suppose that $Z_1, \ldots, Z_n$ are independent identically distributed random variables with mean θ and standard deviation 1 and let $X_n(\theta) = \bar{Z} - \theta$. Then $X_n(\theta) = O_p(n^{-1/2})$. Note however that

$$X_n(\theta+\delta) = \bar{Z} - \theta - \delta = O(1).$$

It is easy to see that $X_n'(\theta) = -1 = O(1)$. On the other hand, if $X_n(\theta) = n^{-1/2}\bar{Z}/\theta$, with $\theta \neq 0$, then $X_n(\theta) = O_p(n^{-1/2})$ and also $X_n(\theta+\delta) = O_p(n^{-1/2})$ for all δ close to 0. In this case,

$$X_n'(\theta) = -n^{-1/2}\bar{Z}/\theta^2 = O_p(n^{-1/2}).$$

A stochastic asymptotic expansion for Y_n can often be used to derive an approximation to the distribution of Y_n by approximating the distribution of the terms in the expansion. For instance, suppose that $Y_n = X_n + O_p(n^{-1/2})$ where X_n is asymptotically distributed according to a standard normal distribution. It follows that Y_n is also asymptotically distributed according to a standard normal distribution.

Now suppose that

$$Y_n = X_n + bX_n^2\frac{1}{\sqrt{n}} + O_p(n^{-1}) \tag{2.26}$$

where the distribution of X_n has an Edgeworth expansion of the form

$$\phi(x)\left[1 + \frac{\rho_3}{6\sqrt{n}}H_3(x) + O(n^{-1})\right].$$

Note that (2.26) implies that

$$X_n = Y_n - bY_n^2\frac{1}{\sqrt{n}} + O_p(n^{-1});$$

one way to show this is to write $X_n = Y_n + cY_n^2/\sqrt{n} + O_p(n^{-1})$, substitute this expression into (2.26) and solve for the constant c. It follows that the distribution of

Y_n may be approximated by

$$\phi(y - by^2/\sqrt{n})\Big[1 + \frac{\rho_3}{6\sqrt{n}}H_3(y - by^2/\sqrt{n})\Big][1 - 2by/\sqrt{n}]$$

with error $O(n^{-1})$. The terms in this expression may be expanded to yield the following expression for the density function of Y_n:

$$\phi(y)\Big[1 + by\frac{1}{\sqrt{n}} + \frac{\rho_3 + 6b}{6\sqrt{n}}H_3(y) + O(n^{-1})\Big]. \qquad (2.27)$$

Finally, to this order of approximation, (2.27) is equivalent to

$$\phi(y - b/\sqrt{n})\Big[1 + \frac{\rho_3 + 6b}{6\sqrt{n}}H_3(y - b/\sqrt{n}) + O(n^{-1})\Big]; \qquad (2.28)$$

that is, a Taylor's series approximation to (2.28) yields (2.27). Hence, the distribution of $Y_n - b/\sqrt{n}$ has a Edgeworth expansion with standardized third cumulant $\rho_3 + 6b$.

It is important to note that this approximation is identical to the one obtained by formally computing the cumulants of the truncated expansion $X + bX^2/\sqrt{n}$ and then using those cumulants as the basis of an Edgeworth expansion. Specifically, the first three cumulants of $X + bX^2/\sqrt{n}$ are $b/\sqrt{n}$, 1, and $\rho_3 + 6b$, respectively, neglecting terms of order $O(n^{-1})$. Using these *formal cumulants* in an Edgeworth expansion leads directly to (2.28). Hence, this approach yields the correct approximation, even though the cumulants of the distribution of Y_n may not exist.

Example 2.15 Log odds ratio
The log odds ratio Y_n may be expanded

$$Y_n = \log\left(\frac{\theta}{1-\theta}\right) + \frac{1}{\theta(1-\theta)}(X_n - \theta) + O_p(n^{-1})$$

so that

$$\sqrt{n}\theta(1-\theta)\Big[Y_n - \log\left(\frac{\theta}{1-\theta}\right)\Big] = \sqrt{n}(X_n - \theta) + O_p(n^{-1/2}).$$

By the central limit theorem, X_n is approximately normally distributed with mean θ and variance $n\theta(1-\theta)$. Hence, Y_n is approximately normally distributed with mean $\log[\theta/(1-\theta)]$ and variance $n/[\theta(1-\theta)]$. More formally,

$$\sqrt{n}(Y_n - \log[\theta/(1-\theta)]) \xrightarrow{\mathcal{D}} N(0, 1). \qquad ∎$$

Example 2.16 Estimator of the rate parameter of an exponential distribution
Let $Z_1, \ldots, Z_n$ denote independent exponential random variables each with rate parameter 1 so that each Z_j has density $\exp\{-z\}$. This distribution has cumulant-generating function $-\log(1-t)$ for $|t| < 1$; the first three cumulants are therefore 1,

1, and 2, respectively. Let $\bar{Z}$ denote the sample mean of the Z_j and let $X_n = \sqrt{n}(\bar{Z} - 1)$; an Edgeworth expansion for the density of X_n is given by

$$\phi(x)\Big[1 + \frac{1}{3\sqrt{n}} H_3(x) + O(n^{-1})\Big].$$

Consider the distribution of $1/\bar{Z}$; this statistic may be viewed as an estimator of the rate parameter of the underlying exponential distribution. Note that

$$\sqrt{n}(\bar{Z}^{-1} - 1) = -\sqrt{n}(\bar{Z} - 1) + n(\bar{Z} - 1)^2 \frac{1}{\sqrt{n}} + O_p(n^{-1}).$$

Let $Y_n = \sqrt{n}(\bar{Z}^{-1} - 1)$. Then

$$Y_n = -X_n + \frac{1}{\sqrt{n}} X_n^2 + O_p(n^{-1}).$$

Following the approach taken in this section, but accounting for the minus sign in the equation relating Y_n to X_n, it follows that an Edgeworth expansion for the density function of Y_n is given by

$$\phi(y - 1/\sqrt{n})\big[1 + \frac{2}{3\sqrt{n}} H_3(y - 1/\sqrt{n}) + O(n^{-1})\big]. \qquad \blacksquare$$

2.10 Approximation of conditional distributions

2.10.1 INTRODUCTION

Let Y denote a random vector of the form $Y = (Y_1, Y_2)$. In general, both Y_1 and Y_2 may be vectors. For simplicity however we assume that both Y_1 and Y_2 are scalars; similar results may be formulated for the vector case.

In this section, we consider the problem of approximating the conditional distribution of Y_2 given Y_1. The conditional density of Y_2 given $Y_1 = y_1$ is, of course, given by the ratio of the joint density of Y_1, Y_2 to the marginal density of Y_1, $p(y_2|y_1) = p(y_1, y_2)/p(y_1)$. Hence, we may approximate $p(y_2|y_1)$ by approximating $p(y_1, y_2)$ and $p(y_1)$.

We will consider two types of approximations, the direct Edgeworth series approximation and the indirect saddlepoint approximation. Often the same type of approximation is used for both $p(y_1, y_2)$ and $p(y_1)$, leading to the direct/direct and indirect/indirect approximations for $p(y_2|y_1)$. This does not necessarily have to be the case however so that a 'mixed' approximation may also be used; for example, a direct approximation may be used for $p(y_1, y_2)$ while an indirect approximation is used for $p(y_1)$.

Let $K(t_1, t_2) \equiv K_n(t_1, t_2)$ denote the cumulant-generating function of (Y_1, Y_2); assume that $K(t_1, t_2)$ and its derivatives are of order $O(n)$; for example, (Y_1, Y_2) may be sums of n independent random vectors.

2.10.2 DIRECT/DIRECT APPROXIMATIONS

We first consider the use of Edgeworth expansions to approximation the density of (Y_1, Y_2) and Y_1. To use an Edgeworth expansion, we must first standardize the variables so that they are of order $O_p(1)$. Hence, let $Z_j = (Y_j - n\mu_j)/(\sqrt{n}\sigma_j)$, $j = 1, 2$ where $n\mu_j$ and $\sqrt{n}\sigma_j$ denote the mean and standard deviation, respectively, of Y_j. Note that μ_j and σ_j are both $O(1)$. We will consider an approximation to the conditional distribution of Z_2 given Z_1; this may be transformed to obtain an approximation to the conditional distribution of Y_2 given Y_1.

Suppose that an Edgeworth expansions exists both for the joint density $p(z_1, z_2)$ as well as for the marginal density $p(z_1)$. We will assume that a cumulant of (Z_1, Z_2) of order j is of order $O(n^{-(j-2)/2})$. We will denote the higher-order cumulants, standardized to be of order $O(1)$, by $\kappa_{ijk}, \kappa_{ijkl}$, and so on; for example,

$$\kappa_{111} = \sqrt{n}\,\text{cum}(Z_1, Z_1, Z_1); \qquad \kappa_{1122} = n\,\text{cum}(Z_1, Z_1, Z_2, Z_2).$$

Using Edgeworth expansions, the approximations for the joint density of Z_1, Z_2 and the marginal density of Z_1 are given by

$$\hat{p}(z_1, z_2) = \phi_2(z; \Sigma)\Big[1 + \frac{1}{6\sqrt{n}}\{\kappa_{111}H_{111}(z, \Sigma) + 3\kappa_{112}H_{112}(z, \Sigma) + 3\kappa_{122}H_{122}(z, \Sigma) + \kappa_{222}H_{222}(z, \Sigma)\} + \cdots\Big]$$

and

$$\hat{p}(z_1) = \phi(z_1)\Big[1 + \frac{1}{6\sqrt{n}}\kappa_{111}H_3(z_1) + \cdots\Big],$$

respectively. Here Σ is of the form

$$\Sigma = \begin{pmatrix} 1 & \rho \\ \rho & 1 \end{pmatrix}$$

and we have used the fact that both κ_{ijk} and h_{ijk} are invariant under permutation of the indices; e.g. $\kappa_{112} = \kappa_{121} = \kappa_{211}$.

An approximation for $p(z_2|z_1)$ may then be obtained by taking the ratio of these approximations:

$$\begin{aligned}\hat{p}(z_2|z_1) &= \frac{\phi_2(z; \Sigma)}{\phi(z_1)}\Big[1 + \frac{1}{6\sqrt{n}}\{\kappa_{111}H_{111}(z, \Sigma) + 3\kappa_{112}H_{112}(z, \Sigma) \\ &\quad + 3\kappa_{122}H_{122}(z, \Sigma) + \kappa_{222}H_{222}(z, \Sigma)\} + \cdots\Big] \\ &\quad \times \Big[1 + \frac{1}{6\sqrt{n}}\kappa_{111}H_3(z_1) + \cdots\Big]^{-1} \\ &= \frac{\phi_2(z; \Sigma)}{\phi(z_1)}\Big[1 + \frac{1}{6\sqrt{n}}\{\kappa_{111}H_{111}(z, \Sigma) + 3\kappa_{112}H_{112}(z, \Sigma) \\ &\quad + 3\kappa_{122}H_{122}(z, \Sigma) + \kappa_{222}H_{222}(z, \Sigma) - \kappa_{111}H_3(z_1)\} + \cdots\Big].\end{aligned}$$

It is straightforward to show that

$$\frac{\phi_2(z; \Sigma)}{\phi(z_1)} = \frac{1}{\tau}\phi((z_2 - \rho z_1)/\tau)$$

where $\tau^2 = 1 - \rho^2$. Hence, the first-order normal approximation to the conditional distribution is a normal distribution with mean ρz_1 and variance $1 - \rho^2$. That is, given $Y_1 = y_1$, Y_2 is approximately normally distributed with mean

$$n\mu_2 + \rho\frac{\sigma_2}{\sigma_1}(y_1 - n\mu_1)$$

and variance $n(1 - \rho^2)\sigma_2^2$.

An approximation with error of order $O(n^{-1})$ is given by

$$\frac{1}{\tau}\phi((z_2 - \rho z_1)/\tau)\left[1 + \frac{1}{6\sqrt{n}}Q(z_2|z_1; \rho)\right]$$

where $Q(z_2|z_1; \rho)$ is a polynomial in (z_1, z_2) given by

$$Q(z_2|z_1; \rho) = \kappa_{111}H_{111}(z, \Sigma) + 3\kappa_{112}H_{112}(z, \Sigma) + 3\kappa_{122}H_{122}(z, \Sigma) + \kappa_{222}H_{222}(z, \Sigma) - \kappa_{111}H_3(z_1).$$

The situation is much simpler whenever $\rho = 0$; this may always be achieved by transforming Z_2 to $Z_2 - \rho Z_1$. Then

$$\hat{p}(z_2|z_1) = \phi(z_2)\left[1 + \frac{1}{6\sqrt{n}}Q(z_2|y_1; 0) + \cdots\right]$$

where

$$Q(z_2|z_1; 0) = \kappa_{222}H_3(z_2) + 3\kappa_{122}z_1H_2(z_2) + 3\kappa_{112}H_2(z_1)z_2.$$

This approximation may be used to derive approximations to the conditional cumulants of Z_2 given Z_1 by simply integrating the approximation. For instance, $\mathrm{E}(Z_2|Z_1 = z_1)$ may be approximated by

$$\int_{-\infty}^{\infty} z_2\hat{p}(z_2|z_1)\,dz_2 = \frac{1}{2\sqrt{n}}\kappa_{112}(z_1^2 - 1).$$

Similarly, the conditional variance of Z_2 given $Z_1 = z_1$ may be approximated by

$$1 + \frac{1}{\sqrt{n}}\kappa_{122}z_1;$$

both of these approximations have error of order $O(n^{-1})$.

There is an important drawback to the direct/direct approximations. Since the approximation for the marginal density of Z_1 is based on an Edgeworth expansion, it holds only for z_1 of the form $z_1 = O(1)$. That is, for approximating the conditional distribution of Y_2 given $Y_1 = y_1$, the direct/direct approximations apply only for y_1 of the form $y_1 = \mathrm{E}(Y_1) + O(\sqrt{n})$.

Example 2.17 Linear functions of exponential random variables
Let $X = (X_1, \ldots, X_n)^T$ where $X_1, \ldots, X_n$ are independent exponential random variables, each with rate parameter 1. Let $a = (a_1, \ldots, a_n)^T$ and $b = (b_1, \ldots, b_n)^T$ denote fixed vectors satisfying $\sum a_j = \sum b_j = 0$ and $\sum a_j^2 = \sum b_j^2 = 1$ and consider the problem of approximating the conditional distribution of $b^T X$ given $a^T X$. Note that $b^T X$ and $a^T X$ both have mean 0 and variance 1; the correlation between $b^T X$ and $a^T X$ is given by $\rho = b^T a$. Assume that $\sum a_j^r b_j^s = O(n^{-(r+s-2)/2})$; this holds, for instance, if a_j and b_j are bounded and of order $O(n^{-1/2})$.

As noted above, it is more convenient to approximate the conditional distribution of $b^T X - \rho a^T X \equiv c^T X$ given $a^T X$; note that $c^T X$ and $a^T X$ are uncorrelated. Hence, let $Z_1 = a^T X$ and $Z_2 = c^T X$. The cumulant-generating function of the standard exponential distribution is $-\log(1-t)$ and it is straightforward to show that the cumulant-generating function of (Z_1, Z_2) is given by

$$K(t_1, t_2) = -\sum \log(1 - t_1 a_j - t_2 c_j).$$

It follows that

$$\kappa_{111} = 2\sqrt{n}\sum a_j^3, \qquad \kappa_{112} = 2\sqrt{n}\sum a_j^2 c_j,$$
$$\kappa_{122} = 2\sqrt{n}\sum a_j c_j^2, \qquad \kappa_{222} = 2\sqrt{n}\sum c_j^3.$$

Hence, the conditional density of Z_2 given $Z_1 = z_1$ may be approximated by

$$\hat{p}(z_2|z_1) = \phi(z_2)\left[1 + \frac{1}{6\sqrt{n}} Q(z_2|z_1; 0)\right]$$

where

$$Q(z_2|z_1; 0) = 2\sum c_j^3\, H_3(z_2) + 6\sum a_j c_j^2\, z_1 H_2(z_2) + 6\sum a_j^2 c_j\, z_2 H_2(z_1);$$

the error of this approximation is of order $O(n^{-1})$. It follows that

$$\mathrm{E}(Z_2|Z_1 = z_1) = \sum a_j^2 c_j (z_1^2 - 1) + O(n^{-1})$$

and

$$\mathrm{Var}(Z_2|Z_1 = z_1) = 1 + 2\sum a_j c_j^2 z_1 + O(n^{-1}).$$

In terms of the original variable $a^T X$ we have

$$\mathrm{E}(a^T X|Z_1 = z_1) = \rho z_1 + \sum a_j^2 (b_j - \rho a_j)\,(z_1^2 - 1) + O(n^{-1})$$

and

$$\mathrm{Var}(a^T X|Z_1 = z_1) = 1 + 2\sum a_j (b_j - \rho a_j)^2\, z_1 + O(n^{-1}).$$

These expressions hold for $z_1 = O(1)$. ∎

2.10.3 INDIRECT/INDIRECT APPROXIMATIONS

Another approach to approximating $p(y_2|y_1)$ is to approximate $p(y_1, y_2)$ and $p(y_1)$ using saddlepoint approximations. Again, we denote the cumulant-generating function of (Y_1, Y_2) by $K(t_1, t_2)$ and assume that $K(t_1, t_2)$ and its derivatives are $O(n)$. Note that t_1 and t_2 both scalars here and that the cumulant-generating function of Y_1 is simply $K(t, 0)$.

For given values of y_1, y_2, let $(\hat{\lambda}_1, \hat{\lambda}_2)$ denote the solution in (t_1, t_2) to the pair of equations $K'(t_1, t_2) = (y_1, y_2)^T$. The saddlepoint approximation for the joint density of (Y_1, Y_2) is therefore given by

$$\frac{\exp\{K(\hat{\lambda}_1, \hat{\lambda}_2) - y_1\hat{\lambda}_1 - y_2\hat{\lambda}_2\}}{(2\pi)|K''(\hat{\lambda}_1, \hat{\lambda}_2)|^{1/2}};$$

it is important to keep in mind that each of $\hat{\lambda}_1$ and $\hat{\lambda}_2$ depends on (y_1, y_2).

Let $\hat{\lambda}_0$ denote the solution to the equation $K_1(\hat{\lambda}_0, 0) = y_1$ where

$$K_1(t_1, 0) = \frac{\partial K(t_1, 0)}{\partial t_1}.$$

Then the saddlepoint approximation to the marginal density of Y_1 is

$$\frac{\exp\{K(\hat{\lambda}_0, 0) - y_1\hat{\lambda}_0\}}{(2\pi)^{1/2}[K_{11}(\hat{\lambda}_0, 0)]^{1/2}}$$

where K_{11} denotes the second derivative of $K(t_1, t_2)$ with respect to t_1.

It follows that an approximation to the conditional density of Y_2 given $Y_1 = y_1$ is given by

$$\frac{\exp\{K(\hat{\lambda}_1, \hat{\lambda}_2) - K(\hat{\lambda}_0, 0) - y_1(\hat{\lambda}_1 - \hat{\lambda}_0) - y_2\hat{\lambda}_2\}}{(2\pi)^{1/2}|K''(\hat{\lambda}_1, \hat{\lambda}_2)|^{1/2}[K_{11}(\hat{\lambda}_0, 0)]^{-1/2}};$$

the error of this approximation is $O(n^{-1})$.

It is important to note that, in contrast to the direct/direct approximations, the indirect/indirect approximation holds for fixed values of y_1.

Example 2.18 Gamma distribution

Let $X_1, \ldots, X_n$ denote independent gamma random variables each with scale parameter 1 and index 1. Let $Y_1 = \sum X_j$ and $Y_2 = \sum \log X_j$ and consider the conditional distribution of Y_2 given $Y_1 = y_1$; this conditional distribution arises when drawing inferences regarding the index of the distribution. The cumulant-generating function of (Y_1, Y_2) is given by

$$n \log \Gamma(1 + t_2) - n(1 + t_2) \log(1 - t_1).$$

It is straightforward to show that $\hat{\lambda}_1$ and $\hat{\lambda}_2$ are given by

$$\frac{1 + \hat{\lambda}_2}{1 - \hat{\lambda}_1} = \frac{y_1}{n}$$

and

$$\Psi(1 + \hat{\lambda}_2) - \log(1 + \hat{\lambda}_2) = \frac{y_2}{n} - \log\left(\frac{y_1}{n}\right), \quad \Psi(t) = \frac{\Gamma'(t)}{\Gamma(t)}$$

and that $\hat{\lambda}_0$ is given by $\hat{\lambda}_0 = 1 - n/y_1$. Note that an explicit expression for $\hat{\lambda}_2$ is not available, although $\hat{\lambda}_2$ is easily found numerically, given values for y_1 and y_2.

The indirect–indirect approximation to the conditional density of Y_2 given y_1 is therefore given by

$$\frac{\Gamma(1 + \hat{\lambda}_2)^n (y_1/n)^{\hat{\lambda}_2} \exp\{(n - y_2)\hat{\lambda}_2\}}{(2\pi)^{1/2}(1 + \hat{\lambda}_2)^{n(1+\hat{\lambda}_2)+1}[(1 + \hat{\lambda}_2)\Psi'(1 + \hat{\lambda}_2) - 1]^{1/2}}.$$

∎

2.10.4 MIXED APPROXIMATIONS

Edgeworth expansions have the advantage that the expansions have a relatively simple form and that the expansions are easy to integrate. On the other hand, saddlepoint approximations generally achieve higher accuracy. Furthermore, with respect conditional distributions, the saddlepoint approximation to the conditional distribution given $Y_1 = y_1$ is valid for all y_1, not just for y_1 satisfying

$$\frac{y_1 - \mathrm{E}(Y_1)}{\sqrt{\mathrm{Var}(Y_1)}} = O(1).$$

In this section, we consider a type of mixed expansion that attempts to realize the benefits of both approaches. Specifically, we will use an indirect approximation for the marginal density of Y_1 and a 'partially tilted' approximation for the joint density of (Y_1, Y_2). This partially tilted approximation is essentially a saddlepoint-type approximation with respect to Y_1 and an Edgeworth series-type approximation with respect to Y_2.

First consider the joint density of (Y_1, Y_2). Define

$$p(y_1, y_2; \lambda) = \exp\{y_1\lambda - K(\lambda, 0)\}\, p(y_1, y_2)$$

where, as above, $K(t_1, t_2)$ denotes the $O(n)$ cumulant-generating function of (Y_1, Y_2) and $K(t_1, 0)$ denotes the cumulant-generating function of Y_1. An approximation to $p(y)$ is given by

$$\hat{p}(y_1, y_2) = \exp\{K(\lambda, 0) - y_1\lambda\}\, \hat{p}(y_1, y_2; \lambda)$$

where $\hat{p}(y_1, y_2; \lambda)$ is an approximation to $p(y_1, y_2; \lambda)$. The value of λ is arbitrary and, hence, we may choose λ to make $p(y_1, y_2; \lambda)$ relatively easy to approximate.

Let $\hat{\lambda}$ denote the solution to $K_1(\hat{\lambda}, 0) = y_1$; then, under $p(y_1, y_2; \hat{\lambda})$, Y_1 has mean y_1. To approximate $p(y_1, y_2; \hat{\lambda})$ we can use an Edgeworth series approximation; note, however, that under the density $p(y_1, y_2; \hat{\lambda})$, (Y_1, Y_2) has cumulant-generating function $K(\hat{\lambda} + t_1, t_2)$. Under this cumulant-generating function, Y_2 has mean $\hat{\mu} = K_2(\hat{\lambda}, 0)$ and variance $\hat{\sigma}^2 = K_{22}(\hat{\lambda}, 0)$; Y_1 has mean y_1 and variance $K_{11}(\hat{\lambda}, 0)$.

The higher-order cumulants of (Y_1, Y_2) are given by the higher-order derivatives of $K(t_1, t_2)$, evaluated at $(t_1, t_2) = (\hat{\lambda}, 0)$ rather than at $(0, 0)$; we will denote these cumulants by $\hat{\kappa}_{ijkl}$ so that, for example,

$$\hat{\kappa}_{112} = \left.\frac{\partial^3 K(t_1, t_2)}{\partial t_1^2\, \partial t_2}\right|_{(t_1,t_2)=(\hat{\lambda},0)}.$$

Recall that the $K(t_1, t_2)$ and, hence, the κ_{ijkl} are of order $O(n)$.

The Edgeworth series approximation to $p(y_1, y_2; \hat{\lambda})$ is therefore given by

$$\phi_2((0, y_2 - \hat{\mu}); \hat{\Sigma})\left[1 + \frac{1}{6}\sum_{i,j,k} \hat{\kappa}_{ijk} H_{ijk}((0, y_2 - \hat{\mu}); \hat{\Sigma}) + \cdots\right]$$

where $\hat{\Sigma} = K''(\hat{\lambda}, 0)$. It follows that an approximation to $p(y_1, y_2)$ with error of order $O(n^{-1})$ is given by

$$\hat{p}(y_1, y_2) = \exp\{K(\hat{\lambda}, 0) - y_1\hat{\lambda}\}\, \phi_2((0, y_2 - \hat{\mu}); \hat{\Sigma}) \times \left[1 + \frac{1}{6}\sum_{i,j,k} \hat{\kappa}_{ijk} H_{ijk}((0, y_2 - \hat{\mu}); \hat{\Sigma})\right].$$

To approximate the marginal density of Y_1 we can use the saddlepoint approximation:

$$\hat{p}(y_1) = \frac{\exp\{K(\hat{\lambda}, 0) - y_1\hat{\lambda}\}}{(2\pi)^{1/2}[\hat{\Sigma}_{11}]^{1/2}};$$

note that the saddlepoint $\hat{\lambda}_0$ used in Section 2.8.2 is the same as $\hat{\lambda}$ defined above.

The resulting approximation to the conditional density of Y_2 given $Y_1 = y_1$ is given by

$$\hat{p}(y_2|y_1) = \phi_2((0, y_2 - \hat{\mu}); \hat{\Sigma})(2\pi)^{1/2}\hat{\Sigma}_{11}^{1/2} \times \left[1 + \frac{1}{6}\sum_{i,j,k} \hat{\kappa}_{ijk} H_{ijk}((0, y_2 - \hat{\mu}); \hat{\Sigma})\right].$$

It is straightforward to show that the leading term in this approximation is a normal distribution with mean $\hat{\mu}$ and variance $\hat{\sigma}^2 = \hat{\Sigma}_{22} - \hat{\Sigma}_{21}\hat{\Sigma}_{11}^{-1}\hat{\Sigma}_{12}$. Note that $\hat{\lambda}$ depends on y_1 so that, in general, both the conditional mean and variance of this normal distribution depend on the value of y_1.

Using this result, we have that

$$\mathrm{E}(Y_2|Y_1 = y_1) = \hat{\mu} + O(1) \tag{2.29}$$

where

$$\hat{\mu} = \frac{\partial}{\partial t_2} K(t_1, t_2)\Big|_{t_1=\hat{\lambda}, t_2=0}. \tag{2.30}$$

This result is easily extended to the multivariate case. Suppose that both Y_1 and Y_2 are vectors. Then (2.29) still holds with $\hat{\mu}$ given by (2.30) and $\hat{\lambda}$ taken to be the solution to the equation

$$\frac{\partial}{\partial t_1} K(t_1, 0) = y_1.$$

Example 2.19 Poisson distribution
Let X_1 and X_2 denote independent Poisson random variables each with mean $n\mu$; for instance, each X_j may be the sum of n independent Poisson random variables with mean μ. Let $Y_1 = X_1 + X_2$ and $Y_2 = X_1$; we will consider approximation of the conditional distribution of Y_2 given $Y_1 = y_1$. It is straightforward to show that this conditional distribution is binomial with index y_1 and success probability $1/2$.

The cumulant-generating function of X_j is

$$n\mu[\exp\{t\} - 1]$$

so that (Y_1, Y_2) has cumulant-generating function

$$K(t_1, t_2) = n\mu[\exp\{t_1\} - 1] + n\mu[\exp\{t_1 + t_2\} - 1].$$

Hence, $\hat{\lambda} = \log(y_1/(n\mu))$, $\hat{\mu} = y_1/2$, and

$$\hat{\Sigma} = y_1 \begin{pmatrix} 1 & 1/2 \\ 1/2 & 1/2 \end{pmatrix}.$$

It follows that the leading term in the approximation is a normal distribution with mean $y_1/2$ and variance $y_1/4$; note that this is simply the normal approximation to the binomial distribution described above.

It is straightforward to show that

$$\hat{\kappa}_{111} = y_1, \qquad \hat{\kappa}_{112} = \hat{\kappa}_{122} = \hat{\kappa}_{222} = y_1/2$$

and that $z_1 = -z_2/2$ where $z = \hat{\Sigma}^{-1}y$. It follows that the $1/\sqrt{n}$ correction term is 0; hence, the error of the normal approximation is $O(n^{-1})$. Hence, using this type of mixed approximation, the conditional density of X_1 given $X_1 + X_2 = s$ is approximately normal with mean $s/2$ and variance $s/4$.

The conditional distribution of X_2 given $X_1 + X_2$ can also be approximated using a direct–direct or an indirect–indirect approximation. First consider an indirect–indirect approximation; take $Y_1 = X_1 + X_2$ and $Y_2 = X_1$. It is straightforward to show that $\hat{\lambda}_1 = \log((y_1 - y_2)/(n\mu))$, $\hat{\lambda}_2 = \log((y_1 - y_2)/(2y_1))$, $\hat{\lambda}_0 = \log(y_1/(2n\mu))$, $|K''(\hat{\lambda}_1, \hat{\lambda}_2)| = y_2(y_1 - y_2)$, and $K''_{11}(\hat{\lambda}_0, 0) = y_1$. It follows that an approximation to the conditional density of Y_2 given $Y_1 = y_1$ is given by

$$\frac{1}{\sqrt{(2\pi)}} \frac{y_1^{y_1+1/2}}{(y_1 - y_2)^{(y_1-y_2)+1/2} y_2^{y_2+1/2}} \left(\frac{1}{2}\right)^{y_1};$$

note that this is simply the density function of a binomial distribution with index y_1 and success probability $1/2$, with the factorials replaced by Stirling's approximation.

The drawback of this approximation, relative to the mixed approximation considered previously, is that, without recognizing the relationship to the binomial distribution, inference based on this approximation is somewhat difficult; for instance, it is difficult to determine approximations to the conditional mean and variance of Y_2.

For the direct–direct approximation, take $Z_1 = (X_1 - n\mu)/[n\mu]^{1/2}$ and $Z_2 = (X_1 + X_2 - 2n\mu)/[2n\mu]^{1/2}$. Then Z_1 and Z_2 both have variance 1 and covariance $1/\sqrt{2}$. Hence, the first-order normal approximation to the conditional density of Z_2 given $Z_1 = z_1$ is normal with mean $z_1/\sqrt{2}$ and variance $1/2$. That is, the conditional density of X_1 given $X_1 + X_2 = s$ is approximately normal with mean $s/2$ and variance $n\mu/2$. Since $X_1 + X_2 = 2n\mu + O_p(\sqrt{n})$, it follows that, to a first order of approximation, X_1 has conditional variance $s/4$, but only for s is the 'normal deviation range' of $s = \mathrm{E}(X_1 + X_2) + O_p(\sqrt{n})$. This may be contrasted with the approximate conditional variance from the mixed approximation which is valid for fixed values of $X_1 + X_2$. Note also that, based on the direct–direct approximation, it appears as if the conditional distribution of X_1 given $X_1 + X_2$ depends on the parameter μ when, in fact, this conditional distribution is independent of μ. ∎

2.11 Laplace approximations

Consider computation of the integral

$$C \equiv \int_a^b h(t) \exp\{-ng(t)\}\, dt$$

where g and h are of order $O(1)$. Laplace's method of approximating an integral of this type is based on the idea that, for large n, most of the contribution to the integral will be from values of t in a relatively small range, near the value of t that maximizes the integrand.

Suppose that $g(t)$ is minimized at $t = \hat{t}$ where $a < \hat{t} < b$, $g'(\hat{t}) = 0$, $g''(\hat{t}) > 0$, and $h(\hat{t}) \neq 0$. Then

$$g(t) \doteq g(\hat{t}) + \tfrac{1}{2} g''(\hat{t})(t - \hat{t})^2$$

and

$$\exp\{-ng(t)\} \doteq \exp\{-ng(\hat{t})\} \exp\left\{-\frac{n}{2} g''(\hat{t})(t - \hat{t})^2\right\}.$$

It follows that

$$\int_a^b h(t) \exp\{-ng(t)\}\, dt \doteq \exp\{-ng(\hat{t})\} \int_a^b h(t) \exp\left\{-\frac{n}{2} g''(\hat{t})(t - \hat{t})^2\right\} dt.$$

The function $h(t)$ may also be approximated by a Taylor's series approximation around $t = \hat{t}$.

Hence,

$$C \doteq \exp\{-ng(\hat{t})\} \int_a^b [h(\hat{t}) + h'(\hat{t})(t - \hat{t}) + \cdots] \exp\left\{-\frac{n}{2} g''(\hat{t})(t - \hat{t})^2\right\} dt.$$

Let $u = \sqrt{ng''(\hat{t})}(t - \hat{t})$; then

$$C \doteq \frac{\exp\{-ng(\hat{t})\}}{\sqrt{ng''(\hat{t})}}$$
$$\times \int_{\sqrt{ng''(\hat{t})}(a-\hat{t})}^{\sqrt{ng''(\hat{t})}(b-\hat{t})} \left[h(\hat{t}) + h'(\hat{t})\frac{t}{\sqrt{ng''(\hat{t})}} + \cdots\right] \exp\left\{-\frac{1}{2}t^2\right\} dt.$$

For large n, the integral in this expression may be approximated by

$$\int_{-\infty}^{\infty} h(\hat{t}) \exp\left\{-\tfrac{1}{2}t^2\right\} dt = h(\hat{t})\sqrt{2\pi}.$$

It follows that

$$C \doteq \frac{\sqrt{2\pi}}{\sqrt{ng''(\hat{t})}} \exp\{-ng(\hat{t})\}h(\hat{t}).$$

A more formal analysis shows that the error in this approximation is $O(n^{-1})$.

Example 2.20 Mean of a gamma distribution
Let Y denote a random variable with a gamma distribution with density of the form

$$\frac{n^n}{\Gamma(n)} y^{n-1} \exp\{-ny\}, \quad y > 0.$$

Consider the problem of approximating $\mathrm{E}(Y)$, which in this case is known to be 1. Note that

$$\mathrm{E}(Y) = \frac{n^n}{\Gamma(n)} \int_0^{\infty} y^n \exp\{-ny\}\, dy$$

so that we may take $h(y) = 1$ and $g(y) = y - \log y$. It is straightforward to show that $g(\cdot)$ is minimized at $y = 1$ and, hence,

$$\mathrm{E}(Y) = \frac{n^n}{\Gamma(n)} \frac{\sqrt{2\pi}}{\sqrt{n}} \exp\{-n\}[1 + O(n^{-1})].$$

Substituting a standard expansion for the gamma function shows that

$$\mathrm{E}(Y) = 1 + O(n^{-1}),$$

in accordance with the general result stated above. ∎

The Laplace method may also be applied in a multivariate setting. Consider approximation of the integral

$$C \equiv \int_D h(t) \exp\{-ng(t)\}\, dt$$

where g and h are $O(1)$ functions defined on $\Re^m$ and D is a subset of $\Re^m$. Assume that $g(t)$ is minimized at $t = \hat{t}$ where $\hat{t}$ is an interior point of D, $g'(\hat{t}) = 0$, $g''(\hat{t})$ is positive definite, and $h(\hat{t}) \neq 0$; here 0 refers to the 0 vector.

The derivation of the approximation closely follows that given for the univariate case. Near $t = \hat{t}$ we have

$$g(t) \doteq g(\hat{t}) + \tfrac{1}{2}(t - \hat{t})^T g''(\hat{t})(t - \hat{t})$$

so that

$$\exp\{-ng(t)\} \doteq \exp\{-ng(\hat{t})\} \exp\left\{-\frac{n}{2}(t - \hat{t})^T g''(\hat{t})(t - \hat{t})\right\}.$$

Using a similar approximation for $h(t)$, we have that

$$\begin{aligned} C &\doteq \exp\{-ng(\hat{t})\}h(\hat{t}) \int_{\Re^m} \exp\left\{-\frac{n}{2}(t - \hat{t})^T g''(\hat{t})(t - \hat{t})\right\} dt \\ &= \exp\{-ng(\hat{t})\}h(\hat{t})(2\pi)^{m/2}|ng''(\hat{t})|^{-1/2} \\ &= \frac{\exp\{-ng(\hat{t})\}h(\hat{t})(2\pi)^{m/2}}{n^{m/2}|g''(\hat{t})|^{1/2}}. \end{aligned}$$

The error of this approximation is of order $O(n^{-1})$.

2.12 Discussion and references

There are a number of books available covering the topics discussed in this chapter in more detail. Barndorff-Nielsen and Cox (1989) discusses all of the topics covered in this chapter with particular emphasis on those topics which are useful in statistical inference.

Ferguson (1996) contains an elementary introduction to large-sample theory based on the central limit theorem, along with applications to statistical inference. A more advanced treatment, including several versions of the central limit theorem, is given by Sen and Singer (1993); in particular, results are given for the case of dependent random variables.

McCullagh (1987, Chapters 5 and 6) contains an introduction to Edgeworth series and saddlepoint approximations. Kolassa (1994) contains a more advanced treatment with detailed discussions of regularity conditions along with many useful results are given for the lattice case; see also Durbin (1980). The rigorous mathematical theory of Edgeworth expansions is given by Bhattacharya and Rao (1976).

Saddlepoint methods are discussed in detail in Jensen (1995); in particular, Sections 5, 6, and 7 of this chapter draw heavily from Chapter 3 of that book. Further discussion of saddlepoint approximations is available in Daniels (1954, 1983), Reid (1988), and Kolassa (1994). Field and Ronchetti (1990) give a detailed account of saddlepoint approximations; this work is particularly useful for its many numerical examples illustrating the accuracy of the approximations. The method of approximating integrals discussed in Section 2.5 is known as Temme's method (Temme, 1982). The saddlepoint approximation to the distribution function, discussed in Section 2.6, is due to Lugannani and Rice (1980). Approximations to distribution functions based on the saddlepoint approximation are discussed in detail in Daniels (1987); see also Skates (1993). The Euler–MacLaurin formula is discussed in Atkinson (1978, Section 5.5).

The method described in Section 2.9, specifically that an Edgeworth expansion for the distribution of a statistic may be based on formal cumulants calculated from a Taylor's series expansion of that statistic, is based on Skovgaard (1981a); see also Skovgaard (1981b) and Barndorff-Nielsen and Cox (1989, Section 4.5). For further details on the notation $O_p(\cdot)$ and $o_p(\cdot)$ see Mann and Wald (1943); a useful summary is given in Azzalini (1996, Section A.8.3).

The material on the approximation of conditional distributions in Section 2.10 is based on Chapter 7 of Barndorff-Nielsen and Cox (1989) where further results are available; see also Michel (1979) and Skovgaard (1987). The mixed approximation considered in Section 2.10.4 is based on Pace and Salvan (1992).

Laplace-type approximations are used in many areas of statistics. For further details, see Barndorff-Nielsen and Cox (1989, Section 3.3), Jensen (1995, Section 3.1) and Kass, Tierney and Kadane (1990).

2.13 Exercises

2.1 Let $Y_1, \ldots, Y_n$ denote independent random variables each with density function

$$\frac{\beta_n^{\alpha_n}}{\Gamma(\alpha_n)} y^{\alpha_n - 1} \exp\{-\beta_n y\}, \quad y > 0$$

where α_n and β_n are known, positive sequences such that $\alpha_n \to \infty$ and $\beta_n \to \infty$ as $n \to \infty$. Find sequences of constants c_n and d_n, and conditions on α_n and β_n, such that $c_n \bar{Y} + d_n$ is asymptotically normally distributed.

2.2 A sequence of real-valued random variables $Y_1, \ldots,$ is said to converge to μ in quadratic mean if

$$\mathrm{E}[|Y_n - \mu|^2] \to 0$$

as $n \to \infty$. Show that, if Y_n converges in quadratic mean to μ, then $Y_n \xrightarrow{p} \mu$. Give an example to show that the converse is not true.

2.3 Let X_n and Y_n denote sequences of real-valued random variables each with mean 0 and standard deviation 1. Suppose that, as $n \to \infty$,

$$X_n \xrightarrow{\mathcal{D}} X; \qquad Y_n \xrightarrow{\mathcal{D}} Y.$$

Let ρ_n denote the correlation of X_n and Y_n. Show that if $\rho_n \to 1$ as $n \to \infty$, then X and Y have the same distribution.

2.4 Let $Y_1, Y_2, \ldots$ denote a sequence of real-valued random variables with mean 0, standard deviation 1, and higher-order cumulants $\kappa_3, \kappa_4, \ldots$. Let $x_1, x_2, \ldots$ denote a sequence of constants. Find conditions on $x_1, x_2, \ldots$ so that

$$\frac{\sum x_j Y_j}{\sqrt{\sum x_j^2}}$$

is asymptotically normally distributed.

2.5 The Edgeworth expansion of the density of a random variable may, for some values of the argument, be negative. To avoid this problem, it is sometimes useful to approximate a density function by expanding the log of the density and then exponentiating the result to obtain an approximation for the density itself. Determine an Edgeworth expansion for the log-density incorporating the $1/\sqrt{n}$ and $1/n$ terms.

2.6 Consider a random variable with mean 0 and standard deviation 1. Suppose the density of this random variable is approximated by an Edgeworth expansion including only the $1/\sqrt{n}$ term. Using this approximation, find approximations for the median and mode of the distribution.

Exercises 2.7 through 2.12 consider the problem of approximating the distribution of X_n, the sample mean of n independent identically distributed exponentially distributed random variables with rate parameter θ (i.e., with mean $1/\theta$). Note that the exact distribution of X_n is available using properties of the gamma distribution. Let $F(x; \theta) = \Pr(X_n \leq x; \theta)$.

2.7 Give an approximation to $F_n(x; \theta)$ based on the central limit theorem.

2.8 Let $X_n, n = 1, 2, \ldots$, a sequence of random variables such that $Y_n = \sqrt{n}(X_n - \mu_n)$ converges in distribution to some random variable Y. In many cases, such as when X_n is the mean of n independent random variables, the third cumulant of Y_n satisfies

$$\kappa_3(Y_n) = O(n^{-1/2}).$$

If h is a continuously differentiable function such that

$$Z_n = \sqrt{n}(h(X_n) - h(\mu_n))$$

has third cumulant

$$\kappa_3(Z_n) = O(n^{-3/2}),$$

then h is called a *skewness-reducing transformation*. In this case it is reasonable to expect that the distribution of Z_n is better approximated by a normal distribution than is the distribution of Y_n, since the skewness of a normal distribution is 0.

Consider the case in which X_n is the sample mean of n independent identically distributed random variables each with an exponential distribution with parameter θ.

(a) Show that

$$\begin{aligned} \mathrm{E}(X_n^r) = \big[1 + \tfrac{1}{2}r(r-1)n^{-1} \\ + \tfrac{1}{8}r(r-1)(r-2)(r-1/3)n^{-2}\big]\lambda^{-r} + O(n^{-3}). \end{aligned}$$

(b) Show that $h(t) = t^{1/3}$ is a skewness-reducing transformation.

(c) Using a normal approximation for $X_n^{1/3}$, approximate $F_n(x; \theta)$.

2.9 In approximating the distribution of $\sqrt{n}(h(X_n) - h(\mu_n))$ by a normal distribution, the approximation can often be improved by replacing $h(\mu_n)$ by an approximate expression for $\mathrm{E}(h(X_n))$.

(a) For the function h given in Exercise 2.8, approximate $\mathrm{E}[h(X_n)]$. Note that $\mathrm{E}(X_n) = \mu_n$ so that this approach is not needed for the approximation given by the central limit theorem.

(b) Approximate $F_n(x; \theta)$ using a normal approximation for $\sqrt{n}(X_n^{1/3} - m)$, where m is your approximation to $\mathrm{E}[X_n^{1/3}]$.

2.10 Approximate $F_n(x; \theta)$ using an Edgeworth expansion that corrects for the third and fourth cumulants.

2.11 Approximate $F_n(x; \theta)$ using a saddlepoint approximation.

2.12 Let X_n denote the sample mean of n independent identically distributed random variables each with an exponential distribution with parameter λ for the case $n = 9$ and $\lambda = 1/2$. For this choice of λ, nX_n has a chi-square distribution with $2n$ degrees of freedom. For $x = 2.5289, 2.8877, 3.2077$ and 3.8672 approximate $F_n(x; \lambda)$ using the approximations derived in Exercises 2.7–2.11. Which approximations appear to be most accurate?

2.13 Let H_j denote the jth Hermite polynomial. Show that

$$\int_{-\infty}^{\infty} H_j(z) H_k(z) \phi(z) \, dz = 0 \quad \text{for } j \neq k.$$

Based on this result, show that if Z is a standard normal random variable and $X_j = H_j(Z)$, $j = 1, \ldots$, then each X_j has mean 0 and X_j and X_k are uncorrelated for $j \neq k$.

2.14 Let $Y_1, \ldots, Y_n$ denote independent random variables each distributed according to the density

$$\theta(1-\theta)^{y-1}, \quad y = 0, 1, \ldots$$

where $0 < \theta < 1$. This is the density of the geometric distribution. Find the saddlepoint approximations to the density and distribution function of $\bar{Y}$. Calculate the approximation to $\Pr(\bar{Y} \leq t)$ for the case $n = 5$ and $\theta = 1/2$ for several choices of t and compare the results to the exact values based on the negative binomial distribution.

2.15 Let Y_n denote a statistic with a density function that has an Edgeworth series approximation of the form (2.2). Let $X_n = Y_n^2$. Find an Edgeworth series approximation to the distribution function of X_n, neglecting terms of order $O(n^{-2})$. Relate the terms in the approximation to the chi-square distribution.

2.16 Consider the saddlepoint approximation to $\Pr(\bar{Y} \geq t)$. Since $\hat{\lambda}_t$ and r are both 0 for $t = \mu$, the mean of $\bar{Y}$, the approximation cannot be used directly when $t = \mu$. Find the limit of

$$\frac{1}{\hat{\lambda}_t K''(\hat{\lambda}_t)^{1/2}} - \frac{1}{r_t}$$

as $t \to \mu$ and, hence, derive an approximation to $\Pr(\bar{Y} \geq \mu)$. Compare this approximation to that based on an Edgeworth series approximation.

2.17 For the model considered in Example 2.17, find an approximation to the conditional mean of $b^T X$ given $a^T X = z_1$ using a mixed approximation. Compare the result to the approximation derived in Example 2.17.

3
Likelihood

3.1 Introduction

Suppose that the observed data y can be modeled as the observed value of a random variable Y distributed according to density $p(y; \theta)$, $\theta \in \Theta$. Consider two possible values for θ, θ_1 and θ_2. Based on the observed data, which value is more likely to be the true value of θ?

First suppose that Y is a discrete random variable so that $p(y; \theta)$ represents the probability that $Y = y$ when θ is the true parameter value. If $p(y; \theta_1) > p(y; \theta_2)$ then the probability of observing what was actually observed, that is, $Y = y$, is greater when the true parameter value is θ_1 than when the true parameter value is θ_2. Hence, it is more likely that $\theta = \theta_1$ than $\theta = \theta_2$. Suppose that

$$\frac{p(y; \theta_1)}{p(y; \theta_2)} = r.$$

We can go further and say that $\theta = \theta_1$ is r times more likely than $\theta = \theta_2$.

The *likelihood function* for θ based on the observation of $Y = y$ is given by

$$L(\theta) \equiv L(\theta; y) = p(y; \theta), \quad \theta \in \Theta. \tag{3.1}$$

Note that the likelihood function is viewed as a function of θ for fixed y; the likelihood function evaluated at θ represents the probability of observing $Y = y$ when θ is the true parameter value. It is often convenient to work with the log of the likelihood function which we will call the log-likelihood function; the log-likelihood function will be denoted by $\ell(\theta)$. The symbol $\ell(\cdot)$ will be used to denote both the log-likelihood function for a given set of data and the log-likelihood function viewed as a random variable; the symbol $L(\cdot)$ will be used in the same manner.

When Y is a continuous random variable we still define the likelihood function by (3.1) where $p(y; \theta)$ denotes the density function of Y. The direct interpretation of the likelihood function in terms of a probability is no longer available, although we may view $p(y; \theta)$ as being proportional to the probability that Y takes values in a small set containing y.

Note that since only relative values of $L(\theta)$ are of interest we may always multiply the likelihood function by a constant independent of θ without changing the value of ratios such as $L(\theta_1)/L(\theta_2)$. It follows that, in a certain sense, likelihood functions $L(\theta)$ and $cL(\theta)$ may be taken to be equivalent. Here we will not distinguish between any two forms of the likelihood function that differ only by a constant factor. Note that in terms of the log-likelihood function $\ell(\theta)$ we will not distinguish between two forms of the log-likelihood function that differ only by a constant additive term.

Example 3.1 Poisson process
Let $Y_1, \ldots, Y_n$ denote the interevent times of a Poisson process with rate parameter λ. Then the density function of $Y = (Y_1, \ldots, Y_n)$ is given by

$$p(y; \lambda) = \lambda^n \exp\{-\lambda(y_1 + \cdots + y_n)\}$$

so that the log-likelihood function for λ based on the observation $Y = y$ is given by

$$\ell(\lambda) = n \log \lambda - \lambda(y_1 + \cdots + y_n).$$

Consider the data in Table A.1 of the Appendix. The observations represent the time between failures of air conditioning in a certain aircraft; the units are hours in service. Here we use the data for aircraft 1, which we model as independent exponential random variables. The log-likelihood function for these data is given by

$$\ell(\lambda) = 6 \log \lambda - 493\lambda.$$

A plot of this log-likelihood function is given in Figure 3.1. In this plot, as in all the likelihood plots in this book, the likelihood function has been standardized to have maximum value 0. ∎

Example 3.2 Logistic regression
Let $Y_1, \ldots, Y_n$ denote independent random variables such that

$$\Pr(Y_j = 1) = 1 - \Pr(Y_j = 0) = \pi_j, \quad j = 1, \ldots, n.$$

Assume that

$$\log \frac{\pi_j}{1 - \pi_j} = x_j \theta$$

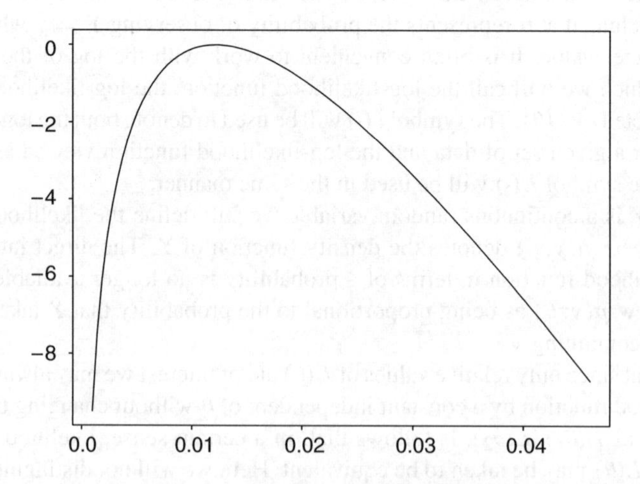

Fig. 3.1 Likelihood function for Example 3.1.

where each x_j is a known $1 \times d$ vector of covariates and θ is an unknown d-dimensional parameter vector. Then, since Y_j has density $\pi_j^y(1-\pi_j)^{1-y}$, the log-likelihood function for θ based on $y_1, \ldots, y_n$ is given by

$$\ell(\theta) = \sum y_j x_j \theta - \sum \log(1 + \exp\{x_j \theta\}).$$

Data of this type are often grouped. Suppose that there are n_j observations with covariate value x_j and that M_j of those observations have a y-value of 1. Then M_j has a binomial distribution with mean

$$n_j \pi_j = n_j \frac{\exp\{x_j \theta\}}{1 + \exp\{x_j \theta\}}$$

and index n_j. The log-likelihood function based on these data is given by

$$\ell(\theta) = \sum m_j x_j \theta - \sum n_j \log(1 + \exp\{x_j \theta\}).$$

Since θ is two-dimensional, a simple plot of the log-likelihood function, as in Example 3.1, is not available. However, $\ell(\theta)$ may be plotted using more sophisticated plotting methods, such a contour plot or a perspective plot. ∎

Example 3.3 Normal-theory nonlinear regression model
Let $Y_1, \ldots, Y_n$ denote independent random variables of the form

$$Y_j = \frac{\beta_1 x_j}{\beta_2 + x_j} + \sigma \epsilon_j, \quad j = 1, \ldots, n$$

where $\epsilon_1, \ldots, \epsilon_n$ are independent standard normal random variables and $x_1, \ldots, x_n$ are fixed scalar constants. The parameter θ is given by $\theta = (\beta_1, \beta_2, \sigma)$. This model is often used in enzyme kinetics where it is known as the Michaelis–Menton model; in that application, Y_j represents the initial velocity of a reaction and x_j represents the substrate concentration.

The log-likelihood function is given by

$$\ell(\theta) = -n \log \sigma - \frac{1}{2\sigma^2} \sum \left(y_j - \frac{\beta_1 x_j}{\beta_2 + x_j} \right)^2.$$ ∎

3.2 Some properties of the likelihood function

Let $L(\theta)$ denote the likelihood function based on the observation of a random variable Y that follows some parametric model. Suppose that θ_1 and θ_2 are two possible values of θ and that $L(\theta_1) > L(\theta_2)$. Why should this lead us to conclude that $\theta = \theta_1$ is 'more likely', in some sense, than $\theta = \theta_2$?

One answer to this question is based on the original motivation for defining the likelihood function: the data support that value of θ that makes what we know to have occurred more probable. In this section however we attempt to give a more formal justification for the use of the likelihood function for statistical inference.

Consider the problem of attempting to choose between two possible values of θ, θ_1 and θ_2, based on the observation of $Y = y$. For certain values of y we choose $\theta = \theta_1$ otherwise $\theta = \theta_2$ is selected. Let $\mathcal{Y}$ denote the set of possible values of Y and let $\mathcal{Y}_1$ denote the set of $y \in \mathcal{Y}$ for which the observation of $Y = y$ leads us to choose $\theta = \theta_1$. How should the set $\mathcal{Y}_1$ be selected?

If we restrict the possible values of θ to the set $\{\theta_1, \theta_2\}$, for a given set $\mathcal{Y}_1$ there are two types of errors that might occur. We might observe $Y \in \mathcal{Y}_1$ whenever $\theta = \theta_2$ or we might observe $Y \notin \mathcal{Y}_1$ whenever $\theta = \theta_1$. Suppose we agree to evaluate the performance of a particular choice for $\mathcal{Y}_1$ by a sum of the probabilities of these two types of error. That is, we consider

$$Q(\mathcal{Y}_1) = \Pr(Y \in \mathcal{Y}_1; \theta_2) + 1 - \Pr(Y \in \mathcal{Y}_1; \theta_1).$$

The 'best' choice for $\mathcal{Y}_1$ is the subset of $\mathcal{Y}$ that minimizes $Q(\mathcal{Y}_1)$ or, equivalently, minimizes

$$\int_{\mathcal{Y}_1} p(y; \theta_2) - \int_{\mathcal{Y}_1} p(y; \theta_1)\, dy = \int_{\mathcal{Y}_1} [p(y; \theta_2) - p(y; \theta_1)]\, dy.$$

Clearly, this will be minimized by choosing

$$\mathcal{Y}_1 = \{y \in \mathcal{Y}\colon p(y; \theta_2) < p(y; \theta_1)\},$$

that is, we should choose $\theta = \theta_1$ whenever $L(\theta_1) > L(\theta_2)$ and choose $\theta = \theta_2$ otherwise. A similar result holds if $Q(\mathcal{Y}_1)$ is of the form

$$Q(\mathcal{Y}_1) = \alpha \Pr(Y \in \mathcal{Y}_1; \theta_2) + \beta[1 - \Pr(Y \in \mathcal{Y}_1; \theta_1)]$$

where α and β are fixed constants. In this case, $Q(\mathcal{Y}_1)$ is minimized by choosing

$$\mathcal{Y}_1 = \{y \in \mathcal{Y}\colon p(y; \theta_2)/p(y; \theta_1) < \beta/\alpha\}.$$

Hence, the likelihood function is optimal, in a certain sense, for choosing between different values of θ.

Furthermore, if the actual parameter value is θ_1, the probability that $L(\theta_2)$ will greatly exceed $L(\theta_1)$ is small. More formally, consider the probability that $L(\theta_2)/L(\theta_1) > k$ when $\theta = \theta_1$. By Markov's inequality,

$$\Pr\left\{\frac{L(\theta_2)}{L(\theta_1)} > k; \theta_1\right\} \le \frac{1}{k}\mathrm{E}\left[\frac{L(\theta_2)}{L(\theta_1)}; \theta_1\right] = \frac{1}{k}.$$

Here we have used the fact that

$$\mathrm{E}\left[\frac{L(\theta_2)}{L(\theta_1)}; \theta_1\right] = \int \frac{p(y; \theta_2)}{p(y; \theta_1)} p(y; \theta_1)\, dy = 1.$$

For example, if θ_1 is the true parameter value, the probability that $L(\theta_2)$ will be ten times as large as $L(\theta_1)$ is at most $1/10$. It should be noted that this probability is, in general, a conservative upper bound; the exact probability will be a function of θ_2

and θ_1. For example, if the underlying probability distribution is normal with mean θ and standard deviation 1, the actual probability that $L(\theta_2)/L(\theta_1) \geq k$ when θ_1 is the true parameter value is

$$1 - \Phi(\log(k)/(\theta_2 - \theta_1) + (\theta_1 + \theta_2)/2), \quad \text{for } \theta_2 > \theta_1.$$

The likelihood function is closely related to the *Kullback–Leibler divergence* between probability distributions. For probability distributions indexed by parameters θ_1 and θ_2, define

$$K(\theta_2, \theta_1) = \mathrm{E}\{[\log p(Y; \theta_1) - \log p(Y; \theta_2)]; \theta_1\}.$$

The quantity $K(\theta_2, \theta_1)$ may be viewed as a measure of the (squared) 'statistical distance' between the distributions with parameters θ_1 and θ_2. In particular, $K(\theta_2, \theta_1) = 0$ if and only if $\theta_1 = \theta_2$. Note, however, that K does not represent a true distance function since it is not symmetric.

Let θ_0 denote the actual parameter value and let θ_1 and θ_2 denote elements of Θ. Then θ_1 is 'closer' in a certain sense to θ_0 than is θ_2 provided that

$$K(\theta_1, \theta_0) < K(\theta_2, \theta_0)$$

that is, provided that

$$\mathrm{E}\{[\log p(y; \theta_1) - \log p(y; \theta_2)]; \theta_0\} > 0. \tag{3.2}$$

Of course, (3.2) cannot be used directly since it requires knowledge of the true parameter value θ_0. However, consider an empirical version of (3.2) which replaces the expectation with respect to the true distribution with expectation with respect to the empirical distribution based on a sample $Y_1, \ldots, Y_n$; for example, $E[\log p(y; \theta_1); \theta_0]$ is replaced by

$$\frac{1}{n} \sum \log p(y_j; \theta_1).$$

Then (3.2) simply becomes $L(\theta_1) > L(\theta_2)$. Hence, the likelihood function evaluated at θ_1 is greater than the likelihood function evaluated at θ_2 if the distribution with parameter θ_1 is closer to the empirical distribution of the data than is the distribution with parameter θ_2, in a certain sense.

3.3 The likelihood principle

In statistical inference, the goal is to draw conclusions regarding the underlying distribution of the random variable Y on the basis of observing $Y = y$. In a parametric model, this is equivalent to drawing conclusions regarding the unknown value of the parameter θ. Since $L(\theta)$ measures how likely different values of θ are to be the true value of θ, the general problem of statistical inference is solved to a large extent by simply examining the likelihood function, at least in principle. The likelihood principle is a somewhat extreme version of this view.

The *likelihood principle* can be stated as follows. Suppose that we have an observation y from a statistical model $\{p(\cdot; \theta) \colon \theta \in \Theta\}$ and an observation x from a statistical model $\{g(\cdot; \theta): \theta \in \Theta\}$ where the parameter θ has the same meaning in both models. Let $L(\theta)$ denote the likelihood function for the model $p(\cdot; \theta)$ and let $\tilde{L}(\theta)$ denote the likelihood function for the model $g(\cdot; \theta)$. If, for given values of x and y, $L(\theta) = \tilde{L}(\theta)$ for all θ, then our conclusions regarding θ based on observing y should be the same as our conclusions regarding θ based on observing x.

Example 3.4 Bernoulli random variables observed under different types of sampling

Suppose that when a certain experiment is performed either a given event occurs or it does not occur and let θ denote the unknown probability that the event occurs. Consider two approaches to gaining information about θ. In the first approach, the experiment is repeated 10 times and Y, the number of times the event occurred, is recorded. The random variable Y has a binomial distribution with index 10. The density of Y is given by

$$\binom{10}{y} \theta^y (1-\theta)^{10-y}, \quad y = 0, \ldots, 10$$

and the likelihood function for θ based on Y is given by

$$L(\theta) = \theta^y (1-\theta)^{10-y}, \quad 0 < \theta < 1.$$

The second approach is to repeat the experiment until 5 events have occurred and to record X, the number of repetitions needed. Then X has a negative binomial distribution. The density of X is given by

$$\binom{x-1}{4} \theta^5 (1-\theta)^{x-5}, \quad x = 5, 6, \ldots$$

and the likelihood function for θ based on X is given by

$$\tilde{L}(\theta) = \theta^5 (1-\theta)^{x-5}, \quad 0 < \theta < 1.$$

Suppose that, using the first approach, 5 events occur so that

$$L(\theta) = \theta^5 (1-\theta)^5.$$

Suppose that, using the second approach, the fifth event occurs on the 10 repetition of the experiment so that $X = 10$; then

$$\tilde{L}(\theta) = \theta^5 (1-\theta)^5.$$

According to the likelihood principle, our conclusions regarding θ should be the same in the two cases; that is, for inference about θ we only need to know that there were 10 repetitions of the experiment and 5 events occurred. Information regarding how the data was collected is not considered. ∎

Under the likelihood principle, quantities that depend on the sampling distribution of a statistic, which is not generally a function of the likelihood function, are irrelevant for statistical inference. Hence, the likelihood principle is in conflict with another important principle of statistical inference, the repeated sampling principle. The *repeated sampling principle* states that statistical procedures should be evaluated on the basis of their behavior in hypothetical repetitions of the experiment that generated the original data. Of course, there is considerable arbitrariness in how this principle is interpreted, in particular, in how the hypothetical repetitions are defined. The basic idea however, which is simply that a given statistical procedure should be evaluated based on how it would perform if adopted for routine use, is certainly reasonable. Many of the commonly used criteria for evaluating statistical procedures, such as bias, variance, and coverage probability, may be justified by the repeated sampling principle.

Hence, the likelihood principle is in disagreement with many of the methods of classical, frequency-based, statistical inference. In particular, the methods described in this book do not, in general, follow the likelihood principle.

The disagreement between the likelihood principle and common statistical practice has led to the development of many counterexamples to the likelihood principle; these counterexamples attempt to show that, at least in some cases, information beyond that provided by the likelihood function is necessary for proper statistical inference.

Example 3.5 A counterexample to the likelihood principle
Let Y denote a random variable taking values in the set $\{1, 2, \dots\}$ with a density function depending on a parameter θ also taking values in the set $\{1, 2, \dots\}$. The form of the density functions depends upon whether θ is even or odd. If θ is even, or $\theta = 1$, then

$$p(y; \theta) = \frac{1}{3} \quad \text{for } y = \left\lceil \frac{\theta}{2} \right\rceil, 2\theta, 2\theta + 1;$$

here $\lceil \cdot \rceil$ denotes the smallest integer greater than or equal to the argument. If θ is an odd number greater than 1 then

$$p(y; \theta) = \begin{cases} \frac{10}{24} & \text{for } y = (\theta - 1)/2 \\ \frac{7}{24} & \text{for } y = 2\theta, 2\theta + 1. \end{cases}$$

It is straightforward, but tedious, to determine the likelihood function based on a given observation $Y = y$. It should be clear however that for any value of y the likelihood function evaluated at $\theta = 2y + 1$ is $10/24$. Clearly, for any other value of θ the likelihood function can be, at most, $1/3$. Hence, for any value of y the likelihood function takes its maximum value at $\theta = s$ where $s = 2y + 1$. Hence, based on the likelihood principle, s is the most likely value for θ based on the observation of $Y = y$. Note, however, that since $2y + 1$ is an odd number greater than 1, whenever θ is even or equal to 1, $\Pr(S = \theta; \theta) = 0$. When θ is an odd number greater than 1, $\Pr(S = \theta; \theta) = 10/24$.

Consider the statistic

$$t \equiv t(y) = \begin{cases} (y - 1)/2 & \text{if } y \text{ is odd and} y > 1 \\ \lceil y/2 \rceil & \text{otherwise.} \end{cases}$$

When θ is 1 then Y is either 1, 2, or 3 and each case $T = \theta$; hence $\Pr(T = \theta; \theta) = 1$. When θ is even and Y is equal to either 2θ or $2\theta + 1$ then $T = \theta$. Hence, in this case, $\Pr(T = \theta; \theta) = 2/3$. Finally, when θ is an odd number greater than 1, $T = \theta$ provided that either $Y = 2\theta$ or $Y = 2\theta + 1$; hence, in this case $\Pr(T = \theta; \theta) = 14/24$. Therefore, for any value of θ,

$$\Pr(T = \theta; \theta) \geq 14/24 > 10/14 \geq \Pr(S = \theta; \theta).$$

The likelihood function evaluated at $\theta = s$ is always 10/24 while the likelihood function evaluated at $\theta = t$ is at most 1/3. Hence, the point with smaller likelihood is more likely to be equal to θ. In particular, if $\theta = 1$, $T = \theta$ with probability 1 while $S \neq \theta$ with probability 1. ∎

This example, like many of the other counterexamples which have been developed, seems far removed from the practical requirements of applied statistics. The main reason that the likelihood principle is not routinely used is most likely due to the fact that many users of statistical methods find other principles, such as the repeated sampling principle, more compelling, rather than the fact that the likelihood function gives misleading inferences in some rare cases.

The likelihood principle, as described above, is sometimes called the strong likelihood principle. A weaker version, called the *weak likelihood principle*, is given by the following. Suppose that we have observations y_1 and y_2 from a statistical model $\{p(\cdot; \theta): \theta \in \Theta\}$ and let $L(\theta; y)$ denote the likelihood function based on the observation of y. If $L(\theta; y_1) = L(\theta; y_2)$ then our conclusions regarding θ based on observing $Y = y_1$ should be the same as our conclusions regarding θ based on observing $Y = y_2$.

3.4 Regular models

Although the emphasis in this book is on the general ideas underlying likelihood-based inference, rather than on precise mathematical results, it is important to note that a general theory of likelihood-based statistical inference requires some regularity conditions on the models under consideration. Generally speaking, we will assume that a model is regular in the sense that the log-likelihood function may be approximated by a polynomial in θ.

More precisely, let $\ell_\theta(\theta)$, $\ell_{\theta\theta}(\theta)$, and so on, denote the derivatives of $\ell(\theta)$ with respect to θ. Hence, $\ell_\theta(\theta)$ is a vector, $\ell_{\theta\theta}(\theta)$ is a matrix, $\ell_{\theta\theta\theta}(\theta)$ is a three-dimensional array, and so on.

For an arbitrary q-dimensional array C define

$$C[t, \ldots, t] = \sum_{i_1, \ldots, i_q} C_{i_1 \cdots i_q} t_{i_1} \cdots t_{i_q}.$$

Property R1

Assume Θ is a subset of $\Re^d$ and that there exists an open subset Θ_0 of Θ with the property that for each $\theta_0 \in \Theta_0$ there exists a neighborhood $N_{\theta_0} \in \Theta_0$ such that for

$\theta \in N_{\theta_0}$ the following expansion holds

$$\ell(\theta) - \ell(\theta_0) = \ell_\theta(\theta_0)^T(\theta - \theta_0) + \tfrac{1}{2}(\theta - \theta_0)^T \ell_{\theta\theta}(\theta_0)(\theta - \theta_0)$$
$$+ \tfrac{1}{6}\ell_{\theta\theta\theta}(\theta_0)[\theta - \theta_0, \theta - \theta_0, \theta - \theta_0]$$
$$+ \tfrac{1}{24}\ell_{\theta\theta\theta\theta}(\theta_0)[\theta - \theta_0, \theta - \theta_0, \theta - \theta_0, \theta - \theta_0] + R_n(\theta)$$

where

$$\frac{\sup_{\theta \in N_{\theta_0}} |R_n(\theta)|}{n||\theta - \theta_0||^5} = O_p(1).$$

Here $||\cdot||$ denotes the Euclidean norm.

Property R2
The log-likelihood derivatives $\ell_\theta(\theta), \ldots, \ell_{\theta\theta\theta\theta}(\theta)$ have joint cumulants up to the fourth order and these cumulants are of order $O(n)$. The first derivative $\ell_\theta(\theta)$ follows the central limit theorem in the sense that $n^{-1/2}\ell_\theta(\theta)$ converges in distribution to a multivariate normal random variable.

Property R3
Let i_1, i_2, i_3, and i_4 denote nonnegative integers such that

$$i_1 + i_2 + i_3 + i_4 \leq 4$$

and let j, k, l, and m take values in the set $\{0, \ldots, d\}$. For $\theta_0 \in \Theta_0$,

$$\frac{\partial^{i_1+i_2+i_3+i_4}}{\partial\theta_j^{i_1}\,\partial\theta_k^{i_2}\,\partial\theta_\ell^{i_3}\,\partial\theta_m^{i_4}} \mathrm{E}\{\exp\{\ell(\theta) - \ell(\theta_0)\}; \theta_0\}\Big|_{\theta=\theta_0}$$
$$= \mathrm{E}\left\{\frac{\partial^{i_1+i_2+i_3+i_4}}{\partial\theta_j^{i_1}\,\partial\theta_k^{i_2}\,\partial\theta_\ell^{i_3}\,\partial\theta_m^{i_4}} \exp\{\ell(\theta) - \ell(\theta_0); \theta_0\}\Big|_{\theta=\theta_0}\right\}.$$

Property R3 is often easily established using standard results on the interchanging of integration and differentiation. For instance, Property R3 holds provided that the support of the underlying random variables does not depend on θ and that $\ell_{\theta\theta\theta\theta}(\theta)$ is a continuous function of θ.

Properties R1–R3 hold in most models of practical interest and verification of these conditions will not be attempted for the models considered in the following chapters. Below are three examples of how the conditions could be verified if there is some question regarding their validity. Example 3.6 considers an independent identically distributed sample from a one-parameter exponential family model. Example 3.7 considers a model in which the underlying observations are not independent. In Example 3.8, a regression-type model is considered in which the random variables are independent, but not identically distributed. All three examples consider the case in which θ is a scalar; the case of a vector parameter can be handled along similar lines, although the details quickly become tedious.

Example 3.6 One-parameter exponential family models
We now show that the one-parameter exponential family model is regular. The same result holds in higher dimensions as well using a similar argument. Let $Y_1, \ldots, Y_n$ denote independent identically distributed random variables each distributed according to a density of the form

$$\exp\{\theta s(y) - k(\theta) + D(y)\}$$

where $\theta \in \Theta$ and Θ is an open subset of $\Re$. In this case the log-likelihood function is given by

$$\ell(\theta) = \theta \sum s(Y_j) - nk(\theta).$$

Since $k(\theta+t) - k(\theta)$ is the cumulant-generating function of Y_1 (see Example 1.15), it follows that $k(\theta)$ has derivatives of all orders and those derivatives are finite. Hence, $k(\cdot)$ may be expanded

$$k(\theta) - k(\theta_0) = k'(\theta_0)(\theta - \theta_0) + \cdots + \tfrac{1}{24}k^{(4)}(\theta_0)(\theta - \theta_0)^4 + O(|\theta - \theta_0|^5).$$

Since

$$\ell(\theta) - \ell(\theta_0) = (\theta - \theta_0)\sum s(Y_j) - n[k(\theta) - k(\theta_0)],$$

Property R1 follows.

To verify Property R2, first note that $\ell_{\theta\theta}(\theta)$, $\ell_{\theta\theta\theta}(\theta)$, and $\ell_{\theta\theta\theta\theta}(\theta)$ are all constants and

$$\ell_\theta(\theta) = \sum S(Y_j) - nk'(\theta).$$

Property R2 now follows from the central limit theorem for independent identically distributed random variables. Property R3 follows since $k^{(4)}(\theta)$ is continuous. ∎

Example 3.7 Time-dependent Poisson process
Consider a stochastic process $N(t)$ as in Example 1.2 such that conditions (a) and (b) given in that example are satisfied and condition (c) is replaced by

$$(\mathrm{c}') \quad \lim_{h \to 0} [\Pr\{N(t+h) - N(t) = 1\} - \lambda(t)h] = 0$$

for some nonnegative function $\lambda(\cdot)$. The stochastic process $\{N(t): t > 0\}$ is a time-dependent or nonhomogeneous Poisson process. For each t, $N(t)$ is a Poisson random variable with mean

$$m(t) = \int_0^t \lambda(s)\, ds.$$

Suppose that the process is observed until n events occur. Let $X_1, \ldots, X_n$ denote the event times and let $Y_1, \ldots, Y_n$ denote the interevent times. In contrast to the stationary Poisson process considered in Example 1.2, $Y_1, \ldots, Y_n$ are, in general, no longer independent or identically distributed.

One way to derive the properties of a time-dependent Poisson process is to transform the time scale. Suppose that $\lambda(t) > 0$ for all t so that $m(\cdot)$ is a strictly increasing, and hence invertible, function. Define $\tilde{N}(t) = N(m^{-1}(t))$. Then

$$\Pr\{\tilde{N}(t) = j\} = \Pr\{N(m^{-1}(t)) = j\}$$

so that $\tilde{N}(t)$ is a Poisson random variable with mean

$$\int_0^{m^{-1}(t)} \lambda(s)\, ds = t$$

that is, $\tilde{N}(t)$ is a stationary Poisson process with rate parameter 1.

Let $\tilde{X}_1, \tilde{X}_2, \ldots, \tilde{X}_n$ denote the event times of the process $\tilde{N}(t)$. Then $\tilde{X}_j = m(X_j)$; it follows from the discussion in Example 1.2 that $\tilde{X}_j - \tilde{X}_{j-1}$, $j = 1, \ldots, n$ are independent exponential random variables with rate parameter 1; here $\tilde{X}_0 = 0$. Hence,

$$m(X_j) - m(X_{j-1}) = \int_{X_{j-1}}^{X_j} \lambda(s)\, ds, \quad j = 1, \ldots, n$$

are independent identically distributed standard exponential random variables. It follows that the density function of $X_1, \ldots, X_n$ is given by

$$\left[\prod \lambda(x_j)\right] \exp\left\{-\int_0^{x_n} \lambda(s)\, ds\right\}.$$

Consider the rate function $\lambda(t) = \theta t$ where $\theta > 0$ is a unknown parameter. Then

$$\ell(\theta) = n \log\theta - \tfrac{1}{2}\theta X_n^2.$$

It is straightforward to show that X_n^2 has a gamma distribution with index n and rate parameter $\theta/2$; that is, $W = X_n^2$ has density function

$$\frac{(\theta/2)^n}{\Gamma(n)} w^{n-1} \exp\left\{-\frac{\theta}{2} w\right\}, \quad w > 0.$$

Hence, X_n^2 has cumulants of all orders. It follows that, by using the usual expansion for $\log\theta$, that condition Property R1 is satisfied; the first part of Property R2 follows easily as well. To verify the remainder of Property R2 it suffices to show that

$$\frac{2}{\sqrt{n}} \ell_\theta(\theta) = \frac{1}{\sqrt{n}}\left[X_n^2 - \frac{2n}{\theta}\right]$$

converges in distribution to a normal distribution. Using the properties of the gamma distribution, it can be shown that this random variable has cumulant-generating function

$$n \log\left(\frac{\theta}{\theta - 2t/\sqrt{n}}\right) - \frac{2\sqrt{n}}{\theta}t = \frac{2}{\theta^2}t^2 + \frac{8}{3\theta^3}t^3\frac{1}{\sqrt{n}} + \cdots.$$

Hence, $2\ell_\theta(\theta)/\sqrt{n}$ converges in distribution to a normal random variable and Property R2 holds. Property R3 follows from the fact that $\ell_{\theta\theta\theta\theta}(\theta)$ is continuous.

It is worth noting that these results continue to hold in the case in which the process is observed for the time interval $(0, T)$ where T is fixed. In this case, in addition to the event times $X_1, \ldots, X_n$, we also observe the fact that no events occurred in the interval (X_n, T) which, given X_n, has probability

$$\exp\left\{-\int_{X_n}^{T} \lambda(s)\, ds\right\}.$$

It follows that the likelihood function is

$$\left[\prod \lambda(x_j)\right] \exp\left\{-\int_0^T \lambda(s)\, ds\right\}.$$

It is important to note that, in this case, n is a random variable.

For the case $\lambda(t) = \theta t$, the log-likelihood function for θ is given

$$\ell(\theta) = n \log\theta - \tfrac{1}{2}\theta T^2$$

where n has a Poisson distribution with mean $\theta T^2/2$. It is straightforward to show that Properties R1–R3 hold. ∎

Example 3.8 Exponential regression model

Let $Y_1, \ldots, Y_n$ denote independent exponential random variables such that Y_j has mean λ_j^{-1} where $\log\lambda_j = \theta x_j$ and $x_1, \ldots, x_n$ are fixed constants. The log-likelihood function for θ is given by

$$\ell(\theta) = \theta \sum x_j - \sum \exp\{\theta x_j\} y_j.$$

It is straightforward to show that $\ell(\theta) - \ell(\theta_0)$ may be expanded in a Taylor's series expansion with remainder term

$$-\frac{1}{120}\sum \exp\{\theta^* x_j\} x_j^5 y_j (\theta - \theta_0)^5$$

where θ^* lies between θ and θ_0. Hence, Property R1 holds provided that

$$\frac{\sum \exp\{\theta^* x_j\}|x_j|^5 y_j}{n} = O_p(1);$$

a sufficient condition for this is that $\sup_j |x_j| \leq C$ for some constant C, i.e., the x_j are bounded.

To verify Property R2, note that for $r = 2, 3, 4$,

$$\frac{\partial^r}{\partial \theta^r} \ell(\theta)\bigg|_{\theta=\theta_0} = -\sum \exp\{\theta_0 x_j\} x_j^r Y_j$$

which has mean $-\sum x_j^r$ and variance $\sum x_j^{2r}$. It follows that the mean and variance of $\ell_{\theta\theta}(\theta_0)$, $\ell_{\theta\theta\theta}(\theta_0)$, and $\ell_{\theta\theta\theta\theta}(\theta_0)$ are all of order $O(n)$ provided that

$$\sum |x_j|^r = O(n) \quad \text{for } r = 2, \ldots, 8. \tag{3.3}$$

Note that this condition prevents the $|x_j|$ from being too small as well as being too large. For instance, if $x_j = 1/j$, then the variance of $\ell_{\theta\theta}(\theta_0)$ is of order $O(1)$. Since

$$\ell_\theta(\theta_0) = \sum (1 - \exp\{\theta_0 x_j\} Y_j) x_j,$$

the first part of Property R2 follows, again provided (3.3) holds.

To verify the remainder of Property R2, we need to show that

$$\frac{1}{\sqrt{n}} \sum (1 - \exp\{\theta_0 x_j\}) Y_j$$

follows the central limit theorem. Let $Z_j = (1 - \exp\{\theta_0 x_j\} Y_j) x_j$, $j = 1, \ldots, n$. Then Z_j has mean 0 and variance x_j^2. By Lyapounov's condition, $\ell_\theta(\theta_0)/\sqrt{n}$ follows the central limit theorem if there exists a $\delta > 0$ such that

$$\lim_{n\to\infty} \frac{\sum \mathrm{E}[|Z_j|^{2+\delta}]}{[\sum x_j^2]^{1+\delta/2}} = 0.$$

Note that $|Z_j| = |x_j||1 - \exp\{\theta_0 x_j\} Y_j|$ and that the distribution of $1 - \exp\{\theta_0 x_j\} Y_j$ does not depend on j. Hence,

$$\sum \mathrm{E}[|Z_j|^{2+\delta}] = c \sum |x_j|^{2+\delta}$$

for some constant $c > 0$ and Lyapounov's condition holds provided that

$$\lim_{n\to\infty} \frac{\sum |x_j|^{2+\delta}}{[\sum x_j^2]^{1+\delta/2}} = 0.$$

This condition follows from (3.3).

Finally, Property R3 follows from the fact that $\ell_{\theta\theta\theta\theta}(\theta)$ is continuous in θ. Hence, this model is regular provided that (3.3) holds. ∎

3.5 Log-likelihood derivatives

3.5.1 INTRODUCTION

A general theory of inference based on the likelihood function relies heavily on the use of Taylor's series approximations of the log-likelihood function and related functions. Hence, log-likelihood derivatives will play a fundamental role in the later chapters. In this section, we consider some basic properties of log-likelihood derivatives.

3.5.2 BARTLETT IDENTITIES

First consider the case in which θ is a scalar. By Property R3,

$$\frac{\partial^j}{\partial\theta^j} \mathrm{E}\{\exp\{\ell(\theta) - \ell(\theta_0)\}; \theta_0\} \Big|_{\theta=\theta_0} = \mathrm{E}\Big\{ \frac{\partial^j}{\partial\theta^j} \exp\{\ell(\theta) - \ell(\theta_0)\} \Big|_{\theta=\theta_0} \Big\};$$

there are a number of important implications of this result. Note that, for all θ,

$$\mathrm{E}[\exp\{\ell(\theta) - \ell(\theta_0)\}; \theta] = 1$$

so that

$$\frac{\partial^j}{\partial\theta^j} \mathrm{E}\{\exp\{\ell(\theta) - \ell(\theta_0)\}; \theta_0\} \Big|_{\theta=\theta_0} = 0$$

and hence

$$\mathrm{E}\Big\{ \frac{\partial^j}{\partial\theta^j} \exp\{\ell(\theta) - \ell(\theta_0)\} \Big|_{\theta=\theta_0} \Big\} = 0. \tag{3.4}$$

This equation is the basis for identities relating the moments of the log-likelihood derivatives. For instance, for $j = 1$,

$$\mathrm{E}[\ell_\theta(\theta); \theta] = 0 \quad \text{for all } \theta;$$

using $j = 2$,

$$\mathrm{E}[\ell_{\theta\theta}(\theta); \theta] + \mathrm{E}[\ell_\theta(\theta)^2; \theta] = 0 \quad \text{for all } \theta.$$

Let

$$\mu_{ijkl} \equiv \mu_{ijkl}(\theta) = \mathrm{E}[\ell_\theta(\theta)^i \ell_{\theta\theta}(\theta)^j \ell_{\theta\theta\theta}(\theta)^k \ell_{\theta\theta\theta\theta}(\theta)^l; \theta];$$

we will write μ_{ij} for μ_{ij00}, and so on. Then

$$\mu_{10} = 0; \qquad \mu_{20} + \mu_{01} = 0.$$

Further identities are possible by taking higher-order derivatives in (3.4):

$$\mu_{001} + 3\mu_{11} + \mu_{30} = 0; \qquad \mu_{0001} + 4\mu_{101} + 3\mu_{02} + 6\mu_{21} + \mu_{40} = 0.$$

These identities are known as the *Bartlett identities*. The Bartlett identities indicate one way in which a log-likelihood function is different from an arbitrary function of θ and Y.

Let ν_{ijkl} denote the (i, j, k, l)th cumulant of $(\ell_\theta(\theta), \ell_{\theta\theta}(\theta), \ell_{\theta\theta\theta}(\theta), \ell_{\theta\theta\theta\theta}(\theta))$. Then, using the relationship between cumulants and moments, we have that

$$\mu_{10} = \nu_{10}, \qquad \mu_{20} = \nu_{20} + \nu_{10}^2, \qquad \mu_{01} = \nu_{01}.$$

Writing the first two Bartlett identities in terms of the cumulants, we have that $\nu_{10} = 0$ and

$$\mu_{20} + \mu_{01} = \nu_{20} + \nu_{10}^2 + \nu_{01} = \nu_{20} + \nu_{01} = 0.$$

Hence, the first two Bartlett identities, written in terms of the cumulants, have exactly the same form as the Bartlett identities written in terms of the moments. The same result applies to all the Bartlett identities so that, for example,

$$\nu_{001} + 3\nu_{11} + \nu_{30} = 0; \qquad \nu_{0001} + 4\nu_{101} + 3\nu_{02} + 6\nu_{21} + \nu_{40} = 0.$$

Let a denote an ancillary statistic. Note that, since

$$p(y|a;\theta) = c(a)p(y;\theta),$$

where $c(a)$ is a constant depending only on a,

$$\ell(\theta) - \ell(\theta_0) = \log p(y|a;\theta) - \log p(y|a;\theta_0).$$

It follows that

$$\mathrm{E}[\exp\{\ell(\theta) - \ell(\theta_0)\}|a;\theta_0] = 1 \quad \text{for all } \theta$$

and, hence, the Bartlett identities hold conditionally on a. That is, if

$$\mu_{ijkl}^{(a)} = \mathrm{E}[\ell_\theta(\theta)^i \ell_{\theta\theta}(\theta)^j \ell_{\theta\theta\theta}(\theta)^k \ell_{\theta\theta\theta\theta}(\theta)^l; a|\theta]$$

then $\mu_{10}^{(a)} = 0$, $\mu_{20}^{(a)} + \mu_{01}^{(a)} = 0$, and so on.

Similar identities apply in the case of a multivariate parameter. Let θ denote a vector parameter of dimension d. Then $\mathrm{E}[\ell_\theta(\theta);\theta] = 0$ and

$$\mathrm{E}[\ell_\theta(\theta)\ell_\theta(\theta)^T;\theta] + \mathrm{E}[\ell_{\theta\theta}(\theta);\theta] = 0;$$

here $\ell_\theta(\theta)$ is a vector of length d and $\ell_{\theta\theta}(\theta)$ is a $d \times d$ matrix. The higher-order identities involve high-dimensional arrays and are not as easily expressed as those in the scalar parameter case. They are however relatively easy to determine since they follow directly from (3.4), except that the differentiation is with respect to the components of θ. For instance, the third Bartlett identity is of the form

$$\mathrm{E}\Big[\frac{\partial^3}{\partial\theta_i\,\partial\theta_j\,\partial\theta_k}\ell(\theta);\theta\Big] + \mathrm{E}\Big[\frac{\partial}{\partial\theta_i}\ell(\theta)\frac{\partial^2}{\partial\theta_j\,\partial\theta_k}\ell(\theta);\theta\Big] + \mathrm{E}\Big[\frac{\partial}{\partial\theta_j}\ell(\theta)\frac{\partial^2}{\partial\theta_i\,\partial\theta_k}\ell(\theta);\theta\Big]$$
$$+\mathrm{E}\Big[\frac{\partial}{\partial\theta_k}\ell(\theta)\frac{\partial^2}{\partial\theta_i\,\partial\theta_j}\ell(\theta);\theta\Big] + \mathrm{E}\Big[\frac{\partial}{\partial\theta_i}\ell(\theta)\frac{\partial}{\partial\theta_j}\ell(\theta);\theta\Big]\mathrm{E}\Big[\frac{\partial}{\partial\theta_k}\ell(\theta);\theta\Big] = 0.$$

Example 3.9 One-parameter exponential family
Let Y denote a random variable with density of the form

$$\exp\{c(\theta)s(y) - k(\theta) + D(y)\};$$

it follows that the log-likelihood function is given by $\ell(\theta) = c(\theta)s - k(\theta)$ where $s = s(y)$. Hence,

$$\ell_\theta(\theta) = c'(\theta)s - k'(\theta), \qquad \ell_{\theta\theta}(\theta) = c''(\theta)S - k''(\theta),$$

and so on. Let $\kappa_1, \kappa_2, \ldots,$ denote the cumulants of S. By the first Bartlett identity, $\kappa_1 = k'(\theta)/c'(\theta)$; by the second Bartlett identity,

$$c'(\theta)^2\kappa_2 + c''(\theta)\kappa_1 - k''(\theta) = 0$$

so that

$$\kappa_2 = \frac{k''(\theta)}{c'(\theta)^2} - \frac{c''(\theta)k'(\theta)}{c'(\theta)^3}.$$

Similarly the third and fourth Bartlett identities may be used to determine expressions for κ_3 and κ_4. ∎

3.5.3 ASYMPTOTIC PROPERTIES

We now briefly consider the asymptotic properties of log-likelihood derivatives under the conditions of Section 3.4; further asymptotic properties will be developed in the subsequent chapters. For simplicity, we consider the case in which θ is a scalar parameter; the vector parameter case may be handled by using these results componentwise, as in Section 3.5.2.

According to Property R2, the mean and variance of the log-likelihood derivatives $\ell_\theta(\theta), \ell_{\theta\theta}(\theta), \ell_{\theta\theta\theta}(\theta), \ell_{\theta\theta\theta\theta}(\theta)$ are of order $O(n)$. Furthermore, $n^{-1/2}\ell_\theta(\theta)$ is asymptotically normally distributed.

The higher-order derivatives are, in general, of order $O_p(n)$ and are equal to a constant of order $O(n)$ plus a term of order $O_p(\sqrt{n})$. For instance,

$$\ell_{\theta\theta}(\theta) = -i(\theta) + O_p(\sqrt{n}).$$

Similar results hold for products of log-likelihood derivatives. For instance, $\ell_\theta(\theta)\ell_{\theta\theta}(\theta)$ is of order $O_p(n)$ and may be written

$$\ell_\theta(\theta)\ell_{\theta\theta}(\theta) = \mathrm{E}[\ell_\theta(\theta)\ell_{\theta\theta}(\theta); \theta] + O_p(\sqrt{n})$$

where $\mathrm{E}[\ell_\theta(\theta)\ell_{\theta\theta}(\theta); \theta]$ is $O(n)$.

3.6 Information

3.6.1 INTRODUCTION

There are a number of important quantities that are derived directly from the likelihood function. For simplicity, assume that in the conditions of Section 3.4, the set Θ_0 may be taken to be the entire parameter space Θ. Recall that the score function, $\ell_\theta(\theta)$, has mean vector 0; the variance of $\ell_\theta(\theta)$ is called the *expected information* and will be denoted by $i(\theta)$, a $d \times d$ matrix. A closely related quantity is the *observed information*, given by $j(\theta) = -\ell_{\theta\theta}(\theta)$. It follows from the second Bartlett identity that

$$\mathrm{E}[j(\theta); \theta] = i(\theta).$$

An important property of both the expected and observed information is that they are additive for independent experiments. More precisely, let Y_1 and Y_2 denote independent random variables, which may be vector valued, and let $\ell^{(k)}(\theta)$ denote the log-likelihood function for a parameter θ based on the observation of Y_k, $k = 1, 2$. Let $i_k(\theta)$ and $j_k(\theta)$ denote the expected information and observed information, respectively, based on $\ell^{(k)}(\theta)$ and let $i(\theta)$ and $j(\theta)$ denote the information matrices based on $\ell^{(1)}(\theta) + \ell^{(2)}(\theta)$, the log-likelihood function based on (Y_1, Y_2). Then

$$i(\theta) = i_1(\theta) + i_2(\theta); \qquad j(\theta) = j_1(\theta) + j_2(\theta).$$

It follows that for independent identically distributed samples of size n, the expected information matrix is simply n times the expected information based on a single observation.

Both types of information depend on the parameterization used. Let $\phi = \phi(\theta)$ denote a one-to-one function of θ and let $\theta(\phi)$ denote the inverse transformation. Let $\tilde{i}(\phi)$ denote the expected information in the ϕ parameterization. Then

$$\tilde{i}(\phi) = \Big(\frac{\partial\theta(\phi)}{\partial\phi}\Big)^T i(\theta(\phi))\Big(\frac{\partial\theta(\phi)}{\partial\phi}\Big).$$

Example 3.10 Full-rank exponential family
Let Y denote a random variable with a density of the form

$$p(y; \theta) = \exp\{s(y)^T c(\theta) - k(\theta) + D(y)\}.$$

Then

$$\ell_\theta(\theta)^T = s(y)^T c'(\theta) - k'(\theta) \tag{3.5}$$

and

$$\ell_{\theta\theta}(\theta) = s(y)^T c''(\theta) - k''(\theta).$$

Note that $c'(\theta)$ is an $d \times d$ matrix and $c''(\theta)$ is a three-dimensional array. The product $s(y)^T c''(\theta)$ is given by

$$s_1(y)c_1''(\theta) + \cdots + s_m(y)c_m''(\theta)$$

where $s(y) = (s_1(y), \ldots, s_d(y))^T$ and $c(\theta) = (c_1(\theta), \ldots, c_d(\theta))^T$.

Since $\ell_\theta(\theta)$ has mean 0, it follows from (3.5) that

$$\mathrm{E}\{s(Y)^T; \theta\} = k'(\theta)c'(\theta)^{-1}$$

and hence

$$i(\theta) = k''(\theta) - k'(\theta)c'(\theta)^{-1}c''(\theta).$$

The observed information is given by

$$j(\theta) = k''(\theta) - s(y)^T c''(\theta).$$

For the special case in which $c(\theta) = \theta$, $c'(\theta) = D_d$, the identity matrix of rank d, and $c''(\theta) = 0$ so that $i(\theta) = j(\theta)$. ∎

Example 3.11 Weibull distribution
Let Y denote a random variable with density function

$$p(y; \theta) = \beta\lambda(\lambda y)^{\beta-1} \exp\{-(\lambda y)^\beta\}, \quad y > 0$$

where $\theta = (\beta, \lambda) \in (0, \infty) \times (0, \infty)$. It follows that

$$-\ell_{\theta\theta}(\theta) = \begin{pmatrix} 1/\beta^2 + (\lambda y)^\beta[\log(\lambda y)]^2 & 1/\lambda(\lambda y)^\beta[\beta \log(\lambda y) + 1] - 1/\lambda \\ 1/\lambda(\lambda y)^\beta[\beta \log(\lambda y) + 1] - 1/\lambda & \beta/\lambda^2[1 + (\beta - 1)(\lambda y)^\beta] \end{pmatrix}$$

and that

$$i(\theta) = \begin{pmatrix} 1/\beta^2[\pi^2/6 + \gamma^2 - 2\gamma] & 1/\lambda(1-\gamma) \\ 1/\lambda(1-\gamma) & \beta^2/\lambda^2 \end{pmatrix};$$

here γ denotes Euler's constant. ∎

3.6.2 INFORMATION INEQUALITY

One justification for the term *information* is the fact that the magnitude of $i(\theta)$ gives some indication of how well it is possible to estimate the parameter θ. Suppose we are interested in estimating a linear function of θ, $c^T\theta$, where c is a known vector with the same length as θ. Let T denote an unbiased estimator of $c^T\theta$ so that $\mathrm{E}\{T; \theta\} = c^T\theta$ for all θ. Then, since T is unbiased,

$$\mathrm{E}\{T(Y); \theta\} = \int T(y)p(y; \theta)\, dy = c^T\theta \quad \text{for all } \theta.$$

Differentiating both sides of this expression with respect to θ leads to

$$\mathrm{E}\{\ell_\theta(\theta)T(Y); \theta\} = c;$$

hence, for any vector d,

$$\mathrm{E}\{\ell_\theta(\theta)^T d\, T(Y); \theta\} = c^T d. \tag{3.6}$$

Since $\mathrm{E}\{\ell_\theta(\theta)^T d; \theta\} = 0$ the left-hand side of (3.6) is simply the covariance of $\ell_\theta(\theta)^T d$ and $T(Y)$ which is bounded by

$$[\mathrm{Var}(\ell_\theta(\theta)^T d; \theta)\mathrm{Var}(T(Y); \theta)]^{1/2}.$$

Since $\mathrm{Var}(\ell_\theta(\theta)^T d; \theta) = d^T i(\theta) d$, it follows that

$$\mathrm{Var}(T(Y); \theta) \geq \frac{(c^T d)^2}{d^T i(\theta) d}. \tag{3.7}$$

Note that (3.7) holds for any d. By the Cauchy–Schwarz inequality, the right-hand side of (3.7) takes its maximum value at $d = i(\theta)^{-1}c$ so that

$$\mathrm{Var}(T(Y); \theta) \geq c^T i(\theta)^{-1} c. \tag{3.8}$$

This result is known as the *information inequality*.

If θ is a scalar and $T(Y)$ is an unbiased estimator of θ, then

$$\mathrm{Var}(T(Y); \theta) \geq \frac{1}{i(\theta)}.$$

3.6.3 PARTIAL INFORMATION

Suppose that the parameter θ may be partitioned into two components (ψ, λ); the information matrix may then be partitioned

$$i(\theta) = \begin{pmatrix} i_{\psi\psi} & i_{\psi\lambda} \\ i_{\lambda\psi} & i_{\lambda\lambda} \end{pmatrix}.$$

For instance, let $\ell_\theta(\theta)^T = (\ell_\psi(\theta) \;\; \ell_\lambda(\theta))^T$ where

$$\ell_\psi(\theta) = \frac{\partial}{\partial \psi}\ell(\theta) \qquad \ell_\lambda(\theta) = \frac{\partial}{\partial \lambda}\ell(\theta);$$

then

$$i_{\psi\psi}(\theta) = \mathrm{E}[\ell_\psi(\theta)\ell_\psi(\theta)^T; \theta].$$

Alternatively,

$$i_{\psi\psi}(\theta) = -\mathrm{E}[\ell_{\psi\psi}(\theta); \theta]$$

where

$$\ell_{\psi\psi}(\theta) = \frac{\partial^2}{\partial \psi\, \partial \psi^T}\ell(\theta).$$

Let $T(Y)$ denote an unbiased estimator of $c_0^T \psi$. Then, by (3.8),

$$\text{Var}(T(Y); \theta) \geq c^T i(\theta)^{-1} c$$

where $c^T = (c_0^T \; 0)$ and 0 denotes the zero vector. Using the formula for the inverse of a partitioned matrix, it can be shown that

$$c^T i(\theta)^{-1} c = c_0^T i_\psi(\theta)^{-1} c_0$$

where

$$i_\psi(\theta) = i_{\psi\psi}(\theta) - i_{\psi\lambda}(\theta) i_{\lambda\lambda}(\theta)^{-1} i_{\lambda\psi}(\theta).$$

Hence, for estimation of ψ, $i_\psi(\theta)$ plays the same role as does $i(\theta)$ in the estimation of the entire parameter vector θ. The matrix $i_\psi(\theta)$ is called the partial information for ψ.

If the parameter λ is known, then the partial information for ψ is simply $i_{\psi\psi}(\theta)$; the difference between $i_\psi(\theta)$ and $i_{\psi\psi}(\theta)$ reflects the loss information about ψ due to the fact that λ is unknown.

Example 3.12 Normal-theory linear regression

Let $Y_1, \ldots, Y_n$ denote independent normally distributed random variables with standard deviation σ and mean $x_j\beta + z_j\lambda$ where x_j is a $1 \times p$ vector of constants, z_j is a $1 \times q$ vector of constants and β and λ are unknown parameters. For simplicity, we take the standard deviation to be known. The information matrix for $\theta = (\beta, \lambda)$, partitioned according to (β, λ), is given by

$$i(\theta) = \frac{1}{\sigma^2} \begin{pmatrix} x^T x & x^T z \\ z^T x & z^T z \end{pmatrix}$$

where x denotes the $n \times p$ matrix with jth row x_j and z denotes the $n \times q$ matrix with jth row z_j.

The partial information for β is therefore

$$i_\beta(\theta) = \frac{1}{\sigma^2}[x^T x - x^T z (z^T z)^{-1} z^T x].$$

If λ is known, the information for β is $(x^T x)/\sigma^2$. The two informations are equal if and only if the rows of x are orthogonal to the rows of z. ∎

A quantity analogous to $i_\psi(\theta)$ may be based on the observed information. The observed information matrix may be partitioned

$$j(\theta) = \begin{pmatrix} j_{\psi\psi}(\theta) & j_{\psi\lambda}(\theta) \\ j_{\lambda\psi}(\theta) & j_{\lambda\lambda}(\theta) \end{pmatrix};$$

the partial observed information for ψ is given by

$$j_\psi(\theta) = j_{\psi\psi}(\theta) - j_{\psi\lambda}(\theta) j_{\lambda\lambda}(\theta)^{-1} j_{\lambda\psi}(\theta).$$

3.6.4 ORTHOGONAL PARAMETERS

Suppose that θ may be written $\theta = (\psi, \lambda)$ and that the log-likelihood function for θ may be written

$$\ell(\psi, \lambda) = \ell^{(1)}(\psi) + \ell^{(2)}(\lambda)$$

where $\ell^{(j)}$ denotes a log-likelihood function depending only on either ψ or λ. In this case, the relative likelihood of different values of ψ does not depend on the value of λ and conversely; hence, likelihood inference regarding ψ may proceed using $\ell^{(1)}(\psi)$, ignoring λ.

The same situation occurs, at least approximately, whenever $i_{\psi\lambda}(\theta) = 0$. In this case, ψ and λ are said to be *orthogonal parameters*. To see this, suppose that $(\psi, \lambda) = (\psi_0, \lambda_0) + O(n^{-1/2})$, where (ψ_0, λ_0) is some fixed parameter value, and consider the following Taylor's series expansion of the log-likelihood function around the point $(\psi, \lambda) = (\psi_0, \lambda_0)$:

$$\begin{aligned}\ell(\psi, \lambda) = \ell(\theta_0) &+ \ell_\psi(\theta_0)^T(\psi - \psi_0) + \ell_\lambda(\theta_0)^T(\lambda - \lambda_0) \\ &+ \tfrac{1}{2}(\psi - \psi_0)^T \ell_{\psi\psi}(\theta_0)(\psi - \psi_0) + \tfrac{1}{2}(\lambda - \lambda_0)^T \ell_{\lambda\lambda}(\theta_0)(\lambda - \lambda_0) \\ &+ (\lambda - \lambda_0)^T \ell_{\lambda\psi}(\theta_0)(\psi - \psi_0) + O_p(n^{-1/2}).\end{aligned}$$

When ψ and λ are orthogonal parameters, then

$$i_{\psi\lambda}(\theta) = -\mathrm{E}\{\ell_{\psi\lambda}(\theta); \theta\} = 0$$

so that $\ell_{\psi\lambda}(\theta_0) = O_p(\sqrt{n})$. Hence,

$$\begin{aligned}\ell(\psi, \lambda) = \ell(\theta_0) &+ \ell_\psi(\theta_0)^T(\psi - \psi_0) + \ell_\lambda(\theta_0)^T(\lambda - \lambda_0) \\ &+ \tfrac{1}{2}(\psi - \psi_0)^T \ell_{\psi\psi}(\theta_0)(\psi - \psi_0) + \tfrac{1}{2}(\lambda - \lambda_0)^T \ell_{\lambda\lambda}(\theta_0)(\lambda - \lambda_0) \\ &+ O_p(n^{-1/2})\end{aligned}$$

so that, to this order of approximation,

$$\ell(\psi, \lambda) = \ell(\psi, \lambda_0) + \ell(\psi_0, \lambda).$$

Suppose that ψ is the parameter of interest while λ is a nuisance parameter. Consider a reparameterization of the model of the form (ψ, ϕ); since the parameter of interest is unchanged, such a reparameterization is said to be *interest respecting*. It is often beneficial to choose ϕ so that it is orthogonal to ψ. This is always possible if ψ is a scalar. Let $\ell(\psi, \lambda)$ and $i(\psi, \lambda)$ denote the log-likelihood function and information matrix, respectively, in the original parameterization and let $\tilde{\ell}(\psi, \phi)$ and $\tilde{i}(\psi, \phi)$ denote the log-likelihood function and information matrix in the (ψ, ϕ) parameterization. Then, $\tilde{\ell}(\psi, \phi) = \ell(\psi, \lambda(\psi, \phi))$ and

$$\tilde{\ell}_\phi(\psi, \phi)^T = \ell_\lambda(\psi, \lambda)^T \frac{\partial \lambda}{\partial \phi},$$

$$\tilde{\ell}_\psi(\psi, \phi) = \ell_\psi(\psi, \lambda) + \ell_\lambda(\psi, \lambda)^T \frac{\partial \lambda}{\partial \psi} = \ell_\psi(\psi, \lambda) + \frac{\partial \lambda}{\partial \psi}^T \ell_\lambda(\psi, \lambda).$$

It follows that

$$\tilde{i}_{\psi\phi}(\psi, \phi) = i_{\psi\lambda}(\psi, \lambda) \frac{\partial \lambda}{\partial \phi} + \frac{\partial \lambda}{\partial \psi}^T i_{\lambda\lambda}(\psi, \lambda) \frac{\partial \lambda}{\partial \phi}.$$

Hence, in order to have $\tilde{i}_{\psi\phi}(\psi, \phi) = 0$, we require that

$$\frac{\partial \lambda}{\partial \psi} = -i_{\lambda\lambda}(\psi, \lambda)^{-1} i_{\lambda\psi}(\psi, \lambda).$$

It is possible, in general, to solve this differential equation for $\phi \equiv \phi(\psi, \lambda)$ although there is not a unique solution. This is to be expected since, if ϕ is orthogonal to ψ, then so is $g(\phi)$ for any smooth function g.

Example 3.13 Exponential family distribution
Suppose that Y is distributed according to a density of the form

$$p(y; \theta) = \exp\{s(y)^T c(\theta) - k(\theta) + D(y)\}.$$

Suppose that θ may be written $\theta = (\psi, \lambda)$ where ψ denotes the parameter of interest, possibly vector valued, and that $c(\theta) = (c_1(\psi)\ c_2(\theta))^T$ for functions c_1, c_2 where $c_1(\cdot)$ is a one-to-one function of ψ. Then, writing $s(y) = (s_1(y) s_2(y))^T$, the log-likelihood function is of the form

$$\ell(\psi, \lambda) = s_1(y)^T c_1(\psi) + s_2(y)^T c_2(\theta) - k(\theta).$$

Let ϕ denote the *complementary mean parameter*, given by

$$\phi \equiv \phi(\theta) = \mathrm{E}\{s_2(Y); \theta\}.$$

Note that, since $\ell_\lambda(\theta)$ has mean 0,

$$\mathrm{E}\left\{s_2(Y)^T \frac{\partial c_2(\theta)}{\partial \lambda}; \theta\right\} = \frac{\partial k(\theta)}{\partial \lambda};$$

it follows that

$$\frac{\partial k(\theta)}{\partial \lambda} = \phi^T \frac{\partial c_2(\theta)}{\partial \lambda}$$

for all θ. Differentiating this expression with respect to ψ yields the following identity:

$$\frac{\partial^2 k(\theta)}{\partial \psi\, \partial \phi} - \phi^T \frac{\partial^2 c_2(\theta)}{\partial \psi\, \partial \phi} = 0. \tag{3.9}$$

The log-likelihood function for (ψ, ϕ) is given by

$$\tilde{\ell}(\psi, \phi) = s_1(y)^T c_1(\psi) + s_2(y)^T c_2(\theta(\psi, \phi)) - k(\theta(\psi, \phi)).$$

Hence,

$$\frac{\partial^2 \tilde{\ell}}{\partial \psi \, \partial \phi}(\psi, \phi) = s_2(y)^T \frac{\partial^2 c_2(\theta)}{\partial \psi \, \partial \phi} - \frac{\partial^2 k(\theta)}{\partial \psi \, \partial \phi}$$

and

$$\tilde{i}_{\psi\phi}(\theta) = \frac{\partial^2 k(\theta)}{\partial \psi \, \partial \phi} - \phi^T \frac{\partial^2 c_2(\theta)}{\partial \psi \, \partial \phi}.$$

It follows from (3.9) that $\tilde{i}_{\psi\phi}(\theta) = 0$ and, hence, ψ and ϕ are orthogonal parameters. ∎

Example 3.14 Weibull distribution

Consider a Weibull distribution as in Example 3.11. Take $\psi = \beta$ to be the parameter of interest and λ to be the nuisance parameter. The information matrix for (ψ, λ) is given by

$$\begin{pmatrix} 1/\psi^2[\pi^2/6 + \gamma^2 - 2\gamma] & 1/\lambda(1-\gamma) \\ 1/\lambda(1-\gamma) & \psi^2/\lambda^2 \end{pmatrix}$$

and, hence, ψ and λ are not orthogonal. Let $\phi \equiv \phi(\psi, \lambda)$ denote an orthogonal nuisance parameter. Then $\lambda \equiv \lambda(\psi, \phi)$ must satisfy

$$\frac{\partial \lambda}{\partial \psi} = -(1-\gamma)\lambda/\psi^2.$$

The general solution to this differential equation is

$$\lambda(\psi) = C \exp\{(1-\gamma)/\psi\}$$

where C is a constant with respect to ψ. Solving for C yields

$$C(\phi) = \lambda \exp\{-(1-\gamma)/\psi\};$$

it follows that ϕ must be of the form

$$\phi = g(\log \lambda + (\gamma - 1)/\psi)$$

where g is an arbitrary smooth function. Taking g to be the identity function yields $\phi = \log \lambda + (\gamma - 1)/\psi$.

The log-likelihood function in terms of this parameterization is given by

$$\psi\phi + \log(\psi) + \psi \log(y) - \exp\{1 - \gamma + \phi\psi + \psi \log(y)\}.$$

The orthogonality of ψ and ϕ may be verified by direct calculation. ∎

3.7 Methods of inference

3.7.1 INTRODUCTION

The remainder of this book considers methods of statistical inference based on the likelihood function. Statistical inference is often divided into three types of methods, construction of point estimates, confidence regions, and hypothesis tests. As an introduction to these methods, this section contains very brief descriptions of the methods of inference commonly used. Many further details are given in the subsequent chapters.

3.7.2 POINT ESTIMATION

Given that the likelihood function $L(\theta)$ measures the relative plausibility of different possible values of θ, the value of θ that maximizes $L(\theta)$ is a reasonable choice for a point estimate of θ; this estimate is called the maximum likelihood estimate of θ. More specifically, a maximum likelihood estimate of θ is any value of $\hat{\theta} \in \Theta$ satisfying

$$L(\hat{\theta}) = \sup_{\theta \in \Theta} L(\theta)$$

or, equivalently,

$$\ell(\hat{\theta}) = \sup_{\theta \in \Theta} \ell(\theta).$$

Note that the symbol $\hat{\theta}$ will be used to denote both the value of the maximum likelihood estimate for a given set of data and the corresponding random variable.

Maximum likelihood estimates may not exist or, if they exist, they may not be unique. We will assume that, with probability approaching 1 as $n \to \infty$, a maximum likelihood exists and is unique. Hence, we will proceed as if there is a unique maximum likelihood estimate.

Maximum likelihood estimation is parameterization invariant in the following sense. Let $\hat{\theta}$ denote the maximum likelihood estimate of θ. If $\phi \equiv \phi(\theta)$ is a one-to-one function of θ, then the maximum likelihood estimate of ϕ is simply $\hat{\phi} = \phi(\hat{\theta})$.

For regular models, $\hat{\theta}$ must satisfy the *likelihood equation*, $\ell_\theta(\hat{\theta}) = 0$; however, not all solutions to the likelihood equation correspond to a maximum likelihood estimate.

Example 3.15 One-parameter exponential family distribution
Let $Y_1, \ldots, Y_n$ be independent random variables each distributed according to a one-parameter exponential family distribution with density of the form

$$\exp\{\theta s(y) - k(\theta) + D(y)\};$$

take Θ to be the natural parameter space. The log-likelihood function is given by

$$\ell(\theta) = \theta \sum s(y_j) - nk(\theta).$$

Since $k(\theta + t) - k(\theta)$ is the cumulant-generating function of $s(Y_1)$, it follows that $k''(\theta) > 0$ and, hence, $\ell(\cdot)$ is a concave function. Therefore, provided that the maximum likelihood estimate exists, it satisfies the equation

$$k'(\hat{\theta}) = \frac{\sum s(y_j)}{n}.$$

The maximum likelihood estimate of any function $h(\theta)$ of θ is given by $h(\hat{\theta})$.

For instance, suppose that each Y_j is a Bernoulli random variable with mean $\Pr(Y_j = 1; \theta) = \theta$ where $0 < \theta < 1$. Then, if the maximum likelihood estimate exists, it satisfies the equation $\hat{\theta} = \bar{y}$; hence, if $0 < \bar{y} < 1$ then $\hat{\theta} = \bar{y}$. If however either $\bar{y} = 0$ or $\bar{y} = 1$, then a maximum likelihood estimate does not exist. ∎

Example 3.16 Laplace distribution

Let $Y_1, \ldots, Y_n$ denote independent observations each distributed according to the distribution with density

$$\frac{1}{2\sigma} \exp\left\{-\frac{1}{\sigma}|y - \mu|\right\}, \quad -\infty < y < \infty$$

so that $\theta = (\mu, \sigma) \in \Re \times \Re^+$. The log-likelihood function is given by

$$\ell(\mu, \sigma) = -\frac{1}{\sigma} \sum |y_j - \mu| - n \log \sigma;$$

a maximum likelihood estimate of μ is the value of μ that minimizes $f(\mu) = \sum |y_j - \mu|$. Let $y_{(1)} < y_{(2)} < \cdots < y_{(n)}$ denote the order statistics of the sample, that is, the ordered observations. Note that $f(\mu)$ is a continuous, piecewise-linear, convex function; however it is not differentiable at $\mu = y_{(j)}$, $j = 1, \ldots, n$.

For $y_{(k)} < \mu < y_{(k+1)}$ the function $f(\cdot)$ may be written

$$f(\mu) = (\mu - y_{(1)}) + \cdots + (\mu - y_{(k)}) + (y_{(k+1)} - \mu) + \cdots + (y_{(n)} - \mu).$$

It follows that, on this interval, $f'(\mu) = 2(k - n/2)$. Hence, if $k < n/2$, then $f(y_{(k)}) > f(y_{(k+1)})$; if $k > n/2$ then $f(y_{(k)}) < f(y_{(k+1)})$. If n is odd, then the minimum must occur at $\mu = y_{(n+1)/2}$ since the function is increasing in both directions. If n is even, then for $y_{(n/2)} < \mu < y_{(n/2)+1}$ we have $f'(\mu) = 0$ so that the minimum value is achieved at any μ in the interval $(y_{(n/2)}, y_{(n/2)+1})$. Hence, a maximum likelihood estimate of μ may be taken to be the sample median.

The maximum likelihood estimate of σ, $\hat{\sigma}$, may be found by maximizing

$$-\frac{1}{\sigma} \sum |y_j - \hat{\mu}| - n \log \sigma$$

where $\hat{\mu}$ denotes the maximum likelihood estimate of μ. It follows that $\hat{\sigma} = \sum |y_j - \hat{\mu}|/n$. ∎

3.7.3 CONFIDENCE REGIONS

There are several possible methods of constructing a confidence region based on the likelihood function. Since the likelihood function indicates which parameter values are most likely, the set of values of θ such that $L(\theta)$ exceeds a given constant may be used as a confidence region. Consider a confidence region of the form

$$\hat{\Theta} = \{\theta \in \Theta \colon L(\theta) \geq c\}$$

where the constant c is chosen so that the coverage probability of the region is approximately $1 - \alpha$. Specifically, we require that

$$\Pr(\theta \in \hat{\Theta}; \theta) = 1 - \alpha + O(n^{-j/2})$$

for some j, typically $j = 1, 2$, or 3. Note that c is a constant with respect to θ but it may, and typically does, depend on the data. This confidence region is known as a likelihood ratio confidence region. It is often written

$$\hat{\Theta} = \{\theta \in \Theta \colon W(\theta) \leq c_1\}$$

for a constant c_1, where

$$W \equiv W(\theta) = 2[\ell(\hat{\theta}) - \ell(\theta)].$$

Other likelihood-based methods of constructing a confidence region are based on the Wald statistic,

$$W_w = (\hat{\theta} - \theta)^T i(\hat{\theta})(\hat{\theta} - \theta)$$

or the score statistic

$$W_s = \ell_\theta(\theta)^T i(\theta)^{-1} \ell_\theta(\theta).$$

For instance, a confidence region based on the Wald statistic is given by

$$\{\theta \in \Theta : (\hat{\theta} - \theta)^T i(\hat{\theta})(\hat{\theta} - \theta) \leq c\}$$

where the constant c is chosen so that the coverage probability of the region is approximately $1 - \alpha$. Hence, a possible value of θ is in the confidence region if it is close to the maximum likelihood estimate $\hat{\theta}$. A closely related approach is to replace $i(\hat{\theta})$ in W_w by the observed information $\hat{j} \equiv j(\hat{\theta})$. When θ is a scalar parameter, the resulting confidence region based on W_w is of the form $\hat{\theta} \pm \sqrt{c}\hat{\sigma}$ where $\hat{\sigma}$ denotes an estimate of the standard error of $\hat{\theta}$ and c is a constant chosen so that the resulting interval has the required coverage probability; it will be shown in Section 4.2.2 that this standard error may be taken to be $1/\sqrt{i(\hat{\theta})}$ or $1/\sqrt{\hat{j}}$.

Example 3.17 Normal distribution
Let $Y_1, \ldots, Y_n$ denote independent observations each distributed according to a normal distribution with mean θ and standard deviation 1. The log-likelihood function for this model is given by

$$\ell(\theta) = -\frac{n}{2}(\hat{\theta} - \theta)^2$$

where $\hat{\theta} = \bar{y}$.

A confidence region for θ may be based on the likelihood ratio statistic

$$W(\theta) = n(\hat{\theta} - \theta)^2;$$

since $\hat{\theta}$ is normally distributed with mean θ and variance n^{-1}, under the distribution with parameter θ, $W(\theta)$ has a chi-square distribution with 1 degree of freedom. Hence, a confidence region for θ is given by

$$\{\theta \in \Re: n(\hat{\theta} - \theta)^2 \leq \chi_1^2(\alpha)\}$$

where $\chi_1^2(\alpha)$ denotes the $1 - \alpha$ quantile of the chi-square distribution with 1 degree of freedom.

Since $i(\theta) = n$, the Wald statistic is given by $W_w = W$; also,

$$\ell_\theta(\theta) = n(\hat{\theta} - \theta)$$

so that $W_s = W$ as well. ∎

Example 3.18 Poisson distribution
Let $Y_1, \ldots, Y_n$ denote independent Poisson random variables each with mean θ. The log-likelihood function is given by

$$\ell(\theta) = n\bar{y} \log \theta - n\theta$$

and $\hat{\theta} = \bar{y}$ is the maximum likelihood estimate of θ.

The likelihood ratio statistic for this model is given by

$$W = 2n[\hat{\theta} \log(\hat{\theta}/\theta) - (\hat{\theta} - \theta)],$$

the Wald statistic is given by

$$W_w = n(\hat{\theta} - \theta)^2/\hat{\theta}$$

and the score statistic is given by

$$W_s = n(\hat{\theta} - \theta)^2/\theta.$$

By the central limit theorem, $\sqrt{n}(\hat{\theta} - \theta)$ is asymptotically normally distributed with mean 0 and variance θ. Based on this fact, it is straightforward to show that W, W_w, and W_s are all asymptotically distributed according to a chi-square distribution with 1 degree of freedom. This result may be used to set approximate confidence limits for θ.

Furthermore, using a Taylor's series expansion for $\log(\hat{\theta}/\theta)$, along with the fact that $\hat{\theta} \equiv \bar{y} = \theta + O_p(n^{-1/2})$, it may be shown that

$$W = W_w + O_p(n^{-1/2}) = W_s + O_p(n^{-1/2})$$

so that, in large samples, there is unlikely to be large difference between the results based on the three methods. For a given small sample size however the confidence limits may differ. ∎

3.7.4 HYPOTHESIS TESTS

Hypothesis tests are closely related to confidence regions and very similar methods are used. Consider a test of the null hypothesis H_0: $\theta = \theta_0$ versus the alternative $H_1 : \theta \neq \theta_0$.

The likelihood ratio test statistic is given by

$$W(\theta_0) = 2[\ell(\hat{\theta}) - \ell(\theta_0)];$$

a large value of W is evidence against the null hypothesis. A critical value for the test or a p-value may be computed using the exact distribution of W under the null hypothesis or, more commonly, using a large-sample approximation. For testing H_0 versus a one-sided alternative, the signed version of the statistic,

$$R = \text{sgn}(\hat{\theta} - \theta_0)\,\{2[\ell(\hat{\theta}) - \ell(\theta_0)]\}^{1/2}$$

may be used.

Hypothesis tests may also be constructed from the Wald statistic or the score statistic, computed using $\theta = \theta_0$. In both cases, a large value of the statistic is evidence against the null hypothesis $\theta = \theta_0$; critical values or p-values are often calculated using a large-sample approximation.

Example 3.19 Normal distribution
Consider the model considered in Example 3.17 and consider a test of the hypothesis $\theta = 0$ versus the alternative $\theta \neq 0$. The likelihood ratio test statistic is given by $W = n\hat{\theta}^2$; under the null hypothesis, W has a chi-square distribution with 1 degree of freedom. As noted in Example 3.17, the Wald and score statistics lead to the same test. ∎

Example 3.20 Poisson distribution
Consider the model considered in Example 3.18 and consider a test of the hypothesis $\theta = \theta_0$ where θ_0 is some specified value. The three test statistics are

$$W = 2n[\hat{\theta}\log(\hat{\theta}/\theta_0) - (\hat{\theta} - \theta_0)];$$

$$W_w = n(\hat{\theta} - \theta_0)^2/\hat{\theta}; \qquad W_s = n(\hat{\theta} - \theta_0)^2/\theta_0.$$

As noted in Example 3.18, under the null hypothesis, each statistic is asymptotically distributed according to a chi-square distribution with 1 degree of freedom. ∎

3.8 Discussion and references

The likelihood function plays a central role in many, if not most, approaches to statistical inference. Many commonly used statistical methods, such as maximum likelihood estimates and likelihood ratio tests, are based on the likelihood function. Early references to statistical methods based on the likelihood function include Fisher (1912, 1922, 1925, 1934), Bartlett (1936), Barnard (1949, 1951), and Barnard, Jenkins and Winsten (1962). A history of likelihood is given by Edward (1974). Many of the points raised in Sections 3.1–3.3 are discussed in greater detail in Royall (1997, Chapter 1); see also Hinkley (1980a). Barndorff-Nielsen and Sørenson (1993) consider likelihood theory for stochastic processes.

The result in Section 3.2 that the likelihood function is optimal for choosing between two values of the parameter is essentially the Neyman–Pearson lemma which states the likelihood ratio test is optimal for the testing $H_0: \theta = \theta_1$ versus $H_1: \theta = \theta_2$; see, for example, Lehmann (1986). The Kullback–Leibler divergence is discussed in detail in Kullback (1959).

Interest in the formal 'likelihood principle' increased with Birnbaum's (1962) proof that the likelihood principle could be deduced from the sufficiency principle and a 'conditionality principle' which states that if the experiment performed is chosen randomly with known probabilities, then the statistical analysis should be conditional on the experiment actually performed. Since this principle is intuitively reasonable and virtually all statistical methods adhere to the sufficiency principle, the conclusion is that the likelihood principle is the 'correct' approach to statistical inference. This important result has resulted in considerable discussion that is still continuing; see, for example, Birnbaum (1970, 1977), Durbin (1970), Savage (1970), Kalbfleisch (1975), Evans, Fraser, and Monette (1986), Severini (1993b), and Kabaila (1998). In particular, although individually the sufficiency and conditionality principles are convincing, additional issues arise when adopting them both (Durbin, 1970; Savage, 1970; Severini, 1993b; Kabaila, 1998).

The likelihood principle, along with many arguments against it, is discussed in detail in Berger and Wolpert (1988). Persuasive arguments for adoption of the likelihood principle have also been given by Basu (1975, 1977); see also Ghosh (1988). The 'counterexample' to the likelihood principle given in Example 3.5 is based on one given in Fraser, Monette, and Ng (1984); see also, Berger and Wolpert (1988, Section 5.3.2), Joshi (1989), Liseo (1992), and Goldstein and Howard (1991).

As noted in Section 3.2, the methods described in this book do not, in general, follow the likelihood principle. Often however they follow the likelihood principle to a high degree of approximation, in a certain sense. See Pierce and Peters (1994).

The Bartlett identities are considered in more detail in McCullagh (1987, Section 7.2) and Barndorff-Nielsen and Cox (1994, Section 5.2); see also Skovgaard (1986).

The properties of the expected information, which is sometimes called the Fisher information, are available in a number of sources; see, for example, Cox and Hinkley (1974), Barndorff-Nielsen and Cox (1994, Section 2.2), and Rao (1973, Section 5a). The information inequality, also known as the Cramér–Rao inequality, is a standard result in mathematical statistics; see, for example, Lehmann (1983, Section 2.6) or Ferguson (1996, Chapter 19). Properties of the observed information are discussed in Barndorff-Nielsen (1991a) and Efron and Hinkley (1978). Properties of the partial information are considered in Bhapkar (1989) and Zhu and Reid (1994). Parameter orthogonality is discussed in Huzurbazar (1950, 1956), Cox and Reid (1987), and Barndorff-Nielsen and Cox (1994, Section 2.7).

The methods discussed in Section 3.7 are considered in detail in many books on statistical theory. See, for example, Lehmann (1983, 1986), Bickel and Doksum

(1977), and Cox and Hinkley (1974, Chapter 9). In many cases, an exact expression for the maximum likelihood estimate is not available. In these cases, computation of the maximum likelihood estimate for a given set of data must be done numerically. A number of methods are available for this; see Thisted (1988, Chapter 4) for an overview of these methods and additional references. Example 3.16 is based on Blyth (1990).

The approach taken here is to use the likelihood function as the basis for frequency-based statistical inference. There are other approaches to inference that are also based on the likelihood function. The most prominent of these is Bayesian inference. In Bayesian inference, the parameter θ is formally treated as a random variable; the distribution of θ represents our beliefs regarding its value. Before the data are observed, we assume that θ is distributed according to its *prior distribution* which must be chosen by the analyst. The model function, $p(y; \theta)$, is interpreted as $p(y|\theta)$, the conditional density of y given θ. After the data are observed, the prior distribution is updated to $p(\theta|y)$, the density of θ given y, using Bayes theorem. This *posterior distribution* represents our beliefs regarding θ based on the observation of $Y = y$ and the prior distribution used. It may be shown that this posterior distribution depends on the data only through the likelihood function and, hence, Bayesian inference respects the likelihood principle.

Bayesian methods have a number of desirable properties and often provide a useful alternative to the methods described in this book. The drawbacks of Bayesian inference are that it requires specification of the prior distribution and that Bayesian methods do not necessarily have good frequency properties, although in many cases they do. For further details of Bayesian methods see, for example, Berger (1985), Bernardo and Smith (1994), and Gelman, Carlin, Stern, and Rubin (1995).

Another approach to statistical inference based on the likelihood function is to interpret the likelihood function directly, without reference to either frequency-based properties such as coverage probabilities or to Bayesian concepts such as a posterior distribution. For instance, a comparison of two parameter values, θ_1 and θ_2, is based directly on the likelihood ratio $L(\theta_1)/L(\theta_2)$. The main drawback of direct likelihood inference is that it is often difficult to interpret quantities such as $L(\theta_1)/L(\theta_2)$ without using either frequency-based or Bayesian concepts. Direct likelihood inference is discussed in detail in Edwards (1972), Kalbfleisch (1985), and Royall (1997).

3.9 Exercises

3.1 Let (Y_{j1}, Y_{j2}), $j = 1, \ldots, n$ denote pairs of real-valued random variables of the form

$$Y_{jk} = \mu + \eta_j + \epsilon_{jk}, \quad k = 1, 2; \quad j = 1, \ldots, n$$

where the ϵ_{jk} are independent normal random variables with mean 0 and standard deviation σ and the η_j are independent normal random variables with mean 0 and standard deviation τ; assume that the ϵ_{jk} and the η_j are also independent. Find the log-likelihood function for (μ, τ, σ).

3.2 Let $Y_1, \ldots, Y_n$ denote independent identically distributed exponential random variables with rate parameter λ. Suppose that there exist constants

$c_1, \ldots, c_n$ such that Y_j is censored at c_j; that is, if $Y_j > c_j$ only c_j is observed. Hence, the observed data are given by

$$x_j = \min(y_j, c_j), \quad j = 1, \ldots, n.$$

Find the log-likelihood function for λ.

3.3 Let $Y_1, \ldots, Y_n$ denote binary random variables taking the values 0, 1. Suppose that there exists a binary latent variable Z such that for each $j = 1, \ldots, n$,

$$\Pr(X_j = 1|Z = z) = \pi_z, \quad z = 0, 1;$$

also, assume that

$$\Pr(Z = 1) = 1 - \Pr(Z = 0) = \theta$$

where θ is an unknown parameter. Find the log-likelihood function for π_0, π_1, θ.

3.4 Let $Y_1, \ldots, Y_n; X_1, \ldots, X_m$ denote independent exponential random variables such that each Y_j has rate parameter λ_1 and each X_j has rate parameter λ_2. Let $\psi = \lambda_1/\lambda_2$ and $\lambda = \lambda_2$. Find the log-likelihood function for (ψ, λ). Find the expected and observed information matrices.

3.5 Let X and Y denote random variables, not necessarily independent, each distributed according to a density depending on a scalar parameter θ. Let $i(\theta; Y)$ denote the expected information for θ based on the marginal distribution of Y and let $i(\theta; X|Y)$ denote the expected information for θ based on the conditional distribution of X given Y. Show that

$$i(\theta) = i(\theta; Y) + E[i(\theta; X|Y); \theta].$$

Does a similar result hold for the observed information?

3.6 Consider a model with parameter θ and consider a reparameterization of the model given by $\phi \equiv \phi(\theta)$. Show that the relationship between the expected information matrix for θ and the expected information matrix for ϕ, given in Section 3.6, holds for the observed information whenever it is evaluated at the maximum likelihood estimate $\hat{\phi}$, but not in general.

3.7 Consider a model with parameter $\theta = (\psi, \lambda)$. Show that if ψ and λ are orthogonal, then $i_{\psi} = i_{\psi\psi}$. Does the converse hold?

3.8 For the model described in Exercise 3.4, find the partial information for ψ.

3.9 For the model described in Exercise 3.4, find a parameter ϕ that is orthogonal to ψ.

3.10 Let $Y_1, \ldots, Y_n$ denote independent identically distributed random variables each having density

$$\frac{\beta^\alpha}{\Gamma(\alpha)} y^{\alpha-1} \exp\{-\beta y\}, \quad y > 0;$$

here $\alpha > 0$ and $\beta > 0$ are unknown parameters. Find the expected information matrix.

3.11 For the model described in Exercise 3.10, find a parameter ϕ that is orthogonal to α and find a parameter η that is orthogonal to β.

3.12 Let Y denote a random variable with a density of the form

$$\exp\{s(y)^T c(\theta) - k(\theta) + D(y)\}$$

where the conditions of a full k-parameter exponential family model are satisfied. Show that $i(\hat{\theta}) = j(\hat{\theta})$.

3.13 Let Y denote a random variable with density $p(y; \psi, \lambda)$ where ψ is a scalar parameter and λ is a vector parameter. Let (ψ, ϕ) denote an interest-respecting reparameterization of the model such that ϕ is orthogonal to ψ. Find the partial information for ψ based on the log-likelihood function for (ψ, ϕ) and compare it to the partial information for ψ based on the log-likelihood function for (ψ, λ).

3.14 Let Y denote a random variable with density $p(y; \theta)$ where $\theta = (\psi, \lambda) \in \Theta \subset \Re^2$ and ψ is the parameter of interest. For a given real-number β, let

$$T \equiv T(\theta) = \ell_{\psi}(\theta) - \beta\ell_{\lambda}(\theta).$$

Find $\beta \equiv \beta(\theta)$ so that T is uncorrelated with $\ell_{\lambda}(\theta)$ for each θ. Show that for this choice of β

$$\mathrm{Var}(T; \theta) = i_{\psi}(\theta) \quad \text{and} \quad -\mathrm{E}\left\{\frac{\partial T}{\partial \psi}(\theta)\right\} = i_{\psi}(\theta)$$

so that T plays the role of a 'score function for ψ'.

3.15 Let X and Y denote independent random variables, possibly vector valued, each distributed according to a distribution depending on a parameter $\theta = (\psi, \lambda)$ where ψ is a scalar. Let $i_{\psi}(\theta; X)$ denote the partial information for ψ based on X, $i_{\psi}(\theta; Y)$ denote the partial information for ψ based on Y, and $i_{\psi}(\theta)$ denote the partial information for ψ based on (X, Y). Give conditions under which

$$i_{\psi}(\theta) = i_{\psi}(\theta; X) + i_{\psi}(\theta; Y).$$

4 First-order asymptotic theory

4.1 Introduction

In Section 3.7 several commonly used likelihood-based statistical methods were described. These methods typically require the determination of the probability distribution of some statistic; for instance, to use the likelihood ratio test, the distribution of the likelihood ratio statistic under the null hypothesis is needed. Although exact results are available in certain special cases, a general theory of likelihood-based statistical inference must rely on asymptotic approximations.

In this chapter, we consider the first-order asymptotic theory of likelihood-based methods. The first-order theory uses approximations based on the central limit theorem. In Chapter 5, higher-order asymptotic theory is considered; the higher-order theory uses approximations based on either an Edgeworth expansion or a saddlepoint approximation. In Chapter 6, the approximation of conditional distributions given an ancillary statistic is discussed.

4.2 Maximum likelihood estimates

4.2.1 CONSISTENCY

An estimate T_n of a parameter θ is said be *consistent* if, under the distribution indexed by θ,

$$T_n \xrightarrow{p} \theta \quad \text{as } n \to \infty.$$

Consistency is a rather weak property which any 'reasonable' estimator should possess. It is important to note that consistency, like all asymptotic properties, refers to the entire sequence of estimates $T_1, T_2, \ldots$, not just the single estimate T_n. Maximum likelihood estimators are typically, but not always, consistent.

Let $\ell(\theta)$, $\theta \in \Theta$ denote the log-likelihood function based on n observations and let $\gamma(\theta) = n^{-1}\mathrm{E}\{\ell(\theta); \theta_0\}$ where θ_0 denotes the true parameter value. Note that $\gamma(\theta_0) - \gamma(\theta)$ is equal to the Kullback–Leibler divergence, $K(\theta_0, \theta)$ discussed in Section 3.2; it follows from the properties of the Kullback–Leibler divergence that $\gamma(\theta)$ is uniquely maximized at θ_0. We also assume that $\gamma(\theta)$ is a continuous function of θ. In broad generality, it follows from the weak law of large numbers that $n^{-1}\ell(\theta)$ converges in probability to $\gamma(\theta)$. Hence, we expect that, under some regularity conditions, $\hat{\theta}$, the maximizer of $n^{-1}\ell(\theta)$, should converge in probability to θ_0, the maximizer of $\gamma(\theta)$.

To make this argument precise, assume that the following two conditions hold:

1. Θ is a compact subset of $\Re^d$
2. $\sup_{\theta\in\Theta} |n^{-1}\ell(\theta) - \gamma(\theta)| \xrightarrow{p} 0$ as $n \to \infty$.

It follows that

$$\sup_{\theta\in\Theta} \frac{1}{n}\ell(\theta) = \frac{1}{n}\ell(\hat{\theta}) \xrightarrow{p} \sup_{\theta\in\Theta} \gamma(\theta) = \gamma(\theta_0).$$

For each $\theta \in \Theta$, $\theta \neq \theta_0$, there exists an $\epsilon > 0$ and an open neighborhood N_θ of θ such that

$$\inf_{\theta_1\in N_\theta} |\gamma(\theta_1) - \gamma(\theta_0)| > \epsilon.$$

It follows that

$$\Pr\{\hat{\theta} \in N_\theta; \theta_0\} \leq \Pr\{|\gamma(\hat{\theta}) - \gamma(\theta_0)| > \epsilon; \theta_0\} \to 0 \quad \text{as } n \to \infty.$$

Let N_0 denote an arbitrary open neighborhood of θ_0 and consider the compact set $\Theta^* = \Theta \setminus N_0$. Then the open sets in $\{N_\theta\colon \theta \in \Theta, \theta \neq \theta_0\}$ cover Θ^*. By compactness, there exists a finite subcover $\{N_{\theta_1}, \ldots, N_{\theta_m}\}$. It follows that

$$\Pr\{\hat{\theta} \notin N_0; \theta_0\} = \Pr\{\hat{\theta} \in \Theta^*; \theta_0\} \leq \sum_1^m \Pr\{\hat{\theta} \in N_{\theta_j}; \theta_0\} \to 0 \quad \text{as } n \to \infty;$$

that is, $\hat{\theta} \xrightarrow{p} \theta_0$ as $n \to \infty$.

Hence, for the case of a compact parameter space, the maximum likelihood estimate is consistent provided that the normalized log-likelihood converges uniformly to its expectation. Note that no local conditions, such as differentiability, are required. Although this is a relatively weak condition that is often satisfied, it is often a nontrivial exercise to verify that it holds. Consistency of the maximum likelihood estimate for a given model is generally easier to establish using an argument based on that particular model, rather than appealing to a general result such as the one given above.

Example 4.1 One-parameter exponential family distribution
Let $Y_1, \ldots, Y_n$ denote independent identically distributed random variables each with density function of the form

$$\exp\{s(y)c(\theta) - k(\theta) + D(y)\}$$

where θ is a real-valued parameter taking values in an open set Θ. We have seen that, if the equation $n^{-1}\sum s(y_j) = k'(\theta)$ has a solution in Θ, then that solution is the maximum likelihood estimate of θ.

Fix θ, the true parameter value. Then $n^{-1}\sum s(Y_j)$ converges in probability to $k'(\theta)$. It follows that, with probability approaching 1 as $n \to \infty$, $n^{-1}\sum s(Y_j)$ takes values in $k'(\Theta)$, the range of the function $k'(\cdot)$. Hence, with probability approaching 1, the maximum likelihood estimate exists and is the solution to $n^{-1}\sum s(y_j) = k'(\hat{\theta})$. It follows that $k'(\hat{\theta}) \xrightarrow{p} k'(\theta)$. As noted previously, $k''(\theta) > 0$ for all θ so that $k'(\cdot)$ is strictly increasing and, hence, invertible. It follows that $\hat{\theta} \xrightarrow{p} \theta$. ∎

Example 4.2 Laplace distribution
Let $Y_1, \ldots, Y_n$ denote independent random variables, each distributed according to the Laplace distribution described in Example 3.16. Recall that maximum likelihood estimate of the location parameter μ may be taken to be the sample median. First suppose that n is odd. Let $\epsilon > 0$ be fixed and for each $j = 1, \ldots, n$, let Z_j denote a random variable taking the value 1 if $Y_j - \mu \geq \epsilon$ and 0 otherwise. Then $\sum Z_j$ is a binomial random variable with index n and success probability $\exp\{-\epsilon/\sigma\}/2$. Note that

$$\begin{aligned}\Pr\{Y_{((n+1)/2)} - \mu \geq \epsilon; \mu\} &= \Pr\Big\{\sum Z_j \geq (n+1)/2\Big\} \\ &= \Pr\{\bar{Z} - 1/(2n) \geq 1/2\}.\end{aligned}$$

Since $\bar{Z} - 1/(2n)$ converges in probability to $\exp\{-\epsilon/\sigma\}/2 < 1/2$, it follows that

$$\Pr\{Y_{((n+1)/2)} - \mu \geq \epsilon; \mu\} \to 0 \quad \text{as } n \to \infty. \tag{4.1}$$

A similar argument can be used to show that

$$\Pr\{Y_{((n+1)/2)} - \mu \leq -\epsilon; \mu\} \to 0 \quad \text{as } n \to \infty. \tag{4.2}$$

In that case, define Z_j to be 1 if $Y_j - \mu \leq -\epsilon$ and 0 otherwise; then Z_j is a binomial random variable with success probability $(1 - \exp\{-\epsilon/\sigma\})/2$. The result (4.2) follows from the fact that

$$\Pr\{Y_{((n+1)/2)} - \mu \leq \epsilon; \mu\} = \Pr\Big\{\sum Z_j \geq (n+1)/2\Big\}.$$

Taking (4.1) and (4.2) together shows that $Y_{((n+1)/2)} \xrightarrow{p} \mu$ as $n \to \infty$.

For the case in which n is even, a similar argument can be used to show that both $Y_{(n/2)}$ and $Y_{(n/2+1)}$ converge in probability to μ. Hence, $\hat{\mu}$ is a consistent estimate of μ. ∎

Example 4.3 Precision of duplicate measurements
Let $(X_1, Y_1), \ldots, (X_n, Y_n)$ denote independent pairs of independent normally distributed random variables such that, for each j, X_j and Y_j each have mean μ_j and variance σ^2. Then

$$\ell(\mu_1, \ldots, \mu_n, \sigma^2) = -\frac{1}{2\sigma^2}\sum_{i=1}^{n}[(X_i - \mu_i)^2 + (Y_i - \mu_i)^2] - 2n\log\sigma^2.$$

It is straightforward to show that the maximum likelihood estimate of σ^2 is given by

$$\hat{\sigma}^2 = \frac{\sum(X_j - Y_j)^2}{4n}.$$

Note that $\hat{\sigma}^2$ has mean $\sigma^2/2$ and variance $\sigma^4/(2n)$ so that $\hat{\sigma}^2 \xrightarrow{p} \sigma^2/2$ as $n \to \infty$; that is, $\hat{\sigma}^2$ is not consistent.

It is important to note, however, that this model falls outside of the general framework we are considering since the dimension of the parameter $(\mu_1, \ldots, \mu_n, \sigma^2)$ depends on the sample size n. ∎

4.2.2 ASYMPTOTIC DISTRIBUTION

For regular models, the asymptotic distribution of the maximum likelihood estimate is easily obtained. By Property R1 the score function evaluated at $\theta = \hat{\theta}$ may be expanded

$$\ell_\theta(\hat{\theta}) - \ell_\theta(\theta) = [\ell_{\theta\theta}(\theta) + O_p(\sqrt{n})](\hat{\theta} - \theta).$$

Solving this equation for $\hat{\theta} - \theta$, and using the fact that $\ell_\theta(\hat{\theta}) = 0$, yields

$$\hat{\theta} - \theta = [-\ell_{\theta\theta}(\theta) + O_p(\sqrt{n})]^{-1}\ell_\theta(\theta)$$

and, hence,

$$\sqrt{n}(\hat{\theta} - \theta) = \Big[-\frac{1}{n}\ell_{\theta\theta}(\theta) + O_p(n^{-1/2})\Big]^{-1} \frac{1}{\sqrt{n}}\ell_\theta(\theta) + O_p(n^{-1/2}). \tag{4.3}$$

Note that

$$-\frac{1}{n}\ell_{\theta\theta}(\theta) = \frac{1}{n}i(\theta) + O_p(n^{-1/2})$$

and since $\ell_\theta(\theta)$ is asymptotically distributed according to a multivariate normal distribution and the covariance matrix of $\ell_\theta(\theta)$ is given by $i(\theta)$,

$$i(\theta)^{-1/2}\ell_\theta(\theta) \xrightarrow{\mathcal{D}} N_d(0, D_d)$$

where $N_d(\mu, \Sigma)$ denotes a multivariate normal random variable of dimension d with mean vector μ and covariance matrix Σ and D_d denotes the identity matrix of rank d. It follows that

$$i(\theta)^{1/2}(\hat{\theta} - \theta) \xrightarrow{\mathcal{D}} N_d(0, D_d); \tag{4.4}$$

that is, the maximum likelihood estimate of θ is approximately distributed according to a multivariate normal distribution with mean 0 and covariance matrix $i(\theta)^{-1}$.

This result may be stated in a number of different ways. As is clear from the argument given above, the expected information in (4.4) may be replaced by the observed information so that

$$j(\theta)^{1/2}(\hat{\theta} - \theta) \xrightarrow{\mathcal{D}} N_d(0, D_d);$$

note, however, that $j(\theta)$ is a random variable whereas $i(\theta)$ is a constant. Also the argument θ in either the observed or expected information may be replaced by $\hat{\theta}$ without changing the asymptotic distribution; for instance,

$$j(\hat{\theta})^{1/2}(\hat{\theta} - \theta) \xrightarrow{\mathcal{D}} N(0, D_d).$$

Hence, to estimate the standard error of a component of $\hat{\theta}$ we may use either $i(\hat{\theta})$ or $j(\hat{\theta})$.

Let $\bar{i}(\theta)$ denote a matrix satisfying

$$\lim_{n\to\infty} \frac{1}{n} i(\theta) = \bar{i}(\theta).$$

In particular, if the likelihood function is based on a sample of independent identically distributed random variables, then $\bar{i}(\theta)$ is simply the expected information based on a single observation; in other cases, $\bar{i}(\theta)$ may be viewed as the limiting average expected information per observation. Then (4.4) may be expressed

$$\sqrt{n}(\hat{\theta} - \theta) \xrightarrow{\mathcal{D}} N_d(0, \bar{i}(\theta)^{-1}).$$

This covariance matrix is the 'smallest' possible in the sense that if T is any estimator such that $\sqrt{n}(T - \theta)$ is asymptotically normally distributed with mean 0 and covariance matrix $V(\theta_0)$ then $V(\theta) - \bar{i}(\theta)^{-1}$ is nonnegative definite, except perhaps for θ in a set with Lebesgue measure 0. If $V(\theta)$ is continuous in θ then $V(\theta) - \bar{i}(\theta)^{-1}$ is nonnegative definite for all θ.

Example 4.4 Exponential distribution
Let $Y_1, \ldots, Y_n$ denote independent random variables each distributed according to an exponential distribution with mean θ. The log-likelihood function for θ is given by

$$\ell(\theta) = -n \log \theta - \sum y_j / \theta;$$

it follows that the maximum likelihood estimate of θ is $\hat{\theta} = \bar{y}$.

It is straightforward to show that

$$\ell_{\theta\theta}(\theta) = \frac{n}{\theta^2} - 2\frac{\sum y_j}{\theta^3}$$

so that $i(\theta) = n/\theta^2$. Hence,

$$\frac{\sqrt{n}(\hat{\theta} - \theta)}{\theta} \xrightarrow{\mathcal{D}} N(0, 1).$$

Using either $i(\hat{\theta})$ or $j(\hat{\theta})$ in place of $i(\theta)$ we have that

$$\frac{\sqrt{n}(\hat{\theta} - \theta)}{\hat{\theta}} \xrightarrow{\mathcal{D}} N(0, 1).$$

Consider the data in Table A.3 of the Appendix. These data represent the time to failure, in hours, of pressure vessels subjected to a constant fixed pressure. Modeling these data as exponential random variables with mean θ, the maximum likelihood estimate of θ is 575.5 hours and the estimated standard error of this estimate is 128.7. ∎

Example 4.5 Normal-theory nonlinear regression model
Consider the nonlinear regression model

$$Y_j = \frac{\beta_1 x_j}{\beta_2 + x_j} + \sigma \epsilon_j$$

considered in Example 3.3; here $x_1, \ldots, x_n$ are fixed constants and $\epsilon_1, \ldots, \epsilon_n$ are independent standard normal random variables. The log-likelihood function is given by

$$\ell(\theta) = -n \log \sigma - \frac{1}{2\sigma^2} \sum \left(y_j - \frac{\beta_1 x_j}{\beta_2 + x_j} \right)^2$$

so that, writing $\theta = (\beta_1, \beta_2, \sigma)$,

$$\ell_\theta(\theta) = \frac{1}{\sigma^2} \begin{pmatrix} \sum \epsilon_j \dfrac{x_j}{\beta_2 + x_j} \\ -\sum \epsilon_j \dfrac{\beta_1 x_j}{(\beta_2 + x_j)^2} \\ \sum (\epsilon_j^2 - \sigma^2)/\sigma \end{pmatrix}$$

where $\epsilon_j = y_j - \beta_1 x_j / (\beta_2 + x_j)$.

It is straightforward to show that the covariance matrix of $\ell_\theta(\theta)$ is given by

$$i(\theta) = \frac{1}{\sigma^2} \begin{pmatrix} \sum \dfrac{x_j^2}{(\beta_2 + x_j)^2} & -\sum \dfrac{\beta_1 x_j^2}{(\beta_2 + x_j)^3} & 0 \\ \cdot & \sum \dfrac{\beta_1^2 x_j^2}{(\beta_2 + x_j)^4} & 0 \\ \cdot & \cdot & 2n \end{pmatrix}.$$

Hence, $\hat{\theta}$ is approximately normally distributed with mean vector θ and covariance matrix $i(\theta)^{-1}$.

The observed information matrix $j(\theta)$ for this model is given by

$$\frac{1}{\sigma^2} \begin{pmatrix} \sum \frac{x_j^2}{(\beta_2+x_j)^2} & \sum \left(y_j - \frac{\beta_1 x_j}{\beta_2+x_j}\right) \frac{x_j}{(\beta_2+x_j)^2} - \sum \frac{\beta_1 x_j^2}{(\beta_2+x_j)^3} & 2 \sum \frac{y_j - \beta_1 x_j}{\beta_2+x_j} \frac{x_j}{\sigma} \\ \cdot & \sum \frac{(\beta_1 x_j)^2}{(\beta_2+x_j)^4} - \sum 2\left(y_j - \frac{\beta_1 x_j}{\beta_2+x_j}\right) \frac{\beta_1 x_j}{(\beta_2+x_j)^3} & \frac{2}{\sigma} \sum \frac{y_j - \beta_1 x_j}{\beta_2+x_j} \frac{\beta_1 x_j}{(\beta_2+x_j)^2} \\ \cdot & \cdot & \frac{1}{\sigma^2} \sum \left(\frac{y_j - \beta_1 x_j}{\beta_2+x_j}\right)^2 + n \end{pmatrix}$$

and the observed information evaluated at $\theta = \hat{\theta}$ is given by

$$j(\hat{\theta}) = \frac{1}{\hat{\sigma}^2} \begin{pmatrix} \sum \dfrac{x_j^2}{(\hat{\beta}_2 + x_j)^2} & -\sum \dfrac{\hat{\beta}_1 x_j^2}{(\hat{\beta}_2 + x_j)^3} & 0 \\ \cdot & \sum \dfrac{(\hat{\beta}_1 x_j)^2}{(\hat{\beta}_2 + x_j)^4} & 0 \\ \cdot & \cdot & 2n \end{pmatrix};$$

here the fact that $\ell_\theta(\hat{\theta}) = 0$ has been used. Note that $i(\hat{\theta}) = j(\hat{\theta})$.

Consider the data in Table A.4 of the Appendix. These data refer to an experiment conducted to study relationship between the 'velocity' of an enzymatic reaction and the concentration of a substrate. Here y_j denotes the velocity of the reaction and x_j denotes the substrate concentration. The maximum likelihood estimates of the parameters are

$$\hat{\beta}_1 = 212.7, \qquad \hat{\beta}_2 = .0641, \qquad \hat{\sigma} = 9.98.$$

Hence,

$$i(\hat{\theta}) = \begin{pmatrix} 0.0599 & -38.511 & 0 \\ \cdot & 42243 & 0 \\ \cdot & \cdot & 0.241 \end{pmatrix}$$

and the asymptotic covariance matrix of $\hat{\theta}$ is given by

$$i(\hat{\theta})^{-1} = \begin{pmatrix} 40.20 & 0.0366 & 0 \\ \cdot & 0.0000571 & 0 \\ \cdot & \cdot & 4.15 \end{pmatrix}.$$

For instance, the estimated standard error of $\hat{\beta}_1$ is $\sqrt{40.20} = 6.34$. ∎

The asymptotic distribution of $\hat{\theta}$ can be used to determine the asymptotic distribution of functions of $\hat{\theta}$. Let $\psi = g(\theta)$ where $g(\cdot)$ is a continuously differentiable function on Θ_0. Then $\hat{\psi} = g(\hat{\theta})$ and

$$\sqrt{n}(\hat{\psi} - \psi) = g'(\theta)\sqrt{n}(\hat{\theta} - \theta) + O_p(n^{-1/2})$$

and

$$[g'(\theta)^T i(\theta)^{-1} g'(\theta)]^{-1/2}(\hat{\psi} - \psi) \xrightarrow{\mathcal{D}} N_q(0, D_q)$$

where q is the dimension of ψ; that is $\hat{\psi}$ is approximately normally distributed with mean vector ψ and covariance matrix $g'(\theta)^T i(\theta)^{-1} g'(\theta)$. Hence, an explicit parameterization of the nuisance parameter of the model is not needed to determine $\hat{\psi}$ or its asymptotic distribution.

Example 4.6 Normal-theory nonlinear regression model

Consider the nonlinear regression model considered in Example 4.5. Take ψ to be the vector of regression parameters, $\psi = (\beta_1, \beta_2)$ so that $g(\beta_1, \beta_2, \sigma) = (\beta_1, \beta_2)^T$. Then

$$g'(\theta) = \begin{pmatrix} 1 & 0 \\ 0 & 1 \\ 0 & 0 \end{pmatrix}$$

and

$$g'(\theta)^T i(\theta)^{-1} g'(\theta) = \sigma^2 \Big/ \beta_1^2 \sum \frac{x_j^2}{(\beta_2 + x_j)^2} \sum \frac{x_j^2}{(\beta_2 + x_j)^4}$$

$$- \beta_1^2 \Big(\sum \frac{x_j^2}{(\beta_2 + x_j)^2} \Big)^2 \begin{pmatrix} \beta_1^2 \sum \frac{x_j^2}{(\beta_2 + x_j)^4} & \beta_1 \sum \frac{x_j^2}{(\beta_2 + x_j)^2} \\ \cdot & \sum \frac{x_j^2}{(\beta_2 + x_j)^2} \end{pmatrix}.$$

If ψ is taken to be β_1/β_2 then

$$g'(\theta) = \begin{pmatrix} 1/\beta_2 \\ -\beta_1/\beta_2^2 \\ 0 \end{pmatrix}$$

and

$$g'(\theta)^T i(\theta)^{-1} g'(\theta)$$

$$= \Big(\frac{1}{\beta_2^2} \sum \frac{x_j^2}{(\beta_2 + x_j)^4} - \frac{2}{\beta_2^3} \sum \frac{x_j^2}{(\beta_2 + x_j)^2} + \frac{1}{\beta_2^4} \sum \frac{x_j^2}{(\beta_2 + x_j)^2} \Big) \Big/$$

$$\Big(\sum \frac{x_j^2}{(\beta_2 + x_j)^2} \sum \frac{x_j^2}{(\beta_2 + x_j)^4} - \Big(\sum \frac{x_j^2}{(\beta_2 + x_j)^2} \Big)^2 \Big).$$

When analyzing a specific set of data the computation of $g'(\hat{\theta})^T i(\hat{\theta})^{-1} g'(\hat{\theta})$ may be done numerically. For instance, using the data considered in Example 4.5, and taking $g(\theta) = \beta_1/\beta_2$,

$$g'(\hat{\theta}) = \begin{pmatrix} 1/\hat{\beta}_2 \\ -\hat{\beta}_1/\hat{\beta}_2^2 \\ 0 \end{pmatrix} = \begin{pmatrix} 15.60 \\ -51767 \\ 0 \end{pmatrix}$$

so that

$$g'(\hat{\theta})^T i(\hat{\theta})^{-1} g'(\hat{\theta}) = (15.60 \quad -51767 \quad 0)$$

$$\times \begin{pmatrix} 40.20 & 0.0366 & 0 \\ \cdot & 0.0000571 & 0 \\ \cdot & \cdot & 4.15 \end{pmatrix} \begin{pmatrix} 15.60 \\ -51767 \\ 0 \end{pmatrix}$$

$$= 103565.$$

Hence, $\hat{\psi} = 3318$ and the estimated standard error of $\hat{\psi}$ is 321.8. ∎

4.3 The likelihood ratio statistic

4.3.1 DISTRIBUTION UNDER THE NULL HYPOTHESIS

First consider the problem of testing the simple null hypothesis H_0: $\theta = \theta_0$ versus the alternative hypothesis H_1: $\theta \neq \theta_0$ where θ_0 is some specified element of Θ. The likelihood ratio test statistic for this case is given by

$$W \equiv W(\theta_0) = 2[\ell(\hat{\theta}) - \ell(\theta_0)]$$

where $\hat{\theta}$ denotes the maximum likelihood estimate of θ. Here we consider the asymptotic distribution of W under the null hypothesis so that θ_0 is the true parameter value.

By expanding $\ell(\hat{\theta}) - \ell(\theta_0)$ around $\hat{\theta} = \theta_0$, it follows that

$$\tfrac{1}{2}W = \ell_\theta(\theta_0)^T(\hat{\theta} - \theta_0) + \tfrac{1}{2}(\hat{\theta} - \theta_0)^T\ell_{\theta\theta}(\theta_0)(\hat{\theta} - \theta_0) + O_p(n^{-1/2}).$$

Substituting the expansion

$$\hat{\theta} - \theta_0 = i(\theta_0)^{-1}\ell_\theta(\theta_0) + O_p(n^{-1}), \tag{4.5}$$

which is based on the results of Section 4.2.2, for $\hat{\theta} - \theta_0$ we have that

$$\begin{aligned}\tfrac{1}{2}W =\ & \ell_\theta(\theta_0)^T i(\theta_0)^{-1}\ell_\theta(\theta_0) + \tfrac{1}{2}\ell_\theta(\theta_0)^T i(\theta_0)^{-1}\ell_{\theta\theta}(\theta_0)i(\theta_0)^{-1}\ell_\theta(\theta_0)\\ &+ O_p(n^{-1/2}).\end{aligned}$$

Finally, since $\ell_{\theta\theta}(\theta_0) = -i(\theta_0) + O_p(\sqrt{n})$, it follows that

$$W = [i(\theta_0)^{-1/2}\ell_\theta(\theta_0)]^T[i(\theta_0)^{-1/2}\ell_\theta(\theta_0)] + O_p(n^{-1/2}); \tag{4.6}$$

since, under the null hypothesis,

$$i(\theta_0)^{-1/2}\ell_\theta(\theta_0)$$

is asymptotically normally distributed with mean vector 0 and identity covariance matrix, it follows that W is asymptotically distributed according to a χ^2 distribution with d degrees of freedom.

Example 4.7 Exponential distribution
Let $Y_1, \ldots, Y_n$ denote independent exponential random variables with mean θ, as in Example 4.4. The likelihood ratio statistic for testing $\theta = \theta_0$ is given by

$$W = 2n\Big[\frac{\hat{\theta}}{\theta_0} - 1 - \log\Big(\frac{\hat{\theta}}{\theta_0}\Big)\Big];$$

under the null hypothesis, W is asymptotically distributed according to χ^2 distribution with 1 degree of freedom. ∎

Now consider the more general case in which the parameter θ may be written (ψ, λ) and the hypothesis to be tested is of the form H_0: $\psi = \psi_0$ versus the alternative H_1: $\psi \neq \psi_0$. The likelihood ratio test statistic for this case is given by

$$W = 2[\ell(\hat{\theta}) - \ell(\hat{\theta}_{\psi_0})]$$

where $\hat{\theta}_\psi$ denotes the maximum likelihood estimate of θ with ψ held fixed. Note that calculation of W does not require explicit parameterization of the nuisance parameter of the model. Let $\psi = \psi(\theta)$; we may write

$$W = 2\left[\sup_{\theta \in \Theta} \ell(\theta) - \sup_{\theta \in \Theta_{\psi_0}} \ell(\theta)\right]$$

where

$$\Theta_{\psi_0} = \{\theta \in \Theta\colon \psi(\theta) = \psi_0\}.$$

We will show that W is asymptotically distributed according to a χ^2 distribution with q degrees of freedom, where q denotes the dimension of ψ. To prove this result, begin by expanding $\ell(\hat{\theta}_{\psi_0})$ around $\hat{\theta}_{\psi_0} = \hat{\theta}$:

$$\ell(\hat{\theta}_{\psi_0}) - \ell(\hat{\theta}) = \ell_\theta(\hat{\theta})^T(\hat{\theta}_{\psi_0} - \hat{\theta}) + \tfrac{1}{2}(\hat{\theta}_{\psi_0} - \hat{\theta})^T \ell_{\theta\theta}(\hat{\theta})(\hat{\theta}_{\psi_0} - \hat{\theta}) + O_p(n^{-1/2}). \tag{4.7}$$

Let θ_0 denote the true value of θ under the null hypothesis. Note that

$$\ell_\theta(\hat{\theta}) = 0; \qquad -\ell_{\theta\theta}(\hat{\theta}) = i(\theta_0) + O_p(\sqrt{n}).$$

Hence,

$$\ell(\hat{\theta}) - \ell(\hat{\theta}_{\psi_0}) = \tfrac{1}{2}(\hat{\theta} - \hat{\theta}_{\psi_0})^T i(\theta_0)(\hat{\theta} - \hat{\theta}_{\psi_0}) + O_p(n^{-1/2}). \tag{4.8}$$

The term $\hat{\theta} - \hat{\theta}_{\psi_0}$ is of the form $(\hat{\psi} - \psi_0, \hat{\lambda} - \hat{\lambda}_{\psi_0})^T$ where $\hat{\lambda}_\psi$ denotes the maximum likelihood estimate of λ for fixed ψ. By (4.3),

$$\begin{pmatrix} \hat{\psi} - \psi_0 \\ \hat{\lambda} - \lambda_0 \end{pmatrix} = i(\theta_0)^{-1} \begin{pmatrix} \ell_\psi(\theta_0) \\ \ell_\lambda(\theta_0) \end{pmatrix} + O_p(n^{-1}); \tag{4.9}$$

here

$$\ell_\psi(\theta) = \frac{\partial}{\partial\psi}\ell(\theta); \qquad \ell_\lambda(\theta) = \frac{\partial}{\partial\lambda}\ell(\theta).$$

For the model with ψ held fixed, a similar result holds, specifically,

$$(\hat{\lambda}_{\psi_0} - \lambda_0) = i_{\lambda\lambda}(\theta_0)^{-1}\ell_\lambda(\theta_0) + O_p(n^{-1}). \tag{4.10}$$

From these two results, it follows that

$$\sqrt{n}(\hat{\lambda} - \hat{\lambda}_{\psi_0}) = \sqrt{n}\tilde{i}_{\lambda\psi}(\theta_0)\ell_\psi(\theta_0) + \sqrt{n}(\tilde{i}_{\lambda\lambda}(\theta_0) - i_{\lambda\lambda}(\theta_0)^{-1})\ell_\lambda(\theta_0) + O_p(n^{-1/2})$$

and

$$\sqrt{n}(\hat{\psi} - \psi_0) = \sqrt{n}\tilde{i}_{\psi\psi}(\theta_0)\ell_\psi(\theta_0) + \tilde{i}_{\psi\lambda}(\theta_0)\ell_\lambda(\theta_0) + O_p(n^{-1/2})$$

where

$$i(\theta)^{-1} = \begin{pmatrix} i_{\psi\psi}(\theta) & i_{\psi\lambda}(\theta) \\ i_{\lambda\psi}(\theta) & i_{\lambda\lambda}(\theta) \end{pmatrix}^{-1} = \begin{pmatrix} \tilde{i}_{\psi\psi}(\theta) & \tilde{i}_{\psi\lambda}(\theta) \\ \tilde{i}_{\lambda\psi}(\theta) & \tilde{i}_{\lambda\lambda}(\theta) \end{pmatrix}.$$

These expressions may be written more concisely as

$$\sqrt{n}\begin{pmatrix} \hat{\psi} - \psi_0 \\ \hat{\lambda} - \hat{\lambda}_{\psi_0} \end{pmatrix} = \sqrt{n}[i(\theta_0)^{-1} - H]\ell_\theta(\theta_0) + O_p(n^{-1/2}) \tag{4.11}$$

where H denotes the $d \times d$ matrix of the form

$$H = \begin{pmatrix} 0 & 0 \\ 0 & i_{\lambda\lambda}(\theta_0)^{-1} \end{pmatrix};$$

recall that $d = \dim(\theta)$.

Substituting this expression into (4.8) yields the following:

$$\begin{aligned} W &= \ell_\theta(\theta_0)^T(i(\theta_0)^{-1} - H)i(\theta_0)(i(\theta_0)^{-1} - H)\ell_\theta(\theta_0) + O_p(n^{-1/2}) \\ &= \Big(\frac{1}{\sqrt{n}}\ell_\theta(\theta_0)\Big)^T(\bar{i}(\theta_0)^{-1} - \bar{H})\bar{i}(\theta_0)(\bar{i}(\theta_0)^{-1} - H)\Big(\frac{1}{\sqrt{n}}\ell_\theta(\theta_0)\Big) \\ &\quad + O_p(n^{-1/2}) \end{aligned}$$

where $\bar{i}(\theta) = \lim n^{-1}i(\theta)$ and

$$\bar{H} = \begin{pmatrix} 0 & 0 \\ 0 & \bar{i}_{\lambda\lambda}(\theta_0)^{-1} \end{pmatrix}.$$

Since, under the null hypothesis, $\ell_\theta(\theta_0)/\sqrt{n}$ is asymptotically distributed according to a multivariate normal distribution with mean 0 and covariance matrix $\bar{i}(\theta_0)$, it follows that the asymptotic distribution of W is the distribution of $Z^T Z$ where Z has a multivariate normal distribution with mean 0 and covariance matrix

$$\Sigma = \bar{i}(\theta_0)^{1/2}(\bar{i}(\theta_0)^{-1} - \bar{H})\bar{i}(\theta_0)(\bar{i}(\theta_0)^{-1} - \bar{H})\bar{i}(\theta_0)^{1/2}.$$

Using the fact that $\bar{H}\bar{i}(\theta_0)\bar{H} = \bar{H}$, it follows that

$$\Sigma = \bar{i}(\theta_0)^{1/2}(\bar{i}(\theta_0)^{-1} - \bar{H})\bar{i}(\theta_0)^{1/2}$$

and that Σ is idempotent. It is straightforward to show Σ has trace, and, hence, rank, equal to q. The result follows.

Example 4.8 Poisson regression
Let $Y_1, \ldots, Y_n$ denote independent Poisson random variables such that Y_j has mean λ_j where

$$\log \lambda_j = \alpha + \beta x_j;$$

here $x_1, \ldots, x_n$ are known constants and α and β are unknown parameters. The log-likelihood function for this model is given by

$$\ell(\theta) = \sum y_j(\alpha + \beta x_j) - \sum \exp\{\alpha + \beta x_j\}.$$

Consider a test of the hypothesis $\beta = 0$; under the null hypothesis Y_j and x_j are not related. Under the null hypothesis, the maximum likelihood estimate of α is $\log \bar{y}$ and the unrestricted estimates $\hat{\alpha}, \hat{\beta}$ satisfy

$$\sum \exp\{\hat{\alpha} + \hat{\beta} x_j\} = \sum y_j; \qquad \sum x_j \exp\{\hat{\alpha} + \hat{\beta} x_j\} = \sum x_j y_j.$$

It follows that the likelihood ratio test statistic may be written

$$W = 2n(\hat{\alpha} - \log \bar{y})\bar{y} + 2\hat{\beta} \sum x_j y_j.$$

Under the null hypothesis, W is asymptotically distributed according to a chi-square distribution with 1 degree of freedom.

Consider the data given in Table A.5 of the Appendix. In these data, y_j denotes the number of cans damaged in a boxcar and x_j denotes the speed of the boxcar at impact. The maximum likelihood estimates of the parameters are $\hat{\alpha} = 3.546$ and $\hat{\beta} = 0.1490$ and $\log \bar{y} = 4.273$. The likelihood ratio statistic for testing $\beta = 0$ is $W = 78.6$. Hence, there is strong evidence against the hypothesis that $\beta = 0$; i.e., there is strong evidence that the number of cans damaged is related to the speed at impact. ∎

Example 4.9 Several exponential distributions
Let $X_1, \ldots, X_d$ denote independent gamma random variables such that X_j has index n_j and rate parameter θ_j. We may view X_j has the sum of n_j independent exponential random variables each with rate parameter θ_j. Consider a test of the hypothesis $\sum \theta_j = c$ where c is a known constant. Note that $\sum \theta_j$ is the rate parameter of the minimum of d independent exponential random variables with rate parameters $\theta_1, \ldots, \theta_d$, respectively.

The log-likelihood function is given by

$$\ell(\theta) = \sum n_j \log \theta_j - \sum \theta_j x_j$$

and hence

$$\ell(\hat{\theta}) = \sum n_j \log(n_j/x_j) - \sum n_j.$$

To find $\sup \ell(\theta)$ under the null hypothesis we may maximize $\ell(\theta)$ subject to the restriction $\sum \theta_j = c$. Using the method of Lagrange multipliers, it is straightforward to show that this maximum is achieved at $\theta_1, \ldots, \theta_d$ of the form $\theta_j = n_j/(x_j + t)$ where t satisfies $\sum n_j/(x_j + t) = c$. It follows that the likelihood ratio statistic is given by

$$W = 2\Big[\sum n_j \log(1 + t/x_j) - tc\Big].$$

Under the null hypothesis, W is asymptotically distributed according to a chi-square random variable with 1 degree of freedom. ∎

When the parameter of interest ψ is a scalar, it is often useful to base inference on the signed square root of W, given by

$$R = \text{sgn}(\hat{\psi} - \psi_0)\sqrt{W};$$

for instance, a test of $\psi = \psi_0$ versus $\psi > \psi_0$ may be based on R. It will be shown in Section 4.4 that R is asymptotically distributed according to a standard normal distribution.

4.3.2 DISTRIBUTION UNDER A LOCAL ALTERNATIVE

We now consider the distribution of the likelihood ratio statistic under an alternative hypothesis of the form $\psi = \psi_n \equiv \psi_0 + \delta/\sqrt{n}$; that is, we want to derive the asymptotic distribution of W when the true parameter value is of the form $\theta_n = (\psi_0 + \delta/\sqrt{n}, \lambda_0)^T$, where ψ_0 is the value of ψ under the null hypothesis and λ_0 denotes the true value of λ.

First note that, since $\psi_n = \psi_0 + O(n^{-1/2})$,

$$\hat{\theta}_{\psi_0} = \hat{\theta}_{\psi_n} + O_p(n^{-1/2}) = \hat{\theta} + O_p(n^{-1/2});$$

hence, the argument leading to (4.8) still holds except that $i(\theta_0)$ must be replaced by $i(\theta_n)$; however since

$$i(\theta_n) = i(\theta_0) + O(\sqrt{n}), \tag{4.12}$$

(4.8) is still valid as it stands.

Equation (4.11) is also valid, except that θ_0 should be replaced by θ_n and ψ_0 should be replaced by ψ_n, that is,

$$\sqrt{n}\begin{pmatrix} \hat{\psi} - \psi_n \\ \hat{\lambda} - \hat{\lambda}_{\psi_n} \end{pmatrix} = \sqrt{n}[i(\theta_n)^{-1} - H_n]\ell_\theta(\theta_n) + O_p(n^{-1/2})$$

where H_n denotes the $d \times d$ matrix of the form

$$H_n = \begin{pmatrix} 0 & 0 \\ 0 & i_{\lambda\lambda}(\theta_n)^{-1} \end{pmatrix}.$$

By (4.12),

$$\sqrt{n}\begin{pmatrix}\hat{\psi}-\psi_n\\ \hat{\lambda}-\hat{\lambda}_{\psi_n}\end{pmatrix} = \sqrt{n}[i(\theta_0)^{-1} - H]\ell_\theta(\theta_n) + O_p(n^{-1/2}).$$

Note that, since $\psi_n = \psi_0 + \delta/\sqrt{n}$, $\hat{\lambda}_{\psi_n} = \hat{\lambda}_{\psi_0} + O_p(n^{-1/2})$. Using the fact that $\ell_\lambda(\psi_0, \hat{\lambda}_{\psi_0}) = \ell_\lambda(\psi_n, \hat{\lambda}_{\psi_n}) = 0$, it follows that

$$\begin{aligned}0 &= \ell_\lambda(\psi_0, \hat{\lambda}_{\psi_0}) - \ell_\lambda(\psi_n, \hat{\lambda}_{\psi_n})\\ &= \ell_{\lambda\psi}(\psi_n, \hat{\lambda}_{\psi_n})(\psi_0 - \psi_n) + \ell_{\lambda\lambda}(\psi_n, \hat{\lambda}_{\psi_n})(\hat{\lambda}_{\psi_0} - \hat{\lambda}_{\psi_n}) + O(1)\\ &= i_{\lambda\psi}(\theta_0)(\psi_0 - \psi_n) + i_{\lambda\lambda}(\theta_0)(\hat{\lambda}_{\psi_0} - \hat{\lambda}_{\psi_n}) + O(1).\end{aligned}$$

Hence,

$$\sqrt{n}(\hat{\lambda}_{\psi_n} - \hat{\lambda}_{\psi_0}) = -i_{\lambda\lambda}(\theta_0)^{-1} i_{\lambda\psi}(\theta_0)\delta + O_p(n^{-1/2}).$$

Since $\sqrt{n}(\hat{\psi} - \psi_n) = \sqrt{n}(\hat{\psi} - \psi_0) - \delta$, it follows that

$$\begin{aligned}\sqrt{n}\begin{pmatrix}\hat{\psi}-\psi_0\\ \hat{\lambda}-\hat{\lambda}_{\psi_0}\end{pmatrix} &= [\bar{i}(\theta_0)^{-1} - \bar{H}]\frac{1}{\sqrt{n}}\ell_\theta(\theta_n)\\ &\quad + \begin{pmatrix}1\\ -\bar{i}_{\lambda\lambda}(\theta_0)^{-1}\bar{i}_{\lambda\psi}(\theta_0)\end{pmatrix}\delta + O_p(n^{-1/2})\\ &= [\bar{i}(\theta_0)^{-1} - \bar{H}]\Big(\frac{1}{\sqrt{n}}\ell_\theta(\theta_n) - \Delta\Big) + O_p(n^{-1/2})\end{aligned}$$

where

$$\Delta = [\bar{i}(\theta_0)^{-1} - \bar{H}]^{-1}\begin{pmatrix}1\\ -\bar{i}_{\lambda\lambda}(\theta_0)^{-1}\bar{i}_{\lambda\psi}(\theta_0)\end{pmatrix}\delta.$$

Hence,

$$\begin{aligned}W = &\Big(\frac{1}{\sqrt{n}}\ell_\theta(\theta_n) - \Delta\Big)^T (\bar{i}(\theta_0)^{-1} - \bar{H})\bar{i}(\theta_0)(\bar{i}(\theta_0)^{-1} - \bar{H})\\ &\times \Big(\frac{1}{\sqrt{n}}\ell_\theta(\theta_n) - \Delta\Big) + O_p(n^{-1/2}).\end{aligned}$$

Under the distribution with parameter θ_n, $\bar{i}(\theta_n)^{-1/2}\ell_\theta(\theta_n)/\sqrt{n}$ is asymptotically distributed according to a multivariate normal distribution with mean vector 0 and identity covariance matrix; also $\ell_\theta(\theta_n)/\sqrt{n} - \Delta$ is asymptotically distributed according to a multivariate normal distribution with mean $-\Delta$ and covariance matrix $\bar{i}(\theta_0)^{-1}$.

It follows that, under θ_n, the asymptotic distribution of W is the distribution of $Z^T Z$ where Z has a multivariate normal distribution with mean

$$\mu \equiv -\bar{i}(\theta_0)^{1/2}[\bar{i}(\theta_0)^{-1} - \bar{H}]\Delta$$

and covariance matrix

$$\Sigma = \bar{i}(\theta_0)^{1/2}(\bar{i}(\theta_0)^{-1} - \bar{H})\bar{i}(\theta_0)(\bar{i}(\theta_0)^{-1} - \bar{H})\bar{i}(\theta_0)^{1/2}.$$

Since Σ is an idempotent matrix of rank q, it follows that W is asymptotically distributed according to a noncentral χ^2 distribution with noncentrality parameter

$$\begin{aligned}\mu^T\mu &= \delta^T \begin{pmatrix} 1 \\ -\bar{i}_{\lambda\lambda}(\theta_0)^{-1}\bar{i}_{\lambda\psi}(\theta_0) \end{pmatrix}^T i(\theta_0) \begin{pmatrix} 1 \\ -\bar{i}_{\lambda\lambda}(\theta_0)^{-1}\bar{i}_{\lambda\psi}(\theta_0) \end{pmatrix} \delta \\ &= \delta^T \bar{i}_\psi(\theta_0)\delta\end{aligned}$$

where

$$\bar{i}_\psi(\theta_0) = \bar{i}_{\psi\psi}(\theta_0) - \bar{i}_{\psi\lambda}(\theta_0)\bar{i}_{\lambda\lambda}(\theta_0)^{-1}\bar{i}_{\lambda\psi}(\theta_0)$$

denotes the limiting average partial information per observation.

Example 4.10 Exponential distribution
Let $Y_1, \ldots, Y_n$ denote independent exponential random variables with mean θ, as in Examples 4.4 and 4.7. The likelihood ratio statistic for testing $\theta = \theta_0$ is given by

$$W = 2n\left[\frac{\hat{\theta}}{\theta_0} - 1 - \log\left(\frac{\hat{\theta}}{\theta_0}\right)\right].$$

Under the null hypothesis, W is approximately distributed according to a χ^2 distribution with 1 degree of freedom; under an exponential distribution with mean $\theta = \theta_0 + \delta/\sqrt{n}$, W is distributed according to a noncentral χ^2 distribution with 1 degree of freedom and noncentrality parameter

$$\delta^2\bar{i}(\theta_0) = \delta^2{\theta_0}^{-2}.$$

Consider the case $\theta_0 = 1$ and $n = 25$ and suppose we want to approximate the probability that the null hypothesis will be rejected when $\theta = 3/2$ using a test with level 0.05. Then

$$\delta = \sqrt{n}(\theta - \theta_0) = 5/2$$

so that, under the alternative hypothesis $\theta = 3/2$, W is approximately distributed according to a noncentral χ^2 distribution with noncentrality parameter $25/4$. Since, under the null hypothesis, W has a χ^2 distribution with 1 degree of freedom, the critical value of the test is approximately 3.841; the probability that the noncentral χ^2 random variable described above exceeds this value is approximately 0.705. ∎

4.4 The score and Wald statistics

Consider the problem of testing a simple null hypothesis $\theta = \theta_0$ versus the alternative $\theta \neq \theta_0$. Two test statistics closely related to the likelihood ratio statistic are the Wald statistic,

$$W_w = (\hat{\theta} - \theta_0)^T i(\hat{\theta})(\hat{\theta} - \theta_0)$$

and the Rao score statistic

$$W_s = \ell_\theta(\theta_0)^T i(\theta_0)^{-1} \ell_\theta(\theta_0).$$

In this section we show that these statistics both have the same first-order asymptotic properties as the likelihood ratio statistic W.

First consider the properties under the null hypothesis. Consider an expansion of $2[\ell(\hat{\theta}) - \ell(\theta_0)]$ around $\theta_0 = \hat{\theta}$:

$$W = -2[\ell_\theta(\hat{\theta})(\theta_0 - \hat{\theta}) + \tfrac{1}{2}(\theta_0 - \hat{\theta})^T \ell_{\theta\theta}(\hat{\theta})(\theta_0 - \hat{\theta})^T + O_p(n^{-1/2})].$$

Since $\ell_\theta(\hat{\theta}) = 0$ and $-\ell_{\theta\theta}(\hat{\theta}) = i(\theta_0) + O_p(\sqrt{n}) = i(\hat{\theta}) + O_p(\sqrt{n})$, it follows that

$$W = W_w + O_p(n^{-1/2}). \tag{4.13}$$

Hence, under the null hypothesis, W_w is asymptotically distributed according to a χ^2 distribution with d degrees of freedom.

The analogous result for W_s follows immediately from (4.6); hence W_s is also asymptotically distributed according to a χ^2 distribution with d degrees of freedom.

The same results hold under a local alternative of the form $\theta = \theta_0 + \delta/\sqrt{n}$. An expansion of $2[\ell(\hat{\theta}) - \ell(\theta_0)]$ around $\theta_0 = \hat{\theta}$ yields

$$W = -2\Big[\tfrac{1}{2}(\theta_0 - \hat{\theta})^T \ell_{\theta\theta}(\hat{\theta})(\theta_0 - \hat{\theta}) + O_p(n^{-1/2})\Big]$$

and since under the distribution with parameter θ_n,

$$\ell_{\theta\theta}(\hat{\theta}) = i(\theta_n) + O_p(\sqrt{n}) = i(\theta_0) + O_p(\sqrt{n}) = i(\hat{\theta}) + O_p(\sqrt{n}),$$

(4.13) still holds. Hence, the test based on W_w has the same local power function as the test based on W. The result for W_s follows along similar lines.

The Wald statistic does have one important disadvantage as compared to either W or W_s; it is not invariant under reparameterization. That is, the value of Wald statistic for a given set of data depends on the parameterization used for the model. This is clearly undesirable.

It is also worth noting that statistics analogous to the Wald and score statistics may be constructed using the observed information in place of the expected information in their definition.

Now consider the case in which the parameter θ may be written (ψ, λ) and the hypothesis to be tested is of the form H_0: $\psi = \psi_0$ versus the alternative H_1: $\psi \neq \psi_0$. The Wald statistic statistic for this case is given by

$$W_w = (\hat{\psi} - \psi_0)^T i_\psi(\hat{\theta})(\hat{\psi} - \psi_0)$$

and the score statistic is given by

$$W_s = \ell_\psi(\hat{\theta}_{\psi_0})^T i_\psi(\hat{\theta}_{\psi_0})^{-1} \ell_\psi(\hat{\theta}_{\psi_0});$$

here $i_\psi(\theta)$ denotes the partial information for ψ.

Using the expansion (4.9) for $\sqrt{n}(\hat{\theta} - \theta)$ we have that

$$\sqrt{n}(\hat{\psi} - \psi_0) = \sqrt{n}\tilde{i}_{\psi\psi}(\theta_0)\ell_\psi(\theta_0) + \sqrt{n}\tilde{i}_{\psi\lambda}(\theta_0)\ell_\lambda(\theta_0) + O_p(n^{-1/2})$$

where we are writing

$$i(\theta)^{-1} = \begin{pmatrix} \tilde{i}_{\psi\psi}(\theta) & \tilde{i}_{\psi\lambda}(\theta) \\ \tilde{i}_{\lambda\psi}(\theta) & \tilde{i}_{\lambda\lambda}(\theta) \end{pmatrix};$$

recall that $i_\psi(\theta) = \tilde{i}_{\psi\psi}(\theta)^{-1}$. This result, together with the fact that $\tilde{i}_{\psi\psi}(\hat{\theta}) = \tilde{i}_{\psi\psi}(\theta_0) + O_p(\sqrt{n})$, shows that

$$W_w = \ell_\theta(\theta_0)^T Q_w \ell_\theta(\theta_0) + O_p(n^{-1/2})$$

where

$$\begin{aligned} Q_w &= (\tilde{i}_{\psi\psi}(\theta_0) \quad \tilde{i}_{\psi\lambda}(\theta_0))^T \tilde{i}_{\psi\psi}(\theta_0)^{-1} (\tilde{i}_{\psi\psi}(\theta_0) \quad \tilde{i}_{\psi\lambda}(\theta_0)) \\ &= \begin{pmatrix} \tilde{i}_{\psi\psi}(\theta_0) & \tilde{i}_{\psi\lambda}(\theta_0) \\ \tilde{i}_{\lambda\psi}(\theta_0) & \tilde{i}_{\lambda\psi}(\theta_0)\tilde{i}_{\psi\psi}(\theta_0)^{-1}\tilde{i}_{\psi\lambda}(\theta_0) \end{pmatrix}. \end{aligned}$$

Note that W has a similar expansion with Q replaced by

$$(i(\theta_0)^{-1} - H)i(\theta_0)(i(\theta_0)^{-1} - H) = i(\theta_0)^{-1} - H(\theta_0).$$

The fact that

$$i(\theta_0) = \begin{pmatrix} \tilde{i}_{\psi\psi}(\theta_0) & \tilde{i}_{\psi\lambda}(\theta_0) \\ \tilde{i}_{\lambda\psi}(\theta_0) & \tilde{i}_{\lambda\lambda}(\theta_0) \end{pmatrix}^{-1}$$

and the formula for the inverse of a partitioned matrix shows that $Q_w = i(\theta_0)^{-1} - H$ so that $W = W_w + O_p(n^{-1/2})$.

A consequence of this result is that, for the case in which ψ is a scalar, R, the signed square root of W, may be written

$$R = \sqrt{i_\psi(\hat{\theta})}(\hat{\psi} - \psi_0) + O_p(n^{-1/2}).$$

It follows that R is asymptotically distributed according to a standard normal distribution.

The result for W_s follows along similar lines. The function $\ell_\psi(\hat{\theta}_{\psi_0}) \equiv \ell_\psi(\psi_0, \hat{\lambda}_{\psi_0})$ can be expanded

$$\ell_\psi(\psi_0, \hat{\lambda}_{\psi_0}) = \ell_\psi(\theta_0) + \ell_{\psi\lambda}(\theta_0)(\hat{\lambda}_{\psi_0} - \lambda_0) + \cdots$$

so that

$$\frac{1}{\sqrt{n}}\ell_\psi(\psi_0, \hat{\lambda}_{\psi_0}) = \frac{1}{\sqrt{n}}\ell_\psi(\theta_0) - \frac{1}{\sqrt{n}}i_{\psi\lambda}(\theta_0)(\hat{\lambda}_{\psi_0} - \lambda_0) + O_p(n^{-1/2}).$$

Substituting the expression for $\hat{\lambda}_{\psi_0} - \lambda_0$ given in (4.10) it follows that

$$\frac{1}{\sqrt{n}}\ell_\psi(\psi_0, \hat{\lambda}_{\psi_0}) = \frac{1}{\sqrt{n}}\ell_\psi(\theta_0) - i_{\psi\lambda}(\theta_0)i_{\lambda\lambda}(\theta_0)^{-1}\frac{1}{\sqrt{n}}\ell_\lambda(\theta_0) + O_p(n^{-1/2})$$

and, hence, that

$$W_s = \ell_\theta(\theta_0)^T Q_s \ell_\theta(\theta_0) + O_p(n^{-1/2})$$

where

$$Q_s = \begin{pmatrix} \tilde{i}_{\psi\psi}(\theta_0) & -\tilde{i}_{\psi\psi}(\theta_0)i_{\psi\lambda}(\theta_0)i_{\lambda\lambda}(\theta_0) \\ -i_{\lambda\lambda}(\theta_0)^{-1}i_{\lambda\psi}(\theta_0)\tilde{i}_{\psi\psi}(\theta_0) & i_{\lambda\lambda}(\theta_0)^{-1}i_{\lambda\psi}(\theta_0)\tilde{i}_{\psi\psi}(\theta_0)i_{\psi\lambda}(\theta_0)i_{\lambda\lambda}(\theta_0)^{-1} \end{pmatrix}.$$

It straightforward to show that $Q_s = i(\theta_0)^{-1} - H$ so that $W = W_s + O_p(n^{-1/2})$.

Example 4.11 Exponential distribution
Let $Y_1, \ldots, Y_n$ denote independent exponential random variables each with mean θ and consider a test of the hypothesis $\theta = \theta_0$ versus the alternative $\theta \neq \theta_0$. Here $\hat{\theta} = \bar{y}$ and $i(\theta) = n/\theta^2$. The Wald statistic is given by

$$W_w = \frac{n(\hat{\theta} - \theta_0)^2}{\hat{\theta}^2} = n\left(\frac{\theta_0}{\hat{\theta}} - 1\right)^2.$$

The score statistic is given by

$$W_s = n\left(\frac{\hat{\theta}}{\theta_0} - 1\right)^2.$$

Recall that the likelihood ratio test statistic for this model is given by

$$W = 2n[\hat{\theta}/\theta_0 - 1 - \log(\hat{\theta}/\theta_0)].$$ ∎

Example 4.12 Poisson regression
Consider the Poisson regression model considered in Example 4.8 and consider a test of the hypothesis $\beta = 0$ versus the alternative $\beta \neq 0$. The expected information

matrix is given by

$$i(\theta) = \begin{pmatrix} \sum \lambda_j & \sum x_j \lambda_j \\ \sum x_j \lambda_j & \sum x_j^2 \lambda_j \end{pmatrix}$$

where $\log \lambda_j = \alpha + \beta x_j$. It follows that the partial information for β is given by

$$i_\beta(\theta) = \sum x_j^2 \lambda_j - \frac{(\sum x_j \lambda_j)^2}{\sum \lambda_j};$$

when evaluated at the maximum likelihood estimate of θ for fixed $\beta = 0$, this becomes

$$\bar{y} \sum (x_j - \bar{x})^2.$$

The Wald statistic is given by

$$W_w = i_\beta(\hat{\theta})\hat{\beta}^2 = \exp\{\hat{\alpha}\} \sum (x_j - \hat{x})^2 \exp\{\hat{\beta} x_j\} \hat{\beta}^2$$

where

$$\hat{x} = \frac{\sum x_j \exp\{\hat{\beta} x_j\}}{\sum \exp\{\hat{\beta} x_j\}}.$$

The score function for β is given by

$$\ell_\beta(\theta) = \sum x_j (y_j - \exp\{\alpha + \beta x_j\});$$

it follows that the score statistic is given by

$$W_s = \frac{[\sum x_j (y_j - \bar{y})]^2}{\bar{y} \sum (x_j - \bar{x})^2}.$$

Consider the data described in Example 4.8. For these data:

$$\bar{y} = 71.77; \qquad \sum x_j (y_j - \bar{y}) = 509.85; \qquad \sum (x_j - \bar{x})^2 = 43.08.$$

It follows that

$$W_w = 82.1; \qquad W_s = 84.1.$$

Recall that the likelihood ratio statistic for these data is $W = 78.6$. ∎

4.5 Confidence regions

As discussed in Section 3.7, any of the test statistics considered in Sections 4.3 and 4.4 may be used as the basis of a confidence region. First consider the case in which we wish to construct a confidence region for the entire parameter θ. Since

$$\Pr\{W(\theta) \leq \chi_d^2(\alpha)\} = 1 - \alpha + O(n^{-1/2})$$

the confidence region of the form

$$\{\theta \in \Theta \colon W(\theta) \leq \chi_d^2(\alpha)\}$$

has coverage probability approximately equal to $1 - \alpha$ for all θ. This confidence region consists of those values of θ for which $L(\theta)$ is larger than a given value.

Alternatively, an approximate confidence region for θ may be based on either the Wald or score statistics. For instance, the set of θ values of the form

$$\{\theta \in \Theta: (\hat{\theta} - \theta)^T i(\hat{\theta})(\hat{\theta} - \theta) \leq \chi^2_d(\alpha)\}$$

has coverage probability approximately equal to $1 - \alpha$; one convenient feature of confidence regions based on the Wald statistic is that they are elliptically shaped. For the case of a scalar parameter, a confidence interval based on the Wald statistic is equivalent to basing a confidence interval on $\hat{\theta}$ together with its standard error estimated using the expected information evaluated at $\hat{\theta}$. As noted previously, the observed information could also be used in this context.

Example 4.13 Exponential distribution
Let $Y_1, \ldots, Y_n$ denote independent exponential random variables each with mean θ. A confidence region for θ based on the likelihood ratio statistic is given by

$$\{\theta \in \Re^+: 2n[\hat{\theta}/\theta - 1 - \log(\hat{\theta}/\theta)] \leq \chi^2_1(\alpha)\}.$$

A confidence region for θ based on the Wald statistic is given by

$$\{\theta \in \Re^+: n(\theta/\hat{\theta} - 1)^2 \leq \chi^2_1(\alpha)\};$$

note that this region is actually an interval of θ values. A confidence region for θ based on the score statistic is given by

$$\{\theta \in \Re^+: n(\hat{\theta}/\theta - 1)^2 \leq \chi^2_1(\alpha)\};$$

this region is also an interval.

Consider the data considered in Example 4.4. For these data, $n = 20$ and $\hat{\theta} = 575.5$. Hence, the 95% likelihood ratio confidence region consists of all values of θ satisfying

$$\frac{575.5}{\theta} + \log\theta \leq 7.451;$$

this leads to the confidence interval (382.7, 922.7) for θ.

The 95% confidence region based on the Wald statistic consists of all values of θ satisfying

$$\left(\frac{\theta}{575.5} - 1\right)^2 \leq 0.1921;$$

this leads to the confidence interval (323.4, 827.7) for θ.

The 95% confidence region based on the score statistic consists of all values of θ satisfying

$$\left(\frac{575.5}{\theta} - 1\right)^2 \leq 0.1921;$$

this leads to the confidence interval (401.7, 1024) for θ. ∎

The same type of procedures are available for constructing confidence regions for a component of θ. Suppose θ may be written $\theta = (\psi, \lambda)$ where ψ is the parameter of interest. A confidence region for ψ may be based on the likelihood ratio statistic for testing $\psi = \psi_0$ versus $\psi \neq \psi_0$; this region has the form

$$\{\psi \in \Psi\colon 2[\ell(\hat{\psi}, \hat{\lambda}) - \ell(\psi, \hat{\lambda}_\psi)] \leq \chi^2_q(\alpha)\}.$$

Here q denotes the dimension of ψ.

An approximate confidence region based on the Wald statistic is given by

$$\{\psi \in \Psi\colon (\hat{\psi} - \psi)^T i_\psi(\hat{\theta})(\hat{\psi} - \psi) \leq \chi^2_q(\alpha)\}$$

and an approximate confidence region based on the score statistic is given by

$$\{\psi \in \Psi\colon \ell_\psi(\hat{\theta}_\psi)^T i_\psi(\hat{\theta}_\psi)^{-1} \ell_\psi(\hat{\theta}_\psi) \leq \chi^2_q(\alpha)\}.$$

Recall that $i_\psi(\theta)$ denotes the partial information for ψ.

Example 4.14 Weibull distribution
Let $Y_1, \ldots, Y_n$ denote independent random variables each distributed according to the density function

$$\psi\lambda(\lambda y)^{\psi-1} \exp\{-(\lambda y)^\psi\}, \quad y > 0.$$

This is a Weibull distribution; see Examples 3.11 and 3.14.

In Example 3.11 it is shown that

$$i(\theta) = n \begin{pmatrix} 1/\psi^2[\pi^2/6 + \gamma^2 - 2\gamma] & 1/\lambda(1-\gamma) \\ 1/\lambda(1-\gamma) & \psi^2/\lambda^2 \end{pmatrix}$$

where γ denotes Euler's constant. It follows that

$$i_\psi(\theta) = \frac{n}{\psi^2}\Big(\frac{\pi^2}{6} - 1\Big).$$

The maximum likelihood estimate of λ for fixed ψ is given by

$$\hat{\lambda}_\psi = \left(\frac{n}{\sum y_j^\psi}\right)^{1/\psi}.$$

It follows that the likelihood ratio confidence region with coverage probability $1-\alpha$ consists of all values of ψ satisfying

$$2n\Big[\log(\hat{\psi}/\psi) + \log\Big(\sum y_j^\psi \Big/ \sum y_j^{\hat{\psi}}\Big) + (\hat{\psi} - \psi)\sum \log y_j/n\Big] \leq \chi^2_1(\alpha).$$

The confidence region for ψ based on the Wald statistic consists of all values of ψ satisfying

$$\Big(\frac{\psi}{\hat{\psi}} - 1\Big)^2 \leq \frac{\chi^2_1(\alpha)}{n(\pi^2/6 - 1)}.$$

The confidence region for ψ based on the score statistic consists of all values of ψ satisfying

$$\left(\frac{\sum y_j^\psi \log(y_j^\psi)}{\sum y_j^\psi} - \frac{\sum \log(y_j^\psi)}{n} - 1\right)^2 \leq \frac{\chi_1^2(\alpha)(\pi^2/6 - 1)}{n}.$$

Consider the data considered in Examples 4.4 and 4.13 where the data as modeled as a sample from an exponential distribution. As a check on the appropriateness of exponential distribution, we can model these data as a sample from a Weibull distribution and estimate the value of ψ; recall that $\psi = 1$ corresponds to an exponential distribution. For these data, $\hat{\psi} = 0.716$. The 95% likelihood ratio confidence interval for ψ is (0.478, 1.013), The 95% confidence interval for ψ based on the Wald statistic is (0.325, 1.107), and the 95% confidence interval for ψ based on the score statistic is (0.528, 0.894). Hence, the conclusion of the analysis depends on which method is used. ∎

4.6 The profile likelihood function

Suppose that θ may be written $\theta = (\psi, \lambda)$ where ψ is the parameter of interest and λ is a nuisance parameter; both ψ and λ may be vectors. Many of the procedures for inference regarding ψ discussed in this chapter may be described in terms of the *profile likelihood function*, given by $L_P(\psi) = L(\hat{\theta}_\psi)$ where $\hat{\theta}_\psi$ denotes the maximum likelihood estimator of θ for fixed ψ; hence, $\hat{\theta}_\psi$ may be written $(\psi, \hat{\lambda}_\psi)$ where $\hat{\lambda}_\psi$ denotes the maximum likelihood estimate of λ treating ψ as fixed. For instance, the maximum likelihood estimator of ψ may be described as the value of ψ that maximizes $L_P(\psi)$ and the likelihood ratio statistic for testing $\psi = \psi_0$ versus $\psi \neq \psi_0$ may be written

$$W = 2[\ell_P(\hat{\psi}) - \ell_P(\psi_0)]$$

where $\ell_P(\psi) = \log L_P(\psi)$ is the profile log-likelihood function.

The profile log-likelihood function has many of the same first-order properties as a genuine log-likelihood function for ψ. First consider the properties of the estimator $\hat{\lambda}_\psi$; to distinguish ψ, the value at which $\hat{\lambda}_\psi$ is computed, from the true value of ψ we will use $\theta_0 = (\psi_0, \lambda_0)$ to denote the true parameter value. Since $\hat{\lambda}_\psi$ is the maximum likelihood estimator of λ with ψ fixed at an 'incorrect' value, in general, $\hat{\lambda}_\psi$ is not a consistent estimator of λ_0. The estimator $\hat{\lambda}_\psi$ maximizes the log-likelihood function $\ell(\psi, \lambda)$ with respect to λ, holding ψ fixed. Since, as $n \to \infty$,

$$\frac{1}{n}\ell(\psi, \lambda) - \frac{1}{n}\mathrm{E}[\ell(\psi, \lambda); \theta_0] \xrightarrow{p} 0,$$

following the proof of consistency of the maximum likelihood estimator discussed in Section 4.2.1, it can be shown that $\hat{\lambda}_\psi - \lambda_\psi \xrightarrow{p} 0$ as $n \to \infty$, where λ_ψ is the value of λ that maximizes $n^{-1}\mathrm{E}\{\ell(\psi, \lambda); \theta_0\}$ holding ψ fixed. Hence, λ_ψ is nonrandom; for the case in which $\ell(\psi, \lambda)$ is based on n independent, identically distributed random variables, λ_ψ does not depend on n. Using an argument similar to the one used in Section 4.2.2 to obtain the asymptotic distribution of $\hat{\theta}$, it can be shown that $\hat{\lambda}_\psi = \lambda_\psi + O_p(n^{-1/2})$.

Example 4.15 Normal distribution

Let $Y_1, \ldots, Y_n$ denote independent random variables each distributed according to a normal distribution with mean ψ and variance λ where $\psi \in \Re$ and $\lambda > 0$; then

$$\ell(\psi, \lambda) = -\frac{1}{2\lambda} \sum (y_j - \psi)^2 - \frac{n}{2} \log \lambda.$$

It is straightforward to show that

$$\hat{\lambda}_\psi = \frac{1}{n} \sum (y_j - \psi)^2$$

and, hence, the profile log-likelihood function is given by

$$\ell_p(\psi) = -\frac{n}{2} \log \sum (y_j - \psi)^2.$$

For this model,

$$\frac{1}{n} \mathrm{E}\{\ell(\psi, \lambda); \theta_0\} = -\frac{1}{2\lambda}[\lambda_0 + (\psi - \psi_0)^2] - \frac{1}{2} \log \lambda.$$

It follows that $\lambda_\psi = \lambda_0 + (\psi - \psi_0)^2$. It is easy to confirm by direct calculation that $\hat{\lambda}_\psi = \lambda_\psi + O_p(n^{-1/2})$. ∎

The profile log-likelihood function may be viewed as an 'estimate' of the genuine log-likelihood function $\tilde{\ell}(\psi) \equiv \ell(\psi, \lambda_\psi)$; here $\tilde{\ell}(\psi)$ is referred to a genuine log-likelihood function since it corresponds to the log-likelihood function for an actual model for the data. Since

$$\begin{aligned} \ell_p(\psi) = \ell(\psi, \hat{\lambda}_\psi) &= \ell(\psi, \lambda_\psi) + \ell_\lambda(\psi, \lambda_\psi)^T (\hat{\lambda}_\psi - \lambda_\psi) \\ &\quad + \tfrac{1}{2} (\hat{\lambda}_\psi - \lambda_\psi)^T \ell_{\lambda\lambda}(\psi, \lambda_\psi)(\hat{\lambda}_\psi - \lambda_\psi) + \cdots \end{aligned}$$

and $\hat{\lambda}_\psi = \lambda_\psi + O_p(n^{-1/2})$, it follows that $\ell_p(\psi) = \ell(\psi, \lambda_\psi) + O_p(1)$.

The log-likelihood derivatives satisfy similar conditions. For instance,

$$\begin{aligned} \ell'_p(\psi_0) &\equiv \left. \frac{\partial \ell_p(\psi)}{\partial \psi} \right|_{\psi=\psi_0} \\ &= \left. \frac{\partial \tilde{\ell}(\psi)}{\partial \psi} \right|_{\psi=\psi_0} + [\ell_{\lambda\psi}(\psi_0, \lambda_0) + \ell_{\lambda\lambda}(\psi_0, \lambda_0)\lambda'_0]^T (\hat{\lambda}_0 - \lambda_0) \\ &\quad + \ell_\lambda(\psi_0, \lambda_0)^T (\hat{\lambda}'_0 - \lambda'_0) + O_p(1). \end{aligned} \tag{4.14}$$

Here

$$\hat{\lambda}_0' = \frac{\partial \hat{\lambda}_\psi}{\partial \psi}\Big|_{\psi=\psi_0}$$

and

$$\lambda_0' = \frac{\partial \lambda_\psi}{\partial \psi}\Big|_{\psi=\psi_0};$$

it is straightforward to show that $\hat{\lambda}_0' = \lambda_0' + O_p(n^{-1/2})$. The quantity λ_0' may be expressed in terms of the expected information matrix $i(\theta_0)$ by noting that λ_ψ must satisfy

$$\mathrm{E}[\ell_\lambda(\psi, \lambda_\psi); \theta_0] = 0 \quad \text{for all } \psi$$

and hence

$$\mathrm{E}[\ell_{\psi\lambda}(\theta_0) + \ell_{\lambda\lambda}(\theta_0)\lambda_0'; \theta_0] = 0$$

so that

$$\lambda_0' = -i_{\lambda\lambda}(\theta_0)^{-1} i_{\lambda\psi}(\theta_0).$$

Hence, the term

$$\ell_{\psi\lambda}(\psi_0, \lambda_0) + \ell_{\lambda\lambda}(\psi_0, \lambda_0)\lambda_0'$$

appearing in (4.14) has mean 0 and is of order $O_p(\sqrt{n})$. It follows that

$$\ell_p'(\psi_0) = \frac{\partial \tilde{\ell}(\psi)}{\partial \psi}\Big|_{\psi=\psi_0} + O_p(1).$$

Using a similar argument, it may be shown that

$$\ell_p''(\psi_0) = \frac{\partial^2 \tilde{\ell}(\psi)}{\partial \psi^2}\Big|_{\psi=\psi_0} + O_p(\sqrt{n}).$$

These results suggest that the first-order properties of procedures based on $\ell_p(\psi)$ will be the same as the first-order properties of procedures based on $\tilde{\ell}(\psi)$. Although procedures based on $\tilde{\ell}(\psi)$ depend on the value of θ_0 and, hence, are not available in practice, the first-order asymptotic properties of those procedures are easily determined since $\tilde{\ell}(\psi)$ is a genuine log-likelihood function.

For instance, consider $\hat{\psi}$, the maximum likelihood estimator of ψ. Since $\hat{\psi}$ satisfies $\ell_p'(\hat{\psi}) = 0$ it is straightforward to show that, as in (4.3),

$$\sqrt{n}(\hat{\psi} - \psi_0) = \Big[-\frac{1}{n}\ell_p''(\psi_0) + o_p(1)\Big]^{-1} \frac{1}{\sqrt{n}}\ell_p'(\psi_0)$$

so that, using the results in this section,

$$\sqrt{n}(\hat{\psi} - \psi_0) = \Big[-\frac{1}{n}\tilde{\ell}''(\psi_0) \Big] \frac{1}{\sqrt{n}} \tilde{\ell}'(\psi_0) + o_p(1).$$

Furthermore,

$$\mathrm{E}[-\tilde{\ell}''(\psi_0); \theta_0] = i_\psi(\theta_0),$$

the partial expected information for ψ, leading to the conclusion that, in general, $\hat{\psi}$ is approximately normally distributed with mean ψ and variance $i_\psi(\theta)^{-1}$.

4.7 Nonregular models

The distributional results presented thus far in this chapter apply to the case in which the regularity conditions of Section 3.4 are satisfied. In this section we consider several examples to illustrate some of the possibilities when these regularity conditions are not satisfied. One example of this type has already been given, Example 4.3, in which the dimension of the parameter space increases with the sample size; models of that type are discussed further in Section 8.2.

Example 4.16 Laplace distribution
Let $Y_1, \ldots, Y_n$ denote independent random variables each distributed according to the distribution with density $\exp\{-|y - \theta|\}/2$, $-\infty < y < \infty$. Then

$$\ell(\theta) = -\sum |y_j - \theta|$$

so that $\ell_\theta(\theta) = -\sum \mathrm{sgn}(y_j - \theta)$ for $\theta \neq y_j$, $j = 1, \ldots, n$. Note that since $\mathrm{E}\{\mathrm{sgn}(Y_j - \theta); \theta\} = 0$ and $\mathrm{sgn}(y_j - \theta)^2 = 1$, it follows that $\mathrm{E}\{\ell_\theta(\theta)^2; \theta\} = n$ while $\ell_{\theta\theta}(\theta) = 0$ with probability 1. Hence, the regularity conditions of Section 3.4 are not satisfied and the large-sample theory described in this section does not necessarily hold.

Recall that $\hat{\theta}$ may be taken to be the sample median. The asymptotic distribution of $\surd(\hat{\theta} - \theta)$ may be obtained using general techniques developed for deriving the asymptotic distributions of sample quantiles. In general, for a sample from a continuous distribution with density $f(\cdot)$, if m denotes the sample median and m_0 denotes the median of the distribution, then

$$\sqrt{n}[2f(m_0)]^{-1}(m - m_0) \xrightarrow{\mathcal{D}} N(0, 1).$$

Hence,

$$\sqrt{n}(\hat{\theta} - \theta_0) \xrightarrow{\mathcal{D}} N(0, 1).$$

Note that, taking $i(\theta) = \mathrm{E}\{\ell_\theta(\theta)^2; \theta\} = n$, the result

$$\sqrt{i(\theta)}(\hat{\theta} - \theta) \xrightarrow{\mathcal{D}} N(0, 1)$$

does hold here, although the proof given in Section 3.2 does not apply.

Consider a test of the null hypothesis $\theta = 0$ versus the alternative $\theta \neq 0$. The likelihood ratio test statistic is given by

$$W = 2[\sum |Y_j| - \sum |Y_j - \hat{\theta}|].$$

The statistic W may be written

$$W = 4 \sum |Y_j|$$

where the sum is over all Y_j falling between 0 and $\hat{\theta}$. It can be shown that W is asymptotically distributed according to a χ_1^2 distribution, although again the proof given in Section 4.3 does not apply. ∎

Example 4.17 Change point in a sequence of normal random variables
Let $Y_1, \ldots, Y_n$ denote a sequence of independent normally distributed random variables such that $Y_1, \ldots, Y_\phi$ have mean μ and variance σ^2 while $Y_{\phi+1}, \ldots, Y_n$ have mean $\mu + \beta$ and variance σ^2. Here μ, β, and ϕ are all unknown parameters with $\mu \in \Re$, $\beta \neq 0$, and ϕ taking values in the set $\{1, \ldots, n\}$. For simplicity, we will assume that σ^2 is known and equal to 1; similar results hold for the case in which $\sigma^2 > 0$ is unknown. Hence, there is a *change point* ϕ such that the mean of all Y_j with $j > \phi$ is increased by β. Consider a test of the hypothesis that there is no change point, which may be written H_0: $\phi = n$ versus the alternative H_1: $\phi < n$.

Several types of nonregularity are exhibited here. The parameter ϕ takes only integer values and the parameter space for ϕ depends on the sample size n; also the value of ϕ under the null hypothesis is on the boundary of this parameter space and, under the null hypothesis, the parameter β is not identifiable. Hence, the asymptotic results given in this section do not necessarily apply.

The likelihood ratio test statistic is given by

$$W = \max_{1 \leq \phi \leq n-1} \left| \sum_{i=1}^{n} (Y_i - \bar{Y})^2 - \sum_{i=1}^{\phi} (Y_i - \bar{Y}_\phi)^2 - \sum_{i=\phi+1}^{n} (Y_i - \bar{Y}'_\phi)^2 \right|$$

where $\bar{Y}$ denotes the sample mean of $Y_1, \ldots, Y_n$, $\bar{Y}_\phi$ denotes the sample mean of $Y_1, \ldots, Y_\phi$, and $\bar{Y}'_\phi$ denotes the sample mean of $Y_{\phi+1}, \ldots, Y_n$. It is straightforward to show that

$$W = \max_{1 \leq \phi \leq n-1} \frac{|S_\phi/\sqrt{n} - (\phi/n)(S_n/\sqrt{n})|}{(\phi/n)(1 - \phi/n)}$$

where $S_t = Y_1 + \cdots + Y_t$.

Let $\{Z(t)\colon 0 \leq t < \infty\}$ denote a standard Brownian motion; then, under H_0, the distribution of $\{(S_t - t\mu)/\sqrt{n}\colon 1 \leq t \leq n\}$ is the same as the distribution of $\{Z(t/n)\colon 1 \leq t \leq n\}$. Hence, the asymptotic null distribution of W may be obtained from properties of the Brownian motion $Z(\cdot)$.

Let $\sqrt{W}$ denote the (unsigned) square root of W. Then it may be shown that for all $-\infty < t < \infty$,

$$\Pr\{\sqrt{2\log\log(n)}\sqrt{W} - 2\log\log(n) - \log\log\log(n)/2 \leq t\}$$
$$\to \exp\{-2\exp\{-t\}/\sqrt{\pi}\}$$

as $n \to \infty$. Hence, the critical value of a test with level 0.05 is given by $[3.09 + 2\log\log(n) + \log\log\log(n)/2]^2/[2\log\log(n)]$. For instance, for $n = 20$ the critical value is 12.9, for $n = 100$ it is 13.2, and for $n = 1000$ it is 13.7. These values may be compared with the critical value based on the usual χ_1^2 approximation which is 2.71. ∎

Example 4.18 Two-parameter exponential distribution
Let $Y_1, \ldots, Y_n$ denote independent observations each distributed according to the distribution with density function

$$p(y; \theta) = \lambda^{-1}\exp\{-(y - \phi)/\lambda\}, \quad y \geq \phi;$$

here $\theta = (\phi, \lambda)$ with $\phi \geq 0$ and $\lambda > 0$.

The nonregularity of this distribution arises from the fact that density function has a discontinuity with location depending on the parameter θ, specifically, ϕ. It follows that, for fixed y, $p(y; \theta)$ is not a continuous function of θ; hence, $\ell(\theta)$, given by

$$\ell(\theta) = -n\log\lambda - n(\bar{y} - \phi)/\lambda, \quad \phi \leq y_{(1)},$$

is not a continuous function of θ. This fact has several consequences. For instance, consider

$$\frac{\partial\ell(\theta)}{\partial\phi} = n/\lambda, \quad \phi < y_{(1)};$$

hence, $\mathrm{E}\{\ell_\theta(\theta); \theta\} \neq 0$.

The maximum likelihood estimator of θ is easily determined and is given by $\hat{\theta} = (\hat{\phi}, \hat{\lambda})$ where $\hat{\phi} = y_{(1)}$ and $\hat{\lambda} = \bar{y} - \hat{\phi}$. Note that

$$\Pr\{Y_{(1)} > y; \theta\} = \Pr\{Y_1 > y, \ldots, Y_n > y; \theta\} = \Pr\{Y_1 > y; \theta\}^n$$
$$= \exp\{-n(y - \phi)/\lambda\}.$$

It follows that $n(\hat{\phi} - \phi)$ has an exponential distribution with mean λ. Hence, not only is $\hat{\phi}$ not asymptotically normally distributed, its variance is of order $O(n^{-2})$ instead of the usual $O(n^{-1})$.

The properties of $\hat{\lambda}$ are similar to those of an estimator based on a regular model; this is not surprising since for fixed ϕ the model does satisfy the usual regularity conditions. Since $\hat{\phi} = O_p(n^{-1})$, $\hat{\lambda} = \bar{y} + O_p(n^{-1})$ and

$$\sqrt{n}(\hat{\lambda} - \lambda) \xrightarrow{\mathcal{D}} N(0, \lambda^2).$$

Hence, the asymptotic variance of $\hat{\lambda}$ is the same as if the parameter ϕ was known.

Consider a test of the null hypothesis $\phi = \phi_0$ versus the alternative $\phi \neq \phi_0$. The likelihood ratio test statistic is given by

$$W = 2n \log\left(1 - \frac{\hat{\phi} - \phi_0}{\bar{y} - \hat{\phi}}\right) = 2\frac{n(\hat{\phi} - \phi_0)}{\lambda} + O_p(n^{-1/2}).$$

Hence, W converges in distribution to an exponential random variable with mean 2. ∎

Example 4.19 Log-normal distribution

Let $Y_1, \ldots, Y_n$ denote independent random variables such that each Y_j has a log-normal distribution with density of the form

$$\frac{1}{\sqrt{(2\pi\sigma^2)}}(y - \phi)^{-1} \exp\left\{-\frac{1}{2\sigma^2}\sum[\log(y - \phi) - \mu]^2\right\}, \quad y > \phi$$

where $\sigma > 0$, ϕ, and μ are unknown parameters. Note that $\log(Y_j - \phi)$ is normally distributed with mean μ and standard deviation σ.

The log-likelihood function is given by

$$\ell(\theta) = -\sum \log(y_j - \phi) - \frac{1}{2\sigma^2}\sum[\log(y_j - \phi) - \mu]^2 - n\log\sigma,$$

$$\phi > y_{(1)}$$

where $y_{(j)}$ denotes the jth order statistic of the sample. For a fixed value of ϕ, the maximum likelihood estimates of μ and σ are given by

$$\hat{\mu}_\phi = \frac{1}{n}\sum \log(y_j - \phi)$$

and

$$\hat{\sigma}^2_\phi = \frac{1}{n}\sum[\log(y_j - \phi) - \hat{\mu}_\phi]^2.$$

Hence, the profile log-likelihood function for ϕ is given by

$$\ell_p(\phi) = -\sum \log(y_j - \phi) - \frac{n}{2}\log\frac{\sum[\log(y_j - \phi) - \hat{\mu}_\phi]^2}{n}.$$

Note that for ϕ close to $y_{(1)}$,

$$\sum \log(y_j - \phi)^2 = \sum \log(y_{(j)} - \phi)^2 \leq n\log(y_{(1)} - \phi)^2$$

since $\log(y_{(1)} - \phi)^2$ tends to ∞ as $\phi \to y_{(1)}$. Hence,

$$\log\frac{\sum[\log(y_j - \phi) - \hat{\mu}_\phi]^2}{n} \leq \log\frac{\sum[\log(y_j - \phi)]^2}{n} \leq \log[\log(y_{(1)} - \phi)]^2.$$

It follows that

$$\ell_p(\phi) \geq -\sum \log(y_j - \phi) - \frac{n}{2} \log[\log(y_{(1)} - \phi)]^2.$$

Let $x = -\log(y_{(1)} - \phi)$. Then

$$\begin{aligned}\ell_p(\phi) &\geq x - \frac{n}{2}\log(x^2) - \sum_{j=2}^{n} \log(y_{(j)} - \phi)\\ &= \log[\exp(x)/|x|^n] - \sum_{j=2}^{n} \log(y_{(j)} - \phi).\end{aligned}$$

Note that as $\phi \to y_{(1)}$, $x \to \infty$,

$$\sum_{j=2}^{n} \log(y_{(j)} - \phi) \to \sum_{j=2}^{n} \log(y_{(j)} - y_{(1)})$$

and

$$\frac{\exp(x)}{|x|^n} \to \infty.$$

It follows that

$$\lim_{\phi \to y_{(1)}} \ell_p(\phi) = \infty$$

for any values of $y_1, \ldots, y_n$.

Since $\hat{\mu}_\phi \to -\infty$ and $\hat{\sigma}_\phi \to \infty$ as $\phi \to y_{(1)}$, strictly speaking the maximum likelihood estimate of θ does not exist for any n and any values of $y_1, \ldots, y_n$. Hence, this model does not satisfy the assumption that, with probability approaching 1, a unique maximum likelihood estimate exists. ∎

4.8 Discussion and references

The first-order asymptotic theory of likelihood-based methods is studied in many texts on statistical theory. See, for example, Cox and Hinkley (1974, Chapter 9), Ferguson (1996, Part 4), Schervish (1997, Chapter 7) and Sen and Singer (1993, Chapter 5). An alternative approach to asymptotic theory that focuses on statistical decision theory is given by LeCam (1986).

The proof of the consistency of maximum likelihood estimates given in Section 4.2 is based on Wald (1949). See Bahadur (1958, 1971), Huber (1967), and Perlman (1970) for detailed discussion of the consistency of maximum likelihood estimates. Example 4.3 is due to Neyman and Scott (1948).

The proof of the asymptotic normality of maximum likelihood estimates given in Section 4.2 is based on Cramér (1946, Section 33) and Lehmann (1983, Chapter 6); alternative approaches are used by LeCam (1970) and Ibragimov and

Has'minskii (1981). A general result on the asymptotic normality of the maximum likelihood estimate, applicable to stochastic process models, is given by Sweeting (1980); see also Sweeting (1992).

The result that maximum likelihood estimates have the minimum possible asymptotic variance has been studied by many authors; see LeCam (1953), Kalianpur and Rao (1955), Bahadur (1964), and Wong (1992). The differences between using the observed and expected information for estimating the standard error of the maximum likelihood estimate are considered in Efron and Hinkley (1978); this issue will be discussed further in Chapter 6. The asymptotic distribution of the likelihood ratio test statistic is considered by Wilks (1938) and Chernoff (1954). The Wald statistic is due to Wald (1943). The score statistic is due to Rao (1947); see also Bartlett (1953a; 1953b).

The function λ_ψ discussed in Section 4.6 is sometimes called a *least favorable curve* in the parameter space and the tangent vector at the true parameter value, λ_0', is called the *least favorable direction*. This terminology is based on the fact that estimation of ψ based on the likelihood function $\ell(\psi, \lambda_\psi)$ is as difficult, asymptotically, as estimation of ψ based on the actual log-likelihood function, $\ell(\psi, \lambda)$ with λ unknown. The family of distributions with parameter (ψ, λ_ψ) is known as Stein's least favorable family and is due to Stein (1956). The least favorable family and related concepts have been applied in many different areas of statistics including bootstrap confidence intervals (DiCiccio and Efron, 1992), empirical likelihood (DiCiccio, Hall and Romano, 1989), and semiparametric estimation (Severini and Wong, 1992).

The asymptotic theory of sample quantiles is discussed in Ferguson (1996) and Rao (1973). The asymptotic results for the likelihood ratio statistic in Example 4.16 are based on Babu and Rao (1992). Example 4.17 is based on Hinkley (1970) and Yao and Davis (1986). Example 4.19 is based on Hill (1963). Other nonregular models are discussed in Barndorff-Nielsen and Cox (1994, Section 3.8).

4.9 Exercises

4.1 Let $Y_1, \ldots, Y_n$ denote the event times of a time-dependent Poisson process with rate function θt, $\theta > 0$ (see Example 3.7). Show that $\hat{\theta}$ is a consistent estimate of θ.

4.2 Let $Y_1, \ldots, Y_n$ denote independent, identically distributed random variables distributed according to a density depending on a scalar parameter θ. Let $\ell(\theta)$ denote the log-likelihood function and define an estimate T as follows. Let $\tilde{\theta}$ denote a preliminary estimate of θ satisfying $\tilde{\theta} - \theta = O_p(n^{-1/2})$. Update $\tilde{\theta}$ to T using one iteration of Newton's method, i.e.,

$$T = \tilde{\theta} - \left[\frac{\partial^2 \ell}{\partial \theta^2}(\tilde{\theta})\right]^{-1} \frac{\partial \ell}{\partial \theta}(\tilde{\theta}).$$

Show that T and $\hat{\theta}$ are asymptotically equivalent in the sense that $T = \hat{\theta} + O_p(n^{-1})$.

4.3 Let Y_j, $j = 1, \ldots, n$, denote independent observations each distributed according to a density

$$\left(\frac{\lambda}{2\pi y^3}\right)^{1/2} \exp\left\{-\frac{1}{2}\left[\phi y + \frac{\lambda}{y} - 2\surd(\phi\lambda)\right]\right\}, \quad y > 0$$

where $\theta = (\lambda, \phi) \in (0, \infty) \times (0, \infty)$. Find the asymptotic covariance matrix of $\hat{\theta}$.

4.4 Let $Y_1, \ldots, Y_n$ denote independent random variables such that Y_j has density function

$$\lambda_j \exp\{-\lambda_j y\}, \quad y > 0$$

where $\log \lambda_j = \alpha + \beta x_j$ and $x_1, \ldots, x_n$ are known constants and α and β are unknown parameters each taking values in $\Re$. Find the large-sample approximation to the distribution of $(\hat{\alpha}, \hat{\beta})$, the maximum likelihood estimate of (α, β).

4.5 Let $Y_1, \ldots, Y_n$ denote independent identically distributed random variables each distributed according to a density depending on a scalar parameter $\theta \in \Theta$. Let $\pi(\cdot)$ denote a probability density function on Θ. Define

$$\tilde{\theta} = \frac{\int_\Theta \theta\pi(\theta)L(\theta)\,d\theta}{\int_\Theta \pi(\theta)L(\theta)\,d\theta}.$$

Using Laplace's method for approximating the integrals in the expression for $\tilde{\theta}$, show that

$$\tilde{\theta} - \hat{\theta} = O_p(n^{-1}).$$

4.6 Let $Y_1, \ldots, Y_n$ denote independent identically distributed random variables each distributed according to a density depending on a scalar parameter $\theta \in \Theta$. Suppose that n is even; let $\hat{\theta}_1$ denote the maximum likelihood estimate of θ based on the first half of the data and let $\hat{\theta}_2$ denote the maximum likelihood estimate of θ based on the second half of the data. Let $T = (\hat{\theta}_1 + \hat{\theta}_2)/2$. Investigate the relationship between T and $\hat{\theta}$, the maximum likelihood estimate based on the entire set of data.

4.7 Let $Y_1, \ldots, Y_n$ denote independent identically distributed random variables each distributed according to a density depending on a scalar parameter $\theta \in \Theta$. Let $\ell^{(j)}(\theta)$ denote the log-likelihood function based on Y_j alone. Let

$$\hat{\sigma}_s^2 = \hat{\jmath}^{-2} \sum \ell_\theta^{(j)}(\hat{\theta})^2$$

here $\hat{\jmath}$ denotes the observed information evaluated at $\hat{\theta}$; $\hat{\sigma}_s^2$ is sometimes called the *sandwich estimate* of the variance of $\hat{\theta}$. Show that $\hat{\sigma}_s^2$ is a consistent estimate of the asymptotic variance of $\hat{\theta}$.
Now suppose that the assumed model does not hold; that is, the Y_j are independent identically distributed random variables each distributed according

to a density $p(y)$, but $p(y)$ is not in the parametric family indexed by θ. Assume that $E[\ell_\theta(\theta)] = 0$, where the expectation is with respect to the true distribution of the data and that (4.3) still holds, where, in this case, θ is defined as the element of Θ that maximizes $E[\ell(\theta)]$. Show that $\hat{\sigma}_s^2$ is a consistent estimate of the asymptotic variance of $\hat{\theta}$.

4.8 Let $Y_1, \ldots, Y_n$ denote random variables of the form

$$Y_j = \theta + \epsilon_j, \quad j = 1, \ldots, n$$

where $\epsilon_1, \ldots, \epsilon_n$ denote independent identically distributed random variables with mean 0, finite variance and a known distribution. Suppose that θ is estimated under the assumption that the ϵ_j are normally distributed. Find the sandwich estimate of the asymptotic variance of $\hat{\theta}$ and show that it is a consistent estimate of the asymptotic variance of $\hat{\theta}$ under any distribution for the ϵ_j.

4.9 Let X_1, X_2 denote independent random variables such that X_j has a binomial distribution with parameters n_j and θ_j, $j = 1, 2$ where n_j is a positive integer and $0 < \theta_j < 1$. Find the likelihood ratio test statistic for testing

$$H_o: \ \theta_1 = \theta_2 \quad \text{vs.} \quad H_1: \ \theta_1 \neq \theta_2.$$

4.10 Let $X_1, \ldots, X_n; Y_1, \ldots, Y_m$ denote independent random variables such that the X_j each have a Poisson distribution with mean λ_1 and the Y_j each have an exponential distribution with mean λ_2^{-1}. Suppose we are interested in testing

$$H_o\colon \lambda_1 = \lambda_2 \quad \text{vs.} \quad H_1\colon \lambda_1 \neq \lambda_2.$$

Find the likelihood ratio test statistic, W, for these hypotheses and give its asymptotic distribution.

4.11 Let $Y_1, \ldots, Y_n$ denote independent identically distributed random variables each distributed according to a Poisson distribution with mean θ. Consider the problem of constructing a confidence region for θ.

(a) Calculate W, W_w and W_s and construct 95% confidence regions for θ based on the asymptotic distribution of each of these statistics.

(b) Now suppose that the model is parameterized by η, the natural parameter of the exponential family. Repeat part (a) for η. Transform the resulting confidence regions into regions for θ and compare the results to the results obtained in part (a).

4.12 Consider a model parameterized by $\theta = (\psi, \lambda)$ where ψ is real valued, let $\ell(\psi, \lambda)$ denote the log-likelihood function and let $j_\psi(\theta)$ denote the partial observed information for ψ. Show that

$$-\ell_p''(\hat{\psi})^{-1} = j_\psi(\hat{\theta}).$$

4.13 Consider a model parameterized by $\theta = (\psi, \lambda)$ where ψ is the parameter of interest. Suppose that $\hat{\lambda}_\psi$ does not depend on ψ so that $\hat{\lambda}_\psi = \hat{\lambda}$. Does it necessarily follow that ψ and λ are orthogonal? Why or why not?

4.14 Consider a full-rank exponential family distribution with log-likelihood function

$$\ell(\psi, \lambda) = s_1(y)^T \psi + s_2(y)^T c(\psi, \lambda) - k(\psi, \lambda)$$

where λ is the complementary mean parameter; see Example 3.13. Show that $\hat{\lambda}_\psi$ does not depend on ψ.

4.15 Let $Y_1, \ldots, Y_n$ denote independent identically distributed random variables each uniformly distributed on the interval $(0, \theta)$ where $\theta > 0$. Consider a test of the hypothesis $\theta = 1$ v versus the alternative $\theta \neq 1$.

Find the distribution of the likelihood ratio test statistic W under the null hypothesis and compare this distribution to the usual chi-square approximation.

5
Higher-order asymptotic theory

5.1 Introduction

In this chapter we consider the problem of constructing higher-order asymptotic approximations to the distribution of the maximum likelihood estimate, the likelihood ratio test statistic, and related quantities. Two basic methods are considered: Edgeworth expansions and saddlepoint approximations. Edgeworth expansions are discussed in Sections 5.3–5.5 and saddlepoint approximations are discussed in Section 5.6; approximations for the conditional distribution of likelihood-based quantities given an ancillary statistic will be considered in Chapter 6.

Edgeworth expansions are based on the following idea. Under mild regularity conditions, an Edgeworth expansion exists for the joint distribution of the log-likelihood derivatives $(\ell_\theta(\theta), \ell_{\theta\theta}(\theta), \ell_{\theta\theta\theta}(\theta), \ell_{\theta\theta\theta\theta}(\theta))$, suitably normalized. Many likelihood-based statistics, such as the maximum likelihood estimate, can be written as a function of these log-likelihood derivatives to a high degree of approximation. The Edgeworth expansion for the statistic under consideration may then be derived by transforming the Edgeworth expansion for the the distribution of the log-likelihood derivatives. As noted in Section 2.9, the resulting expansion is identical to the one obtained by formally computing the cumulants of a stochastic asymptotic expansion for the statistic in question.

5.2 Some preliminary results

5.2.1 INTRODUCTION

We begin by describing a few basic results that will be useful in these derivations. Let $H(\cdot)$ denote a smooth real-valued random function defined on the real line such that H and its derivatives are of order $O_p(n)$. Let X denote the maximizer of $H(\cdot)$; assume that $H'(X) = 0$ and that $X = O_p(n^{-1/2})$. Suppose the derivatives $H'(0), H''(0), \ldots,$ satisfy

$$\frac{H^{(j)}(0)}{n} = V_j + \frac{Z_j}{\sqrt{n}}, \quad j = 1, \ldots, 4 \tag{5.1}$$

where V_j is an $O(1)$ constant and Z_j is an $O_p(1)$ random variable. Since $H'(X) = 0$ and $X = O_p(n^{-1/2})$, it follows that we may take $V_1 = 0$.

5.2.2 AN EXPANSION FOR X

We first consider an expansion for the statistic X that maximizes $H(\cdot)$. By expanding $H'(X)$ around $X = 0$ and using the fact that $H'(X) = 0$ we have that

$$H'(0) + H''(0)X + \tfrac{1}{2}H'''(0)X^2 + \tfrac{1}{6}H^{(4)}(0)X^3 = O_p(n^{-1}). \tag{5.2}$$

Using (5.1), (5.2) may be written

$$Z_1 + (V_2 + Z_2/\sqrt{n})(\sqrt{n}X) + \tfrac{1}{2}(V_3 + Z_3/\sqrt{n})(\sqrt{n}X)^2/\sqrt{n}$$
$$+ \tfrac{1}{6}(V_4 + Z_4/\sqrt{n})(\sqrt{n}X)^3/n = O_p(n^{-3/2});$$

rearranging the terms, this expression may be written

$$[Z_1 + V_2(\sqrt{n}X)] + [Z_2(\sqrt{n}X) + \tfrac{1}{2}V_3(\sqrt{n}X)^2]/\sqrt{n}$$
$$+ [\tfrac{1}{2}Z_3(\sqrt{n}X)^2 + \tfrac{1}{6}V_4(\sqrt{n}X)^3]/n + O_p(n^{-3/2}). \tag{5.3}$$

This expression can be inverted to yield an expansion for $\sqrt{n}X$. Perhaps the simplest way to do this is to write

$$\sqrt{n}X = B + \frac{C}{\sqrt{n}} + \frac{D}{n} + O_p(n^{-3/2}), \tag{5.4}$$

where B, C, D are all $O_p(1)$, substitute (5.4) into (5.3), and then solve for B, C, D.

Equation (5.3) with the expansion in (5.4) replacing $\sqrt{n}X$ becomes

$$(Z_1 + V_2 B) + (Z_2 B + \tfrac{1}{2}V_3 B^2 + V_2 C)/\sqrt{n}$$
$$+ (\tfrac{1}{2}Z_3 B^2 + \tfrac{1}{6}V_4 B^3 + Z_2 C + V_3 BC + V_3 D)/n = O_p(n^{-3/2}).$$

We may solve for B, C, D by setting each of the three terms in this expression equal to 0. This leads to the following expansion for X:

$$\sqrt{n}X = -\frac{Z_1}{V_2} + \left(\frac{Z_1 Z_2}{V_2^2} - \frac{1}{2}\frac{V_3}{V_2^3}Z_1^2\right)\bigg/\sqrt{n}$$
$$+ \left[\left(\frac{1}{6}\frac{V_4}{V_2^4} - \frac{1}{2}\frac{V_3^2}{V_2^5}\right)Z_1^3 - \frac{1}{2}\frac{Z_1^2 Z_3}{V_2^3} - \frac{Z_1 Z_2^2}{V_2^3} + \frac{3}{2}\frac{V_3}{V_2^4}Z_1^2 Z_2\right]\bigg/n$$
$$+ O_p(n^{-3/2}). \tag{5.5}$$

5.2.3 AN EXPANSION FOR $H(X) - H(0)$

The expansion given above for X may be used to derive an expansion for $H(X) - H(0)$. Expanding $H(X) - H(0)$ around $X = 0$ yields

$$H(X) - H(0) = H'(0)X + \tfrac{1}{2}H''(0)X^2 + \tfrac{1}{6}H'''(0)X^3$$
$$+ \tfrac{1}{4}H^{(4)}(0)X^4 + O_p(n^{-3/2}).$$

Using (5.1) and (5.5) leads to the following expansion

$$H(X) - H(0) = -\frac{1}{2}\frac{Z_1^2}{V_2} + \left(\frac{1}{2}\frac{Z_1^2 Z_2}{V_2^2} - \frac{1}{6}\frac{V_3}{V_2^3}Z_1^3\right)\Big/\sqrt{n}$$

$$+\left(-\frac{1}{8}\frac{V_3^2}{V_2^5}Z_1^4 - \frac{1}{6}\frac{Z_1^3 Z_3}{V_2^3} + \frac{1}{24}\frac{V_4}{V_2^4}Z_1^4 + \frac{1}{2}\frac{V_3}{V_2^4}Z_1^3 Z_2 - \frac{1}{2}\frac{Z_1^2 Z_2^2}{V_2^3}\right)\Big/n$$

$$+ O_p(n^{-3/2}). \tag{5.6}$$

5.2.4 CUMULANTS OF EXPANSIONS

Suppose that a statistic X may be expanded as

$$X = Y_1 + Y_2/\sqrt{n} + Y_3/n + O_p(n^{-3/2})$$

where Y_1, Y_2, Y_3 are $O_p(1)$ random variables with cumulants of all orders. Assume that the (i, j, k)th cumulant of (Y_1, Y_2, Y_3) is of order $O(n^{-(i+j+k-2)/2})$ as would be the case if the Y_j were standardized sample means.

The formal cumulants of X can be approximated by the cumulants of

$$\hat{X} = Y_1 + Y_2/\sqrt{n} + Y_3/n$$

and, hence, may be expressed in terms of the cumulants of Y_1, Y_2, Y_3, at least to a given order of approximation.

One way to derive these expressions is to use the method of generating functions. Let $K_Y(t_1, t_2, t_3)$ denote the joint cumulant-generating function of (Y_1, Y_2, Y_3). Since $K_{\hat{X}}(t)$, the cumulant-generating function of $\hat{X}$, is given by

$$K_{\hat{X}}(t) = \log \mathrm{E}[\exp\{t(Y_1 + Y_2/\sqrt{n} + Y_3/n)\}] = K_Y(t, t/\sqrt{n}, t/n),$$

the cumulants of $\hat{X}$ may be obtained by differentiating K_Y.

For instance,

$$K'_{\hat{X}}(0) = \frac{\partial K_Y}{\partial t_1}(0,0,0) + \frac{\partial K_Y}{\partial t_2}(0,0,0)\frac{1}{\sqrt{n}} + \frac{\partial K_Y}{\partial t_3}(0,0,0)\frac{1}{n}$$

so that the mean of X may be expanded

$$\begin{aligned}\mathrm{cum}(X) &= \mathrm{cum}(\hat{X}) + O(n^{-3/2}) \\ &= \mathrm{cum}(Y_1) + \mathrm{cum}(Y_2)/\sqrt{n} + \mathrm{cum}(Y_3)/n + O(n^{-3/2}).\end{aligned}$$

The same approach can be used for the higher-order cumulants leading to the expressions

$$\begin{aligned}\mathrm{cum}(X, X) &= \mathrm{cum}(Y_1, Y_1) + 2\,\mathrm{cum}(Y_1, Y_2)/\sqrt{n}\\ &\quad + [2\,\mathrm{cum}(Y_1, Y_3) + \mathrm{cum}(Y_2, Y_2)]/n + O(n^{-3/2})\\ \mathrm{cum}(X, X, X) &= \mathrm{cum}(Y_1, Y_1, Y_1) + 3\,\mathrm{cum}(Y_1, Y_1, Y_2)/\sqrt{n}\\ &\quad + 3\,\mathrm{cum}(Y_1, Y_1, Y_3)/n + 3\,\mathrm{cum}(Y_1, Y_2, Y_2)/n + O(n^{-3/2})\\ \mathrm{cum}(X, X, X, X) &= \mathrm{cum}(Y_1, Y_1, Y_1, Y_1) + 4\,\mathrm{cum}(Y_1, Y_1, Y_1, Y_2)/\sqrt{n}\\ &\quad + 4\,\mathrm{cum}(Y_1, Y_1, Y_1, Y_3)/n + 4\,\mathrm{cum}(Y_1, Y_1, Y_2, Y_2)/n\\ &\quad + O(n^{-3/2}).\end{aligned}$$

5.3 Maximum likelihood estimates

5.3.1 THE SINGLE PARAMETER CASE

First consider the case in which θ is a real-valued parameter. Taking $H(t) = \ell(\theta + t)$, a stochastic asymptotic expansion for the maximum likelihood estimate $\hat{\theta}$ may be obtained from the results in Section 5.2.2. The quantities appearing in (5.5) may be expressed in terms of the log-likelihood derivatives.

Let $n\nu_{ijk\ell}$ denote the (i, j, k, ℓ)th cumulant of $(\ell_\theta(\theta), \ell_{\theta\theta}(\theta), \ell_{\theta\theta\theta}(\theta), \ell_{\theta\theta\theta\theta}(\theta))$, calculated under the distribution with parameter θ; note that $\nu_{ijk\ell}$ is of order $O(1)$. In using this notation, trailing 0s will be omitted so that, for example, the mean of $\ell_{\theta\theta}(\theta)$ will be written as $n\nu_{01}$ instead of as $n\nu_{0100}$. Hence, $V_2 = \nu_{01}$, $V_3 = \nu_{001}$, and $V_4 = \nu_{0001}$. The random variable Z_j is a standardized log-likelihood derivative; for instance,

$$Z_1 = (\ell_\theta(\theta) - \mathrm{E}[\ell_\theta(\theta); \theta])/\sqrt{n} = \ell_\theta(\theta)/\sqrt{n}$$

and

$$Z_2 = (\ell_{\theta\theta}(\theta) - \mathrm{E}[\ell_{\theta\theta}(\theta); \theta])/\sqrt{n}.$$

Hence, we obtain immediately from (5.5) that

$$\begin{aligned}\sqrt{n}(\hat{\theta} - \theta) &= -\frac{Z_1}{\nu_{01}} + \left(\frac{Z_1 Z_2}{\nu_{01}^2} - \frac{1}{2}\frac{\nu_{001}}{\nu_{01}^3} Z_1^2\right)\Big/\sqrt{n}\\ &\quad + \left[\left(\frac{1}{6}\frac{\nu_{0001}}{\nu_{01}^4} - \frac{1}{2}\frac{\nu_{001}^2}{\nu_{01}^5}\right) Z_1^3 - \frac{1}{2}\frac{Z_1^2 Z_3}{\nu_{01}^3} - \frac{Z_1 Z_2^2}{\nu_{01}^3} + \frac{3}{2}\frac{\nu_{001}}{\nu_{01}^4} Z_1^2 Z_2\right]\Big/ n\\ &\quad + O_p(n^{-3/2}).\end{aligned} \tag{5.7}$$

Let

$$T = \sqrt{n\nu_2}(\hat{\theta} - \theta) = \sqrt{i(\theta)}(\hat{\theta} - \theta)$$

denote the standardized maximum likelihood estimate. Using the expansion (5.7), the results in Section 5.2.4 may now be used to obtain the formal cumulants of T; these cumulants will be denoted $\hat{\kappa}_j \equiv \hat{\kappa}_j(T)$; these formal cumulants may be used in an Edgeworth expansion for the distribution of T.

For instance, $\mathrm{E}[-Z_1/\nu_{01};\theta] = 0$,

$$\mathrm{E}\left[\frac{Z_1 Z_2}{\nu_{01}^2} - \frac{1}{2}\frac{\nu_{001}}{\nu_{01}^3}Z_1^2;\theta\right] = \frac{\nu_{11}}{\nu_{01}^2} - \frac{1}{2}\frac{\nu_{001}\nu_2}{\nu_{01}^3}$$

and

$$\mathrm{E}\left[\left(\frac{1}{6}\frac{\nu_{0001}}{\nu_{01}^4} - \frac{1}{2}\frac{\nu_{001}^2}{\nu_{01}^5}\right)Z_1^3 - \frac{1}{2}\frac{Z_1^2 Z_3}{\nu_{01}^3} - \frac{Z_1 Z_2^2}{\nu_{01}^3} + \frac{3}{2}\frac{\nu_{001}}{\nu_{01}^4}Z_1^2 Z_2;\theta\right]$$

$$= \left(\frac{1}{6}\frac{\nu_{0001}}{\nu_{01}^4} - \frac{1}{2}\frac{\nu_{001}^2}{\nu_{01}^5}\right)\nu_3/\sqrt{n} - \frac{1}{2}\frac{1}{\nu_{01}^3}\nu_{201}/\sqrt{n} - \frac{1}{\nu_{01}^3}\nu_{12}/\sqrt{n}$$

$$+ \frac{3}{2}\frac{\nu_{001}}{\nu_{01}^4}\nu_{21}/\sqrt{n}.$$

It follows that the first formal cumulant of $\sqrt{i(\theta)}(\hat{\theta} - \theta)$ is given by

$$\hat{\kappa}_1(T) = \left(\frac{\nu_{11}}{\nu_2^{3/2}} + \frac{1}{2}\frac{\nu_{001}}{\nu_2^{3/2}}\right)\frac{1}{\sqrt{n}};$$

here the fact that $-\nu_{01} = \nu_2$ has been used to simplify the expression. This result may be used to construct a bias-corrected version of the maximum likelihood estimate.

To determine $\hat{\kappa}_2(T)$ we first multiply the expansion (5.7) by $\sqrt{i(\theta)}$, then square the expansion and take its expected value; this yields an expansion for the second moment of T. Subtracting $\hat{\kappa}_1(T)^2$ yields the following result:

$$\hat{\kappa}_2(T) = 1 + \left(\frac{\nu_{001}\nu_{30}}{\nu_2^3} + 2\frac{\nu_{21}}{\nu_2^2} + \frac{\nu_{0001}}{\nu_2^2} + \frac{7}{2}\frac{\nu_{001}^2}{\nu_2^3} + 3\frac{\nu_{101}}{\nu_2^2}\right.$$

$$\left. + 3\frac{\nu_{02}}{\nu_2^2} + 5\frac{\nu_{11}^2}{\nu_2^3} - 7\frac{\nu_{001}\nu_{11}}{\nu_2^3}\right)\frac{1}{n}.$$

The same approach can be used to show that

$$\hat{\kappa}_3(T) = \left(\frac{\nu_{30}}{\nu_2^{3/2}} + 6\frac{\nu_{11}}{\nu_2^{3/2}} + 3\frac{\nu_{001}}{\nu_2^{3/2}}\right)\frac{1}{\sqrt{n}},$$

and

$$\hat{\kappa}_4(T) = \frac{\nu_{40}}{\nu_2^2}\frac{1}{n};$$

all of these expressions have an error of $O(n^{-3/2})$ or smaller.

An Edgeworth expansion for the distribution function of $\sqrt{i(\theta)}(\hat{\theta} - \theta)$ may now be derived as discussed in Section 2.3. The resulting Edgeworth expansion for the distribution function of T is given by

$$\Phi(t) - \phi(t)\Big\{\hat{\kappa}_1(T) + \tfrac{1}{6}\hat{\kappa}_3(T)H_2(t) + \tfrac{1}{2}\big[\hat{\kappa}_1(T)^2 + \hat{\kappa}_2(T) - 1\big]t$$
$$+ \Big[\tfrac{1}{24}\hat{\kappa}_4(T)H_3(t) + \tfrac{1}{72}\hat{\kappa}_3(T)^2 H_5(t) + \tfrac{1}{6}\hat{\kappa}_1(T)\hat{\kappa}_3(T)H_3(t)\Big] + O(n^{-3/2})\Big\}.$$

Example 5.1 Exponential distribution

Let $Y_1, \ldots, Y_n$ denote independent exponential random variables each with rate parameter θ. The maximum likelihood estimate of θ is given by $1/\bar{y}$; we consider an Edgeworth expansion for the distribution of $\sqrt{n}(\hat{\theta} - \theta)$. The log-likelihood function for this model is given by

$$\ell(\theta) = n \log \theta - n\theta\bar{y}.$$

It follows that

$$\ell_\theta(\theta) = \frac{n}{\theta} - n\bar{y}; \qquad \ell_{\theta\theta}(\theta) = -\frac{n}{\theta^2},$$
$$\ell_{\theta\theta\theta}(\theta) = \frac{2n}{\theta^3}; \qquad \ell_{\theta\theta\theta\theta}(\theta) = -\frac{6n}{\theta^4}.$$

It is straightforward to show that $\nu_{11}, \nu_{21}, \nu_{101}, \nu_{02}$ are all 0 and that

$$\nu_{01} = -\frac{1}{\theta^2}; \qquad \nu_{001} = \frac{2}{\theta^3}; \qquad \nu_{0001} = -\frac{6}{\theta^4};$$
$$\nu_{30} = -\frac{2}{\theta^3}; \qquad \nu_{40} = \frac{6}{\theta^4}.$$

It follows that

$$\hat{\kappa}_1 = \frac{1}{\sqrt{n}}; \qquad \hat{\kappa}_2 = 1 + \frac{4}{n};$$
$$\hat{\kappa}_3 = \frac{4}{\sqrt{n}}; \qquad \hat{\kappa}_4 = \frac{6}{n}.$$

Based on these results, the bias of $\hat{\theta}$ may be approximated by θ/n so that a bias-corrected estimate is given by

$$\hat{\theta} - \hat{\theta}/n = \frac{n-1}{n}\hat{\theta};$$

it is straightforward to show that this estimator is exactly unbiased.

The probability $\Pr[\sqrt{n}\theta^{-1}(\hat{\theta} - \theta) \le t; \theta]$ may be approximated by

$$\Phi(t) - \phi(t)\Big[\Big(1 + \frac{2}{3}H_2(t)\Big)\frac{1}{\sqrt{n}} + \Big(\frac{5}{2}t + \frac{11}{12}H_3(t) + \frac{2}{9}H_5(t)\Big)\frac{1}{n}\Big].$$

For this model, the exact distribution of $\hat{\theta}$ is easy to determine. Note that $n\theta\bar{y}$ has a standard gamma distribution with index n. Hence,

$$\Pr[\sqrt{n}\theta^{-1}(\hat{\theta} - \theta) \le t; \theta] = \Pr\Big[Z \ge \frac{n}{1 + t/\sqrt{n}}\Big]$$

where Z denotes a standard gamma random variable with index n. This result can be used to evaluate the accuracy of the Edgeworth series approximation. Table 5.1 contains the the normal approximation, along with two Edgeworth series approximations for several values of t, for the case $n = 10$.

The results in Table 5.1 illustrate two important properties of Edgeworth series approximations. The first is that the approximations may be negative for extreme values of t. The second is that, for a given value of t and a given value of n, the approximation based on one correction term may be more accurate than the approximation based on two correction terms. ∎

Example 5.2 Linear regression for Poisson random variables

Let $Y_1, \ldots, Y_n$ denote independent Poisson random variables such that Y_j has mean θx_j where $x_1, \ldots, x_n$ are known constants. Then

$$\ell(\theta) = \log\theta \sum y_j - \theta \sum x_j$$

so that $\hat{\theta} = \sum y_j / \sum x_j$,

$$\ell_\theta(\theta) = \frac{1}{\theta}\sum y_j - \sum x_j; \qquad \ell_{\theta\theta}(\theta) = -\frac{\sum y_j}{\theta^2};$$

$$\ell_{\theta\theta\theta}(\theta) = \frac{2\sum y_j}{\theta^3}; \qquad \ell_{\theta\theta\theta\theta}(\theta) = -\frac{6\sum y_j}{\theta^4}.$$

Table 5.1 Comparison of probability approximations in Example 5.1

			Edgeworth series	
t	Exact	Normal	1 term	2 terms
−1.48	0.01	0.070	−0.006	0.019
−1.15	0.05	0.125	0.046	0.088
−0.94	0.10	0.175	0.100	0.151
1.92	0.90	0.973	0.917	0.901
2.67	0.95	0.996	0.978	0.963
4.49	0.99	1.00	1.00	0.999

It follows that

$$\nu_{01} = -\frac{\bar{x}}{\theta}; \qquad \nu_2 = \frac{\bar{x}}{\theta}; \qquad \nu_{11} = -\frac{\bar{x}}{\theta^2};$$
$$\nu_{001} = \frac{2\bar{x}}{\theta^2}; \qquad \nu_3 = \frac{\bar{x}}{\theta^2}.$$

The first three formal cumulants of $i(\theta)^{1/2}(\hat{\theta} - \theta)$ are therefore 0, 1, and $[\theta \sum x_j]^{-1/2}$; these expressions have error of order $O(n^{-1})$. Since $\sum Y_j$ has a Poisson distribution with mean $\theta \sum x_j$, these expressions are exact for this example. ∎

In general $i(\theta)$ depends on the unknown value of the parameter θ. Hence, it is useful to have an approximation to the distribution of $(\hat{\theta} - \theta)/\hat{\sigma}$, where $\hat{\sigma}$ denotes an estimate of the standard deviation of $\hat{\theta}$; that is, $\hat{\sigma}$ is an estimate of $i(\theta)^{-1/2}$. For instance, consider $\sqrt{\hat{\jmath}}(\hat{\theta} - \theta)$ where $\hat{\jmath}$ denotes the observed information evaluated at $\theta = \hat{\theta}$. Note that

$$\frac{\hat{\jmath}}{n} = \frac{-\ell_{\theta\theta}(\theta)}{n} - \frac{\ell_{\theta\theta\theta}(\theta)}{n}\sqrt{n}(\hat{\theta} - \theta)\frac{1}{\sqrt{n}} + O_p(n^{-1})$$

which may be written

$$\frac{\hat{\jmath}}{n} = \nu_2 - Z_2\frac{1}{\sqrt{n}} - \nu_{001}\frac{Z_1}{\nu_2}\frac{1}{\sqrt{n}} + O_p(n^{-1}).$$

Combining this expansion with the expansion (5.7) for $\sqrt{n}(\hat{\theta} - \theta)$ yields the following expansion for $\sqrt{\hat{\jmath}}(\hat{\theta} - \theta)$:

$$\sqrt{\hat{\jmath}}(\hat{\theta} - \theta) = (\hat{\jmath}/n)^{1/2}\sqrt{n}(\hat{\theta} - \theta) = \frac{Z_1}{\sqrt{\nu_2}} + \frac{1}{2}\frac{Z_1 Z_2}{\nu_2^{3/2}}\frac{1}{\sqrt{n}} + O_p(n^{-1}).$$

Based on this expansion, the first three formal cumulants of $\sqrt{\hat{\jmath}}(\hat{\theta} - \theta)$ are given by

$$\frac{1}{2}\frac{\nu_{11}}{\nu_2^{3/2}}\frac{1}{\sqrt{n}}, \qquad 1, \qquad \left(3\frac{\nu_{11}}{\nu_2^{3/2}} + \frac{\nu_3}{\nu_2^{3/2}}\right)\frac{1}{\sqrt{n}},$$

respectively. The error in these expressions is of order $O(n^{-1})$ or smaller.

Example 5.3 Exponential distribution
Let $Y_1, \ldots, Y_n$ denote independent exponential random variables each with rate parameter θ, as in Example 5.1. Consider an Edgeworth expansion for the distribution of $\sqrt{\hat{\jmath}}(\hat{\theta} - \theta)$. From Example 5.1, we have that $\nu_2 = \theta^{-2}$, $\nu_3 = -2\theta^{-3}$, and $\nu_{11} = 0$. Hence, the first three formal cumulants of $\sqrt{\hat{\jmath}}(\hat{\theta} - \theta)$ are 0, 1, and $-2/\sqrt{n}$, respectively. It follows that

$$\Pr[\sqrt{\hat{\jmath}}(\hat{\theta} - \theta) \leq t; \theta] = \Phi(t) - \phi(t)\frac{1}{6\sqrt{n}} + O(n^{-1}).$$

Note that $\hat{j} = n/\hat{\theta}^2$ so that

$$\sqrt{\hat{j}}(\hat{\theta} - \theta) = \sqrt{n}(1 - \theta/\hat{\theta}) = -\sqrt{n}(\theta\bar{y} - 1).$$

The expressions for the cumulants given above can be verified directly by noting that the first three cumulants of $n\theta\bar{Y}$ are n, n, and $2n$, respectively. ∎

5.3.2 VECTOR PARAMETER CASE

We now consider the case in which the parameter θ is a vector. One approach is to derive an Edgeworth expansion for the vector-valued random variable $\sqrt{n}(\hat{\theta}-\theta)$ using a generalization of the approach used in Section 5.3.1 for the scalar parameter case. Although it is possible to take this approach, the resulting multivariate Edgeworth expansion is complicated and difficult to use. An alternative approach is to consider the distribution of $\sqrt{n}(\hat{\psi} - \psi)$ where $\hat{\psi}$ is a scalar function of the parameter θ. This approach is appropriate whenever the relationship between two or more components of $\hat{\theta}$ is not of interest.

Assume that θ may be written (ψ, λ) where ψ is a real-valued parameter of interest and λ is a nuisance parameter. We will derive an Edgeworth expansion for the distribution of $\sqrt{n}(\hat{\psi} - \psi)$ that includes the $1/\sqrt{n}$ term; the $1/n$ term may, in principle, be derived using the same approach.

The symbol μ will be used to denote the average expected value of certain log-likelihood derivatives with the subscripts of μ indicating the derivatives under consideration. For example,

$$\mu_{\psi\lambda\lambda}(\theta) = \frac{1}{n}\mathrm{E}[\ell_{\psi\lambda\lambda}(\theta); \theta]; \qquad \mu_{\lambda,\lambda\psi}(\theta) = \frac{1}{n}\mathrm{E}[\ell_{\lambda}(\theta)\ell_{\lambda\psi}(\theta)^T; \theta].$$

As noted in Section 4.6, $\hat{\psi}$ may be described as the maximizer of the profile log-likelihood function $\ell_p(\psi)$. Hence, the expansion derived in Section 5.2.2 applies here, defining the function H by $H(t) = \ell_p(\psi + t)$. It follows that

$$\sqrt{n}(\hat{\psi} - \psi) = -\frac{Z_1}{V_2} + \left(\frac{Z_1 Z_2}{V_2^2} - \frac{1}{2}\frac{V_3}{V_2^3} Z_1^2\right)\Big/\sqrt{n} + O_p(n^{-1}) \qquad (5.8)$$

where the Z_j and V_j are defined in terms of $\ell_p(\psi)$. Specifically, $Z_1 = \ell'_p(\psi)/\sqrt{n}$, V_2 satisfies $n^{-1}\ell''_p(\psi) = V_2 + o_p(1)$,

$$Z_2 = [\ell''_p(\psi) - V_2]/\sqrt{n}$$

and V_3 satisfies $n^{-1}\ell'''_p(\psi) = V_3 + o_p(1)$.

Let $i_\psi(\theta)$ denote the partial information for ψ. In some cases, the formal cumulants of $\sqrt{i_\psi(\theta)}(\hat{\psi} - \psi)$ may be obtained as in Section 5.3.1 by computing the cumulants of the terms in the expansion (5.8). In doing this, it is important to keep in mind that, unlike the case in Section 5.3.1, here the Z_j are not log-likelihood derivatives. In particular, although $V_1 = 0$, Z_1 does not have mean exactly 0; it will be shown later in this section that the mean of Z_1 is $O(n^{-1/2})$. Also, recall that $i_\psi(\theta)/n = -V_2 + O(n^{-1})$.

Using this approach, the first three formal cumulants of $\sqrt{i_\psi(\theta)}(\hat{\psi} - \psi)$ are

$$\frac{\text{cum}(Z_1)}{(-V_2)^{1/2}} + \Big[\frac{\text{cum}(Z_1, Z_2)}{(-V_2)^{3/2}} + \frac{1}{2}\frac{V_3}{(-V_2)^{5/2}}\,\text{cum}(Z_1, Z_1)\Big]\frac{1}{\sqrt{n}},$$

$-\text{cum}(Z_1, Z_1)/V_2$, and

$$\frac{\text{cum}(Z_1, Z_1, Z_1)}{(-V_2)^{3/2}} + \frac{6\text{cum}(Z_1, Z_1)\,\text{cum}(Z_1, Z_2)}{(-V_2)^{5/2}}\frac{1}{\sqrt{n}} + \frac{3V_3}{(-V_2)^{7/2}}\,\text{cum}(Z_1, Z_1)^2\frac{1}{\sqrt{n}};$$

these expressions have error of order $O(n^{-1})$ or smaller.

Example 5.4 Standard deviation of a normal distribution

Let $Y_1, \ldots, Y_n$ denote independent normally distributed random variables each with mean λ and standard deviation ψ. Then

$$\ell(\psi, \lambda) = -\frac{1}{2\psi^2}\sum(y_j - \lambda)^2 - n\log\psi.$$

The maximum likelihood estimate of λ for fixed ψ is simply $\bar{y}$ so that

$$\ell_p(\psi) = -\frac{1}{2\psi^2}\sum(y_j - \bar{y})^2 - n\log\psi.$$

Hence,

$$\ell_p'(\psi) = \frac{1}{\psi^3}\sum(y_j - \bar{y})^2 - \frac{n}{\psi}; \qquad \ell_p''(\psi) = -\frac{3}{\psi^4}\sum(y_j - \bar{y})^2 + \frac{n}{\psi^2};$$

$$\ell_p'''(\psi) = \frac{12}{\psi^5}\sum(y_j - \bar{y})^2 - \frac{2n}{\psi^3}.$$

Note that $\sum(Y_j - \bar{Y})^2/\psi^2$ has a χ^2 distribution with $n - 1$ degrees of freedom and, hence, has cumulants $n - 1$, $2(n - 1)$ and $8(n - 1)$. It follows that $V_2 = -2/\psi^2$, $V_3 = 10/\psi^3$,

$$\text{cum}(Z_1) = -\frac{1}{\psi\sqrt{n}}; \qquad \text{cum}(Z_1, Z_2) = -\frac{6}{\psi^3} + O(n^{-1}).$$

Using these results, it is straightforward to show that the first three formal cumulants of $\sqrt{i_\psi(\theta)}(\hat{\psi} - \psi)$ are $-3/(2^{3/2}\sqrt{n})$, 1, and $1/\sqrt{(2n)}$, respectively; these expressions have error of order $O(n^{-1})$, or smaller.

For instance, a bias-corrected maximum likelihood estimate is given by

$$\hat{\psi} + \frac{3}{4n}\hat{\psi} = \frac{4n+3}{4n}\hat{\psi}. \qquad \blacksquare$$

Example 5.5 Index of a gamma distribution
Let $Y_1, \ldots, Y_n$ denote independent gamma random variables each with index ψ and rate parameter λ. The log-likelihood function for this model is given by

$$n\psi \log \lambda - n \log \Gamma(\psi) + \psi \sum \log y_j - n\lambda \bar{y}.$$

The maximum likelihood estimate of λ for fixed ψ is given by $\hat{\lambda}_\psi = \psi/\bar{y}$; hence,

$$\ell_p(\psi) = n\psi \log \psi - n\psi \log \bar{y} - n \log \Gamma(\psi) + \psi \sum \log y_j - n\psi.$$

It follows that

$$\ell'_p(\psi) = n[\log \psi - \log \sum y_j + n^{-1} \sum \log y_j - \Psi(\psi) + \log n];$$

$$\ell''_p(\psi) = n[1/\psi - \Psi'(\psi)]; \qquad \ell'''_p(\psi) = -n[1/\psi^2 + \Psi''(\psi)].$$

Here $\Psi(t) = d \log \Gamma(t)/dt$. Hence, $V_2 = 1/\psi - \Psi'(\psi)$ and $V_3 = -[1/\psi^2 + \Psi''(\psi)]$.

Using facts about the gamma distribution, it may be shown that $\mathrm{E}[\log Y_j] = \Psi(\psi) - \log \lambda$ and that $\mathrm{E}[\log \sum Y_j] = \Psi(n\psi) - \log \lambda$. Hence,

$$\mathrm{E}[\ell'_p(\psi)] = n[\log \psi + \log n - \Psi(n\psi)]$$

so that

$$\mathrm{cum}(Z_1) = \sqrt{n}[\log(n\psi) - \Psi(n\psi)];$$

using standard asymptotic expansions for the functions $\log(\cdot)$ and $\Psi(\cdot)$,

$$\mathrm{cum}(Z_1) = \frac{1}{2\psi} \frac{1}{\sqrt{n}} + O(n^{-3/2}).$$

Note that, since $\ell''_p(\psi)$ is a constant, $Z_2 = 0$.

To approximate the cumulants of $\hat{\psi}$, we need to calculate, either exactly or approximately, $\mathrm{cum}(Z_1, Z_1)$ and $\mathrm{cum}(Z_1, Z_1, Z_1)$. Note that

$$\begin{aligned} Z_1 &= \sqrt{n}[\log \psi - \Psi(\psi) + n^{-1} \log y_j - \log \bar{y}] \\ &= \sqrt{n}[\log \psi - \Psi(\psi) - \log \mu] + \frac{1}{\sqrt{n}} \sum \log y_j - \frac{\sqrt{n}}{\mu}(\bar{y} - \mu) + O_p(n^{-1/2}) \end{aligned}$$

where $\mu = \mathrm{E}(Y_1; \theta) = \psi/\lambda$; the leading term in an expansion of the $O_p(n^{-1/2})$ term is a constant times $(\bar{Y} - \mu)^2$. Hence, in calculating the second and third cumulants of Z_1, we may use

$$\sqrt{n}[\log \psi - \Psi(\psi) - \log \mu] + \frac{1}{\sqrt{n}} \sum \log y_j - \frac{\sqrt{n}}{\mu}(\bar{y} - \mu)$$

in place of Z_1 without changing the order of the approximation; note that this is not the case when calculating $\mathrm{cum}(Z_1)$. Let $Z = \log y_1 - y_1/\mu$. Then

$$\mathrm{cum}(Z_1, Z_1) = \mathrm{cum}(Z, Z) + O(n^{-1});$$

$$\mathrm{cum}(Z_1, Z_1, Z_1) = \mathrm{cum}(Z, Z, Z) \frac{1}{\sqrt{n}} + O(n^{-1}).$$

A direct calculation shows that cumulant-generating function of Z is given by

$$\log \Gamma(\psi + t) - (\psi + t) \log(\lambda + t/\mu) + \psi \log \lambda - \log \Gamma(\psi)$$

so that

$$\text{cum}(Z, Z) = \Psi'(\psi) - 1/\psi; \qquad \text{cum}(Z, Z, Z) = \Psi''(\psi) + 1/\psi^2.$$

It follows that the first three formal cumulants of $\sqrt{i_\psi(\theta)}(\hat{\psi} - \psi)$ are given by

$$\frac{1}{2} \frac{\Psi'(\psi)/\psi - \Psi''(\psi) - 2/\psi^2}{(\Psi'(\psi) - 1/\psi)^{3/2}} \frac{1}{\sqrt{n}},$$

1, and

$$-2 \frac{\Psi''(\psi) + 1/\psi^2}{(\Psi'(\psi) - 1/\psi)^{3/2}} \frac{1}{\sqrt{n}};$$

these expressions have error $O(n^{-1})$ or smaller. ∎

As in previous example, it is often difficult to calculate the cumulants of Z_1 and Z_2 without further approximation, due to the fact that the derivatives $\ell_p'(\psi)$ and $\ell_p''(\psi)$ are functions of $\hat{\lambda}_\psi$, the maximum likelihood estimate of λ for fixed ψ. In the remainder of this section, we develop some general approximation results along these lines. When deriving these results, it is important to distinguish between the value at which ψ is held fixed, and the value of ψ at which probabilities and expectations are computed, that is, the 'true' value of ψ. Hence, in these derivations we will use $\theta_0 = (\psi_0, \lambda_0)$ to denote the true parameter value; all expectations will be assumed to be calculated at the distribution with parameter θ_0.

The key idea is that $\ell_p(\psi)$ may be approximated by $\tilde{\ell}(\psi)$ where $\tilde{\ell}(\psi) = \ell(\psi, \lambda_\psi)$ and λ_ψ denotes the value of λ that maximizes $n^{-1}\text{E}[\ell(\psi, \lambda)]$ with respect to λ for fixed ψ.

The relationship between $\ell_p(\psi)$ and $\tilde{\ell}(\psi)$ was considered in Section 4.6. This relationship is simplified whenever ψ and λ are orthogonal parameters so that $i_{\psi\lambda} = 0$. Hence, we make this assumption throughout the remainder of this section. Let $\lambda_\psi' = d\lambda_\psi/d\psi$ and $\lambda_0' = \lambda_{\psi_0}'$. When ψ and λ are orthogonal, then $\lambda_0' = 0$.

Let $\tilde{Z}_1, \tilde{Z}_2, \tilde{Z}_3$ denote approximations to Z_1, Z_2, Z_3, respectively, based on $\tilde{\ell}(\psi)$; for instance, $\tilde{Z}_1$ and $\tilde{V}_1$ satisfy

$$\frac{1}{n}\tilde{\ell}_\psi(\psi_0) = \tilde{V}_1 + \frac{\tilde{Z}_1}{\sqrt{n}}.$$

Since $\hat{\lambda}_\psi \xrightarrow{p} \lambda_\psi$, $\tilde{V}_j = V_j$ as defined previously; however, to emphasize the fact that we are using the approximation based on $\tilde{\ell}(\psi)$, we will continue to write $\tilde{V}_j$.

Note that

$$Z_j = \tilde{Z}_j + B_j/\sqrt{n} + O_p(n^{-1}), \quad j = 1, 2, 3$$

where the B_j are $O_p(1)$. Since $\tilde{\ell}(\cdot)$ is a genuine log-likelihood function $\mathrm{E}[\tilde{Z}_1] = 0$, $\mathrm{E}[\tilde{Z}_1^2] + \mathrm{E}[\tilde{Z}_2] = 0$, and so on. The B_j represent the effect of using $\ell_p(\psi)$ instead of the genuine log-likelihood $\tilde{\ell}(\psi)$.

It follows that equation (5.8) may be written

$$\sqrt{n}(\hat{\psi} - \psi_0) = -\frac{\tilde{Z}_1}{\tilde{V}_2} + \left(\frac{\tilde{Z}_1 \tilde{Z}_2}{\tilde{V}_2^2} - \frac{1}{2}\frac{\tilde{V}_3}{\tilde{V}_2^3}\tilde{Z}_1^2 - \frac{B_1}{\tilde{V}_2}\right)\Big/\sqrt{n} + O_p(n^{-1}). \quad (5.9)$$

The formal cumulants of $\hat{\psi}$ may then be determined by computing the formal cumulants of the expansion in (5.9).

Using the results in Section 5.2.4, the formal cumulants of $\sqrt{i_\psi}(\hat{\psi} - \psi_0)$ may be written

$$\hat{\kappa}_1 = \left[\frac{\mathrm{cum}(\tilde{Z}_1, \tilde{Z}_2)}{(-\tilde{V}_2)^{3/2}} + \frac{1}{2}\frac{\tilde{V}_3}{(-\tilde{V}_2)^{3/2}} + \frac{\mathrm{cum}(B_1)}{(-\tilde{V}_2)^{1/2}}\right]\frac{1}{\sqrt{n}} + O(n^{-3/2});$$

$$\hat{\kappa}_2 = 1 + \frac{2\mathrm{cum}(\tilde{Z}_1, B_1)}{-\tilde{V}_2}\frac{1}{\sqrt{n}} + O(n^{-1});$$

$$\hat{\kappa}_3 = \left[\frac{\mathrm{cum}(\tilde{Z}_1, \tilde{Z}_1, \tilde{Z}_1)}{(-\tilde{V}_2)^{3/2}} + 3\frac{\mathrm{cum}(\tilde{Z}_1, \tilde{Z}_1, \tilde{Z}_1\tilde{Z}_2)}{(-\tilde{V}_2)^{7/2}} + \frac{1}{2}\frac{\tilde{V}_3}{(-\tilde{V}_2)^{7/2}}\mathrm{cum}(\tilde{Z}_1, \tilde{Z}_1^2) + \frac{\mathrm{cum}(\tilde{Z}_1, B_1)}{(-\tilde{V}_2)^{3/2}}\right]\frac{1}{\sqrt{n}} + O(n^{-3/2}).$$

Note that since the $\tilde{Z}_j$ are based on a genuine log-likelihood function, we have used the fact that $\mathrm{cum}(\tilde{Z}_1, \tilde{Z}_1) = -\tilde{V}_2 + O(n^{-1})$.

All of the cumulants appearing in these expressions may be written in terms of moments of log-likelihood derivatives. The function $\ell(\psi, \hat{\lambda}_\psi)$ may be expanded

$$\begin{aligned}\ell(\psi, \hat{\lambda}_\psi) &= \ell(\psi, \lambda_\psi) + \ell_\lambda(\psi, \lambda_\psi)^T(\hat{\lambda}_\psi - \lambda_\psi) \\ &\quad + \tfrac{1}{2}(\hat{\lambda}_\psi - \lambda_\psi)^T \ell_{\lambda\lambda}(\psi, \lambda_\psi)(\hat{\lambda}_\psi - \lambda_\psi) + \cdots. \end{aligned} \quad (5.10)$$

Hence,

$$\begin{aligned}\ell'_p(\psi_0) &= \tilde{\ell}_\psi(\psi_0) + \left[\frac{d}{d\psi}\ell_\lambda(\psi, \lambda_\psi)\Big|_{\psi=\psi_0}\right](\hat{\lambda}_0 - \lambda_0) + \ell_\lambda(\psi_0, \lambda_0)(\hat{\lambda}'_0 - \lambda'_0) \\ &\quad + (\hat{\lambda}_0 - \lambda_0)^T \ell_{\lambda\lambda}(\psi_0, \lambda_0)(\hat{\lambda}'_0 - \lambda'_0) \\ &\quad + \frac{1}{2}(\hat{\lambda}_0 - \lambda_0)^T \ell_{\lambda\lambda\psi}(\lambda_0, \psi_0)(\hat{\lambda}_0 - \lambda_0) + O_p(n^{-1}).\end{aligned}$$

Note that

$$\sqrt{n}(\hat{\lambda}_0 - \lambda_0) = \bar{i}_{\lambda\lambda} \frac{1}{\sqrt{n}} \ell_\lambda(\psi_0, \lambda_0) + O_p(n^{-1/2}),$$

which follows immediately from the results of Section 4.2.2 applied to the model with ψ held fixed at ψ_0, and that

$$-\ell_{\lambda\lambda}(\psi_0, \lambda_0)/n = \bar{i}_{\lambda\lambda} + O_p(n^{-1/2}).$$

Hence,

$$\ell_\lambda(\psi_0, \lambda_0)(\hat{\lambda}_0' - \lambda_0') + (\hat{\lambda}_0 - \lambda_0)^T \ell_{\lambda\lambda}(\psi_0, \lambda_0)(\hat{\lambda}_0' - \lambda_0') = O_p(n^{-1/2})$$

and

$$\begin{aligned} \ell_p'(\psi_0) &= \tilde{\ell}_\psi(\psi_0) + \frac{d}{d\psi} \ell_\lambda(\psi, \lambda_\psi) \Big|_{\psi=\psi_0} (\hat{\lambda}_0 - \lambda_0) \\ &\quad + \frac{1}{2} (\hat{\lambda}_0 - \lambda_0)^T \ell_{\lambda\lambda\psi}(\lambda_0, \psi_0)(\hat{\lambda}_0 - \lambda_0) + O_p(n^{-1}). \end{aligned}$$

Since $\lambda_0' = 0$,

$$\frac{d}{d\psi} \ell_\lambda(\psi, \lambda_\psi) \Big|_{\psi=\psi_0} = \ell_{\psi\lambda}(\psi_0, \lambda_0),$$

and

$$\begin{aligned} \frac{1}{\sqrt{n}} \ell_p'(\psi_0) &= \frac{1}{\sqrt{n}} \tilde{\ell}_\psi(\psi_0) + \Big[\frac{1}{\sqrt{n}} \ell_{\psi\lambda}(\psi_0, \lambda_0) \bar{i}_{\lambda\lambda}^{-1} \frac{1}{\sqrt{n}} \ell_\lambda(\psi_0, \lambda_0) \frac{1}{\sqrt{n}} \\ &\quad + \frac{1}{2} \Big(\frac{1}{\sqrt{n}} \ell_\lambda(\psi_0, \lambda_0) \Big)^T \bar{i}_{\lambda\lambda}^{-1} \mu_{\lambda\lambda\psi} \bar{i}_{\lambda\lambda}^{-1} \Big(\frac{1}{\sqrt{n}} \ell_\lambda(\psi_0, \lambda_0) \Big) \Big] \frac{1}{\sqrt{n}} \\ &\quad + O_p(n^{-1}). \end{aligned} \tag{5.11}$$

In computing $\hat{\kappa}_j(T)$, $j = 1, 2, 3$, we may therefore calculate $\tilde{Z}_j$ using $\tilde{\ell}(\psi)$; for instance,

$$\tilde{Z}_1 = \tilde{\ell}_\psi'(\psi_0)/\sqrt{n}.$$

Also,

$$\begin{aligned} B_1 &= \frac{1}{\sqrt{n}} \ell_{\psi\lambda}(\psi_0, \lambda_0) \bar{i}_{\lambda\lambda}^{-1} \frac{1}{\sqrt{n}} \ell_\lambda(\psi_0, \lambda_0) \\ &\quad + \frac{1}{2} \Big(\frac{1}{\sqrt{n}} \ell_\lambda(\psi_0, \lambda_0) \Big)^T \bar{i}_{\lambda\lambda}^{-1} \mu_{\lambda\lambda\psi} \bar{i}_{\lambda\lambda}^{-1} \Big(\frac{1}{\sqrt{n}} \ell_\lambda(\psi_0, \lambda_0) \Big). \end{aligned}$$

The derivatives of $\tilde{\ell}(\psi_0)$ may be expanded in terms of the usual log-likelihood derivatives. Specifically,

$$\tilde{\ell}_\psi(\psi_0) = \frac{d}{d\psi}\ell(\psi, \lambda_\psi)\Big|_{\psi=\psi_0} = \ell_\psi(\psi_0, \lambda_0) + \ell_\lambda(\psi_0, \lambda_0)\lambda_0' = \ell_\psi(\psi_0, \lambda_0);$$

$$\tilde{\ell}_{\psi\psi}(\psi_0) = \ell_{\psi\psi}(\psi_0, \lambda_0) + \ell_\lambda(\psi_0, \lambda_0)\lambda_0''$$

$$= \ell_{\psi\psi}(\psi_0, \lambda_0) + \ell_\lambda(\psi_0, \lambda_0)\bar{i}_{\lambda\lambda}^{-1}\mu_{\psi\psi\lambda};$$

$$\tilde{\ell}_{\psi\psi\psi}(\psi_0) = \ell_{\psi\psi\psi}(\psi_0, \lambda_0) + 3\ell_{\psi\lambda}(\psi_0, \lambda_0)\lambda_0'' + \ell_\lambda(\psi_0, \lambda_0)\lambda_0'''.$$

Note that the expression $\lambda_0'' = \bar{i}_{\lambda\lambda}^{-1}\mu_{\psi\psi\lambda}$ may be obtained by differentiating the expression $\mathrm{E}[\ell_\lambda(\psi, \lambda_\psi); \theta_0] = 0$ with respect to ψ; a similar expression may be obtained for λ_0''', although it will not be needed.

From these results we obtain that $\tilde{Z}_1 = \ell_\psi(\psi_0, \lambda_0)/\sqrt{n}$, and

$$\tilde{Z}_2 = \frac{\ell_\lambda(\psi_0, \lambda_0)}{\sqrt{n}}\bar{i}_{\lambda\lambda}^{-1}\mu_{\psi\psi\lambda} + \frac{\ell_{\psi\psi}(\psi_0, \lambda_0)/n + \bar{i}_{\psi\psi}}{\sqrt{n}}.$$

It follows that

$$\tilde{V}_2 = -\bar{i}_{\psi\psi}; \qquad \tilde{V}_3 = \mu_{\psi\psi\psi}; \qquad \mathrm{cum}(\tilde{Z}_1, \tilde{Z}_1) = \bar{i}_{\psi\psi};$$

$$\mathrm{cum}(\tilde{Z}_1, \tilde{Z}_1, \tilde{Z}_1) = \mu_{\psi,\psi,\psi}\frac{1}{\sqrt{n}}; \qquad \mathrm{cum}(\tilde{Z}_1, \tilde{Z}_2) = \mu_{\psi,\psi\psi};$$

$$\mathrm{cum}(B_1) = \mathrm{tr}(\mu_{\lambda,\lambda\psi}\bar{i}_{\lambda\lambda}^{-1}) + \tfrac{1}{2}\mathrm{tr}(\mu_{\lambda\lambda\psi}\bar{i}_{\lambda\lambda}^{-1}); \qquad \mathrm{cum}(\tilde{Z}_1, B_1) = O(n^{-1/2});$$

$$\mathrm{cum}(\tilde{Z}_1, \tilde{Z}_1, B_1) = O(n^{-1/2}).$$

Also,

$$\mathrm{cum}(\tilde{Z}_1, \tilde{Z}_1, \tilde{Z}_1\tilde{Z}_2) = 2\mathrm{cum}(\tilde{Z}_1, \tilde{Z}_1)\mathrm{cum}(\tilde{Z}_1, \tilde{Z}_2) + O(n^{-1})$$

$$= 2\bar{i}_{\psi\psi}\mu_{\psi,\psi\psi} + O(n^{-1});$$

$$\mathrm{cum}(\tilde{Z}_1, \tilde{Z}_1, \tilde{Z}_1^2) = 2\mathrm{cum}(\tilde{Z}_1, \tilde{Z}_1)^2 + O(n^{-1}) = 2\bar{i}_{\psi\psi}^2 + O(n^{-1}).$$

Using these results, the first three formal cumulants of $\sqrt{i_\psi}(\hat{\psi} - \psi)$ are given by

$$\hat{\kappa}_1 = \left[\frac{\mu_{\psi,\psi\psi}}{\bar{i}_{\psi\psi}^{3/2}} + \frac{1}{2}\frac{\mu_{\psi\psi\psi}}{\bar{i}_{\psi\psi}^{3/2}} + \frac{\mathrm{tr}(\mu_{\lambda,\lambda\psi}\bar{i}_{\lambda\lambda}^{-1}) + \mathrm{tr}(\mu_{\lambda\lambda\psi}\bar{i}_{\lambda\lambda}^{-1})/2}{\bar{i}_{\psi\psi}^{1/2}}\right]\frac{1}{\sqrt{n}} + O(n^{-3/2});$$

$$\hat{\kappa}_2 = 1 + O(n^{-1});$$

$$\hat{\kappa}_3 = \frac{\mu_{\psi,\psi,\psi} + 3\mu_{\psi\psi\psi} + 6\mu_{\psi,\psi\psi}}{\bar{i}_{\psi\psi}^{3/2}}\frac{1}{\sqrt{n}} + O(n^{-3/2}).$$

Note that, using the Bartlett identities, these expressions may be written in different, but equivalent, forms. For instance,

$$\hat{\kappa}_3 = \frac{\mu_{\psi\psi\psi} - \mu_{\psi,\psi,\psi}}{\bar{i}_{\psi\psi}^{3/2}} \frac{1}{\sqrt{n}} + O(n^{-3/2}).$$

Further simplification is possible using the fact that ψ and λ are orthogonal. Orthogonality implies that

$$E[\ell_{\psi\lambda}(\theta); \psi, \lambda] = 0$$

for all ψ, λ. Differentiating this expression with respect to λ, for example, shows that

$$\mu_{\psi\lambda\lambda} + \mu_{\lambda,\lambda\psi} = 0.$$

Example 5.6 Standard deviation of a normal distribution
Let $Y_1, \ldots, Y_n$ denote independent normally distributed random variables each with mean λ and standard deviation ψ as in Example 5.4. The log-likelihood function is given by

$$\ell(\psi, \lambda) = -\frac{1}{2\psi^2} \sum (y_j - \lambda)^2 - n \log \psi.$$

Note that ψ and λ are orthogonal parameters. It is straightforward to show that

$$\bar{i}_{\psi\psi} = \frac{2}{\psi^2}; \quad \bar{i}_{\lambda\lambda} = \frac{1}{\psi^2}; \quad \mu_{\psi,\psi,\psi} = \frac{8}{\psi^3}; \quad \mu_{\psi\psi\psi} = \frac{10}{\psi^3};$$

$$\mu_{\psi,\psi\psi} = -\frac{6}{\psi^3}; \quad \mu_{\lambda,\lambda\psi} = -\frac{2}{\psi^3}; \quad \mu_{\lambda\lambda\psi} = \frac{2}{\psi^3}.$$

Hence, the first three formal cumulants of $\sqrt{i_\psi}(\hat{\psi} - \psi)$ are given by

$$\hat{\kappa}_1 = -\frac{3}{2\sqrt{2}} \frac{1}{\sqrt{n}}; \quad \hat{\kappa}_2 = 1; \quad \hat{\kappa}_3 = \frac{1}{\sqrt{2}} \frac{1}{\sqrt{n}}:$$

the same results obtained in Example 5.4. ∎

Example 5.7 Index of a gamma distribution
Let $Y_1, \ldots, Y_n$ denote independent gamma random variables each with index ψ and rate parameter λ, as in Example 5.5. The log-likelihood function for this model is given by

$$n\psi \log \lambda - n \log \Gamma(\psi) + n(\psi - 1) \sum \log y_j - n\lambda\bar{y}.$$

To use the results derived in this section, we need an orthogonal parameterization. Since this is an exponential family model and ψ is a natural parameter, a parameter orthogonal to ψ is given by the complementary mean parameter

$$\phi = \mathrm{E}[\bar{Y}; \psi, \lambda] = \frac{\psi}{\lambda}.$$

The log-likelihood function in terms of (ψ, ϕ) is given by

$$\ell(\psi, \phi) = n\psi \log \psi - n\psi \log \phi - n \log \Gamma(\psi) \\ + (\psi - 1) \sum \log y_j - \frac{\psi}{\phi} \sum y_j.$$

It follows that

$$\mu_{\psi,\psi,\psi} = \Psi''(\psi) + \frac{1}{\psi^2}; \qquad \mu_{\psi\psi\psi} = -\Psi''(\psi) - \frac{1}{\psi^2}; \qquad \mu_{\phi,\psi\phi} = \frac{1}{\phi^2};$$

$$\mu_{\psi,\psi\psi} = 0; \qquad \mu_{\phi\phi\psi} = -\frac{1}{\phi^2}; \qquad \bar{i}_{\phi\phi} = \frac{\psi}{\phi^2}; \qquad \bar{i}_{\psi\psi} = \Psi'(\psi) - \frac{1}{\psi}.$$

It follows that the first three formal cumulants of $\sqrt{i_\psi}(\theta)(\hat{\psi} - \psi)$ are given by

$$\frac{1}{2} \frac{\Psi'(\psi)/\psi - \Psi''(\psi) - 2/\psi^2}{(\Psi'(\psi) - 1/\psi)^{3/2}} \frac{1}{\sqrt{n}},$$

1, and

$$-2 \frac{\Psi''(\psi) + 1/\psi^2}{(\Psi'(\psi) - 1/\psi)^{3/2}} \frac{1}{\sqrt{n}}.$$

These agree with results of Example 5.5. ∎

5.4 Likelihood ratio statistic

5.4.1 SINGLE PARAMETER CASE

Suppose that θ is a real-valued parameter. For a given value of θ, the likelihood ratio statistic is given by

$$W \equiv W(\theta) = 2[\ell(\hat{\theta}) - \ell(\theta)].$$

The statistic $W(\theta_0)$ can be used to test the null hypothesis $\theta = \theta_0$ versus the alternative $\theta \neq \theta_0$ or $W(\theta)$ may be used to set confidence limits for θ. In studying the asymptotic properties of the likelihood ratio statistic, it is often convenient to also work with the square-root version of the statistic, given by

$$R \equiv R(\theta) = \mathrm{sgn}(\hat{\theta} - \theta)\{2[\ell(\hat{\theta}) - \ell(\theta)]\}^{1/2}.$$

A stochastic asymptotic expansion for W may be obtained immediately from the result in Section 5.2.3, taking $H(\cdot)$ to be $2\ell(\cdot)$. As in Section 5.3, let $\nu_{ijk\ell}$ denote the standardized cumulants of the log-likelihood derivatives. Then

$$
\begin{aligned}
W = {} & \frac{Z_1^2}{\nu_{20}} + \left(\frac{Z_1^2 Z_2}{\nu_{20}^2} + \frac{1}{3} \frac{\nu_{001}}{\nu_{20}^3} Z_1^3 \right) \frac{1}{\sqrt{n}} \\
& + \left(\frac{1}{4} \frac{\nu_{001}^2}{\nu_{20}^5} Z_1^4 + \frac{1}{3} \frac{Z_1^3 Z_3}{\nu_{20}^3} + \frac{1}{12} \frac{\nu_{0001}}{\nu_{20}^4} Z_1^4 + \frac{\nu_{001}}{\nu_{20}^4} Z_1^3 Z_2 + \frac{Z_1^2 Z_2^2}{\nu_{20}^3} \right) \frac{1}{n} \\
& + O(n^{-3/2}).
\end{aligned}
$$

Using a Taylor's series expansion for the square-root function, an expansion for the signed square root of W is given by

$$
\begin{aligned}
R = {} & \frac{Z_1}{\sqrt{\nu_{20}}} \left[1 + \left(\frac{1}{6} \frac{\nu_{001}}{\nu_{20}^2} Z_1 + \frac{1}{2} \frac{Z_2}{\nu_{20}} \right) \frac{1}{\sqrt{n}} \right. \\
& \left. + \left(\frac{1}{6} \frac{Z_1 Z_3}{\nu_{20}^3} - \frac{1}{8} \frac{\nu_{0001}}{\nu_{20}^3} Z_1^2 - \frac{7}{18} \frac{\nu_{001}^2}{\nu_{20}^4} Z_1^2 - \frac{7}{12} \frac{\nu_{001}}{\nu_{20}^3} Z_1 Z_2 + \frac{3}{8} \frac{Z_2^2}{\nu_{20}^2} \right) \frac{1}{n} \right] \\
& + O(n^{-3/2}).
\end{aligned}
$$

From these expressions it is straightforward to derive the following expressions for the formal cumulants of R under the null hypothesis.

$$
\hat{\kappa}_1(R) = \left(\frac{1}{6} \frac{\nu_{001}}{\nu_{20}^{3/2}} + \frac{1}{2} \frac{\nu_{11}}{\nu_{20}^{3/2}} \right) \frac{1}{\sqrt{n}} + O(n^{-3/2});
$$

$$
\begin{aligned}
\hat{\kappa}_1(R)^2 + \hat{\kappa}_2(R) & = \hat{\kappa}_1(W) \\
& = 1 + \left[\frac{\nu_{21}}{\nu_{20}^2} + \frac{1}{3} \frac{\nu_{001} \nu_{30}}{\nu_{20}^3} + \frac{3}{4} \frac{\nu_{001}^2}{\nu_{20}^3} + \frac{\nu_{101}}{\nu_{20}} \right. \\
& \qquad \left. + \frac{1}{4} \frac{\nu_{0001}}{\nu_{20}^2} + 3 \frac{\nu_{001} \nu_{11}}{\nu_{20}^3} + \frac{\nu_{02}}{\nu_{20}^2} + 2 \frac{\nu_{11}^2}{\nu_{20}^3} \right] \frac{1}{n} + O(n^{-3/2});
\end{aligned}
$$

$$
\hat{\kappa}_3(R) = \frac{\nu_{30} + \nu_{001} + 3\nu_{11}}{\nu_{20}^{3/2}} \frac{1}{\sqrt{n}} + O(n^{-3/2});
$$

$$
\begin{aligned}
\hat{\kappa}_4(R) = {} & \left[\frac{\nu_{40} + 3\nu_{02} + 6\nu_{21} + 4\nu_{101} + \nu_{0001}}{\nu_{20}^2} \right. \\
& \left. + \frac{(4\nu_{001} + 6\nu_{11})(\nu_{30} + 3\nu_{11} + \nu_{001})}{\nu_{20}^3} \right] \frac{1}{n} + O(n^{-2}).
\end{aligned}
$$

Recall that the third and fourth Bartlett identities are

$$
\nu_{30} + 3\nu_{11} + \nu_{001} = 0; \qquad \nu_{40} + 3\nu_{02} + 6\nu_{21} + 4\nu_{101} + \nu_{0001} = 0.
$$

It follows that $\hat{\kappa}_3(R) = 0$ and $\hat{\kappa}_4(R) = 0$, neglecting terms of order $O(n^{-3/2})$. Hence, under the null hypothesis,

$$\hat{\kappa}_1(R) = \frac{\mu}{\sqrt{n}} + O(n^{-3/2}), \qquad \hat{\kappa}_2(R) = 1 + \frac{\tau}{n} + O(n^{-3/2}),$$

$\hat{\kappa}_3(R) = \hat{\kappa}_4(R) = O(n^{-3/2})$ where μ and τ are $O(1)$ constants.

Hence,

$$\Pr(R \le r) = \Phi(r) - \phi(r)\Big\{\frac{\mu}{\sqrt{n}} + \frac{1}{2}\Big(\frac{\mu^2}{n} + \frac{\tau}{n}\Big)r + O(n^{-3/2})\Big\}.$$

Using a similar expansion for $\Pr(R \le -r)$ and subtracting shows that, under the null hypothesis,

$$\begin{aligned}\Pr(W \le r^2; \theta_0) &= \Pr(-r \le R \le r)\\ &= 2\Phi(r) - 1 - \phi(r)[(\mu^2 + \tau)r]\frac{1}{n} + O(n^{-2})\end{aligned}$$

where

$$\mu^2 + \tau = \Bigg[\frac{\nu_{21}}{\nu_{20}^2} + \frac{1}{3}\frac{\nu_{001}\nu_{30}}{\nu_{20}^3} + \frac{3}{4}\frac{\nu_{001}^2}{\nu_{20}^3} + \frac{\nu_{101}}{\nu_{20}^2} + \frac{1}{4}\frac{\nu_{0001}}{\nu_{20}^2} + 3\frac{\nu_{001}\nu_{11}}{\nu_{20}^3} + \frac{\nu_{02}}{\nu_{20}^2} + 2\frac{\nu_{11}^2}{\nu_{20}^3}\Bigg]. \tag{5.12}$$

Note that $1+(\mu^2+\tau)/n$ is the formal second moment of R under the null hypothesis or, equivalently, the formal first moment of W. The remainder term in the above expansion is $O(n^{-2})$, rather than $O(n^{-3/2})$, as might be expected, because the $O(n^{-3/2})$ term is an even function of r and, hence the terms corresponding to r and $-r$ cancel.

Let $b = \mu^2 + \tau$. Consider the adjusted likelihood ratio statistic

$$W_B = \frac{W}{1 + b/n}.$$

Then

$$\begin{aligned}\Pr(W_B \le r^2) &= \Pr[W \le r^2(1 + b/n)]\\ &= \Pr\{W \le [r(1 + b/(2n))]^2\} + O(n^{-2}).\end{aligned}$$

It follows that

$$\Pr(W_B \le r^2) = 2\Phi(r) - 1 + O(n^{-2});$$

that is, W_B has a chi-square distribution to order $O(n^{-2})$. The same result holds if W_B is given by

$$W_B = \frac{W}{1 + \hat{b}/n}$$

where $\hat{b} = b + O_p(n^{-1/2})$. The statistic W_B is called the Bartlett-adjusted likelihood ratio statistic.

Example 5.8 Exponential distribution
Let $Y_1, \dots, Y_n$ denote independent random variables each with an exponential distribution with rate parameter θ and consider inference about θ.

The likelihood ratio statistic is given by

$$W = 2n[\theta\bar{y} - \log(\theta\bar{y}) - 1].$$

Using the fact that, under the distribution with parameter θ, $n\theta\bar{Y}$ has a standard gamma distribution with index n, it is straightforward to show that

$$\mathrm{E}[W; \theta] = 2n[\log(n) - \Psi(n)].$$

Hence, the Bartlett-corrected likelihood ratio statistic is given by

$$W_B = \frac{\theta\bar{y} - \log(\theta\bar{y}) - 1}{\log(n) - \Psi(n)};$$

this statistic can be used to set confidence limits for θ or as a test statistic.

Consider the problem of setting confidence limits for θ. Confidence limits based on W consist of all θ satisfying

$$\theta\bar{y} - \log(\theta\bar{y}) \leq 1 + \frac{\chi_1^2(\alpha)}{2n};$$

these inequalities can be written in terms of inequalities for $\theta\bar{y}$. For instance, for $n = 10$ and $\alpha = 0.05$,

$$0.36 \leq \theta\bar{y} \leq 2.15.$$

The same is true for confidence limits based on W_B.

Since $n\theta\bar{y}$ has a standard gamma distribution with index n, it is straightforward to determine the exact coverage probability of confidence limits based on either W or W_B. The coverage probabilities of approximate 95% confidence limits based on these methods are given in Table 5.2 for several choices of n. These results indicate that, while the coverage probability of confidence limits based on W are reasonably close to the nominal value 0.95, the coverage probability of those based on the Bartlett-adjusted likelihood ratio statistic W_B is essentially exactly equal to 0.95.

Table 5.2 Coverage probabilities of confidence intervals in Example 5.8

n	W	W_B
1	0.932	0.951
2	0.941	0.950
3	0.944	0.950
5	0.946	0.950
10	0.948	0.950

∎

Example 5.9 Logistic distribution
Let $Y_1, \ldots, Y_n$ denote independent random variables each with a logistic distribution with location parameter θ and scale parameter 1. Hence, each Y_j has density

$$\frac{\exp\{-(y-\theta)\}}{[1+\exp\{-(y-\theta)\}]^2}, \quad -\infty < y < \infty$$

where $-\infty < \theta < \infty$. Hence,

$$\ell(\theta) = -\sum (y_j - \theta) - 2\sum \log[1 + \exp\{-(y_j - \theta)\}].$$

Consider the problem of testing $\theta = \theta_0$. Without loss of generality we may take $\theta_0 = 0$.

There is not a closed-form expression for the maximum likelihood estimate $\hat{\theta}$ and, hence, not for the log-likelihood ratio statistic W. The Bartlett adjustment factor may be calculated using the expression (5.12). The log-likelihood derivatives are given by

$$\ell_\theta(\theta_0) = \sum \frac{1-\exp\{-y_j\}}{1+\exp\{-y_j\}}; \qquad \ell_{\theta\theta}(\theta_0) = -2\sum \frac{\exp\{-y_j\}}{(1+\exp\{-y_j\})^2};$$

$$\ell_{\theta\theta\theta}(\theta_0) = -2\sum \frac{\exp\{-y_j\}(1-\exp\{-y_j\})}{(1+\exp\{-y_j\})^3};$$

$$\ell_{\theta\theta\theta\theta}(\theta_0) = -2\sum \frac{\exp\{-y_j\}(1-4\exp\{-y_j\}+\exp\{-2y_j\})}{(1-\exp\{-y_j\})^4}.$$

The cumulants required for (5.12) are given by $\nu_{11} = \nu_{001} = \nu_{30} = 0$, $\nu_{20} = 1/3$, $\nu_{21} = \nu_{101} = -1/15$, $\nu_{02} = 2/15$, $\nu_{0001} = 1/15$. Hence, the Bartlett-corrected likelihood ratio statistic is given by $W/[1 + 2/(5n)]$. ∎

Example 5.10 Poisson distribution
Let $Y_1, \ldots, Y_n$ denote independent Poisson random variables each with mean θ. The log-likelihood function is given by

$$\ell(\theta) = \log\theta \sum y_j - n\theta$$

so that

$$W = 2n[\bar{y}\log(\bar{y}/\theta) - (\bar{y} - \theta)].$$

Using a Taylor's series expansion for the log function,

$$W = n\left[\frac{1}{\theta}(\bar{y}-\theta)^2 - \frac{1}{3\theta^2}(\bar{y}-\theta)^3 + \frac{1}{6\theta^3}(\bar{y}-\theta)^4\right] + O_p(n^{-3/2}).$$

Using the fact that all cumulants of the Poisson distribution are equal to the mean, the first formal cumulant of W may be written

$$1 + \frac{1}{6\theta}\frac{1}{n} + O(n^{-2});$$

the Bartlett-adjusted likelihood ratio statistic is therefore given by

$$\frac{2n[\bar{y}\log(\bar{y}/\theta) - (\bar{y} - \theta)]}{1 + 1/(6\theta n)}.$$

Note, however, that W has a discrete distribution (although it is nonlattice); hence, it does not necessarily follow that the distribution of the Bartlett-adjusted likelihood ratio statistic is better approximated by a χ^2 distribution than is the distribution of W itself. Suppose that we are interested in testing the hypothesis that $\theta = 1$ versus the alternative $\theta \neq 1$. For the case $n = 5$ the acceptance region of the likelihood ratio test with nominal level 0.05 is

$$2 \leq \sum y_j \leq 9.$$

For the Bartlett-adjusted likelihood ratio test, the acceptance region is also

$$2 \leq \sum y_j \leq 9.$$

This is due to the fact that $\sum y_j$ can only take integer values. The actual level of this test is 0.072. Hence, in this case, the Bartlett-adjusted likelihood ratio statistic offers no improvement over the unadjusted statistic.

For the case $n = 2$, the acceptance region of the likelihood ratio test with nominal level 0.05 is

$$1 \leq \sum y_j \leq 5;$$

for the corresponding test based on the Bartlett-adjusted likelihood ratio statistic, the acceptance region is

$$0 \leq \sum y_j \leq 5.$$

The likelihood ratio test has actual level 0.152 while the Bartlett-adjusted likelihood ratio test has actual level 0.017. Hence, in this case, the Bartlett adjustment is beneficial. ∎

It is important to note that, although W_B has a chi-square distribution with error $O(n^{-2})$, it does not follow that, in general, the signed square root of W_B, R_B has a standard normal distribution to this same order. This is due to the fact that

$$\hat{\kappa}_1(R_B) = \hat{\kappa}_1(R) + O(n^{-3/2}) = \left(\frac{1}{6}\frac{\nu_{001}}{\nu_{20}^{3/2}} + \frac{1}{2}\frac{\nu_{11}}{\nu_{20}^{3/2}}\right)\frac{1}{\sqrt{n}} + O(n^{-3/2}).$$

The higher-order asymptotic properties of R, as well as the problem of adjusting R to have a standard normal distribution with error of order $O(n^{-3/2})$, is the subject of Chapter 7.

Example 5.11 Exponential distribution
As in Example 5.8, let $Y_1, \ldots, Y_n$ denote independent random variables each with an exponential distribution with rate parameter θ. Recall that, for the case $n = 5$,

95% confidence limits for θ based on W_B have exact coverage probability 0.950. However the 97.5% upper confidence limit for θ based on R_B has exact coverage probability 0.983 while the 97.5% lower confidence limit based on R_B has exact coverage probability 0.967. Taken together, these limits yield a confidence interval for θ with coverage probability

$$0.983 + 0.967 - 1 = 0.950.$$

The individual upper and lower confidence limits however do not have this same accuracy in terms of coverage probability.

Viewed another way, the confidence interval based on W_B does not have the symmetry properties usually associated with a confidence interval. ∎

5.4.2 SCALAR PARAMETER OF INTEREST WITH A NUISANCE PARAMETER

We now consider the case in which θ may be written $\theta = (\psi, \lambda)$ where ψ is a real-valued parameter of interest and λ is a nuisance parameter. The likelihood ratio test statistic for testing $\psi = \psi_0$ is given by

$$W = 2[\ell_p(\hat{\psi}) - \ell_p(\psi_0)]$$

where ℓ_p denotes the profile log-likelihood function. Define the Z_j and V_j in terms of ℓ_p and, let $\hat{\nu}_{ijk\ell}$ denote the standardized cumulants of $(\ell'_p(\psi_0), \ldots, \ell_p^{(4)}(\psi_0))$ under the distribution with parameter θ_0. For instance,

$$\frac{1}{n}\ell'_p(\psi_0) = V_1 + Z_1/\sqrt{n}$$

and

$$\hat{\nu}_2 = \mathrm{E}[\ell''_p(\psi_0); \theta_0].$$

Using this approach, many of the results derived in the previous section continue to hold. One major difference however is that we do not have $\hat{\nu}_1$ and $\hat{\nu}_{20} + \hat{\nu}_{01}$ both exactly equal to zero, although both will be approximately zero.

Let $\delta = \sqrt{n}\hat{\nu}_1$ and $\tilde{Z}_1 = Z_1 - \delta/\sqrt{n}$; hence, $\tilde{Z}_1$ is simply a mean-corrected version of Z_1. Hence, the first cumulant of $\tilde{Z}_1$ is 0 and all other cumulants involving $\tilde{Z}_1$ are the same as those involving Z_1. Note that $\delta = O(1)$. The statistic W has the following expansion:

$$\begin{aligned} W = {} & \frac{\tilde{Z}_1^2}{(-\hat{\nu}_{01})} + \left(\frac{\tilde{Z}_1^2 Z_2}{(-\hat{\nu}_{01})^2} + \frac{1}{3}\frac{\hat{\nu}_{001}}{(-\hat{\nu}_{01})^3}\tilde{Z}_1^3 + 2\frac{\delta}{(-\hat{\nu}_{01})}\tilde{Z}_1 \right)\frac{1}{\sqrt{n}} \\ & + \left(\frac{1}{4}\frac{\hat{\nu}_{001}^2}{(-\hat{\nu}_{01})^5}\tilde{Z}_1^4 + \frac{1}{3}\frac{\tilde{Z}_1^3 Z_3}{(-\hat{\nu}_{01})^3} + \frac{1}{12}\frac{\hat{\nu}_{0001}}{(-\hat{\nu}_{01})^4}\tilde{Z}_1^4 + \frac{\hat{\nu}_{001}}{(-\hat{\nu}_{01})^4}\tilde{Z}_1^3 Z_2 \right. \\ & \left. + \frac{\tilde{Z}_1^2 Z_2^2}{(-\hat{\nu}_{01})^3} + \frac{\delta^2}{(-\hat{\nu}_{01})} + 2\frac{\delta}{(-\hat{\nu}_{01})^2}\tilde{Z}_1 Z_2 + \frac{\hat{\nu}_{001}\delta}{(-\hat{\nu}_{01})^3}\tilde{Z}_1^2 \right)\frac{1}{n} + O_p(n^{-3/2}). \end{aligned}$$

A similar expansion for R, the signed square root of W is also easily obtained.

The formal cumulants of R may be obtained using the same approach as in the previous section. First consider the relationship between $\hat{\nu}_{20}$ and $\hat{\nu}_{01}$. Since the log-likelihood ratio statistic W is invariant under reparameterizations that leave ψ unchanged, we may assume without loss of generality that λ is orthogonal to ψ.

Recall that the profile log-likelihood function $\ell_p(\psi)$ may be approximated by $\tilde{\ell}(\psi) = \ell(\psi, \lambda_\psi)$ and that $\tilde{\ell}(\psi)$ is a genuine log-likelihood function. Let $\tilde{\nu}_{ijk\ell}$ denote the standardized cumulants of $(\tilde{\ell}_\psi(\psi), \tilde{\ell}_{\psi\psi}(\psi), \tilde{\ell}_{\psi\psi\psi}(\psi), \tilde{\ell}_{\psi\psi\psi\psi}(\psi))$. Using (5.11) it follows that

$$\hat{\nu}_1 = \tilde{\nu}_1 + O(n^{-1}); \qquad \hat{\nu}_2 = \tilde{\nu}_2 + O(n^{-1}).$$

Using (5.10), it can be shown that

$$\frac{1}{n}\ell_p''(\psi) = \frac{1}{n}\tilde{\ell}_{\psi\psi}(\psi) + O_p(n^{-1}) \tag{5.13}$$

and, hence,

$$\hat{\nu}_{01} = \tilde{\nu}_{01} + O(n^{-1}).$$

It follows that

$$\hat{\nu}_2 + \hat{\nu}_{01} = \tilde{\nu}_2 + \tilde{\nu}_{01} + O(n^{-1}) = O(n^{-1}).$$

The third and fourth Bartlett identities may be handled using the same approach used for the second Bartlett identity. It may be shown that

$$\hat{\nu}_3 + 3\hat{\nu}_{11} + \hat{\nu}_{001} = O(n^{-1/2})$$

and that

$$\hat{\nu}_{40} + 3\hat{\nu}_{02} + 6\hat{\nu}_{21} + 4\hat{\nu}_{101} + \hat{\nu}_{0001} = O(n^{-1/2}).$$

Using these results, it follows that the formal cumulants of R are of the form

$$\hat{\kappa}_1(R) = \frac{\mu}{\sqrt{n}} + O(n^{-3/2}), \qquad \hat{\kappa}_2(R) = 1 + \frac{\tau}{n} + O(n^{-2}),$$

$\hat{\kappa}_3(R) = \hat{\kappa}_4(R) = O(n^{-3/2})$. Hence, $\Pr(W \leq r^2)$ is of the form

$$\Pr(W \leq r^2) = 2\Phi(r) - 1 - \phi(r)[(\mu^2 + \tau)r]\frac{1}{n} + O(n^{-2}).$$

Write

$$\hat{\kappa}_1(W) = \hat{\kappa}_1(R)^2 + \hat{\kappa}_2(R) = 1 + b/n;$$

then

$$W_B = \frac{W}{1 + b/n}$$

has a chi-square distribution with 1 degree of freedom, with error $O(n^{-2})$.

Example 5.12 Comparison of exponential distributions
Let $Y_1, \ldots, Y_n$ and $X_1, \ldots, X_m$ denote independent exponential random variables such that $Y_1, \ldots, Y_n$ each have rate parameter $\psi\lambda$ and $X_1, \ldots, X_m$ each have rate parameter λ. Hence, ψ represents the ratio of the rate parameters of the Y_j and that of the X_j. Consider the problem of setting confidence limits for ψ. The log-likelihood function for this model is given by

$$\ell(\psi, \lambda) = n \log \psi + (n+m) \log \lambda - \lambda(n\psi\bar{y} + m\bar{x}).$$

The log-likelihood ratio statistic for testing $\psi = \psi_0$ is therefore given by

$$W = 2\left[(n+m)\log\left(\frac{n\psi_0\bar{y} + m\bar{x}}{n+m}\right) - (n\log(\psi_0\bar{y}) + m\log\bar{x})\right].$$

Note that, when ψ_0 and λ_0 represent the true parameter values, then $n\psi_0\lambda_0\bar{y}$ and $m\lambda_0\bar{x}$ are independent gamma random variables with indices n and m, respectively. Hence, we may write

$$\begin{aligned} W = 2[&(n+m)\log(Y_n + Y_m) - n\log Y_n - m\log Y_m \\ &- (n+m)\log(n+m) + n\log(n) + m\log(m)] \end{aligned}$$

where Y_n and Y_m denote independent gamma random variables with indices n and m, respectively. Using properties of the gamma distribution, it is straightforward to show that

$$\begin{aligned} \mathrm{E}[W; \theta_0] = 2[&(n+m)\Psi(n+m) - n\Psi(n) - m\Psi(m) \\ &- (n+m)\log(n+m) + n\log(n) + m\log(m)]. \end{aligned}$$

As $n \to \infty$,

$$n[\log(n) - \Psi(n)] = \frac{1}{2} + \frac{1}{12n} + O(n^{-3})$$

and, hence,

$$\mathrm{E}[W; \theta_0] = 1 + \frac{1}{6}\left(\frac{1}{n} + \frac{1}{m} - \frac{1}{n+m}\right) + O(n^{-3}) + O(m^{-3}).$$

Let $Q \equiv Q_\psi = \psi\bar{y}/\bar{x}$ and let $\pi = n/(n+m)$. Then confidence limits for ψ based on the chi-square approximation to the distribution of R^2 are given by the set of all ψ satisfying

$$\pi\log(\pi + (1-\pi)/Q) + (1-\pi)\log(\pi Q + (1-\pi)) \le \frac{\chi_1^2(\alpha)}{2(n+m)}.$$

Confidence limits based on the chi-square approximation to the distribution of the Bartlett-corrected likelihood ratio statistic are given by

$$\pi \log(\pi + (1-\pi)/Q) + (1-\pi)\log(\pi Q + (1-\pi))$$
$$\leq \frac{\chi_1^2(\alpha)}{2(n+m)}\Big(1 + \frac{1}{6}\Big(\frac{1}{n} + \frac{1}{m} - \frac{1}{n+m}\Big)\Big).$$

Note that Q has an F-distribution with $(2n, 2m)$ degrees of freedom and, hence, the exact coverage probabilities of these confidence limits may be calculated.

For instance, for the case $n = 4, m = 2$, the coverage probability of confidence limits based on W is 0.939 while the coverage probability of confidence limits based on W_B is 0.950. Similar calculations for other choice of n and m indicate that confidence limits based on W_B are very accurate in terms of coverage probability. ∎

5.5 Saddlepoint approximations

Another approach to approximating the distribution of estimates and test statistics is to use approximations based on the saddlepoint method. Unlike methods based on Edgeworth expansions, the saddlepoint approximation requires knowledge of the entire cumulant-generating function of the statistic under consideration. Hence, the use of the saddlepoint approximation is limited to certain special cases. Here we consider the case of a one-parameter exponential family distribution.

5.5.1 DISTRIBUTION OF THE MAXIMUM LIKELIHOOD ESTIMATOR

Let $Y_1, \ldots, Y_n$ denote independent random variables each distributed according to a density of the form

$$\exp\{c(\theta)T(y) - k(\theta) + S(y)\}.$$

The cumulant-generating function of $T(Y_1)$ is given by $k_0(\eta + s) - k_0(\eta)$ where

$$k_0(\eta + s) = k(c^{-1}(c(\theta) + s))$$

and, hence, the saddlepoint approximation to the density of $\bar{T} = n^{-1}\sum T(Y_j)$ is given by

$$\frac{1}{\sqrt{2\pi}}\frac{\sqrt{n}}{[k_0''(\eta + \hat{s})]^{1/2}}\exp\{n[k_0(\eta + \hat{s}) - k_0(\eta) - \hat{s}t]\}$$

where $\hat{s}$ satisfies $k_0'(\eta + \hat{s}) = t$.

The log-likelihood function for this model is given by

$$\ell(\theta) = c(\theta)\sum T(y_j) - nk(\theta)$$

so that the maximum likelihood estimate $\hat{\theta}$ satisfies

$$\frac{k'(\hat{\theta})}{c'(\hat{\theta})} = \bar{T}.$$

Using the usual change-of-variable formula, it follows that the saddlepoint approximation to the density of $\hat{\theta}$ is given by

$$\hat{p}(\hat{\theta}; \theta) = \frac{1}{\sqrt{2\pi}} \frac{\sqrt{n}}{[k_0''(\eta + \hat{s})]^{1/2}} \exp\{n[k_0(\eta + \hat{s}) - k_0(\eta) - \hat{s}k'(\hat{\theta})/c'(\hat{\theta})]\}$$
$$\times \left| \frac{k''(\hat{\theta})}{c'(\hat{\theta})} - \frac{k'(\hat{\theta})}{c'(\hat{\theta})} \frac{c''(\hat{\theta})}{c'(\hat{\theta})} \right|$$

where $\hat{s}$ satisfies

$$k_0'(\eta + \hat{s}) = \frac{k'(\hat{\theta})}{c'(\hat{\theta})}.$$

By the definition of k_0, it follows that

$$k_0'(\eta) = k'(c^{-1}(\eta)) \frac{dc^{-1}(\eta)}{d\eta} = \frac{k'(c^{-1}(\eta))}{c'(c^{-1}(\eta))}$$

so that $\hat{s}$ satisfies

$$\frac{k'(c^{-1}(\eta + \hat{s}))}{c'(c^{-1}(\eta + \hat{s}))} = \frac{k'(\hat{\theta})}{c'(\hat{\theta})}$$

which holds provided that $\eta + \hat{s} = c(\hat{\theta})$. Hence,

$$\hat{p}(\hat{\theta}; \theta) = \frac{\sqrt{n}}{\sqrt{2\pi}} \frac{|k''(\hat{\theta})c'(\hat{\theta}) - (k'(\hat{\theta})/c'(\hat{\theta}))(c''(\hat{\theta})/c'(\hat{\theta}))|}{[k_0''(c(\hat{\theta}))]^{1/2}}$$
$$\times \exp\{n[k_0(c(\hat{\theta})) - k_0(c(\theta)) - (c(\hat{\theta}) - c(\theta))k'(\hat{\theta})/c'(\hat{\theta})]\}.$$

The log-likelihood function for θ may be written

$$\ell(\theta) = c(\theta) \sum T(y_j) - nk(\theta) = nc(\theta) \frac{k'(\hat{\theta})}{c'(\hat{\theta})} - nk_0(c(\theta));$$

hence,

$$n[k_0(c(\hat{\theta})) - k_0(c(\theta)) - (c(\hat{\theta}) - c(\theta))k'(\hat{\theta})/c'(\hat{\theta})] = \ell(\theta) - \ell(\hat{\theta})$$

and

$$nk_0''(c(\hat{\theta})) = \frac{-\ell_{\theta\theta}(\hat{\theta})}{c'(\hat{\theta})^2}.$$

It follows that

$$\hat{p}(\hat{\theta};\theta) = \frac{1}{\sqrt{2\pi}} \frac{n|c'(\hat{\theta})|}{[-\ell_{\theta\theta}(\hat{\theta})]^{1/2}} \exp\{\ell(\theta) - \ell(\hat{\theta})\} \left| \frac{k''(\hat{\theta})}{c'(\hat{\theta})} - \frac{k'(\hat{\theta})}{c'(\hat{\theta})} \frac{c''(\hat{\theta})}{c'(\hat{\theta})} \right|.$$

Finally, note that

$$\ell_{\theta\theta}(\hat{\theta}) = n\Big[c''(\hat{\theta})\frac{k'(\hat{\theta})}{c'(\hat{\theta})} - k''(\hat{\theta})\Big]$$

so we may write

$$\hat{p}(\hat{\theta};\theta) = \frac{1}{\sqrt{2\pi}}[-\ell_{\theta\theta}(\hat{\theta})]^{1/2} \exp\{\ell(\theta) - \ell(\hat{\theta})\}. \tag{5.14}$$

In evaluating this expression it is important to keep in mind that that the log-likelihood function $\ell(\theta)$ is to be interpreted as a function of θ and $\hat{\theta}$, that is, $\ell(\theta) \equiv \ell(\theta; \hat{\theta})$.

In general, (5.17) will not integrate exactly to 1. Hence, we may wish to use a renormalized version of this approximation,

$$p^*(\hat{\theta};\theta) = c[-\ell_{\theta\theta}(\hat{\theta})]^{1/2} \exp\{\ell(\theta) - \ell(\hat{\theta})\},$$

where c is determined numerically. This approximation is a special case of the *likelihood ratio* approximation for the distribution of $\hat{\theta}$; it is also called *Barndorff-Nielsen's formula* or the p^* formula. A more general version of this approximation is given in Chapter 6.

To approximate the distribution function of $\hat{\theta}$, we may begin by approximating the distribution function of $\bar{T}$. Using the result in Section 2.6, $\Pr[\bar{T} \leq t]$ may be approximated by

$$\Phi(r) + \Big(\frac{1}{r} - \frac{1}{v}\Big)\phi(r) \tag{5.15}$$

where

$$r = \sqrt{n}\ \mathrm{sgn}(\hat{s})[2(\hat{s}k_0'(\eta + \hat{s}) - k_0(\eta + \hat{s}))]^{1/2}$$

and

$$v = \hat{s}[nk_0''(\eta + \hat{s})]^{1/2}.$$

Recall that $\bar{T} = k_0'(c(\hat{\theta}))$ so that

$$\Pr[c(\hat{\theta}) \leq z] = \Pr[\bar{T} \leq k_0'(z)].$$

Also, using the form of the log-likelihood function for an exponential family model, r and v may be written

$$r = \mathrm{sgn}(c(\hat{\theta}) - c(\theta))\ [2(\ell(\hat{\theta}) - \ell(\theta))]^{1/2}$$

$$v = (c(\hat{\theta}) - c(\theta))[-\ell_{\theta\theta}(\hat{\theta})]^{1/2}/|c'(\hat{\theta})|;$$

in approximating $\Pr[c(\hat{\theta}) \leq z]$ both of these expressions are to be evaluated at $\hat{\theta} = c^{-1}(z)$.

Example 5.13 Exponential distribution
Consider n independent identically distributed random variables each with an exponential distribution with rate parameter θ; see Example 5.1. Then

$$\ell(\theta) = n \log(\theta) - n\theta/\hat{\theta}$$

and $c(\theta) = \theta$. Hence, $\Pr[\hat{\theta} \leq z; \theta]$ may be approximated by (5.15) with

$$r = \text{sgn}(z - \theta)\, [2(n\theta/z - n - n\log(\theta/z))]^{1/2}$$

and

$$v = \frac{\sqrt{n}(z - \theta)}{|z|}.$$

In order to compare this approximation with those based on the Edgeworth expansion-type approximations, it is convenient to consider an approximation for $\Pr[\sqrt{n}\,(\hat{\theta} - \theta)/\theta \leq t]$. Since

$$\Pr[\sqrt{n}(\hat{\theta} - \theta)/\theta \leq t; \theta] = \Pr[\hat{\theta} \leq \theta(1 + t/\sqrt{n}); \theta],$$

this probability may be approximated by (5.15) with

$$r = \text{sgn}(t)\, [2(n\log(1 + t/\sqrt{n}) - \sqrt{n}t/(1 + t/\sqrt{n}))]^{1/2}$$

and $v = t/(1 + t/\sqrt{n})$.

For this model, it is well known that $\theta/(n\hat{\theta})$ is exactly distributed according to a standard gamma distribution with index n; hence exact values of $\Pr[\sqrt{n}(\hat{\theta} - \theta)/\theta \leq t; \theta]$ are available. For the values of t used in Table 5.1, the probabilities based on the approximation (5.15) are exact to three decimal places. ∎

The above discussion is for continuous random variables. Now consider the case in which $T_1, \ldots, T_n$ are lattice random variables which, for simplicity, we assume take values in the set of nonnegative integers. Then $\bar{T}$ is a lattice random variable, taking values in the set $\{0, 1/n, 2/n, \ldots\}$. The saddlepoint method discussed in Section 2.7 may then be applied, yielding the approximation

$$\frac{1}{\sqrt{2\pi}} \frac{1}{[nk_0''(\eta + \hat{s})]^{1/2}} \exp\{n[k_0(\eta + \hat{s}) - k_0(\eta) - \hat{s}t]\},$$

where $\hat{s}$ satisfies $k_0'(\eta + \hat{s}) = t$.

Since $\bar{T} = k_0'(\hat{\theta})/c'(\hat{\theta})$, the saddlepoint approximation to the density of $\hat{\theta}$ is given by

$$\frac{1}{\sqrt{2\pi}} \frac{1}{[nk_0''(\eta + \hat{s})]^{1/2}} \exp\{n[k_0(\eta + \hat{s}) - k_0(\eta) - \hat{s}k_0'(\hat{\theta})/c'(\hat{\theta})]\}.$$

Writing this expression in terms of the log-likelihood function, we have that the approximate density of $\hat{\theta}$ is given by

$$\frac{|c'(\hat{\theta})|}{\sqrt{2\pi}}[-\ell_{\theta\theta}(\hat{\theta})]^{-1/2}\exp\{\ell(\theta)-\ell(\hat{\theta})\}$$

which differs from the approximation (5.14). This expression is valid at the support points of the distribution of $\hat{\theta}$.

Example 5.14 Poisson distribution
Let $Y_1, \ldots, Y_n$ denote independent random variables each distributed according to a Poisson distribution with mean μ. Take as the parameter of the model $\theta = \log\mu$, the natural parameter. The maximum likelihood estimate of θ is given by $\hat{\theta} = \log\bar{y}$; hence,

$$\ell(\theta) = n(\theta\exp\{\hat{\theta}\} - \exp\{\theta\}).$$

It follows that $-\ell_{\theta\theta}(\hat{\theta}) = n\exp\{\hat{\theta}\}$ and that the approximate density of $\hat{\theta}$ is given by

$$\frac{1}{\sqrt{2\pi}}\exp\{-\hat{\theta}/2\}\frac{(n\exp\{\theta\})^{n\exp\{\hat{\theta}\}}\exp\{-n\exp\{\theta\}\}}{(n\exp\{\hat{\theta}\})^{n\exp\{\hat{\theta}\}}\exp\{-n\exp\{\hat{\theta}\}\}}.$$

Let $x = n\exp\{z\}$ and $\lambda = n\exp\{\theta\}$. Then for values of z in the support of $\hat{\theta}$, $\Pr[\hat{\theta} = z; \theta]$ is approximated by

$$\frac{\lambda^x\exp\{-\lambda\}}{\sqrt{2\pi}x^{x+1/2}\exp\{-x\}}.$$

Note that $n\exp\{\hat{\theta}\} = \sum y_j$ and, therefore, has a Poisson distribution with mean λ. Hence, the exact value of $\Pr(\hat{\theta} = z; \theta)$ is given by

$$\frac{\lambda^x\exp\{-\lambda\}}{x!}, \quad x = n\exp(z).$$

The saddlepoint approximation is simply the exact density with $x!$ replaced by Stirling's approximation. ∎

Now consider approximation of the distribution function of $\hat{\theta}$. As in the case in which the T_j have a continuous distribution, we do this by approximating the distribution function of $\bar{T}$ and then using that approximation as the basis for an approximation to the distribution of $\hat{\theta}$. Using the results in Section 2.7, together with the results for the continuous case given in this section, we have that the tail probability $\Pr[c(\hat{\theta}) \geq z] = \Pr[\bar{T} \geq k_0'(z)]$ may be approximated by

$$1 - \Phi(r) + \left(\frac{1}{v} - \frac{1}{r}\right)\phi(r)$$

where

$$r = \text{sgn}(c(\hat{\theta}) - c(\theta))[2(\ell(\hat{\theta}) - \ell(\theta))]^{1/2}$$
$$v = (1 - \exp\{c(\theta) - c(\hat{\theta})\})[-\ell_{\theta\theta}(\hat{\theta})]^{1/2}/|c'(\hat{\theta})|;$$

both of these expressions are to be evaluated at $\hat{\theta} = c^{-1}(z)$ and it is assumed that $c^{-1}(z)$ is a support point of the distribution of $\hat{\theta}$. An approximation for the lower tail probability may be derived along similar lines.

Example 5.15 Poisson distribution

As in Example 5.14, let $Y_1, \ldots, Y_n$ denote independent Poisson random variables each with natural parameter θ. Then

$$r = \text{sgn}(\hat{\theta} - \theta)\,(2n)^{1/2}[\exp\{\hat{\theta}\}\,(\hat{\theta} - \theta) + \exp\{\theta\} - \exp\{\hat{\theta}\}]^{1/2}$$

and

$$v = (1 - \exp\{\theta - \hat{\theta}\})(n\exp\{\hat{\theta}\})^{1/2}$$

where $\hat{\theta} = z$.

Consider the case $n = 10$ and $\theta = 0$; hence, $10\exp\{\hat{\theta}\}$ has a Poisson distribution with mean 10. For this case,

$$r = \text{sgn}(z)[20(z\exp\{z\} - \exp\{z\} + 1)]^{1/2}$$

and

$$v = (10\exp\{z\})^{1/2}(1 - \exp\{-z\}).$$

In Table 5.3 the resulting approximation to $\Pr(\hat{\theta} \geq z)$ for several values of z corresponding to support points of the distribution of $\hat{\theta}$, together with the exact probability, a normal approximation based on the fact that $\sqrt{n}\hat{\theta}$ is asymptotically distributed according to a standard normal distribution, and a normal approximation based on the distribution of $\bar{Y}$ that incorporates a continuity correction. Clearly, the saddlepoint approximation is extremely accurate.

Table 5.3 Comparison of probability approximations in Example 5.15

z	Exact	Normal-$\hat{\theta}$	Normal-$\bar{Y}$	Saddlept
−0.916	0.9897	0.9981	0.9801	0.9896
−0.693	0.9707	0.9858	0.9590	0.9707
−0.511	0.9329	0.9469	0.9226	0.9329
0.470	0.0487	0.0686	0.0410	0.0487
0.588	0.0143	0.0315	0.0089	0.0143
0.693	0.0035	0.0142	0.0013	0.0035

∎

5.5.2 DISTRIBUTION OF THE LIKELIHOOD RATIO STATISTIC

Consider the likelihood ratio test statistic for testing $\theta = \theta_0$ versus $\theta \neq \theta_0$. Since the likelihood ratio test statistic does not depend on the parameterization used, we may assume without loss of generality that the model is parameterized by the natural parameter of the model, which we will denote by η. The saddlepoint approximation to the density of $\hat{\eta}$, the maximum likelihood estimate of η, is given by

$$p^*(\hat{\eta}; \eta) = c[-\ell_{\theta\theta}(\hat{\eta})]^{1/2} \exp\{\ell(\eta) - \ell(\hat{\eta})\}$$

where

$$\ell(\eta) = n[\eta k_0'(\hat{\eta}) - k_0(\eta)].$$

Note that

$$W = 2[\ell(\hat{\eta}) - \ell(\eta)] = 2n[(\hat{\eta} - \eta)k_0'(\hat{\eta}) + k_0(\eta) - k_0(\hat{\eta})].$$

Hence, an approximation to the density function of W may be obtained from $p^*(\hat{\eta}; \eta)$ using the usual change of variable formula. In carrying out this change-of-variable it is important to realize that there is, in general, more than one value of $\hat{\eta}$ that yields a given value of W. For this class of models,

$$\frac{\partial W}{\partial \hat{\eta}} = 2(\hat{\eta} - \eta)[-\ell_{\theta\theta}(\hat{\eta})] \tag{5.16}$$

so that $W \equiv W(\hat{\eta})$ is a strictly increasing function of $\hat{\eta}$ for $\hat{\eta} > \eta$ and a strictly decreasing function of $\hat{\eta}$ for $\hat{\eta} < \eta$. It follows that, for a given value of W, say w, there are two values of $\hat{\eta}$ such that $W(\hat{\eta}) = w$, one greater than η and one less than η. Hence, we must perform the change of variable once in each of these two regions and add the results.

First consider the case in which $\hat{\eta} > \eta$. Using (5.16) it follows that the approximate density of W is given by

$$c \frac{1}{(\hat{\eta} - \eta)[-\ell_{\theta\theta}(\hat{\eta})]^{1/2}} \exp\{-w/2\}$$

where $\hat{\eta}$ is viewed as a function of w. Hence, we need to determine $\hat{\eta}(w)$, the inverse of the transformation from $\hat{\eta}$ to W. To do this, we assume that $\hat{\eta} - \eta = O(n^{-1/2})$. Using a Taylor's series expansion, we have that

$$W = n\Big[k_0''(\eta)(\hat{\eta} - \eta)^2 + \tfrac{2}{3}k_0'''(\eta)(\hat{\eta} - \eta)^3 + \tfrac{1}{4}k_0^{(4)}(\eta)(\hat{\eta} - \eta)^4 + O(n^{-5/2})\Big].$$

Inverting this expansion yields the following expansion for $\hat{\eta}$ in terms of W:

$$\begin{aligned}\sqrt{n}(\hat{\eta} - \eta) = \frac{\sqrt{W}}{\sqrt{k_0''(\eta)}}\Bigg[1 &- \frac{1}{3}\frac{k_0'''(\eta)}{(k_0''(\eta))^{3/2}}\frac{\sqrt{W}}{\sqrt{n}} \\ &+ \left\{\frac{5}{18}\frac{(k_0'''(\eta))^2}{(k_0''(\eta))^3} - \frac{1}{3}\frac{k_0^{(4)}(\eta)}{(k_0''(\eta))^2}\right\}\frac{W}{n} + O(n^{-3/2})\Bigg].\end{aligned}$$

It follows that

$$\frac{1}{(\hat{\eta}-\eta)[-\ell_{\theta\theta}(\hat{\eta})]^{1/2}} = W^{-1/2}\Bigg[1 - \frac{2}{3}\frac{k_0'''(\eta)}{(k_0''(\eta))^{3/2}}\frac{\sqrt{W}}{\sqrt{n}} + \left\{\frac{5}{24}\frac{(k_0'''(\eta))^2}{(k_0''(\eta))^3} - \frac{1}{8}\frac{k_0^{(4)}(\eta)}{(k_0''(\eta))^2}\right\}\frac{W}{n} + O(n^{-3/2})\Bigg].$$

Hence, the contribution to the density of W from $\hat{\eta} > \eta$ is given by

$$\frac{c}{2}w^{-1/2}\exp\{-w/2\}\Bigg[1 - \frac{2}{3}\frac{k_0'''(\eta)}{k_0''(\eta)^{3/2}}\frac{\sqrt{w}}{\sqrt{n}} + \left\{\frac{5}{24}\frac{k_0'''(\eta)^2}{k_0''(\eta)^3} - \frac{1}{8}\frac{k_0^{(4)}(\eta)}{k_0''(\eta)^2}\right\}\frac{w}{n} + O(n^{-3/2})\Bigg]. \tag{5.17}$$

The derivation for $\hat{\eta} < \eta$ is very similar, except that the $1/\sqrt{n}$ term in (5.17) is of the opposite sign. When these two approximations are combined, they yield the following approximation for the density of W:

$$p^*(w;\eta) = cw^{-1/2}\exp\{-w/2\}\left[1 + \left\{\frac{5}{24}\frac{k_0'''(\eta)^2}{k_0''(\eta)^3} - \frac{1}{8}\frac{k_0^{(4)}(\eta)}{k_0''(\eta)^2}\right\}\frac{w}{n}\right].$$

Based on the analysis given above, the error of this approximation is $O(n^{-3/2})$; however the first omitted term in the expansion will be an odd power of $\sqrt{w}$ and, hence, the contributions to the density of W from $\hat{\eta} > \eta$ and $\hat{\eta} < \eta$ will cancel. It follows that error in $p^*(w;\eta)$ is $O(n^{-2})$, provided that it is renormalized; that is, the constant c may be chosen so that $p^*(w;\eta)$ integrates to 1.

Note that the lead term in $p^*(w;\eta)$ is the χ_1^2 density function. Furthermore, the $1/n$ term in the expansion may be eliminated by considering a scaled version of W. Let $W_B = W/(1+b/n)$ denote the Bartlett-corrected likelihood ratio statistic where

$$b = \left\{\frac{5}{12}\frac{k_0'''(\eta)^2}{k_0''(\eta)^3} - \frac{1}{4}\frac{k_0^{(4)}(\eta)}{k_0''(\eta)^2}\right\}.$$

Then the density function of W_B may be approximated by

$$p^*(w;\eta) = cw^{-1/2}\exp\{-w/2\};$$

the error of this approximation is $O(n^{-2})$. Again, this density may be renormalized; it is straightforward to show that this leads to $c = 1/\sqrt{(2\pi)}$.

Example 5.16 Exponential distribution

For the exponential distribution considered in Example 5.8, the natural parameter is $\eta = -\theta$ and $k_0(\eta) = -\log(-\eta)$. Hence

$$k_0''(\eta) = \eta^{-2}; \qquad k_0'''(\eta) = -2\eta^{-3}; \qquad k_0^{(4)}(\eta) = 6\eta^{-4}.$$

The Bartlett correction factor is given by

$$b = \left\{ \frac{5}{12} \frac{(k_0'''(\eta))^2}{(k_0''(\eta))^3} - \frac{1}{4} \frac{k_0^{(4)}(\eta)}{(k_0''(\eta))^2} \right\} = \frac{1}{6}.$$

Recall that in Example 5.8 it was shown that $\mathrm{E}[W; \theta_0] = 2n[\log(n) - \Psi(n)]$; using the usual asymptotic expansion for $\Psi(n)$ it can be shown that

$$2n[\log(n) - \Psi(n)] = 1 + \frac{1}{6n} + O(n^{-3}).$$

∎

5.6 Discussion and references

Edgeworth expansions for the distribution of the maximum likelihood estimate are considered by Bhattacharya and Ghosh (1978) and Skovgaard (1981b); see also Pfanzagl (1979). Amari (1985) considers the case of a curved exponential family using methods from differential geometry. Approximations to the higher moments of the maximum likelihood estimate are given by Haldane and Smith (1956), McCullagh (1987, Chapter 7), Peers and Iqbal (1985) and Shenton and Bowman (1977, Chapter 3).

A related area of higher-order asymptotic theory, not discussed here, is the theory of higher-order asymptotic efficiency. See Ghosh (1994) for a general overview of this field and Amari (1985) and Kass and Vos (1997) for results for the case of a curved exponential family model.

The Bartlett adjustment to the likelihood ratio statistic originated with Bartlett (1937), who considered the problem of testing the equality of variances of normal distributions. The same approach was used for different problems by Bartlett (1938) and Box (1949). Lawley (1956) gave a formal demonstration of the fact that the Bartlett adjustment is generally available for likelihood ratio statistics, along with an expression for the correction factor in terms of cumulants of log-likelihood derivatives. In particular, the Bartlett adjustment is available for the likelihood ratio statistic for inference regarding a vector parameter of interest, a case not considered here. Further proofs of the validity of the Bartlett adjustment are given by Hayakawa (1977), Barndorff-Nielsen and Cox (1984) and Bickel and Ghosh (1990); see also Barndorff-Nielsen and Hall (1988), Cordeiro (1987) and McCullagh and Cox (1986). The effect of the Bartlett adjustment on the properties of likelihood ratio statistics based on discrete data is considered by Frydenberg and Jensen (1989).

Saddlepoint approximations for the distribution of the maximum likelihood estimate in full-rank exponential family models follow directly from the saddlepoint approximation for the sample mean; see, for example, Barndorff-Nielsen and Cox (1989, Chapter 6) and Jensen (1995). Saddlepoint approximations for curved exponential family models are considered in Hougaard (1985) and Barndorff-Nielsen and Cox (1994, Chapter 7). Saddlepoint approximations to the distribution of M-estimates in general models are considered by Field (1982), Field and Ronchetti (1990) and Jensen (1995, Chapter 4). The saddlepoint approximation approach to deriving the Bartlett adjustment is closely related to Barndorff-Nielsen and Cox (1984).

5.7 Exercises

5.1 Let $Y_1, \ldots, Y_n$ denote independent random variables each distributed according to the density

$$\frac{\theta}{y^{\theta+1}}, \quad y > 1$$

where $\theta > 1$. Find the first three formal cumulants of $\sqrt{i(\theta)}(\hat{\theta} - \theta)$.

5.2 Let $Y_1, \ldots, Y_n$ denote independent random variables of the form

$$Y_j = f_j(\lambda) + \sqrt{\psi}\epsilon_j, \quad j = 1, \ldots, n$$

where $\epsilon_1, \ldots, \epsilon_n$ are independent standard normal random variables and for each j, $f_j(\cdot)$ is a known function. Find the first three formal cumulants of $\hat{\psi}$ and compare these to the cumulants of $\hat{\psi}$ for the case in which $f_j(\lambda) \equiv \lambda$ for each j.

5.3 Let $Y_1, \ldots, Y_n$ denote independent random variables each distributed according to a density depending on a scalar parameter θ. Let

$$h(\theta) = \frac{1}{3}\frac{3\nu_{11}(\theta) + 2\nu_{001}(\theta)}{\nu_{20}(\theta)}.$$

Consider a parameterization of the model by $\phi(\theta)$ where $\phi'(\theta)$ satisfies

$$\phi'(\theta) = \exp\left\{\int^{\theta} h(t)\,dt\right\}.$$

Show that the third formal cumulant of $\sqrt{n}(\hat{\phi} - \phi)$ is of order $O(n^{-3/2})$.

5.4 Let $Y_1, \ldots, Y_n$ denote independent normal random variables each with mean 0 and standard deviation σ^2. Find a parameterization $\phi \equiv \phi(\sigma^2)$ such that the third formal cumulant of $\sqrt{n}(\hat{\phi} - \phi)$ is of order $O(n^{-3/2})$.

5.5 Let $Y_1, \ldots, Y_n$ denote independent random variables each distributed according to the density

$$\frac{\theta}{y^{\theta+1}}, \quad y > 1$$

where $\theta > 1$. Find the Bartlett adjustment factor the the likelihood ratio statistic for testing $\theta = \theta_0$ versus $\theta \neq \theta_0$.

5.6 Let $Y_1, \ldots, Y_n$ and $X_1, \ldots, X_m$ denote independent normal random variables such that each Y_j has mean 0 and standard deviation σ_1 and each X_j has mean 0 and standard deviation σ_2. Let $\psi = \sigma_1/\sigma_2$. Show how to construct a confidence interval for ψ using the Bartlett-adjusted likelihood ratio statistic and compare that result to an exact confidence interval based on the F-distribution.

5.7 Let X and Y denote independent Poisson random variables such that X has mean $n\lambda_1$ and Y has mean $m\lambda_2$; here λ_1 and λ_2 are unknown parameters and n and m are known positive integers. Find the Bartlett correction factor for the likelihood ratio statistic for testing the hypothesis $\lambda_1 = \lambda_2$.

5.8 Let $Y_1, \ldots, Y_n$ denote independent Bernoulli random variables with $\Pr(Y_j = 1) = \pi$, $j = 1, \ldots, n$ and let $\theta = \log[\pi/(1-\pi)]$. Using the results in Section 5.3, find an approximation to the bias of $\hat{\theta}$, the maximum likelihood estimate of θ. Using this approximation, construct a bias-corrected maximum likelihood estimate, $\tilde{\theta}$. Show that

$$\tilde{\theta} = \log\left(\frac{y + 1/2}{n - y + 1/2}\right) + O_p(n^{-3/2})$$

where $y = \sum y_j$.

5.9 Consider a sample of size n from a distribution in the one-parameter exponential family and let η denote the canonical parameter. Based on the results in Section 5.5, the density of $\hat{\eta}$, the maximum likelihood estimate of η may be approximated by

$$p^*(\hat{\eta}; \eta) = c[-\ell_{\eta\eta}(\hat{\eta})]^{1/2} \exp\{\ell(\eta) - \ell(\hat{\eta})\}$$

and this approximation may be transformed into the following approximation for the density of W, the likelihood ratio statistic:

$$cw^{-1/2} \exp\{-w/2\} \left[1 + \frac{1}{2}\frac{b}{n}w + O(n^{-3/2})\right].$$

Here b is the term appearing in the Bartlett correction factor. Hence,

$$\int p^*(\hat{\eta}; \eta)\, d\hat{\eta} = c \int w^{-1/2} \exp\{-w/2\} \left[1 + \frac{1}{2}\frac{b}{n}w\right] dw + O(n^{-3/2}).$$

Using this fact, show that c and b are related by

$$c = \frac{1}{\sqrt{(2\pi)}} \exp\left\{-\frac{1}{2}\frac{b}{n}\right\} [1 + O(n^{-3/2})].$$

5.10 Let $Y_1, \ldots, Y_n$ denote independent random variables each distributed according to the density

$$p(y; \theta) = \frac{\theta^y / y}{-\log(1-\theta)}, \quad y = 1, 2, \ldots$$

where $0 < \theta < 1$. Find the saddlepoint approximation to the density of $\hat{\theta}$, the maximum likelihood estimate of θ.

5.11 Let $Y_1, \ldots, Y_n$ denote independent random variables each distributed according to the density

$$p(y; \theta) = \frac{\sqrt{\theta}}{\sqrt{(2\pi)}} \exp\{\theta\} y^{-3/2} \exp\left\{-\frac{1}{2}(y + 1/y)\right\}, \quad y > 0$$

where $\theta > 0$. Find the saddlepoint approximation to the density of $\hat{\theta}$, the maximum likelihood estimate of θ.

5.12 Let $Y_1, \ldots, Y_n$ denote independent identically distributed integer random variables each distributed according to a density in the one-parameter exponential family with natural parameter $c(\theta)$. Determine the saddlepoint approximation for the lower tail probability $\Pr(c(\hat{\theta}) \leq z)$. Calculate the approximation for the Poisson distribution described in Example 5.15 using the values of z given in Table 5.3. Compare the approximations to the exact values.

5.13 Consider the same type of model considered in the previous exercise. Determine the continuity-corrected saddlepoint approximation for the lower tail probability $\Pr(c(\hat{\theta}) \leq z)$. Calculate the approximation for the Poisson distribution described in Example 5.15 using the values of z given in Table 5.3. Compare the approximation to the exact values.

5.14 Consider a one-parameter exponential family distribution with density of the form

$$\exp\{\eta y - k_0(\eta) + S(y)\}.$$

Let $g(\eta) = \log k_0''(\eta)$. Express the Bartlett adjustment factor for the likelihood ratio test statistic in terms of the function g and its derivatives. Find the form of g for which the Bartlett adjustment factor is 0.

5.15 In Section 5.5.1, the distribution function of $\hat{\theta}$ is approximated by approximating the probability $\Pr(\bar{T} \leq t)$. Another approach is to approximate the integral of $p^*(\hat{\theta}; \theta)$ directly using the methods described in Section 2.5. Derive this approximation and compare it to the one given in Section 5.5.1.

6
Asymptotic theory and conditional inference

6.1 Introduction

The asymptotic theory developed in the preceding sections was concerned with the marginal distributions of various likelihood-based quantities. In this chapter we consider the conditional distributions of likelihood-based quantities given an ancillary statistic.

There are at least two motivations for such conditioning. One is that, by keeping certain functions of the underlying random variable fixed at their observed values, it makes the probability calculations more relevant to the data at hand; this is discussed in Section 1.6 and illustrated in Example 1.23. Another motivation is that conditioning simplifies the stochastic structure of the model, in a certain sense. For instance, consider the case in which the sufficient statistic may be written $(\hat{\theta}, a)$ where $\hat{\theta}$ is the maximum likelihood estimator of θ and a is an ancillary statistic. In the conditional model based on the conditional distribution of the data given a, $\hat{\theta}$ itself is sufficient. It follows that, in considering probability calculations, one only needs to consider the conditional distribution of $\hat{\theta}$ or some one-to-one function of $\hat{\theta}$. Of course, determination of this conditional distribution may be difficult and, hence, large-sample approximations play an important role.

The approximations considered in this chapter are based on the following model. Let $Y_1, \ldots, Y_n$ denote independent random variables distributed according to a distribution depending on a parameter θ, which may be vector valued. Suppose that the sufficient statistic for the model may be written $(\hat{\theta}, a)$ where $\hat{\theta}$ denotes the maximum likelihood estimator of θ and a is an ancillary statistic.

In this case, the log-likelihood function $\ell(\theta)$ may be written $\ell(\theta; \hat{\theta}, a)$ to emphasize that it is a function of the sufficient statistic $(\hat{\theta}, a)$ as well as the usual argument θ.

Example 6.1 Exponential family distribution
Let $Y_1, \ldots, Y_n$ denote independent random variables each distributed according to a density of the form

$$\exp\left\{y^T\theta - k(\theta) + D(y)\right\}$$

where $\theta \in \Theta \subset \Re^d$; this is a full d-parameter exponential family model with natural parameter θ. The log-likelihood function for the model is given by

$$\ell(\theta) = \sum y_j^T\theta - nk(\theta).$$

Since $\hat{\theta}$ satisfies the likelihood equation

$$\ell_\theta(\theta) = \sum y_j - nk'(\theta) = 0,$$

the log-likelihood function may be written

$$\ell(\theta) = nk'(\hat{\theta})^T\theta - nk(\theta).$$

∎

Example 6.2 Location model

Let $Y_1, \ldots, Y_n$ denote independent random variables following the location model

$$Y_j = \theta + \epsilon_j, \quad j = 1, \ldots, n$$

where $\epsilon_1, \ldots, \epsilon_n$ are independent random variables each having known density function $\exp\{g(\cdot)\}$. The log-likelihood function for the model is given by

$$\ell(\theta) = \sum g(y_j - \theta).$$

Let $a = (a_1, \ldots, a_n)$ where $a_j = y_j - \hat{\theta}$; it has been shown that a is an ancillary statistic.

Each observation Y_j may be written $Y_j = a_j + \hat{\theta}$ so that the log-likelihood function may be written

$$\ell(\theta) = \sum g(a_j + \hat{\theta} - \theta).$$

∎

6.2 Log-likelihood derivatives

6.2.1 INTRODUCTION

As in the previous chapters, the log-likelihood derivatives will play a central role in deriving approximations to the conditional distributions. In addition to the standard log-likelihood derivatives, such as

$$\ell_\theta(\theta) = \frac{\partial \ell}{\partial \theta}(\theta); \qquad \ell_{\theta\theta}(\theta) = \frac{\partial^2 \ell}{\partial \theta \partial \theta^T}(\theta),$$

we may also calculate derivatives with respect to $\hat{\theta}$, holding a fixed. These derivatives are often called *sample space derivatives* since the differentiation is with respect to certain functions of the sample, holding other functions of the sample fixed.

Derivatives with respect to functions of the data, such as $\hat{\theta}$, will be denoted by a semicolon followed by the function(s) of the data used in the differentiation. For instance,

$$\ell_{\theta;\hat{\theta}}(\theta; \hat{\theta}, a) = \frac{\partial}{\partial \hat{\theta}} \ell_\theta(\theta; \hat{\theta}, a);$$

$$\ell_{\theta;\hat{\theta}\hat{\theta}}(\theta; \hat{\theta}, a) = \frac{\partial^2}{\partial \hat{\theta} \partial \hat{\theta}^T} \ell_\theta(\theta; \hat{\theta}, a).$$

It follows that $\ell_{\theta;\hat{\theta}}(\theta; \hat{\theta}, a)$ is a matrix, $\ell_{\theta\theta;\hat{\theta}}(\theta; \hat{\theta}, a)$ and $\ell_{\theta;\hat{\theta}\hat{\theta}}(\theta; \hat{\theta}, a)$ are three-dimensional arrays, and so on. Hence, we assume that $\ell(\theta; \hat{\theta}, a)$ is four-times continuously differentiable in $(\theta, \hat{\theta})$.

In addition, these derivatives may be evaluated at $\hat{\theta} = \theta$ where θ is an arbitrary element of the parameter space. For instance,

$$\ell_{\theta;\hat{\theta}}(\theta; \theta, a) = \frac{\partial^2}{\partial\hat{\theta}\,\partial\theta^T}\ell(\theta; \hat{\theta}, a)\Big|_{\hat{\theta}=\theta}.$$

Note that, in a conditional analysis considering a to be fixed, $\ell_{\theta;\hat{\theta}}(\theta; \theta, a)$ is a constant; we will see that derivatives of this type play the role played by cumulants of log-likelihood derivatives in the unconditional analysis of Chapter 5.

Example 6.3 Location model
Consider the location model considered in Example 6.2. The log-likelihood function is given by

$$\ell(\theta; \hat{\theta}, a) = \sum g(a_j + \hat{\theta} - \theta).$$

Hence,

$$\ell_{\theta;\hat{\theta}}(\theta; \hat{\theta}, a) = -\sum g''(a_j + \hat{\theta} - \theta)$$

and

$$\ell_{\theta;\hat{\theta}}(\theta; \theta, a) = -\sum g''(a_j). \qquad \blacksquare$$

6.2.2 THE SCALAR PARAMETER CASE

We begin by considering the case in which θ is a scalar parameter. As noted above, in the conditional model the only random variable is the maximum likelihood estimate $\hat{\theta}$. Alternatively, we may work with the standardized score function $\ell_\theta(\theta)/\sqrt{n}$ which, for fixed a, will be a one-to-one function of $\hat{\theta}$, at least for θ near $\hat{\theta}$. The advantage of working with the score function is that, because of the Bartlett identities, it is possible to determine the cumulants of the score function to a high-degree of approximation.

Consider the likelihood equation $\ell_\theta(\hat{\theta}) = 0$ which may be expanded in the usual way:

$$\ell_\theta(\hat{\theta}) = \ell_\theta(\theta) + \ell_{\theta\theta}(\theta)(\hat{\theta} - \theta) + \tfrac{1}{2}\ell_{\theta\theta\theta}(\theta)(\hat{\theta} - \theta)^2 + \tfrac{1}{6}\ell_{\theta\theta\theta\theta}(\theta)(\hat{\theta} - \theta)^3 + \cdots.$$

Each of the log-likelihood derivatives appearing in this expansion may be written as a function of $\hat{\theta}$ and, hence, may be expanded around $\hat{\theta} = \theta$; for instance,

$$\ell_{\theta\theta}(\theta) \equiv \ell_{\theta\theta}(\theta; \hat{\theta}, a) = \ell_{\theta\theta}(\theta; \theta, a) + \ell_{\theta\theta;\hat{\theta}}(\theta; \theta, a)(\hat{\theta} - \theta) + \tfrac{1}{2}\ell_{\theta\theta;\hat{\theta}\hat{\theta}}(\theta; \theta, a)(\hat{\theta} - \theta)^2 + \cdots.$$

Carrying out an expansion of this type for each of $\ell_{\theta\theta}(\theta)$, $\ell_{\theta\theta\theta}(\theta)$ and $\ell_{\theta\theta\theta\theta}(\theta)$ leads to the following expansion for $\ell_\theta(\hat{\theta})$:

$$\begin{aligned}\ell_\theta(\hat{\theta}) &= \ell_\theta(\theta) + \ell_{\theta\theta}(\theta;\theta,a)(\hat{\theta}-\theta)\\ &\quad + \left[\ell_{\theta\theta;\hat{\theta}}(\theta;\theta,a) + \tfrac{1}{2}\ell_{\theta\theta\theta}(\theta;\theta,a)\right](\hat{\theta}-\theta)^2\\ &\quad + \left[\tfrac{1}{2}\ell_{\theta\theta;\hat{\theta}\hat{\theta}}(\theta;\theta,a) + \tfrac{1}{2}\ell_{\theta\theta\theta;\hat{\theta}}(\theta;\theta,a) + \tfrac{1}{6}\ell_{\theta\theta\theta\theta}(\theta;\theta,a)\right](\hat{\theta}-\theta)^3 + \cdots .\end{aligned}$$

Using an approach similar to that used in Section 5.2.2, this expansion may be inverted to give an expansion for $\sqrt{n}(\hat{\theta}-\theta)$:

$$\begin{aligned}\sqrt{n}(\hat{\theta}-\theta) &= \frac{Z}{\bar{j}_o} + \frac{\ell_{\theta\theta;\hat{\theta}}(\theta;\theta,a) + \ell_{\theta\theta\theta}(\theta;\theta,a)/2}{j_o}\left[\frac{Z}{\bar{j}_o}\right]^2\frac{1}{\sqrt{n}}\\ &\quad + \left[\frac{\ell_{\theta\theta;\hat{\theta}\hat{\theta}}(\theta;\theta,a)/2 + \ell_{\theta\theta\theta;\hat{\theta}}(\theta;\theta,a)/2 + \ell_{\theta\theta\theta\theta}(\theta;\theta,a)/6}{j_o}\right.\\ &\quad \left.\times\frac{2\ell_{\theta\theta;\hat{\theta}}(\theta;\theta,a) + \ell_{\theta\theta\theta}(\theta;\theta,a)}{j_o}\,\frac{\ell_{\theta\theta;\hat{\theta}}(\theta;\theta,a) + \ell_{\theta\theta\theta}(\theta;\theta,a)/2}{j_o}\right]\\ &\quad \times\left[\frac{Z}{\bar{j}_o}\right]^3\frac{1}{n} + O_p(n^{-3/2});\end{aligned} \tag{6.1}$$

here $Z = \ell_\theta(\theta)/\sqrt{n}$, $j_o = -\ell_{\theta\theta}(\theta;\theta,a)$, and $\bar{j}_o = j_o/n$. Note that in this expansion, only Z is a random variable; all other quantities depend on the data only through a and, hence, may be viewed as constants in a conditional analysis given a.

We now derive approximations to the conditional cumulants of Z. First recall that the Bartlett identities hold conditionally as well. That is

$$E[\ell_\theta(\theta)|a;\theta] = 0; \qquad E[\ell_{\theta\theta}(\theta)|a;\theta] + E[\ell_\theta(\theta)^2|a;\theta] = 0;$$

and so on.

Let $\nu_j(a)/n^{(j-2)/2}$, $j = 1, 2, \ldots$, denote the standardized conditional cumulants of Z. Then, as noted above, $\nu_1(a) = 0$, and

$$\nu_2(a) = -E[\ell_{\theta\theta}(\theta)/n|a;\theta].$$

Expanding $\ell_{\theta\theta}(\theta) \equiv \ell_{\theta\theta}(\theta;\theta,a)$ as

$$\ell_{\theta\theta}(\theta) = \ell_{\theta\theta}(\theta;\theta,a) + \ell_{\theta\theta;\hat{\theta}}(\theta;\theta,a)(\hat{\theta}-\theta) + \cdots$$

and using the expansion for $\sqrt{n}(\hat{\theta}-\theta)$ given above yields

$$\begin{aligned}-\frac{1}{n}\ell_{\theta\theta}(\theta) &= -\frac{1}{n}\ell_{\theta\theta}(\theta;\theta,a) - \frac{\ell_{\theta\theta;\hat{\theta}}(\theta;\theta,a)}{j_o}Z\frac{1}{\sqrt{n}}\\ &\quad - \frac{\ell_{\theta\theta;\hat{\theta}}(\theta;\theta,a)(\ell_{\theta\theta;\hat{\theta}}(\theta;\theta,a) + \ell_{\theta\theta\theta}(\theta;\theta,a)/2)}{j_o^3}Z^2\\ &\quad + O_p(n^{-3/2}).\end{aligned}$$

Hence,

$$\nu_2(a) = -\frac{1}{n}\ell_{\theta\theta}(\theta;\theta,a) - \frac{\ell_{\theta\theta;\hat\theta}(\theta;\theta,a)(\ell_{\theta\theta;\hat\theta}(\theta;\theta,a) + \ell_{\theta\theta\theta}(\theta;\theta,a)/2)}{j_o^3}\nu_2(a) + O(n^{-3/2})$$

so that

$$\nu_2(a) = -\frac{1}{n}\ell_{\theta\theta}(\theta;\theta,a) - \frac{\ell_{\theta\theta;\hat\theta}(\theta;\theta,a)(\ell_{\theta\theta;\hat\theta}(\theta;\theta,a) + \ell_{\theta\theta\theta}(\theta;\theta,a)/2)}{j_o^2}\frac{1}{n} - \frac{1}{2}\frac{\ell_{\theta\theta;\hat\theta\hat\theta}(\theta;\theta,a)}{j_o}\frac{1}{n} + O(n^{-3/2}).$$

Using the same approach, it can be shown that

$$\nu_3(a) = -\frac{3\ell_{\theta\theta;\hat\theta}(\theta;\theta,a) + \ell_{\theta\theta\theta}(\theta;\theta,a)}{n} + O(n^{-1}).$$

Example 6.4 Exponential family distribution
Let $Y_1, \ldots, Y_n$ denote independent random variables each distributed according to a density of the form

$$\exp\{y\theta - k(\theta) + D(y)\}$$

where $\theta \in \Theta \subset \Re$. The log-likelihood function for this model may be written

$$\ell(\theta) = nk'(\hat\theta)\theta - nk(\theta);$$

see Example 6.1. Hence,

$$\ell_{\theta\theta}(\theta;\theta,a) = -nk''(\theta); \qquad \ell_{\theta\theta\theta}(\theta;\theta,a) = -nk'''(\theta);$$
$$\ell_{\theta\theta;\hat\theta}(\theta;\theta,a) = 0.$$

It follows that Z has conditional cumulants

$$\nu_1(a) = 0; \qquad \nu_2(a) = k''(\theta) + O(n^{-3/2}); \qquad \nu_3(a) = k'''(\theta) + O(n^{-1}).$$

For this model, $\hat\theta$ itself is sufficient so that a is null. Also $Z = \sqrt{n}(\bar{Y} - k'(\theta))$ and, hence,

$$\nu_1(a) = 0; \qquad \nu_2(a) = k''(\theta); \qquad \nu_3(a) = k'''(\theta)$$

hold exactly. ∎

Example 6.5 Two measuring instruments
Let $(Y_1, A_1), \ldots, (Y_n, A_n)$ denote independent identically distributed random variables such that A_j takes the value 0 or 1, each with probability 1/2 and, given $A_j = a_j$, Y_j is normally distributed with mean θ and variance $\sigma_{a_j}^2$ where σ_0^2 and σ_1^2 are known. The log-likelihood function for this model may be written

$$\ell(\theta) = -\frac{1}{2}\sum(y_j - \theta)^2/\sigma_{a_j}^2 = \left(\theta\hat{\theta} - \frac{1}{2}\theta^2\right)\left[(n-a)\frac{1}{\sigma_0^2} + a\frac{1}{\sigma_1^2}\right]$$

where $a = \sum a_j$ is an ancillary statistic and

$$\hat{\theta} = \frac{\sum y_j/\sigma_{a_j}^2}{\sum 1/\sigma_{a_j}^2}.$$

Then

$$Z = \frac{1}{\sqrt{n}}\sum(Y_j - \theta)/\sigma_{a_j}^2.$$

Since

$$\ell(\theta) = \left(\theta\hat{\theta} - \frac{1}{2}\theta^2\right)\left[(n-a)\frac{1}{\sigma_0^2} + a\frac{1}{\sigma_1^2}\right];$$

$$\ell_{\theta\theta}(\theta;\theta,a) = -\left[(n-a)\frac{1}{\sigma_0^2} + a\frac{1}{\sigma_1^2}\right];$$

$$\ell_{\theta\theta\theta}(\theta) = \ell_{\theta\theta\theta\theta}(\theta) = \ell_{\theta\theta;\hat{\theta}}(\theta) = \ell_{\theta\theta;\hat{\theta}\hat{\theta}}(\theta) = \ell_{\theta\theta\theta;\hat{\theta}}(\theta) = 0,$$

it follows that the approximations to the first three conditional cumulants of Z given a given in this section are given by

$$0; \qquad \frac{1}{n}\left[(n-a)\frac{1}{\sigma_0^2} + a\frac{1}{\sigma_1^2}\right]; \qquad 0,$$

respectively.

In this example, the conditional cumulants of Z may be calculated exactly. Since, given a_j, $(Y_j - \theta)/\sigma_{a_j}^2$ is normally distributed with mean 0 and variance $1/\sigma_{a_j}^2$, the conditional moment-generating function of $(Y_j - \theta)/\sigma_{a_j}^2$ is $\exp\{t^2/(2\sigma_{a_j}^2)\}$. It follows that the conditional moment-generating function of Z given $a_1, \ldots, a_n$ is given by

$$\exp\left\{\frac{1}{2n}t^2\sum 1/\sigma_{a_j}^2\right\} = \exp\left\{\frac{1}{2n}t^2\left[(n-a)\frac{1}{\sigma_0^2} + a\frac{1}{\sigma_1^2}\right]\right\}.$$

Since this function depends on $a_1, \ldots, a_n$ only through $a = \sum a_j$, it is also the conditional moment-generating function given a. Hence, the conditional cumulant-generating function of Z given a is given by

$$\frac{1}{2n}t^2\left[(n-a)\frac{1}{\sigma_0^2} + a\frac{1}{\sigma_1^2}\right].$$

It follows that the approximations to the first three conditional cumulants of Z are exact in this example. ∎

Example 6.6 Location model
Consider the location model considered in Examples 6.2 and 6.3. Then

$$\ell_{\theta\theta}(\theta;\theta,a) = \sum g''(a_j); \qquad \ell_{\theta\theta;\hat{\theta}}(\theta;\theta,a) = \sum g'''(a_j);$$
$$\ell_{\theta\theta\theta}(\theta;\theta,a) = -\sum g'''(a_j); \qquad \ell_{\theta\theta;\hat{\theta}\hat{\theta}}(\theta;\theta,a) = -\sum g^{(4)}(a_j).$$

Hence, the first three conditional cumulants of Z are given by

$$\nu_1(a) = 0;$$
$$\nu_2(a) = -\frac{1}{n}\sum g''(a_j) - \frac{1}{2}\left[\frac{(\sum g'''(a_j))^2}{(\sum g''(a_j))^2} + \frac{\sum g^{(4)}(a_j)}{\sum g''(a_j)}\right]\frac{1}{n} + O(n^{-3/2});$$
$$\nu_3(a) = -\frac{2}{n}\sum g'''(a_j) + O(n^{-1}).$$ ∎

6.2.3 VECTOR PARAMETER CASE

Now consider the case in which θ is a vector; hence, $\ell_\theta(\theta)$ is a vector, $\ell_{\theta\theta}(\theta)$ is a matrix, and $\ell_{\theta\theta\theta}(\theta)$ is a three-dimensional array. Results analogous to those presented in Section 6.2.2 may be derived in this case as well, although the derivations are considerably more complicated. Let $\nu_j(a)$ denote the jth conditional cumulant of Z. Then $\nu_1(a) = 0$,

$$\nu_2(a) = -E_\theta\left[\frac{1}{n}\ell_{\theta\theta}(\theta)|a\right]$$

and since

$$\frac{1}{n}\ell_{\theta\theta}(\theta) \equiv \frac{1}{n}\ell_{\theta\theta}(\theta;\hat{\theta},a) = \frac{1}{n}\ell_{\theta\theta}(\theta;\theta,a) + \frac{1}{n}\ell_{\theta\theta;\hat{\theta}}(\theta;\theta,a)(\hat{\theta}-\theta) + \cdots$$

it is easy to show that

$$\nu_2(a) = -\frac{1}{n}\ell_{\theta\theta}(\theta;\theta,a) + O(n^{-1}).$$

Some care is needed when deriving the third cumulant of Z since it is a three-dimensional array. A relatively simple approach is to proceed component-wise. For instance, consider $\mathrm{cum}(\ell_{\theta_i}(\theta), \ell_{\theta_j}(\theta), \ell_{\theta_k}(\theta))$ where $\theta_i, \theta_j, \theta_k$ are components of the vector θ. Using the multivariate form of the third Bartlett identity given in Section 3.5, we have that

$$E[\ell_{\theta_i}(\theta)\ell_{\theta_j}(\theta)\ell_{\theta_k}(\theta)|a;\theta] = -E[\ell_{\theta_i\theta_j\theta_k}(\theta)|a;\theta] - E[\ell_{\theta_i}(\theta)\ell_{\theta_j\theta_k}(\theta)|a;\theta]$$
$$- E[\ell_{\theta_j}(\theta)\ell_{\theta_i\theta_k}(\theta)|a;\theta] - E[\ell_{\theta_k}(\theta)\ell_{\theta_i\theta_j}(\theta)|a;\theta].$$

Each term in the right-hand side of this expression may now be handled using the approach used in Section 6.2.2. This leads to the expression

$$E[\ell_{\theta_i}(\theta)\ell_{\theta_j}(\theta)\ell_{\theta_k}(\theta)|a;\theta] = -[\ell_{\theta_i\theta_j;\hat{\theta}_k}(\theta;\theta,a) + \ell_{\theta_i\theta_k;\hat{\theta}_j}(\theta;\theta,a)$$
$$+ \ell_{\theta_j\theta_k;\hat{\theta}_i}(\theta;\theta,a) + \ell_{\theta_i\theta_j\theta_k}(\theta;\theta,a)] + O(1).$$

The standardized third cumulant array of Z may be written

$$\nu_3(a) = -\frac{\ell_{\theta\theta;\hat{\theta}}(\theta;\theta,a)[3] + \ell_{\theta\theta\theta}(\theta;\theta,a)}{n} + O(n^{-3/2})$$

where [3] indicates summation over the three permutations of the indices. For instance, let $\text{cum}(Z_i, Z_j, Z_k)$ denote the cumulant of ith, jth, and kth components of Z. Then

$$\text{cum}(Z_i, Z_j, Z_k) = -\frac{1}{n}[\ell_{\theta_i\theta_j;\hat{\theta}_k}(\theta;\theta,a) + \ell_{\theta_i\theta_k;\hat{\theta}_j}(\theta;\theta,a) + \ell_{\theta_j\theta_k;\hat{\theta}_i}(\theta;\theta,a)$$
$$+ \ell_{\theta_i\theta_j\theta_k}(\theta;\theta,a)] + O(n^{-3/2}).$$

Example 6.7 Location-scale model
Let $Y_1, \ldots, Y_n$ denote independent random variables following the location-scale model

$$Y_j = \mu + \sigma\epsilon_j, \quad j = 1, \ldots, n$$

where $\epsilon_1, \ldots, \epsilon_n$ are independent random variables each having known density function $\exp\{g(\cdot)\}$ and μ and σ are unknown parameters with $\sigma > 0$. The log-likelihood function for the model is given by

$$\ell(\theta) = \sum g[(y_j - \mu)/\sigma] - n\log\sigma.$$

Let $a = (a_1, \ldots, a_n)$ where $a_j = (y_j - \hat{\theta})/\hat{\sigma}$; it has been shown that a is an ancillary statistic. Then $\ell(\theta)$ may be written

$$\ell(\theta;\hat{\theta},a) = \sum g\left(\frac{\hat{\sigma}a_j + \hat{\mu} - \mu}{\sigma}\right) - n\log\sigma.$$

It follows that

$$-\ell_{\theta\theta}(\theta;\theta,a) = -\frac{1}{\sigma^2}\begin{pmatrix} \sum g''(a_j) & \sum g''(a_j)a_j \\ \sum g''(a_j)a_j & \sum g''(a_j)a_j^2 + n \end{pmatrix};$$

note that $\sum g'(a_j)a_j = \sum g'(a_j) = 0$.

The standardized third cumulant array of Z may be calculated component-wise using the approach described above. For instance,

$$\ell_{\mu\mu\sigma}(\theta;\theta,a) = -\frac{1}{\sigma^3}\sum g'''(a_j)a_j; \qquad \ell_{\mu\mu;\hat{\sigma}}(\theta;\theta,a) = \frac{1}{\sigma^3}\sum g'''(a_j)a_j;$$

$$\ell_{\mu\sigma;\hat{\mu}}(\theta;\theta,a) = \frac{1}{\sigma^3}\Big[\sum g'''(a_j)a_j + 2\sum g''(a_j)\Big].$$

It follows that the (μ, μ, σ) element of the third cumulant array of Z is given by

$$-\frac{1}{\sigma^3}\Big[2\sum g'''(a_j)a_j + 4\sum g''(a_j)\Big]\frac{1}{n} + O(n^{-1}).$$

∎

6.3 Conditional distribution of maximum likelihood estimates

6.3.1 SCALAR PARAMETER CASE

We begin by calculating the formal conditional cumulants of $\sqrt{n}(\hat{\theta} - \theta)$, the standardized maximum likelihood estimator; we will denote these cumulants by $\hat{\kappa}_j(a)$. Using the stochastic asymptotic expansion given for $\hat{\theta}$ in Section 6.2.2, together with the expansion for the conditional cumulants of Z given in Section 6.2.2, it is straightforward to show that

$$\hat{\kappa}_1(a) = \frac{\ell_{\theta\theta;\hat{\theta}}(\theta;\theta,a) + \ell_{\theta\theta\theta}(\theta;\theta,a)/2}{n\bar{\jmath}_o{}^2}\frac{1}{\sqrt{n}} + O(n^{-3/2});$$

$$\hat{\kappa}_2(a) = \frac{1}{\bar{\jmath}_o} + O(n^{-1});$$

$$\hat{\kappa}_3(a) = \frac{[2\ell_{\theta\theta\theta}(\theta;\theta,a) + 3\ell_{\theta\theta;\hat{\theta}}(\theta;\theta,a)]/n}{\bar{\jmath}_o{}^3}\frac{1}{\sqrt{n}} + O(n^{-3/2}).$$

Example 6.8 Poisson distribution
Let $Y_1, \ldots, Y_n$ denote independent Poisson random variables each with mean θ. The log-likelihood function for this model may be written

$$\ell(\theta) = n\hat{\theta}\log\theta - n\theta;$$

note that $\hat{\theta}$ is sufficient so that no ancillary statistic is needed. Using the approximations derived in this section, it is straightforward to show that

$$\hat{\kappa}_1(a) = O(n^{-3/2}); \qquad \hat{\kappa}_2(a) = \theta + O(n^{-1});$$

$$\hat{\kappa}_3(a) = \theta\frac{1}{\sqrt{n}} + O(n^{-3/2}).$$

A direct calculation shows that these approximations are exact for this model. ∎

Example 6.9 Location model
Consider the location model considered in Example 6.6. Then

$$\hat{\kappa}_1(a) = \frac{1}{2}\frac{\sum g'''(a_j)}{\left[\sum g''(a_j)\right]^2}\sqrt{n} + O(n^{-3/2}); \quad \hat{\kappa}_2(a) = -\frac{n}{\sum g''(a_j)} + O(n^{-1});$$

$$\hat{\kappa}_3(a) = -\frac{\sum g'''(a_j)}{\left[\sum g''(a_j)\right]^3} n^{3/2} + O(n^{-3/2}).$$

■

Using the approximations $\hat{\kappa}_j(a)$ for the conditional cumulants, it is possible to construct an Edgeworth expansion for the conditional distribution of $\sqrt{n}(\hat{\theta} - \theta)$.

Another approach to approximating the conditional distribution of $\hat{\theta}$ is to use a method closely related to the saddlepoint method. Let $p(\hat{\theta}, a; \theta)$ denote the density function of $(\hat{\theta}, a)$, $p(\hat{\theta}|a; \theta)$ denote the conditional density function of $\hat{\theta}$ given a, and $p(a)$ denote the density function of a which, because of the ancillarity of a, does not depend on θ. It follows that for any value of θ,

$$p(a) = \frac{p(\hat{\theta}, a; \theta)}{p(\hat{\theta}|a; \theta)}$$

for all $\hat{\theta}, a$. Hence,

$$p(\hat{\theta}|a; \theta) = \frac{p(\hat{\theta}, a; \theta)}{p(a)} = \frac{p(\hat{\theta}, a; \theta)}{p(\hat{\theta}, a; \theta_0)} p(\hat{\theta}|a; \theta_0)$$

for any θ_0. Note that

$$\frac{p(\hat{\theta}, a; \theta)}{p(\hat{\theta}, a; \theta_0)} = \exp\{\ell(\theta; \hat{\theta}, a) - \ell(\theta_0; \hat{\theta}, a)\}$$

so that, given an approximation $\hat{p}(\hat{\theta}|a; \theta_0)$ to $p(\hat{\theta}|a; \theta_0)$, an approximation to $p(\hat{\theta}|a; \theta)$ is given by

$$\exp\{\ell(\theta; \hat{\theta}, a) - \ell(\theta_0; \hat{\theta}, a)\}\, \hat{p}(\hat{\theta}|a; \theta_0).$$

The idea behind this approximation method is to choose the value of θ_0 so that $p(\hat{\theta}|a; \theta_0)$ is easy to accurately approximate; note the similarity between this approach and the saddlepoint method.

Consider an Edgeworth expansion for the density of $T = \sqrt{j_o}(\hat{\theta} - \theta)$ under the distribution with parameter θ where, as above, $j_o = -\ell_{\theta\theta}(\theta; \theta, a)$. This expansion is of the form

$$\phi(t)\Big[1 + \Big[\mu(\theta)t + \frac{1}{6}\rho_3(\theta)H_3(t)\Big]\frac{1}{\sqrt{n}}\Big] + O(n^{-1})$$

where $\mu(\theta)/\sqrt{n}$ denotes the mean of T and $\rho_3(\theta)$ denotes the standardized third cumulant of T. It follows that an expansion for the density of $\hat{\theta}$ is given by

$$\sqrt{j_o}\phi(\sqrt{j_o}(\hat{\theta}-\theta))\Big[1+\Big[\mu(\theta)\sqrt{j_o}(\hat{\theta}-\theta) + \frac{1}{6}\rho_3(\theta)H_3(\sqrt{j_o}(\hat{\theta}-\theta))\Big]\frac{1}{\sqrt{n}}+O(n^{-1})\Big]. \tag{6.2}$$

If we approximate the density of $\hat{\theta}$ using the parameter value $\hat{\theta}$, the approximation becomes $\sqrt{\hat{\jmath}}\phi(0)$ and the error of this approximation is $O(n^{-1})$. Hence, we may approximate the density of $\hat{\theta}$ by

$$\sqrt{\hat{\jmath}}\frac{1}{\sqrt{(2\pi)}}\exp\{\ell(\theta)-\ell(\hat{\theta})\}.$$

Note that in evaluating this expression $\hat{\jmath} = -\ell_{\theta\theta}(\hat{\theta};\hat{\theta},a)$; $\ell(\hat{\theta}) = \ell(\hat{\theta};\hat{\theta},a)$; $\ell(\theta) = \ell(\theta;\hat{\theta},a)$. This approximation may be renormalized so that it integrates exactly to 1; in this case the approximation is of the form

$$p^*(\hat{\theta}|a;\theta) = c\sqrt{\hat{\jmath}}\exp\{\ell(\theta)-\ell(\hat{\theta})\}$$

where c is a constant that may depend on a and θ; c is often determined numerically. This result is known as the *likelihood ratio approximation*, *Barndorff-Nielsen's formula* or the p^* formula. It may also be written

$$p^*(\hat{\theta}|a;\theta) = \frac{\bar{c}}{\sqrt{(2\pi)}}\sqrt{\hat{\jmath}}\exp\{\ell(\theta)-\ell(\hat{\theta})\}$$

where $\bar{c} = 1 + O(n^{-1})$.

The derivation of p^* shows that it approximates the true conditional density of $\hat{\theta}$ given a with relative error $O(n^{-1})$. Like the saddlepoint approximation, the Edgeworth expansion used in this derivation is evaluated only at the value 0; hence, the relative error $O(n^{-1})$ applies to any fixed point at which p^* is evaluated. That is, the relative error is $O(n^{-1})$ in the large-deviation range.

For $\hat{\theta}$ in the moderate-deviation range, $\hat{\theta} = \theta + O(n^{-1/2})$, higher accuracy is achieved. To see this, note that the $O(n^{-1})$ term in (6.2) is of the form

$$\Big[\frac{1}{24}\rho_4(\theta)H_4(\sqrt{j_o}(\hat{\theta}-\theta)) + \frac{1}{72}\rho_3(\theta)^2H_6(\sqrt{j_o}(\hat{\theta}-\theta))\Big]\frac{1}{n} + O(n^{-3/2}).$$

Evaluating this using $\theta = \hat{\theta}$, together with the fact that $\hat{\theta} = \theta + O(n^{-1/2})$, shows that the $O(n^{-1})$ term may be written

$$\Big[\frac{1}{8}\rho_4(\hat{\theta}) - \frac{5}{24}\rho_3(\hat{\theta})^2\Big]\frac{1}{n} + O(n^{-3/2}) = \Big[\frac{1}{8}\rho_4(\theta) - \frac{5}{24}\rho_3(\theta)^2\Big]\frac{1}{n} + O(n^{-3/2})$$
$$\equiv \frac{\bar{c}}{n} + O(n^{-3/2}).$$

Hence, in the moderate-deviation range, p^* has relative error $O(n^{-3/2})$.

Example 6.10 Exponential family distribution
Let $Y_1, \ldots, Y_n$ denote independent random variables each distributed according to a density of the form

$$\exp\{y\theta - k(\theta) + D(y)\};$$

this is a one-parameter exponential family model with natural parameter θ. As in Example 6.1, the log-likelihood function for this model may be written

$$\ell(\theta) = nk'(\hat{\theta})\theta - nk(\theta);$$

also $\hat{\jmath} = nk''(\hat{\theta})$. An approximation to the density of $\hat{\theta}$ is given by

$$p^*(\hat{\theta};\theta) = [nk''(\hat{\theta})]^{1/2}\frac{\bar{c}}{\sqrt{2\pi}}\exp\{nk'(\hat{\theta})(\theta - \hat{\theta}) + n[k(\hat{\theta}) - k(\theta)]\}.$$

As noted in Section 5.5, this is identical to the renormalized saddlepoint approximation to the density of $\hat{\theta}$. ∎

Example 6.11 Normal distribution with known coefficient of variation
Let $Y_1, \ldots, Y_n$ denote independent normally distributed random variables each with mean θ and standard deviation $r\theta$ where $\theta > 0$ and the coefficient of variation r is known; for simplicity take $r = 1$. The sufficient statistic for the model may be written $(\hat{\theta}, a)$ where

$$a = \sqrt{n}\frac{[\sum y_j^2]^{1/2}}{\sum y_j}$$

is an ancillary statistic and

$$\hat{\theta} = \frac{(\sum y_j^2)^{1/2}}{\sqrt{n}}\frac{2|a|}{(1+4a^2)^{1/2} + \text{sgn}(a)}$$

is the maximum likelihood estimator of θ. Assume that $a > 0$ which occurs with probability rapidly approaching 1 as $n \to \infty$.

The log-likelihood function may be written

$$\ell(\theta;\hat{\theta}, a) = -\frac{n}{2\theta^2}\left[q^2\hat{\theta}^2 - \frac{2q\theta\hat{\theta}}{a}\right] - n\log\theta$$

where

$$q = \frac{(1+4a^2)^{1/2}+1}{2a}.$$

It follows that

$$p^*(\hat{\theta};\theta|a) = \frac{\sqrt{n\bar{c}}}{\sqrt{(2\pi)}\pi}\left(\frac{\hat{\theta}}{\theta}\right)^{n-1}\frac{1}{\theta}(1+q^2)^{1/2}$$
$$\times \exp\left\{-\frac{n}{2}\left[\frac{q^2}{\theta^2}(\hat{\theta}^2 - \theta^2) - \frac{2\theta}{aq}(\hat{\theta} - \theta)\right]\right\}.$$

This expression may be rewritten as

$$p^*(\hat{\theta};\theta|a) = \frac{\sqrt{n}\bar{c}}{\sqrt{(2\pi)}}\exp\Big\{\frac{n}{2}(q-1/a)^2\Big\}(1+q^2)^{1/2}\Big(\frac{\hat{\theta}}{\theta}\Big)^{n-1}\frac{1}{\theta}$$
$$\times\exp\Big\{-\frac{n}{2}q^2(\hat{\theta}/\theta-1/(aq))^2\Big\}.$$

It may be shown that the exact conditional density of $\hat{\theta}$ given a is of the form

$$p(\hat{\theta};\theta|a) = b(a)\Big(\frac{\hat{\theta}}{\theta}\Big)^{n-1}\frac{1}{\theta}\exp\Big\{-\frac{n}{2}q^2(\hat{\theta}/\theta-1/(aq))^2\Big\}$$

where $b(a)$ is a normalizing constant depending on a. Hence, the likelihood ratio approximation is exact for this model. ∎

To approximate tail probabilities regarding $\hat{\theta}$, we can approximate the integral of p^* using the approach discussed in Section 2.5. Consider the tail probability

$$\int_{\hat{\theta}}^{\infty} p^*(s|a;\theta)ds = \int_{\hat{\theta}}^{\infty}[-\ell_{\theta\theta}(s;s,a)]^{1/2}\frac{\bar{c}}{\sqrt{2\pi}}$$
$$\times\exp\{\ell(\theta;s,a)-\ell(s;s,a)\}\,ds. \tag{6.3}$$

The first step is to transform the integrand to make the integral of the form

$$\bar{c}\int_t^{\infty} h(z)\sqrt{n}\phi(\sqrt{n}z)\,dz.$$

Let

$$z = \operatorname{sgn}(s-\theta)[2(\ell(s;s,a)-\ell(\theta;s,a))]^{1/2}/\sqrt{n};$$

then

$$\frac{dz}{ds} = \frac{\ell_{;\hat{\theta}}(s;s,a)-\ell_{;\hat{\theta}}(\theta;s,a)}{nz}$$

and, assuming the transformation from s to z is one to one, the integral in (6.2) may be written

$$\bar{c}\int_{r/\sqrt{n}}^{\infty}\Big[\frac{-\ell_{\theta\theta}(s;s,a)}{n}\Big]^{1/2}\frac{z}{\ell_{;\hat{\theta}}(\theta;s,a)-\ell_{;\hat{\theta}}(s;s,a)}\sqrt{n}\phi(\sqrt{n}z)\,dz$$

where

$$r = \operatorname{sgn}(\hat{\theta}-\theta)[2(\ell(\hat{\theta})-\ell(\theta))]^{1/2}.$$

Let

$$h(z) = \frac{[-\ell_{\theta\theta}(s;s,a)]^{1/2}z}{\ell_{;\hat{\theta}}(s;s,a)-\ell_{;\hat{\theta}}(\theta;s,a)}$$

where s is a function of z. Note that $z = 0$ is equivalent to $s = \theta$ and that for $z \doteq 0$,

$$z \doteq [-\ell_{\theta\theta}(\theta; \theta, a)]^{1/2}(s - \theta);$$
$$\ell_{;\hat{\theta}}(\theta; s, a) - \ell_{;\hat{\theta}}(s; s, a) \doteq \ell_{\theta;\hat{\theta}}(s; s, a)(\theta - s).$$

By differentiating the likelihood equation $\ell_\theta(s; s, a) = 0$ with respect to s we have that

$$\ell_{\theta;\hat{\theta}}(s; s, a) = -\ell_{\theta\theta}(s; s, a)$$

and hence

$$\ell_{;\hat{\theta}}(\theta; s, a) - \ell_{;\hat{\theta}}(s; s, a) \doteq -\ell_{\theta;\hat{\theta}}(s; s, a)(\theta - s).$$

It follows that $h(0) = 1$. Note that $z = r/\sqrt{n}$ is equivalent to $s = \hat{\theta}$ so that

$$h(r/\sqrt{n}) = \frac{[-\ell_{\theta\theta}(\hat{\theta})]^{1/2} r}{\sqrt{n}[\ell_{;\hat{\theta}}(\hat{\theta}; \hat{\theta}, a) - \ell_{;\hat{\theta}}(\theta; \hat{\theta}, a)]}.$$

The approximation (2.8) now shows that the probability in (6.3) may be approximated by

$$1 - \Phi(r) + \phi(r)\left[\frac{1}{v} - \frac{1}{r}\right]$$

where

$$v = [-\ell_{\theta\theta}(\hat{\theta})]^{-1/2}[\ell_{;\hat{\theta}}(\hat{\theta}; \hat{\theta}, a) - \ell_{;\hat{\theta}}(\theta; \hat{\theta}, a)].$$

The relative error of this approximation is $O(n^{-3/2})$ for $r = O(1)$, corresponding to $\hat{\theta} = \theta + O(n^{-1/2})$ and $O(n^{-1})$ for $r = O(\sqrt{n})$, corresponding to fixed values of $\hat{\theta}$.

Example 6.12 Normal distribution with known coefficient of variation
Let $Y_1, \ldots, Y_n$ denote independent normally distributed random variables each with mean θ and standard deviation θ, $\theta > 0$; see Example 6.11.

For this model,

$$r = \text{sgn}(\hat{\theta} - \theta)\sqrt{n}\{q^2[(\hat{\theta}/\theta)^2 - 1] - 2(q/a)[\hat{\theta}/\theta - 1] - 2\log(\hat{\theta}/\theta)\}^{1/2};$$
$$v = \frac{\sqrt{n}q^2}{(1+q^2)}\left[(\hat{\theta}/\theta)^2 - 1 - \frac{1}{aq}(\hat{\theta}/\theta - 1)\right]$$

where q and a are as given in Example 6.11.

Consider the case in which $n = 10$, $a = 1$, and $\theta = 1$; suppose that $\hat{\theta} = 3/2$ is observed. The standard error of $\hat{\theta}$ based on the observed information is approximately 0.249 so that, based on a normal approximation, the probability that $\hat{\theta}$ exceeds 3/2 is approximately 0.022.

For $\hat{\theta} = 3/2$, $r \doteq 2.904$ and $v \doteq 4.096$. Hence, the approximate tail probability based on the approximation derived in this section is approximately 0.001253. The exact probability, obtained by using the exact density function of $\hat{\theta}$ and numerical integration, is 0.001247. ∎

6.3.2 VECTOR PARAMETER CASE

The case of a vector parameter is similar. The cumulants of $\sqrt{n}(\hat{\theta} - \theta)$ may be determined from the cumulants of the score statistic. In particular, the first formal cumulant of $\sqrt{n}(\hat{\theta} - \theta)$ is of order $O(n^{-1/2})$, the second formal cumulant is given by $\bar{j_o}^{-1} + O(n^{-1})$ and the third and fourth formal cumulants are of order $O(n^{-1/2})$ and $O(n^{-1})$, respectively.

To approximate the density of $\hat{\theta}$ we may use the same approach as in the previous section. The conditional density of $\hat{\theta}$ given a, $p(\hat{\theta}|a; \theta)$, is given by

$$\exp\{\ell(\theta; \hat{\theta}, a) - \ell(\theta_0; \hat{\theta}, a)\} p(\hat{\theta}|a; \theta_0)$$

where θ_0 is an arbitrary parameter value, which may be chosen so that $p(\hat{\theta}|a; \theta)$ may be accurately approximated. A natural choice for θ_0 is $\hat{\theta}$, the point at which the density is to be evaluated. An Edgeworth expansion for the density of $\hat{\theta}$ with $\theta_0 = \hat{\theta}$ is given by $|\hat{j}|^{1/2}\phi_d(0)$, where $\phi_d(\cdot)$ denotes the density function of the standard multivariate normal density of dimension d; the error of this approximation is $O(n^{-1})$. Hence, we may approximate the density of $\hat{\theta}$ by

$$|\hat{j}|^{1/2}(2\pi)^{-d/2} \exp\{\ell(\theta) - \ell(\hat{\theta})\}.$$

The renormalized form of this approximation is given by

$$p^*(\hat{\theta}|a; \theta) = c|\hat{j}|^{1/2} \exp\{\ell(\theta) - \ell(\hat{\theta})\}$$

or, equivalently,

$$p^*(\hat{\theta}|a; \theta) = |\hat{j}|^{1/2} \frac{\bar{c}}{(2\pi)^{d/2}} \exp\{\ell(\theta) - \ell(\hat{\theta})\}.$$

The asymptotic properties of p^* in the vector parameter case are the same as those in the scalar parameter case. Specifically, the relative error of the approximation is $O(n^{-1})$ for fixed $\hat{\theta}$ and $O(n^{-3/2})$ for $\hat{\theta}$ of the form $\theta + O(n^{-1/2})$.

Example 6.13 Linear model
Consider the linear model in which Y is a $n \times 1$ vector of the form

$$Y = \mu + \sigma\epsilon$$

where μ is an unknown $n \times 1$ parameter vector taking values in a linear space $\mathcal{M}$, ϵ is an $n \times 1$ vector of unobservable random variables with known density function $\exp\{g(\epsilon)\}$, and $\sigma > 0$ is an unknown scalar parameter. The parameter μ may be written $\mu(\beta) = x\beta$ where x is an $n \times p$ matrix such that the columns of x span the linear space $\mathcal{M}$ and β is a $p \times 1$ parameter. Let $a = (y - \hat{\mu})/\hat{\sigma}$ where $\hat{\mu}$ and $\hat{\sigma}$ denote the maximum likelihood estimates of μ and σ, respectively; recall that a is an ancillary statistic.

The log-likelihood function is given by

$$\ell(\theta) = g[(\hat{\sigma}a + \mu(\hat{\beta}) - \mu(\beta))/\sigma] - n \log \sigma.$$

It is straightforward to show that

$$\hat{\jmath} = \hat{\sigma}^{-2} \begin{pmatrix} \mu'(\hat{\beta})^T g''(a)\mu'(\hat{\beta}) & \mu'(\hat{\beta})^T g''(a)a \\ ag''(a)\mu'(\hat{\beta})^T & a^T g''(a)a + g'(a)^T a \end{pmatrix};$$

note that $\mu'(\hat{\beta})$ is a constant and that $g''(a)$ is an $n \times n$ matrix depending only on a. Hence,

$$|\hat{\jmath}|^{1/2} = c_1(a)\hat{\sigma}^{-(p+1)}$$

where $c_1(a)$ depends only on a. Since

$$\ell(\hat{\theta}) = c_2(a) - n \log \hat{\sigma}$$

where $c_2(a)$ is a function of a alone, it follows that

$$p^*(\hat{\beta}, \hat{\sigma}|a) = c(a)\frac{\hat{\sigma}^{n-p-1}}{\sigma^n} \exp\{g[(\hat{\sigma}a + \mu(\hat{\beta}) - \mu(\beta))/\sigma]\}.$$

For a model of this type, it is possible to calculate the exact conditional density of the maximum likelihood estimators. The density of Y is given by

$$\sigma^{-n} \exp\{g[(\hat{\sigma}a + \mu(\hat{\beta}) - \mu(\beta))/\sigma]\}.$$

To determine the conditional density of $(\hat{\beta}, \hat{\sigma})$ given a we might first attempt to determine the density of $(\hat{\beta}, \hat{\sigma}, a)$ by transforming y to $(\hat{\beta}, \hat{\sigma}, a)$. Note, however, that since a is of dimension n, the transformation from y to $(\hat{\beta}, \hat{\sigma}, a)$ is not one to one. The components of a satisfy certain functional relationships; for instance, if $\mu(\beta) = \beta$ is a location parameter and ϵ is normally distributed with identity covariance matrix, then $a_j = (y_j - \bar{y})/\hat{\sigma}$ and, hence, the sample mean of the elements of a is identically 0.

In general, there exists an ancillary statistic a_0 of dimension $n - (p + 1)$ such that $a = h(a_0)$ for some function h and the transformation from y to $(\hat{\beta}, \hat{\sigma}, a_0)$ is one to one; a_0 may often be taken to be the first $n - (p + 1)$ components of a. The density of $(\hat{\beta}, \hat{\sigma}, a_0)$ is given by

$$\sigma^{-n} \exp\{g[(\hat{\sigma}h(a_0) + \mu(\hat{\beta}) - \mu(\beta))/\sigma]\} |J|$$

where $|J|$ denotes the Jacobian of the transformation from y to $(\hat{\beta}, \hat{\sigma}, a_0)$. This Jacobian is given by

$$|J| = \left|h(a_0) \quad \mu'(\hat{\beta}) \quad \hat{\sigma}h'(a_0)\right|.$$

Note that $\mu'(\hat{\beta})$ does not depend on $(\hat{\beta}, \hat{\sigma})$ so that we may write

$$|J| = \left| Q \quad \hat{\sigma} h'(a_0) \right|$$

where Q is a matrix depending only on a. It follows that

$$|J|^2 = |J^T J| = \begin{vmatrix} Q^T Q & \hat{\sigma} Q^T h'(a_0) \\ \hat{\sigma} h'(a_0)^T Q & \hat{\sigma}^2 h'(a_0)^T h'(a_0) \end{vmatrix}.$$

Using the usual the formula for the determinant of a partitioned matrix, it can be shown that

$$|J|^2 = c_3(a) \hat{\sigma}^{2(n-p-1)}$$

where $c_3(a)$ is a constant depending only on a; recall that the dimension of $h'(a_0)$ is $n \times (n-p-1)$.

Thus, the exact joint density of $(\hat{\beta}, \hat{\sigma}, a_0)$ is of the form

$$c_4(a) \frac{\hat{\sigma}^{n-p-1}}{\sigma^n} \exp\{g[(\hat{\sigma} a + \mu(\hat{\beta}) - \mu(\beta))/\sigma]\}.$$

To compute the conditional density of $(\hat{\beta}, \hat{\sigma})$ given a_0 we must divide this joint density by the marginal density of a_0; since this marginal density depends only on a, it follows that the exact conditional density of $(\hat{\beta}, \hat{\sigma})$ given a is of the form

$$c(a) \frac{\hat{\sigma}^{n-p-1}}{\sigma^n} \exp\{g[(\hat{\sigma} a + \mu(\hat{\beta}) - \mu(\beta))/\sigma]\}$$

where $c(a)$ depends only on a; this is exactly the p^* formula. Hence, for this class of models, the likelihood ratio approximation is exact. This same result holds for all transformation models. ∎

The p^* formula gives an expression for the density of the entire vector $\hat{\theta}$. In practice, it is often useful to have an approximation for the density of a single component of $\hat{\theta}$. This may be obtained by integrating out the unwanted components of $\hat{\theta}$ from p^* using either exact integration, approximate analytic integration, or numerical integration.

Example 6.14 Comparison of exponential distributions

Consider two independent samples of independent exponential random variables, each of size n, and with rate parameters $\psi\lambda$ and λ, respectively. Hence, ψ denotes the ratio of the rate parameters of the two distributions. Let $\bar{y}$ and $\bar{x}$ denote the sample means; then $\hat{\psi} = \bar{x}/\bar{y}$ and $\hat{\lambda} = 1/\bar{x}$. The log-likelihood function may be written

$$\ell(\theta) = 2n \log \lambda + n \log \psi - n \left[\frac{\lambda \psi}{\hat{\lambda} \hat{\psi}} + \frac{\lambda}{\hat{\lambda}} \right];$$

it follows that

$$\hat{j} = n\begin{pmatrix} \hat{\psi}^{-2} & (\hat{\psi}\hat{\lambda})^{-1} \\ (\hat{\psi}\hat{\lambda})^{-1} & 2\hat{\lambda}^{-2} \end{pmatrix}.$$

The p^* approximation to the density of $(\hat{\psi}, \hat{\lambda})$ is given by

$$p^*(\hat{\psi}, \hat{\lambda}) = \frac{\bar{c}}{2\pi}\frac{n}{\hat{\psi}\hat{\lambda}}\left(\frac{\lambda}{\hat{\lambda}}\right)^{2n}\left(\frac{\psi}{\hat{\psi}}\right)^{n} \exp\left\{-n\left(\frac{\lambda\psi}{\hat{\lambda}\hat{\psi}} + \frac{\lambda}{\hat{\lambda}} - 2\right)\right\}.$$

Suppose that we are primarily interested in the parameter ψ. To obtain the marginal density of $\hat{\psi}$ we need to evaluate the integral

$$\int p^*(\hat{\psi}, \hat{\lambda})\, d\hat{\lambda} = \bar{c}\frac{n}{2\pi}\int \frac{1}{\hat{\psi}\hat{\lambda}}\left(\frac{\lambda}{\hat{\lambda}}\right)^{2n}\left(\frac{\psi}{\hat{\psi}}\right)^{n} \exp\left\{-n\left(\frac{\lambda\psi}{\hat{\lambda}\hat{\psi}} + \frac{\lambda}{\hat{\lambda}} - 2\right)\right\} d\hat{\lambda}.$$

Let $t = \lambda/\hat{\lambda}$. The desired integral may be written

$$\begin{aligned} &\bar{c}\frac{n}{2\pi}\left(\frac{\psi}{\hat{\psi}}\right)^{2n+1}\frac{1}{\psi}\exp\{2n\}\int_0^\infty t^{2n-1}\exp\{-n(\psi/\hat{\psi}+1)t\}\, dt \\ &= \bar{c}\frac{n}{2\pi}\left(\frac{\psi}{\hat{\psi}}\right)^{2n+1}\frac{1}{\psi}\exp\{2n\}\frac{\Gamma(2n)}{(\psi/\hat{\psi}+1)^{2n}}. \end{aligned}$$

Hence, an approximation to the marginal density of $\hat{\psi}$ given by

$$\bar{c}\frac{\exp\{2n\}\Gamma(2n)}{2\pi n^{2n-1}}\frac{1}{\psi}\left(\frac{\hat{\psi}}{\psi}\right)^{n-1}\left(\frac{\hat{\psi}}{\psi}+1\right)^{-2n}.$$

The exact distribution of $\hat{\psi}/\psi$ is an F-distribution with degrees of freedom $(2n, 2n)$. It follows that the exact density of $\hat{\psi}$ is given by

$$\frac{\Gamma(2n)}{\Gamma(n)^2}\frac{1}{\psi}\left(\frac{\hat{\psi}}{\psi}\right)^{n-1}\left(\frac{\hat{\psi}}{\psi}+1\right)^{-2n};$$

hence, the approximation derived here is exact, aside from the normalization factor. ∎

Example 6.15 Scale parameter of a linear model

Consider the linear model described in Example 6.13. The p^* formula for the conditional density of $(\hat{\beta}, \hat{\sigma})$ given $a = (y - \hat{\mu})/\hat{\sigma}$ was shown to be

$$p^*(\hat{\beta}, \hat{\sigma}\,|a) = c(a)\frac{\hat{\sigma}^{n-p-1}}{\sigma^n}\exp\{g[(\hat{\sigma}a + x\hat{\beta} - x\beta)/\sigma]\}.$$

The conditional density of the scale parameter $\hat{\sigma}$ is therefore given by

$$c(a)\frac{\hat{\sigma}^{n-p-1}}{\sigma^n}\int_{\Re^p}\exp\{g[(\hat{\sigma}a + x\hat{\beta} - x\beta)/\sigma]\}\,d\hat{\beta}.$$

This integral may be approximated using Laplace's method. Let $t = (\hat{\beta} - \beta)/\sigma$. Then

$$\int_{\Re^p}\exp\{g[(\hat{\sigma}a + x\hat{\beta} - x\beta)/\sigma]\}\,d\hat{\beta} = \sigma^p\int_{\Re^p}\exp\{g(a\hat{\sigma}/\sigma + xt)\}\,dt.$$

Let $\hat{t} \equiv \hat{t}(a\hat{\sigma}/\sigma; x)$ denote the maximizer of $g(a\hat{\sigma}/\sigma + xt)$. Then

$$\int_{\Re^p}\exp\{g(a\hat{\sigma}/\sigma + xt)\}\,dt = c_1\frac{\exp\{g(a\hat{\sigma}/\sigma + x\hat{t})\}}{|x^T g''(a\hat{\sigma}/\sigma + x\hat{t})x|^{1/2}}[1 + O(n^{-1})]$$

for some constant c_1.

Hence, the approximate conditional density of $\hat{\sigma}$ is given by

$$\frac{c_2(a)}{\sigma}\left(\frac{\hat{\sigma}}{\sigma}\right)^{n-p-1}\frac{\exp\{g(a\hat{\sigma}/\sigma + x\hat{t})\}}{|-x^T g''(a\hat{\sigma}/\sigma + x\hat{t})x|^{1/2}}.$$

Consider the case of a normal-theory linear model so that

$$g(y) = -\tfrac{1}{2}y^T y.$$

Then $\hat{t}$ minimizes

$$\left(a\frac{\hat{\sigma}}{\sigma} + xt\right)^T\left(a\frac{\hat{\sigma}}{\sigma} + xt\right).$$

Since $x^T a = 0$, this is equivalent to minimizing $t^T(x^T x)t$; hence, $\hat{t} = 0$. It follows that the approximate conditional density of $\hat{\sigma}$ is given by

$$\frac{c_2(a)}{\sigma}\left(\frac{\hat{\sigma}}{\sigma}\right)^{n-p-1}\exp\{-n(\hat{\sigma}/\sigma)^2/2)\};$$

here we have used the fact that $a^T a = n$.

Recall that for this model $n\hat{\sigma}^2/\sigma^2$ has a chi-square density with $n - p$ degrees of freedom and that $\hat{\sigma}$ and a are independent. It follows that the exact conditional density of $\hat{\sigma}$ is given by

$$\frac{n^{(n-p)/2}}{\Gamma((n-p)/2)2^{(n-p)/2-1}}\frac{1}{\sigma}\left(\frac{\hat{\sigma}}{\sigma}\right)^{n-p-1}\exp\{-n(\hat{\sigma}/\sigma)^2/2)\};$$

hence, the approximation derived above is exact for this model. ∎

6.3.3 THE LATTICE CASE

The results discussed thus far in this section apply to the case in which the underlying random variables have a continuous distribution. We now consider in which these random variables have a lattice distribution. Note that, in general, the distribution of $\hat{\theta}$ will not be a lattice distribution. Hence, it is important to keep track of the lattice random variables. For simplicity, we consider only the scalar parameter case; a similar approach may be used for vector parameters.

The difficulties with p^* in the discrete case arise from the use of an Edgeworth expansion for approximating the density of $\hat{\theta}$ under the parameter value $\theta_0 = \hat{\theta}$. If $\hat{\theta}$ has a lattice distribution, then the argument given above continues to hold with only minor modifications and p^*, which in this case is a density with respect to counting measure, provides a valid approximation to the conditional density of $\hat{\theta}$ given a with the usual properties.

In many cases however in which the underlying data have a lattice distribution, the distribution of $\hat{\theta}$ is discrete, but nonlattice. In these cases, an Edgeworth series approximation for the density of $\hat{\theta}$ is not available. Here we consider an approximation to the conditional distribution of $\hat{\theta}$ given a under the assumption that there exists a one-to-one differentiable function g such that $g(\hat{\theta})$ takes values in the set $\{0, 1, 2, \ldots\}$. Note that the function g may depend on both n and a.

We use will use the following approach. Let $\hat{\phi} = g(\hat{\theta})$; since $\hat{\phi}$ is a lattice variable, p^* may be used to approximate the conditional density of $\hat{\phi}$ given a. This approximation may be transformed to obtain an approximation to the conditional density of $\hat{\theta}$ given a; note that, since these densities are with respect to counting measure, no Jacobian term is needed. This leads to the following approximation:

$$c|\hat{j}|^{1/2}\exp\{\ell(\theta) - \ell(\hat{\theta})\}\left|\frac{\partial\hat{\theta}}{\partial\hat{\phi}}\right|.$$

The differential term in this expression arises from writing the observed information for ϕ in terms of the observed information for θ.

Let

$$m(\hat{\theta}) \equiv m(\hat{\theta}; \hat{\Theta}) = \left|\frac{\partial\hat{\theta}}{\partial\hat{\phi}}\right| = \left|\frac{\partial g(y)}{\partial y}\bigg|_{y=\hat{\theta}}\right|^{-1}$$

and let $\mu^*(\cdot) \equiv \mu^*(\cdot; \hat{\Theta})$ denote the discrete measure placing mass $m(\hat{\theta})$ at each point $\hat{\theta}$ of $\hat{\Theta}$. Note that μ^* does not depend on the choice of g since if for two functions g_1, g_2 satisfy the required conditions then g_1 and g_2 must differ by a constant; also, since the spacing of elements of $\hat{\Theta}$ is $O(n^{-1})$, $m(\hat{\theta})$ is of order $O(n^{-1})$.

We may then approximate the density of $\hat{\theta}$ by

$$p^*(\hat{\theta}|a; \theta)d\mu^*(\hat{\theta}) = c|\hat{j}|^{1/2}\exp\{\ell(\theta) - \ell(\hat{\theta})\}\,d\mu^*(\hat{\theta}),$$

which is valid at the support points of the distribution of $\hat{\theta}$. To approximate the probability that $\hat{\theta} = s$, we use the approximation $p^*(s|a; \theta)m(s)$. The asymptotic properties of p^* are the same as in the continuous case. As in the continuous case, an unnormalized version of the approximation is also available in which the constant c is replaced by $(2\pi)^{-1/2}$.

Example 6.16 Geometric distribution
Let $Y_1, \ldots, Y_n$ denote independent random variables each with density of the form

$$(1-\theta)\theta^y, \quad y = 0, 1, \ldots$$

where $0 < \theta < 1$; this is a geometric distribution. Then

$$\ell(\theta) = n\bar{y}\log\theta + n\log(1-\theta)$$

and $\hat{\theta} = \bar{y}/(1+\bar{y})$. Note that

$$\hat{\Theta} = \left\{0, \frac{1}{n+1}, \frac{2}{n+2}, \ldots\right\}$$

so that $g(\hat{\theta}) = n\hat{\theta}/(1-\hat{\theta})$ and $m(\hat{\theta}) = (1-\hat{\theta})^2/n$. It is straightforward to show that

$$p^*(\hat{\theta};\theta) = c\frac{\theta^{n\hat{\theta}/(1-\hat{\theta})}(1-\theta)^n}{\hat{\theta}^{n\hat{\theta}/(1-\hat{\theta})+1/2}(1-\hat{\theta})^{n+1}}.$$

Table 6.1 contains a comparison of probabilities based on p^* with the exact probabilities based on the fact that $\sum Y_j$ has a negative binomial distribution for the case $\theta = 1/2$ and $n = 10$; results are also given for the unnormalized approximation, which is denoted by $p^\dagger$. ∎

Note that, in some cases, the expression for p^* is not valid for values of $\hat{\theta}$ at the boundary of its support; this occurs whenever the variance of $\hat{\theta}$ is 0 for certain values of θ. For instance, for a Poisson distribution with mean θ, the variance of $\hat{\theta}$ is 0 for $\theta = 0$; hence, p^* does not provide a valid approximation to the probability that $\hat{\theta} = 0$. Since these extreme value of $\hat{\theta}$ occur only for extreme values of the data, it is often possible to determine the exact probability of that occurrence; for instance, in the Poisson example, $\hat{\theta} = 0$ occurs only if all data values are 0, the probability of which is easily determined.

As in the continuous case, p^* may be used to construct an approximation to tail probabilities. Using the method described in Section 2.5, it follows that $\Pr(\hat{\theta} \leq t|a;\theta)$

Table 6.1 Probability approximations in Example 6.16

$\hat{\theta}$	Exact	p^*	$p^\dagger$
0.167	0.0134	0.0138	0.0140
0.333	0.0611	0.0614	0.0623
0.500	0.0881	0.0879	0.0892
0.600	0.0390	0.0388	0.0394
0.667	0.00933	0.00928	0.00928

Table 6.2 Distribution function approximations in Example 6.17

$\hat\theta$	Exact	LR	r
0.091	0.00586	0.00631	0.00173
0.167	0.0193	0.0199	0.00791
0.333	0.151	0.152	0.0962
0.600	0.885	0.886	0.842
0.667	0.9786	0.9787	0.9674
0.722	0.99803	0.99804	0.99668

may be approximated by

$$\Phi(r) + \phi(r)\left(\frac{1}{v} + \frac{1}{r}\right)$$

where r denotes the signed likelihood ratio statistic,

$$v = \frac{1 - \exp[\{\ell_{;\hat\theta}(\hat\theta) - \ell_{;\hat\theta}(\theta)\}m(\hat\theta)]}{|\hat\jmath|^{1/2}m(\hat\theta)},$$

and $\hat\theta$ is taken to be t.

Example 6.17 Geometric distribution

Consider the geometric distribution model considered in Example 6.16. For this model,

$$r = \text{sgn}(\hat\theta - \theta)\left\{2n\left[\frac{\hat\theta}{1-\hat\theta}\log(\hat\theta/\theta) - \log\frac{\hat\theta}{1-\hat\theta}\right]\right\};$$

$$v = \frac{\sqrt{n}(\hat\theta - \theta)}{[\hat\theta(1-\hat\theta)]^{1/2}}.$$

Table 6.2 contains the distribution function approximation discussed in this section for the case $\theta = 1/2$, $n = 10$; this approximation, based on the Lugannani and Rice formula, is denoted by 'LR'.

For comparison, the exact probabilities are presented as well as the probabilities based on a standard normal approximation to r, denoted simply by 'r'. For this example, the distribution function approximation is highly accurate. ∎

6.4 Stable inference

6.4.1 INTRODUCTION

In Section 6.2 expressions for the conditional cumulants of the normalized score statistic Z were given. In general, these conditional cumulants will differ from the unconditional cumulants of Z derived in Section 5.3. However certain cumulants or combinations of cumulants will be unaffected by conditioning, either exactly or approximately; these combinations of cumulants will be said to be *stable*. For instance, we have seen that

$$\mathrm{E}[Z;\theta] = \mathrm{E}[Z|a;\theta] = 0.$$

Hence, the first cumulant of Z is (exactly) stable; the value of the cumulant does not depend on whether it is calculated conditionally or unconditionally.

One way to view this result is that we may calculate the conditional expected value of Z given a by carrying out the unconditional calculation. Not only do we not need to know the conditional distribution of the data given a, we do not even need to specify the ancillary a; the result holds for all ancillary statistics. Of course, this case is particularly simple since the cumulant in question is exactly zero. All of the cumulant combinations given by the Bartlett identities have this same property.

The same ideas may be applied to the distribution of a statistic T. If the conditional distribution given an arbitrary ancillary statistic a and the unconditional distribution of T are approximately the same, then the distribution of T will be said to be stable. More precisely, we will say that the distribution of T is stable to first order if the first-order asymptotic approximation to the conditional distribution of T given a does not depend on a and stable to second order if an asymptotic expansion for the conditional distribution of T given a does not depend on a, neglecting terms of order $O(n^{-1})$. Inference based on a statistic with a distribution that is stable is particularly convenient since the goals of conditioning on an ancillary statistic may be achieved using an unconditional procedure. Furthermore, the conclusions reached do not depend on the ancillary statistic used and precise specification of the ancillary is not needed. This is particularly important if it is difficult to determine an appropriate ancillary statistic or the conditional analysis is difficult to carry out.

In this section, we consider the stability of the distribution of the score statistic, the maximum likelihood estimate, and the likelihood ratio statistic for the case of a scalar parameter. Similar results hold when θ is a vector.

6.4.2 THE RELATIONSHIP BETWEEN CONDITIONAL AND UNCONDITIONAL CUMULANTS OF LOG-LIKELIHOOD DERIVATIVES

Consider the distribution of the score statistic Z. The first four conditional cumulants of Z were derived in Section 6.2 and are given by $\nu_1(a)$; $\nu_2(a)$; $\nu_3(a)/\sqrt{n}$; $\nu_4(a)/n$, respectively, where $\nu_1(a) = 0$; $\nu_2(a) = \bar{\jmath}_o + O(n^{-1})$;

$$\nu_3(a) = -\frac{3\ell_{\theta\theta;\hat{\theta}}(\theta;\theta,a) + \ell_{\theta\theta\theta}(\theta;\theta,a)}{n} + O(n^{-1});$$

$$\nu_4(a) = -\frac{\ell_{\theta\theta\theta\theta}(\theta;\theta,a) + 4\ell_{\theta\theta\theta;\hat{\theta}}(\theta;\theta,a) + 6\ell_{\theta\theta;\hat{\theta}\hat{\theta}}(\theta;\theta,a)}{n}$$

$$+ 3\left(\frac{\ell_{\theta\theta;\hat{\theta}}(\theta;\theta,a)}{n}\right)^2 \frac{1}{\bar{\jmath}_o} + O(n^{-1}),$$

respectively. In order to compare these to the unconditional standardized cumulants of Z which, in the notation of Section 5.3, are denoted by $\nu_1, \nu_2, \nu_3, \nu_4$, we need to consider the relationship between mixed log-likelihood derivatives evaluated at $\hat{\theta} = \theta$ and the unconditional cumulants of $(\ell_\theta(\theta), \ell_{\theta\theta}(\theta), \ell_{\theta\theta\theta}(\theta), \ell_{\theta\theta\theta\theta}(\theta))$.

For instance, consider $\ell_{\theta\theta}(\theta)$. Using a Taylor's series expansion around $\hat{\theta} = \theta$, we have that

$$\begin{aligned}\frac{1}{n}\ell_{\theta\theta}(\theta_1) &\equiv \frac{1}{n}\ell_{\theta\theta}(\theta_1; \hat{\theta}, a) \\ &= \frac{1}{n}\ell_{\theta\theta}(\theta_1; \theta, a) + \frac{1}{n}\ell_{\theta\theta;\hat{\theta}}(\theta_1; \theta, a)(\hat{\theta} - \theta) + \cdots.\end{aligned}$$

It follows that

$$\frac{1}{n}\ell_{\theta\theta}(\theta_1) = \frac{1}{n}\ell_{\theta\theta}(\theta_1; \theta, a) + O_p(n^{-1/2})$$

and that

$$\frac{1}{n}\mathrm{E}[\ell_{\theta\theta}(\theta_1); \theta] = \frac{1}{n}\mathrm{E}[\ell_{\theta\theta}(\theta_1; \theta, a); \theta] + O(n^{-1}) \tag{6.4}$$

for any value of θ_1. Since

$$\frac{1}{n}\ell_{\theta\theta}(\theta) = \frac{1}{n}\mathrm{E}[\ell_{\theta\theta}(\theta); \theta] + O_p(n^{-1/2})$$

it follows that

$$\frac{1}{n}\ell_{\theta\theta}(\theta; \theta, a) = \frac{1}{n}\mathrm{E}[\ell_{\theta\theta}(\theta; \theta, a); \theta] + O_p(n^{-1/2})$$

and that

$$\nu_{01} = \frac{1}{n}\mathrm{E}[\ell_{\theta\theta}(\theta); \theta] = \frac{1}{n}\ell_{\theta\theta}(\theta; \theta, a) + O(n^{-1/2}).$$

Similar results hold for other derivatives as well, so that, neglecting terms of order $O(n^{-1/2})$,

$$\nu_{001} = \frac{1}{n}\ell_{\theta\theta\theta}(\theta; \theta, a); \qquad \nu_{0001} = \frac{1}{n}\ell_{\theta\theta\theta\theta}(\theta; \theta, a).$$

Joint cumulants may be handled in a similar manner. Differentiating (6.4) with respect to θ and evaluating the result at $\theta_1 = \theta$, shows that

$$\begin{aligned}\frac{1}{n}\mathrm{E}[\ell_{\theta\theta}(\theta)\ell_\theta(\theta); \theta] = \frac{1}{n}&\mathrm{E}[\ell_{\theta\theta;\hat{\theta}}(\theta; \theta, a); \theta] \\ &+ \mathrm{E}[\ell_{\theta\theta}(\theta; \theta, a)\ell_\theta(\theta); \theta] + O(n^{-1}).\end{aligned}$$

Note that

$$\mathrm{E}[\ell_{\theta\theta}(\theta; \theta, a)\ell_\theta(\theta)|a; \theta] = 0$$

so that

$$\frac{1}{n}\mathrm{E}[\ell_{\theta\theta}(\theta)\ell_\theta(\theta); \theta] = \frac{1}{n}\mathrm{E}[\ell_{\theta\theta;\hat{\theta}}(\theta; \theta, a); \theta] + O(n^{-1}).$$

Since

$$\frac{1}{n}\ell_{\theta\theta;\hat{\theta}}(\theta) = \frac{1}{n}\mathrm{E}[\ell_{\theta\theta;\hat{\theta}}(\theta);\theta] + O_p(n^{-1/2}),$$

$$\frac{1}{n}\mathrm{E}[\ell_{\theta\theta}(\theta)\ell_{\theta}(\theta);\theta] = \frac{1}{n}\ell_{\theta\theta;\hat{\theta}}(\theta;\theta,a) + O(n^{-1/2}).$$

It follows that, neglecting terms of order $O(n^{-1/2})$,

$$\nu_{11} = \frac{1}{n}\ell_{\theta\theta;\hat{\theta}}(\theta;\theta,a);$$

similarly,

$$\nu_{101} = \frac{1}{n}\ell_{\theta\theta\theta;\hat{\theta}}(\theta;\theta,a).$$

The same approach can be used with second derivatives so that

$$\nu_{21} + \nu_{02} = \frac{1}{n}\ell_{\theta\theta;\hat{\theta}\hat{\theta}}(\theta;\theta,a) + O(n^{-1/2}).$$

Note that since

$$\ell_{\theta}(\hat{\theta}) = \ell_{\theta}(\hat{\theta};\hat{\theta},a) = 0,$$

it follows that $\ell_{\theta}(\theta;\theta,a) = 0$; this expression may be differentiated with respect to θ to obtain identities for certain mixed log-likelihood derivatives; these identities are analogous to the Bartlett identities in a certain sense. For instance,

$$\ell_{\theta\theta}(\theta;\theta,a) + \ell_{\theta;\hat{\theta}}(\theta;\theta,a) = 0.$$

Using these identities together with the Bartlett identities we may obtain expressions for the remaining ν_{ijk} that are needed:

$$\nu_{02} = \frac{1}{n}\ell_{\theta\theta;\hat{\theta}\hat{\theta}}(\theta;\theta,a) + \frac{1}{n}\ell_{\theta\theta;\hat{\theta}}(\theta;\theta,a);$$

$$\nu_{30} = -\left[\frac{1}{n}\ell_{\theta\theta\theta}(\theta;\theta,a) + \frac{3}{n}\ell_{\theta\theta;\hat{\theta}}(\theta;\theta,a)\right];$$

$$\nu_{40} = -\left[\frac{1}{n}\ell_{\theta\theta\theta\theta}(\theta;\theta,a) + \frac{4}{n}\ell_{\theta\theta\theta;\hat{\theta}}(\theta;\theta,a) + \frac{3}{n}\ell_{\theta\theta;\hat{\theta}\hat{\theta}}(\theta;\theta,a) - \frac{3}{n}\ell_{\theta\theta;\hat{\theta}}(\theta;\theta,a)\right].$$

These expressions neglect terms of order $O(n^{-1/2})$.

Using these results, the expressions for the standardized conditional cumulants of Z given above may be expressed in terms of the standardized unconditional cumulants:

$$\nu_2(a) = \nu_{20} + O(n^{-1/2}); \qquad \nu_3(a) = \nu_3 + O(n^{-1/2});$$

$$\nu_4(a) = [\nu_{40} - 3(\nu_{02} - \nu_{11}^2/\nu_{20})] + O(n^{-1/2}).$$

Hence, the leading terms in the expansions for the first three conditional cumulants of Z agree with the leading terms in the expansions for the unconditional cumulants. It follows that the distribution of Z is stable to first order; it is not, in general, stable to second order since the conditional and unconditional variance agree only to $O(n^{-1/2})$.

Even the first terms in the conditional and unconditional expansions for the the standardized fourth cumulant of Z do not agree. Hence, the value of the fourth cumulant of Z depends heavily on whether it is calculated conditionally or unconditionally. The difference between the conditional and unconditional standardized fourth cumulant of Z may be written as

$$\nu_4(a) - \nu_4 = -3\gamma^2\nu_2^2 + O(n^{-1/2})$$

where

$$\gamma^2 = \frac{\nu_{02} - \nu_{11}^2/\nu_{20}}{\nu_{20}^2}.$$

The quantity γ is called the *statistical curvature* of the model. The statistical curvature is a measure of how close a given model is to exponential family form; if the model is a one-parameter exponential family model then $\gamma = 0$.

Example 6.18 Two measuring instruments
Consider the two measuring instruments example considered in Example 6.5. The log-likelihood function for this model may be written

$$\ell(\theta) = -\frac{1}{2}\sum(y_j - \theta)^2/\sigma_{a_j}^2 = \left(\theta\hat{\theta} - \frac{1}{2}\theta^2\right)\left[(n-a)\frac{1}{\sigma_0^2} + a\frac{1}{\sigma_1^2}\right]$$

where $a = \sum a_j$ is an ancillary statistic and

$$\hat{\theta} = \frac{\sum y_j/\sigma_{a_j}^2}{\sum 1/\sigma_{a_j}^2}.$$

Then

$$Z = \frac{1}{\sqrt{n}}\sum(Y_j - \theta)/\sigma_{a_j}^2.$$

The first four standardized unconditional cumulants of $(Y_j - \theta)/\sigma_{a_j}^2$ are given by

$$0; \qquad \frac{1}{2}\left(\frac{1}{\sigma_0^2} + \frac{1}{\sigma_1^2}\right); \qquad 0; \qquad \frac{3}{4}\left(\frac{1}{\sigma_0^2} - \frac{1}{\sigma_1^2}\right)^2,$$

respectively. It follows that the first four unconditional cumulants of Z are given by

$$0; \qquad \frac{1}{2}\left(\frac{1}{\sigma_0^2}+\frac{1}{\sigma_1^2}\right); \qquad 0; \qquad \frac{3}{4}\left(\frac{1}{\sigma_0^2}-\frac{1}{\sigma_1^2}\right)^2\frac{1}{n}.$$

Recall that first four conditional cumulants of Z given a are given by

$$0; \qquad \frac{1}{n}\left[(n-a)\frac{1}{\sigma_0^2}+a\frac{1}{\sigma_1^2}\right]; \qquad 0; \qquad 0,$$

respectively. Since $a = 1/2 + O_p(n^{-1/2})$, the second cumulants differ by $O(n^{-1/2})$; the difference in the conditional and unconditional fourth cumulants is $O(n^{-1})$, the same order as the cumulants themselves. ∎

6.4.3 MAXIMUM LIKELIHOOD ESTIMATES

Consider the distribution of the normalized maximum likelihood estimator, $T = \sqrt{i(\theta)}(\hat{\theta}-\theta)$. Using the results from Section 6.2, it follows that the first three conditional cumulants of T are given by

$$\frac{\ell_{\theta\theta;\hat{\theta}}(\theta;\theta,a)+\ell_{\theta\theta\theta}(\theta;\theta,a)/2}{n\bar{j_O}^{3/2}}\frac{1}{\sqrt{n}}+O(n^{-1}); \qquad \frac{\nu_2}{\bar{j_O}}+O(n^{-1});$$

$$-\frac{3}{2}\frac{\ell_{\theta\theta\theta}(\theta;\theta,a)}{n\bar{j_O}^{3/2}}\frac{1}{\sqrt{n}}+O(n^{-1}),$$

respectively. In Section 5.3 it was shown that the first three unconditional cumulants of T are given by

$$\frac{\nu_{11}+\nu_{001}/2}{\nu_{01}^{3/2}}\frac{1}{\sqrt{n}}+O(n^{-1}); \qquad 1+O(n^{-1});$$

$$\frac{\nu_{001}-\nu_{30}}{\nu_{20}^{3/2}}\frac{1}{\sqrt{n}}+O(n^{-1}).$$

It follows that the distribution of T is stable to first, but not to second order.

Now consider the distribution of the maximum likelihood estimator normalized by the observed information, $T_O = \sqrt{\hat{j}}(\hat{\theta}-\theta)$. First consider the conditional distribution of T_O. Recall that $\sqrt{n}(\hat{\theta}-\theta)$ may be expanded

$$\sqrt{n}(\hat{\theta}-\theta) = \frac{Z}{\bar{j_O}} + \frac{\ell_{\theta\theta;\hat{\theta}}(\theta;\theta,a)+\ell_{\theta\theta\theta}(\theta;\theta,a)/2}{j_O}\left[\frac{Z}{\bar{j_O}}\right]^2\frac{1}{\sqrt{n}}+O_p(n^{-1}).$$

Since

$$\ell_{\theta\theta}(\hat{\theta}) = \ell_{\theta\theta}(\theta;\theta,a)+(\ell_{\theta\theta;\hat{\theta}}(\theta;\theta,a)+\ell_{\theta\theta\theta}(\theta;\theta,a))(\hat{\theta}-\theta)+O_p(1),$$

it follows that

$$[-\ell_{\theta\theta}(\hat{\theta})/n]^{1/2} = \sqrt{\bar{j}_O}\left[1 - \frac{1}{2}\frac{\ell_{\theta\theta;\hat{\theta}}(\theta;\theta,a) + \ell_{\theta\theta\theta}(\theta;\theta,a)}{n\bar{j}_O{}^2}\frac{1}{\sqrt{n}} + O_p(n^{-1})\right].$$

Hence,

$$T_O = \frac{Z}{\sqrt{\bar{j}_O}} + \frac{\ell_{\theta\theta;\hat{\theta}}(\theta;\theta,a)/2}{n\bar{j}_O{}^{5/2}}Z^2\frac{1}{\sqrt{n}} + O_p(n^{-1}).$$

From this expansion it is straightforward to show that the first three conditional cumulants of T_O are

$$\frac{\ell_{\theta\theta;\hat{\theta}}(\theta;\theta,a)/2}{n\bar{j}_O{}^{3/2}}\frac{1}{\sqrt{n}} + O(n^{-1}); \qquad 1 + O(n^{-1});$$

$$\left(\frac{3}{2}\frac{\ell_{\theta\theta;\hat{\theta}}(\theta;\theta,a)}{n\bar{j}_O{}^{3/2}} - \frac{\ell_{\theta\theta\theta}(\theta;\theta,a)}{n\bar{j}_O{}^{3/2}}\right)\frac{1}{\sqrt{n}} + O(n^{-1}).$$

Now consider the unconditional distribution of T_O. Recall that $\sqrt{n}(\hat{\theta} - \theta)$ may be expanded

$$\sqrt{n}(\hat{\theta} - \theta) = \frac{Z}{\nu_2} + \left(\frac{ZZ_2}{\nu_2^2} + \frac{1}{2}\frac{\nu_{001}}{\nu_2^3}Z^2\right)\frac{1}{\sqrt{n}} + O_p(n^{-1}).$$

The observed information may be expanded

$$-\ell_{\theta\theta}(\hat{\theta}) = n\nu_2 - \sqrt{n}Z_2 - n\nu_{001}\sqrt{n}(\hat{\theta} - \theta)\frac{1}{\sqrt{n}} + O_p(n^{-1});$$

it follows that

$$[-\ell_{\theta\theta}(\hat{\theta})/n]^{1/2} = \sqrt{\nu_2}\left[1 - \frac{1}{2}\left(\frac{Z_2}{\nu_2} + \frac{\nu_{001}}{\nu_2^2}Z\right)\frac{1}{\sqrt{n}}\right] + O_p(n^{-1}).$$

Hence,

$$T_O = \frac{Z}{\sqrt{\nu_{20}}} + \frac{1}{2}\frac{ZZ_2}{\nu_2^{3/2}}\frac{1}{\sqrt{n}} + O_p(n^{-1}).$$

The first three formal cumulants of T_O are therefore given by

$$\frac{1}{2}\frac{\nu_{11}}{\nu_2^{3/2}}\frac{1}{\sqrt{n}} + O(n^{-1}); \qquad 1 + O(n^{-1}); \qquad \frac{\nu_{30} + 9\nu_{11}/2}{\nu_2^{3/2}}\frac{1}{\sqrt{n}} + O(n^{-1}),$$

respectively. Comparing these expressions to the expressions for the conditional cumulants given above using the results of Section 6.4.2 we see that the first two conditional and unconditional cumulants agree to order $O(n^{-1})$. The third conditional cumulant may be written

$$\frac{3\nu_{11}/2 - \nu_{001}}{\nu_2^{3/2}}\frac{1}{\sqrt{n}} + O(n^{-1}).$$

By the third Bartlett identity, $-\nu_{001} = \nu_{30} + 3\nu_{11}$ so that the third cumulants agree to order $O(n^{-1})$ as well. It follows that the distribution of T_O is stable to second order.

Example 6.19 Two measuring instruments
Consider the two measuring instruments example considered in Examples 6.5 and 6.18. The maximum likelihood estimator of $\hat{\theta}$ is given by

$$\hat{\theta} = \frac{\sum y_j/\sigma_{a_j}^2}{\sum 1/\sigma_{a_j}^2}$$

so that

$$T = \sqrt{i(\theta)}(\hat{\theta} - \theta) = \frac{[\sigma_0^2/2 + \sigma_1^2/2]^{1/2}}{(1 - a/n)(1/\sigma_0^2) + (a/n)(1/\sigma_1^2)} Z.$$

The first three conditional cumulants of T are

$$0; \qquad \frac{[\sigma_0^2/2 + \sigma_1^2/2]}{[(1 - a/n)(1/\sigma_0^2) + (a/n)(1/\sigma_1^2)]}; \qquad 0.$$

Since $a/n = 1/2 + O(n^{-1/2})$,

$$\frac{[\sigma_0^2/2 + \sigma_1^2/2]}{[(1 - a/n)(1/\sigma_0^2) + (a/n)(1/\sigma_1^2)]} = 1 + O(n^{-1/2}).$$

Now consider

$$T_0 = \sqrt{\hat{j}}(\hat{\theta} - \theta) = \frac{Z}{[(1 - a/n)(1/\sigma_0^2) + (a/n)(1/\sigma_1^2)]^{1/2}}.$$

The first three conditional cumulants of T_0 are 0, 1, and 0. Since T_0 has conditional mean 0, the first three unconditional cumulants are simply the expected values of the conditional cumulants. Hence, the conditional and unconditional cumulants agree and T_0 is stable to second order. ∎

6.4.4 LIKELIHOOD RATIO STATISTIC

Consider the distribution of the signed likelihood ratio statistic,

$$R = \text{sgn}(\hat{\theta} - \theta)[2(\ell(\hat{\theta}) - \ell(\theta))]^{1/2}.$$

We first consider the unconditional formal cumulants of R. In Section 5.4 it was shown that

$$\hat{\kappa}_1(R) = \left(\frac{1}{6}\frac{\nu_{001}}{\nu_2^{3/2}} + \frac{1}{2}\frac{\nu_{11}}{\nu_2^{3/2}}\right)\frac{1}{\sqrt{n}} + O(n^{-3/2});$$

$$\hat{\kappa}_2(R) = 1 + O(n^{-1}); \qquad \hat{\kappa}_3(R) = 0 + O(n^{-3/2}).$$

To determine the conditional cumulants of R we may derive a stochastic asymptotic expansion of R in terms of mixed log-likelihood derivatives and the score statistic Z:

$$R = \frac{Z}{\sqrt{\bar{j}_o}}\left[1 + \frac{1}{2}\frac{\ell_{\theta\theta;\hat{\theta}}(\theta;\theta,a) + \ell_{\theta\theta\theta}(\theta;\theta,a)/3}{n}\frac{Z}{\bar{j}_o{}^2}\frac{1}{\sqrt{n}}\right] + O_p(n^{-1}).$$

From this expression, together with the conditional cumulants of Z, it follows easily that the conditional cumulants of R are given by

$$\hat{\kappa}_1(R|a) = \left[\tfrac{1}{2}\ell_{\theta\theta;\hat{\theta}}(\theta;\theta,a) + \tfrac{1}{6}\ell_{\theta\theta\theta}(\theta;\theta,a)\right] j_o^{-3/2} + O(n^{-3/2});$$

$$\hat{\kappa}_2(R|a) = 1 + O(n^{-1}); \qquad \hat{\kappa}_3(R|a) = O(n^{-3/2}).$$

Using the results in Section 6.4.2 it follows that the distribution of R is stable to second order.

Example 6.20 Two measuring instruments
Since

$$\ell(\theta) = \left[(n-a)\frac{1}{\sigma_0^2} + a\frac{1}{\sigma_1^2}\right]\left(\theta\hat{\theta} - \frac{1}{2}\theta^2\right),$$

$$R = \left[(n-a)\frac{1}{\sigma_0^2} + a\frac{1}{\sigma_1^2}\right]^{1/2}(\hat{\theta} - \theta) = \sqrt{\hat{j}}(\hat{\theta} - \theta).$$

It now follows from Example 6.19 that the distribution of R is stable to second order. ∎

6.5 Approximation of the conditional model

One of the main drawbacks to the use of the methods discussed in this chapter is that they require that the conditional model, given in Section 6.1, holds. Specifically, they require that the minimal sufficient statistic for the model may be written $(\hat{\theta}, a)$ where a is an ancillary statistic. Although there are a number of important examples in which this model holds, such as transformation models in which the ancillary statistic may be taken to be the maximal invariant, or full exponential family models in which an ancillary statistic is not needed, in general, it may be difficult to establish that the conditional model holds. In particular, construction of an appropriate ancillary statistic is often difficult.

There are at least three situations in which it may be difficult to show that the conditional model holds. We may be able to show that the minimal sufficient statistic is of the form $(\hat{\theta}, a)$, but we may not be able to show that a is ancillary. Alternatively, we may be able to identify an ancillary statistic a, but $(\hat{\theta}, a)$ is not sufficient. Finally, $(\hat{\theta}, a)$ may be sufficient, but not minimal sufficient. The goal of this section is to develop appropriate definitions of approximate sufficiency and approximate ancillarity that allow the use of the results of this chapter even when exact sufficiency or exact ancillarity does not hold. In particular, we consider definitions of approximate sufficiency and approximate ancillarity that are strong enough for the likelihood ratio approximation to the conditional density of $\hat{\theta}$ to be valid to order $O(n^{-1})$ for values of the argument $\hat{\theta}$ of the form $\hat{\theta} = \theta + O(n^{-1/2})$. We consider only the case in which θ is a scalar; similar conditions may be formulated for the vector parameter case.

First note that minimal sufficiency is only required for ancillary a to be relevant to the statistical analysis. If $(\hat{\theta}, a)$ is sufficient, but not minimal sufficient, the results of this chapter continue to hold.

Consider the case in which $(\hat{\theta}, a)$ is sufficient, but a is not necessarily ancillary. There are two consequences of the ancillarity of a that are used in the derivation of p^*. The first is that

$$\frac{p(\hat{\theta}, a; \theta_0)}{p(\hat{\theta}|a; \theta_0)} = p(a; \theta_0)$$

does not depend on the value of θ_0. If this does not hold, then the conditional density of $\hat{\theta}$ is given by

$$\frac{p(a; \hat{\theta})}{p(a; \theta)} \exp\{\ell(\theta) - \ell(\hat{\theta})\}\, p(\hat{\theta}|a; \hat{\theta}).$$

Hence, we need to derive conditions under which

$$\frac{p(a; \hat{\theta})}{p(a; \theta)}$$

is approximately independent of $\hat{\theta}$.

Let $\ell(\theta; a)$ denote the log-likelihood function based on a alone. Then

$$\frac{p(a; \hat{\theta})}{p(a; \theta)} = \exp\{\ell(\hat{\theta}; a) - \ell(\theta; a)\}.$$

Suppose that

$$\ell(\theta + \delta/\sqrt{n}; a) - \ell(\theta; a) = O(n^{-1}). \tag{6.5}$$

Then

$$\frac{p(a; \hat{\theta})}{p(a; \theta)} = 1 + O(n^{-1})$$

and the approximate conditional density of $\hat{\theta}$ is given by

$$\exp\{\ell(\theta) - \ell(\hat{\theta})\}\, p(\hat{\theta}|a; \hat{\theta})[1 + O(n^{-1})].$$

The second way in which ancillarity is used in the derivation of p^* is in deriving the conditional cumulants used in constructing an approximation to $p(\hat{\theta}|a; \hat{\theta})$. Specifically, we use the fact that the Bartlett identities hold conditionally given a.

In order that $p(\hat{\theta}|a; \hat{\theta})$ can be approximated by

$$\frac{1}{\sqrt{j(\hat{\theta})}}\phi(0)$$

with error $O(n^{-1})$ we require that

$$\mathrm{E}[\sqrt{n}(\hat{\theta} - \theta)|a; \theta] = O(n^{-1/2})$$

and

$$\text{Var}[\sqrt{n}(\hat{\theta} - \theta)|a; \theta] = \frac{1}{\bar{j}_o} + O(n^{-1})$$

where $\bar{j}_o = -\ell_{\theta\theta}(\theta; \theta, a)/n$. Using expansion (6.1), it follows that these conditions hold provided that

$$\text{E}[Z|a; \theta] = O(n^{-1/2}); \qquad \text{E}[Z^2|a; \theta] = \bar{j} + O(n^{-1})$$

where $Z = \ell_\theta(\theta)/\sqrt{n}$. Hence, these conditions may be written

$$\text{E}[\ell_\theta(\theta)|a; \theta] = O(1); \qquad \text{E}[\ell_\theta(\theta)^2 + \ell_{\theta\theta}(\theta)|a; \theta] = O(1). \tag{6.6}$$

Let $\ell(\theta; \hat{\theta}|a)$ denote the log-likelihood function based on the conditional distribution of $\hat{\theta}$ given a. It follows from the Bartlett identities applied to this conditional distribution that

$$\text{E}[\ell_\theta(\theta; \hat{\theta}|a)|a; \theta] = 0; \qquad \text{E}[\ell_\theta(\theta; \hat{\theta}|a)^2 + \ell_{\theta\theta}(\theta; \hat{\theta}|a)|a; \theta] = 0.$$

Note that

$$\ell(\theta; \hat{\theta}|a) = \ell(\theta) - \ell(\theta; a)$$

so that

$$\ell_\theta(\theta; \hat{\theta}|a) = \ell_\theta(\theta) - \ell_\theta(\theta; a)$$

and

$$\ell_{\theta\theta}(\theta; \hat{\theta}|a) = \ell_{\theta\theta}(\theta) - \ell_{\theta\theta}(\theta; a).$$

Hence,

$$\text{E}[\ell_\theta(\theta)|a; \theta] = \ell_\theta(\theta; a)$$

and

$$\text{E}[\ell_\theta(\theta)^2 + \ell_{\theta\theta}(\theta)|a; \theta] = \ell_\theta(\theta; a)^2 + \ell_{\theta\theta}(\theta; a).$$

It follows that sufficient conditions for (6.6) are

$$\ell_\theta(\theta; a) = O(1); \qquad \ell_{\theta\theta}(\theta; a) = O(1).$$

Suppose (6.5) holds. Then

$$\ell_\theta(\theta; a)\delta/\sqrt{n} + \ell_{\theta}\theta(\theta; a)\delta^2/n + \ell_{\theta\theta\theta}(\theta; a)\delta^3/n^{3/2} = O(n^{-1}).$$

It follows that

$$\ell_\theta(\theta; a) = O(n^{-1/2}); \qquad \ell_{\theta\theta}(\theta; a) = O(1)$$

so that (6.6) holds. Hence, for the p^* formula to hold with error $O(n^{-1})$, it is sufficient that (6.5) holds.

For a statistic a of fixed dimension, sufficient conditions for (6.5) may be given in terms of the cumulants of a. Suppose a is an $O_p(1)$ scalar statistic such that the mean of a is constant to order $O(n^{-1/2})$ and the variance of a is constant to order $O(n^{-1})$. Without loss of generality, we may assume that leading terms in the expansions for the mean and variance of a are 0 and 1, respectively. Let $\mu(\theta)/\sqrt{n} + O(n^{-3/2})$ denote an expansion for the mean of a and let $1 + O(n^{-1})$ denote an expansion for the variance of a; assume that $\kappa_j(a; \theta)$, the jth cumulant of a is of order $O(n^{(j-1)/2})$, $j = 3, 4, \ldots$.

Then, using a Edgeworth expansion for the density of a, the likelihood function based on a is given by

$$\ell(\theta; a) = \frac{\mu(\theta)}{\sqrt{n}} a + \frac{\kappa_3(a; \theta)}{6\sqrt{n}} H_3(a) + O(n^{-1}).$$

Hence,

$$\ell(\theta + \delta/\sqrt{n}; a) - \ell(\theta; a) = O(n^{-1});$$

i.e., (6.5) is satisfied.

In general, sufficient conditions for (6.5) are that

$$\kappa_j(a; \theta + \delta/\sqrt{n}) - \kappa_j(a; \theta) = O(n^{-1})$$

for $j = 1, 2, \ldots$.

Now suppose that a is ancillary, or at least satisfies (6.5), but that $(\hat{\theta}, a)$ is not necessarily sufficient. In deriving p^*, sufficiency was used to show that

$$\frac{p(\hat{\theta}, a; \theta)}{p(\hat{\theta}, a; \theta_0)} = \exp\{\ell(\theta) - \ell(\theta_0)\}.$$

In the general case, in which $(\hat{\theta}, a)$ is not necessarily sufficient,

$$\frac{p(\hat{\theta}, a; \theta)}{p(\hat{\theta}, a; \theta_0)} = \exp\{\bar{\ell}(\theta) - \bar{\ell}(\theta_0)\}$$

where $\bar{\ell}(\theta) \equiv \bar{\ell}(\theta; \hat{\theta}, a)$ denotes the log-likelihood function based on the distribution of $(\hat{\theta}, a)$. The symbol $\bar{\ell}(\theta; \hat{\theta}, a)$ is used rather than $\ell(\theta; \hat{\theta}, a)$ to indicate that in this context $(\hat{\theta}, a)$ is not sufficient.

Using the same approach used in Section 6.3, an approximation for the conditional distribution is given by

$$\frac{[-\ell_{\theta\theta}(\hat{\theta})]^{1/2}}{\sigma(\hat{\theta}; a)\sqrt{(2\pi)}} \exp\{\bar{\ell}(\theta) - \bar{\ell}(\hat{\theta})\}.$$

Hence, the p^* approximation holds provided that

$$\bar{\ell}(\theta) - \bar{\ell}(\hat{\theta}) = \ell(\theta) - \ell(\hat{\theta}) + O(n^{-1}).$$

Suppose that there exists a function $h(\theta) \equiv h(\theta; \hat{\theta}, a)$, depending on the data only through $(\hat{\theta}, a)$, such that

$$\ell(\theta + \delta/\sqrt{n}) - \ell(\theta) = h(\theta + \delta/\sqrt{n}) - h(\theta) + O(n^{-1}).$$

Then

$$\frac{p(y; \theta + \delta/\sqrt{n})}{p(y; \theta)} = \exp\{h(\theta + \delta/\sqrt{n}) - h(\theta)\}[1 + R_n(y; \theta)],$$

where $R_n(y; \theta) = O(n^{-1})$, and

$$p(\hat{\theta}, a; \theta + \delta/\sqrt{n}) = \exp\{h(\theta + \delta/\sqrt{n}) - h(\theta)\}\left[1 + \int p(y; \theta) R_n(y; \theta)\, dy\right]$$

where the integral is over all y with $(\hat{\theta}, a)$ fixed. It follows that

$$\log\left(\frac{p(\hat{\theta}, a; \theta + \delta/\sqrt{n})}{p(\hat{\theta}, a; \theta)}\right) = h(\theta + \delta/\sqrt{n}) - h(\theta) + O(n^{-1})$$

so that

$$\ell(\theta + \delta/\sqrt{n}) - \ell(\theta) = \bar{\ell}(\theta + \delta/\sqrt{n}) - \bar{\ell}(\theta) + O(n^{-1}),$$

which is what is required for the p^* approximation to have error $O(n^{-1})$.

The results of this section may be stated as follows. Let a be a statistic satisfying the following properties:

(S) There exists a function $h(\theta)$ depending on y only through $(\hat{\theta}, a)$ such that

$$\ell(\theta + \delta/\sqrt{n}) - \ell(\theta) = h(\theta + \delta/\sqrt{n}) - h(\theta) + O(n^{-1}).$$

(A) The log-likelihood function based on a, $\ell(\theta; a)$, satisfies

$$\ell(\theta + \delta/\sqrt{n}; a) - \ell(\theta; a) = O(n^{-1}).$$

In this case, we will call a a second-order locally ancillary statistic.

Under these conditions, the conditional density of $\hat{\theta}$ given a is given by

$$\frac{[-\ell_{\theta\theta}(\hat{\theta})]^{1/2}}{\sqrt{(2\pi)}} \exp\{\ell(\theta) - \ell(\hat{\theta})\}\,[1 + O(n^{-1})].$$

It should be noted that the error of this approximation cannot be reduced to $O(n^{-3/2})$ by renormalization without further conditions on $(\hat{\theta}, a)$ due to the fact that the $O(n^{-1})$ term depends, in general, on $\hat{\theta}$; for instance,

$$[\bar{\ell}(\theta) - \bar{\ell}(\hat{\theta})] - [\ell(\theta) - \ell(\hat{\theta})]$$

depends on $\hat{\theta}$. For the renormalized form of the approximation to hold with error $O(n^{-3/2})$ conditions (S) and (A) given above must hold to order $O(n^{-3/2})$, rather than to $O(n^{-1})$.

6.6 Approximate ancillarity

6.6.1 INTRODUCTION

In this section we consider the construction of approximately ancillary statistics that satisfy the conditions given in the previous section.

We will consider four approaches to constructing an approximately ancillary statistic. The first approach considers a *local model* which is valid near the point $\theta = \theta_0$; in this model $\ell_\theta(\theta_0)$ and $\ell_{\theta\theta}(\theta_0)$ are approximately sufficient and an approximately ancillary statistic may be constructed as a function of $(\ell_\theta(\theta_0), \ell_{\theta\theta}(\theta_0))$. When θ is a location parameter, the statistic $a = y - \hat{\theta}$ may be used as an exactly ancillary statistic. The second approach is to approximate a general model with parameter θ as a location model and then use the maximal invariant statistic from the approximating location model as an approximately ancillary statistic. Suppose there exist a pivotal quantity $P(y; \theta)$, a function of the data y and the parameter θ that has a distribution not depending on θ. Clearly, $P(y; \theta)$ itself cannot be used as an ancillary statistic since it depends on θ; however, $P(y; \hat{\theta})$ is, in general, approximately ancillary. The third approach is based on this idea. The fourth approach is applicable whenever the given model may embedded in a 'larger' model with parameter (θ, δ) such that the original model corresponds to a specific value of δ, say $\delta = \delta_0$. If in the larger model an ancillary statistic is available, or no ancillary statistic is needed, an additional ancillary statistic may be constructed from the likelihood ratio test statistic for testing $\delta = \delta_0$.

These four methods will be illustrated with the following examples.

Example 6.21 Exponential hyperbola
Let $(X_1, Y_1), \ldots, (X_n, Y_n)$ denote independent pairs of independent exponential random variables such that each X_j has mean $1/\theta$ and each Y_j has mean θ. The log-likelihood function is given by

$$\ell(\theta) = -n(\theta\bar{x} + \bar{y}/\theta).$$

The sufficient statistic for the model may be written $(\hat{\theta}, a)$ where $\hat{\theta} = (\bar{y}/\bar{x})^{1/2}$ denotes the maximum likelihood estimator of θ and $a = (\bar{x}\bar{y})^{1/2}$ is an ancillary statistic. The log-likelihood function may be written

$$\ell(\theta) = -na(\theta/\hat{\theta} + \hat{\theta}/\theta).$$

■

Example 6.22 Normal-theory nonlinear regression model
Let $Y_1, \ldots, Y_n$ denote independent normally distributed random variables such that Y_j has mean $m_j(\theta)$ and standard deviation 1. For each j, $m_j(\theta)$ is a known function of the unknown parameter θ. If each $m_j(\theta)$ is a linear function of θ, then this is simply a normal-theory linear model and no ancillary statistic is needed; that is, $\hat{\theta}$ itself is sufficient. When the $m_j(\theta)$ are nonlinear functions of θ, $\hat{\theta}$ is not sufficient and it is difficult, if not impossible, to construct an exactly ancillary statistic in many cases.

■

6.6.2 A LOCAL MODEL

Fix a parameter value θ and consider a model with parameter $\theta + \delta/\sqrt{n}$. The log-likelihood function for this model may be written

$$\ell(\theta + \delta/\sqrt{n}) - \ell(\theta) = Z\delta + \tfrac{1}{2}(\nu_{01} + Z_2/\sqrt{n})\delta^2 + \cdots$$

where

$$Z = \frac{1}{\sqrt{n}}\ell_\theta(\theta); \qquad Z_2 = \frac{1}{\sqrt{n}}(\ell_{\theta\theta}(\theta) - n\nu_{01})$$

and cumulants such as ν_{01}, are calculated under the distribution with parameter θ. Hence, the log-likelihood derivatives $Z, Z_2, \ldots$ may be viewed as the sufficient statistic for the model and they may be used to construct an approximately ancillary statistic.

We have seen that

$$\sqrt{n}(\hat{\theta} - \theta) = \frac{Z}{\nu_{20}} + \left(\frac{1}{2}\frac{\nu_{001}}{\nu_{20}^3}Z^2 + \frac{ZZ_2}{\nu_{20}^2}\right)\frac{1}{\sqrt{n}} + O_p(n^{-1}).$$

Consider a function of the form

$$A_0 = Z_2 + \alpha Z + R/\sqrt{n}$$

where R is an $O_p(1)$ statistic that is a function of Z, Z_2 and the higher-order log-likelihood derivatives. The constant α and the statistic R may be chosen so that the resulting statistic is approximately ancillary. Recall that we require that

$$\mathrm{E}[A_0; \theta + \delta/\sqrt{n}] = \mathrm{E}[A_0; \theta] + O(n^{-1}) \tag{6.7}$$

and

$$\mathrm{Var}[A_0; \theta + \delta/\sqrt{n}] = \mathrm{Var}[A_0; \theta] + O(n^{-1}).$$

Since

$$\frac{\partial}{\partial \delta} \mathrm{E}[A_0; \theta + \delta/\sqrt{n}] \Big|_{\delta=\delta_0} = \mathrm{E}[A_0 Z; \theta] = \nu_{11} + \alpha \nu_2,$$

it follows that we must take $\alpha = -\nu_{11}/\nu_2$. Hence, we consider A_0 of the form

$$A_0 = Z_2 - \frac{\nu_{11}}{\nu_2} Z + R/\sqrt{n}.$$

It is convenient to standardize the statistic to have standard deviation approximately 1 under the distribution with parameter θ. Let

$$\sigma^2(\theta) = \mathrm{Var}\Big(Z_2 - \frac{\nu_{11}}{\nu_2} Z_1; \theta\Big).$$

It is straightforward to show that

$$\sigma^2(\theta) = \nu_2(\theta)^2 \gamma^2(\theta)$$

where $\gamma(\theta)$ denotes the statistical curvature of the model. Hence, we consider a statistic of the form

$$A = \frac{Z_2 - (\nu_{11}/\nu_2) Z}{\nu_2(\theta)\gamma(\theta)} + R/\sqrt{n}.$$

Since ν_{11}, ν_2, Z, and Z_2 are all evaluated at θ, the term

$$Z_2 - \frac{\nu_{11}}{\nu_2} Z$$

depends on θ. To use this statistic in practice, we need to be able to approximate this function by a statistic depending only on the data. Using the facts that

$$Z = \nu_2 \sqrt{n}(\hat{\theta} - \theta) + O_p(n^{-1/2})$$

and

$$Z_2 = -\frac{\hat{j} - i(\hat{\theta}) + \sqrt{n}(2\nu_{11} + \nu_3)\sqrt{n}(\hat{\theta} - \theta)}{\sqrt{n}} - \nu_{001}\sqrt{n}(\hat{\theta} - \theta) + O_p(n^{-1/2}), \tag{6.8}$$

it follows that

$$Z_2 - \frac{\nu_{11}}{\nu_2} Z = \frac{i(\hat{\theta}) - \hat{j}}{\sqrt{n}} + O_p(n^{-1/2}).$$

Since

$$\nu_2 = i(\hat{\theta})/n + O_p(n^{-1/2}); \qquad \gamma(\theta) = \gamma(\hat{\theta}) + O_p(n^{-1/2}),$$

we may write

$$A = -\frac{\sqrt{n}[\hat{j} - i(\hat{\theta})]}{i(\hat{\theta})\gamma(\hat{\theta})} + R/\sqrt{n}$$

where R is an arbitrary $O_p(1)$ statistic chosen so that a is a second-order locally ancillary statistic.

In fact, it has been shown that the statistic

$$\frac{\sqrt{n}[\hat{\jmath} - i(\hat{\theta})]}{i(\hat{\theta})\gamma(\hat{\theta})}$$

itself is a second-order locally ancillary statistic. This approximate ancillary statistic was first proposed by Efron and Hinkley (1978) and is referred to as the Efron–Hinkley ancillary.

Example 6.23 Exponential hyperbola
In this example, $i(\theta) = 2n/\theta^2$, $\gamma(\theta)^2 = 1/2$, and $\hat{\jmath} = 2n\bar{y}/\hat{\theta}^3$. It follows that the Efron–Hinkley ancillary is given by

$$\sqrt{(2n)}(\bar{y}/\hat{\theta} - 1) = \sqrt{(2n)}((\bar{x}\bar{y})^{1/2} - 1)$$

which is exactly ancillary. ∎

Example 6.24 Normal-theory nonlinear regression model
Here

$$\ell(\theta) = -\frac{1}{2}\sum(y_j - m_j(\theta))^2.$$

It follows that $i(\theta) = \sum m'_j(\theta)^2$,

$$\hat{\jmath} = i(\hat{\theta}) - \sum(y_j - m_j(\hat{\theta}))m''_j(\hat{\theta})$$

and

$$\gamma(\hat{\theta})^2 = n\frac{\sum m''_j(\hat{\theta})^2 - \left[\sum m'_j(\hat{\theta})m''_j(\hat{\theta})\right]^2 / \sum m'_j(\hat{\theta})^2}{\left[\sum m'_j(\hat{\theta})^2\right]^2}.$$

It follows that the Efron–Hinkley ancillary statistic is given by

$$-\frac{\sqrt{n}\sum(y_j - m_j(\hat{\theta}))m''_j(\hat{\theta})}{i(\hat{\theta})\gamma(\hat{\theta})}.$$ ∎

6.6.3 APPROXIMATION BY A LOCATION MODEL

Another approach to deriving an approximate ancillary statistic is to approximate the model by a location model and then use the maximal invariant statistic of the location model as an approximate ancillary statistic in the original model.

Consider a real-valued random variable Y with distribution function $F(y;\theta)$. We will assume that Y is stochastically increasing in θ near a given point θ_0 so that

$$\frac{\partial}{\partial\theta}F(y;\theta)\Big|_{\theta=\theta_0} < 0.$$

Our goal is to transform Y to a new variable X that has a distribution function $G(x;\theta)$ that is approximately of location model form. A distribution function $G(x;\theta)$ is of location model form if it may be written

$$G(x;\theta) = G_0(b(x) - \mu(\theta)) \tag{6.9}$$

where G_0 is a distribution function on the real line and b and μ are one-to-one functions. In this case, given a change in θ to $\theta + \delta$, there is a change in x to $x + \epsilon$ such that

$$G(x + \epsilon; \theta + \delta) \doteq G(x;\theta).$$

Since

$$G(x + \epsilon; \theta + \delta) \doteq G(x;\theta) + \frac{\partial}{\partial x}G(x;\theta)\epsilon + \frac{\partial}{\partial \theta}G(x;\theta)\delta$$

and

$$\frac{\partial}{\partial x}G(x;\theta) = G_0'(b(x) - \mu(\theta))b'(x);$$
$$\frac{\partial}{\partial \theta}G(x;\theta) = -G_0'(b(x) - \mu(\theta))\mu'(\theta),$$

ϵ is given by

$$-\frac{(\partial/\partial\theta)G(x;\theta)}{(\partial/\partial x)G(x;\theta)}\delta = \frac{\mu'(\theta)}{b'(x)}\delta.$$

In the special case in which $\mu(\theta) = \theta$ and $b(x) = x$, $\epsilon = \delta$.

To approximate the distribution of Y by a location model we proceed by transforming Y to X such that the distribution function of X satisfies

$$-\frac{(\partial/\partial\theta)G(x;\theta)}{(\partial/\partial x)G(x;\theta)} = 1$$

at $\theta = \theta_0$. In this case,

$$G(x + \delta; \theta_0 + \delta) = G(x;\theta_0) + O(\delta^2).$$

Since $G(x;\theta) = F(y(x);\theta)$, it follows that $y(x)$ must satisfy

$$y'(x) = -\frac{(\partial/\partial\theta)F(y(x);\theta)}{(\partial/\partial y)F(y(x);\theta)}\bigg|_{\theta=\theta_0}$$

or, equivalently,

$$x'(y) = -\frac{(\partial/\partial y)F(y;\theta)}{(\partial/\partial\theta)F(y;\theta)}\bigg|_{\theta=\theta_0}.$$

It follows that

$$x = -\int^y \frac{(\partial/\partial t)F(t;\theta)}{(\partial/\partial\theta)F(t;\theta)}dt\,\bigg|_{\theta=\theta_0}.$$

The distribution function of x is approximately of location model form (6.9) with $b(x) = x$ and $\mu(\theta) = \theta$, locally near $\theta = \theta_0$.

Let $Y_1, \ldots, Y_n$ denote n independent random variables such that Y_j has distribution function $F_j(y; \theta)$. Each Y_j may be transformed to a random variable X_j such that X_j approximately follows a location model with parameter θ; hence, $X_1, \ldots, X_n$ are independent random variables each approximately following a location model with location parameter θ.

The ancillary statistic corresponding this local location model is given by

$$a = (H_2(y_2) - H_1(y_1), H_3(y_3) - H_1(y_1), \ldots, H_n(y_n) - H_1(y_1))$$

where

$$H_j(s) = -\int^s \frac{(\partial/\partial t)F_j(t; \theta)}{(\partial/\partial\theta)F_j(t; \theta)}\, dt \,\Big|_{\theta=\theta_0};$$

this statistic may be taken as an approximately ancillary statistic in the original model for $Y_1, \ldots, Y_n$.

Let $X_j = H_j(Y_j)$, $j = 1, \ldots, n$. Then the distribution function of X_j, $G_j(x; \theta)$, satisfies

$$\frac{\partial}{\partial\theta} G_j(x; \theta) \,\Big|_{\theta=\theta_0} = -\frac{\partial}{\partial x} G_j(x; \theta_0).$$

It follows that the density of X_j, $g_j(x; \theta)$, satisfies

$$\frac{\partial}{\partial\theta} g_j(x; \theta) \,\Big|_{\theta=\theta_0} = -\frac{\partial}{\partial x} g_j(x; \theta_0).$$

Hence,

$$g_j(x; \theta_0 + \delta/\sqrt{n}) = g_j(x - \delta/\sqrt{n}; \theta_0) + O(n^{-1}).$$

It can be shown that the mean of X_j satisfies

$$\mathrm{E}[X_j; \theta_0 + \delta/\sqrt{n}] = \mu_j(\theta_0) + \frac{\delta}{\sqrt{n}} + O(n^{-1})$$

and the variance of X_j satisfies

$$\mathrm{Var}[X_j; \theta_0 + \delta/\sqrt{n}] = \sigma_j^2(\theta_0) + O(n^{-1});$$

here $\mu_j(\theta_0)$ and $\sigma_j^2(\theta_0)$ are constants not depending on δ. Let $A_j = X_{j+1} - X_1$. Then a_j has mean

$$\mu_{j+1}(\theta_0) - \mu_1(\theta_0) + O(n^{-1})$$

and variance

$$\sigma_{j+1}^2(\theta_0) + \sigma_1^2(\theta_0) + O(n^{-1}).$$

Hence, the cumulants of A satisfy the conditions given in Section 6.5 for local ancillarity. Note, however, that it does not necessarily follow that A is second-order locally ancillary since the dimension of A increases with n. The establishment of the local ancillarity of A may be done on a case-by-case basis.

In some cases, the approximate location ancillary is equivalent to an ancillary statistic that does not depend on θ_0; in other cases, θ_0 may be replaced by $\hat{\theta}$.

Example 6.25 Exponential hyperbola
In this model, there are two distributions to be considered. For each X_j,

$$F_j(x;\theta) = 1 - \exp\{-\theta x\}$$

so that

$$H_j(s) = -\theta_0 \log(s);$$

for each Y_j,

$$F_j(y;\theta) = 1 - \exp\{-y/\theta\}$$

so that

$$H_j(s) = \theta_0 \log(s).$$

Taking y_1 to be the first observation, the approximate location ancillary is given by

$$\theta_0(\log(y_2/y_1), \log(y_3/y_1), \ldots, -\log(x_1 y_1), -\log(x_2 y_1), \ldots, -\log(x_n y_1))$$

which is equivalent to

$$(y_2/y_1, \ldots, y_n/y_1, x_1 y_1, \ldots, x_n y_1).$$

It is straightforward to show that this statistic is exactly ancillary. ∎

Example 6.26 Normal-theory nonlinear regression model
For this model,

$$H_j(s) = \int^s \frac{\phi(t - m_j(\theta_0))}{\phi(t - m_j(\theta_0)) m'_j(\theta_0)}\, dt = \frac{s}{m'_j(\theta_0)}.$$

Hence, the local location ancillary is given by

$$\left(\frac{y_2}{m'_2(\hat{\theta})} - \frac{y_1}{m'_1(\hat{\theta})}, \ldots, \frac{y_n}{m'_n(\hat{\theta})} - \frac{y_1}{m'_1(\hat{\theta})}\right).$$ ∎

6.6.4 APPROXIMATELY ANCILLARY STATISTICS BASED ON PIVOTAL QUANTITIES

Let $P_1(y;\theta), \ldots, P_m(y;\theta)$ denote functions of y and θ such that, under the distribution with parameter θ, $P_1(Y;\theta), \ldots, P_m(Y;\theta)$ are independent random variables with known distributions. Of course, we cannot use the $P_j(y;\theta)$ as ancillary statistics since they depend on the parameter θ. We may however replace θ by $\hat{\theta}$ to obtain an approximately ancillary statistic.

One particularly simple approach is to use pivotals based on the distribution functions of the observations. Suppose that $Y_1, \ldots, Y_n$ are independent random variables and that Y_j had distribution function $F_j(\cdot;\theta)$. Then $F_j(Y_j;\theta)$ has a uniform distribution. Hence, we may consider $a = (a_1, \ldots, a_n)$, where $a_j = F_j(y_j;\hat{\theta})$ as an approximate ancillary statistic.

We now consider the distributional properties of a. Let $\hat{\theta}_j$ denote the maximum likelihood estimate of θ based on all the data except y_j. Then

$$a_j = F_j(y_j;\hat{\theta}) = F_j(y_j;\hat{\theta}_j) + O_p(n^{-1}).$$

It follows that

$$\begin{aligned}
&\Pr[F(Y_j;\hat{\theta}) \leq t;\theta] \\
&\quad = \Pr[F(Y_j;\hat{\theta}_j) \leq t;\theta] + O(n^{-1}) \\
&\quad = \mathrm{E}\{\Pr[F(Y_j;\hat{\theta}_j) \leq t \mid y_1, \ldots, y_{j-1}, y_{j+1}, \ldots, y_n;\theta];\theta\} + O(n^{-1}) \\
&\quad = t + O(n^{-1}).
\end{aligned}$$

It follows that each a_j is approximately ancillary to second order. The same approach can be used to show that any fixed number of $a_1, \ldots, a_n$ are also independent to the same order. However, as in the case of the local location ancillary, it does not necessarily follow that a is a second-order locally ancillary statistic since the dimension of a increases with n.

Example 6.27 Exponential hyperbola
In this model, there are two distributions to be considered. For each X_j,

$$F_j(x;\theta) = 1 - \exp\{-\theta x\}$$

so that the corresponding approximately ancillary statistic is

$$1 - \exp\{-\hat{\theta} x_j\}.$$

For each Y_j,

$$F_j(y;\theta) = 1 - \exp\{-y/\theta\}$$

so that the corresponding approximately ancillary statistic is

$$1 - \exp\{-y_j/\hat{\theta}\}.$$

It follows that the pivotal-based approximately ancillary statistic is equivalent to

$$(y_1/\hat{\theta}, \ldots, y_n/\hat{\theta}, x_1\hat{\theta}, \ldots, x_n\hat{\theta}).$$

It is straightforward to show that this statistic is exactly ancillary. ∎

Example 6.28 Normal-theory nonlinear regression model
The simplest pivotals to use in this example are those of the form $y_j - m_j(\theta)$, $j = 1, \ldots, n$; clearly, these are equivalent to those based on the distribution function of the Y_j. The corresponding approximately ancillary statistic is given by the residual vector

$$(y_1 - m_1(\hat{\theta}), \ldots, y_n - m_n(\hat{\theta})).$$

■

6.6.5 LIKELIHOOD RATIO ANCILLARY STATISTICS

Suppose that the data vector Y has a density $p(y; \theta)$ that may be embedded in a larger model with parameter (θ, δ), where δ is a scalar parameter, such that the actual model for the data corresponds to a specific value of δ, say $\delta = \delta_0$. Let $\tilde{p}(y; \theta, \delta)$ denote the model function of the more general model and let $\tilde{\ell}(\theta, \delta)$ denote the corresponding log-likelihood function. We assume that the minimal sufficient statistic in the general model is of the form $(\hat{\theta}^*, \hat{\delta}, a_0)$ where $\hat{\theta}^*$ and $\hat{\delta}$ are the maximum likelihood estimates of θ and δ, respectively, in the general model, and a_0 is a statistic that is ancillary in the general model. In many cases, a_0 may be taken to be null and, for simplicity, we assume this is the case in the remainder of this section.

The signed likelihood ratio statistic for testing $\delta = \delta_0$ is given by

$$a = \text{sgn}(\hat{\delta} - \delta_0)\, [2(\tilde{\ell}(\hat{\theta}^*, \hat{\delta}) - \tilde{\ell}(\hat{\theta}, \delta_0))]^{1/2}$$

where $\hat{\theta}$ denotes the maximum likelihood estimate of θ based on the log-likelihood function $\ell(\theta)$. Under the assumption that the original model holds, a is distributed approximately according to a standard normal distribution. The results in Section 5.4 show that a is second-order locally ancillary.

Example 6.29 Exponential hyperbola
Consider a 'larger' model in which the Y_j have mean θ and the X_j have mean $1/(\delta\theta)$; hence, the exponential hyperbola model corresponds to $\delta = 1$. The log-likelihood function for this model is given by

$$\tilde{\ell}(\theta, \delta) = -n(\theta\delta\bar{x} + \bar{y}/\theta) + n \log \delta;$$

it follows that $\hat{\theta}^* = \bar{y}$ and $\hat{\delta} = 1/(\bar{x}\bar{y})$. Therefore write

$$\tilde{\ell}(\theta, \delta) = -n\left(\frac{\theta\delta}{\hat{\theta}^*\hat{\delta}} + \frac{\hat{\theta}^*}{\theta}\right).$$

In this model $(\hat{\theta}^*, \hat{\delta})$ is sufficient so that no ancillary statistic is needed.

The signed likelihood ratio statistic for testing $\delta = 1$ is given by

$$\text{sgn}(\bar{x}\bar{y} - 1)2\surd n(\surd(\bar{x}\bar{y}) - \log \surd(\bar{x}\bar{y}) - 1)^{1/2}.$$

This statistic is equivalent to $\bar{x}\bar{y}$, which is exactly ancillary. ■

6.7 Approximation of sample space derivatives

6.7.1 INTRODUCTION

Many likelihood-based methods rely on the use of sample space derivatives, derivatives of the likelihood function, or a closely related function, with respect to certain functions of the data. For instance, the approximation to the tail probability of a scalar maximum likelihood estimate $\hat{\theta}$ derived in Section 6.3.1 uses $\ell_{;\hat{\theta}}(\hat{\theta}) - \ell_{;\hat{\theta}}(\theta)$. In Chapters 7 and 9 it is shown that sample space derivatives arise in other contexts as well.

Consider a parametric statistical model for data y, of dimension n, with parameter θ, which is d-dimensional. Suppose that the sufficient statistic for the model may be written $(\hat{\theta}, a)$, where $\hat{\theta}$ denotes the maximum likelihood estimate of θ and a is an ancillary statistic. Calculation of a sample space derivative such as $\ell_{;\hat{\theta}}(\theta)$ requires specification of the ancillary statistic a and, hence, is only easily calculated for certain special classes of models, such as full-rank exponential family models and transformation models. In general models, calculation of these sample space derivatives may be difficult or impossible. To deal with this difficulty, several approximations to sample space derivatives have been proposed. In this section, a number of these approximations are reviewed and compared.

Specifically, we will be concerned with approximating either $\ell_{\theta;\hat{\theta}}(\theta)$ or $\ell_{;\hat{\theta}}(\hat{\theta}) - \ell_{;\hat{\theta}}(\theta)$. When evaluating the accuracy of these approximations we will distinguish between two cases, the case in which the argument θ is fixed, the large-deviation case, and the case in which θ is of the form $\theta = \hat{\theta} + O(n^{-1/2})$, the moderate-deviation case. Since $\hat{\theta}$ is within $O(n^{-1/2})$ of the true parameter value, and we are generally concerned with values of θ within $O(n^{-1/2})$ of the true parameter value, the moderate-deviation case is the more important of the two. However, good approximation properties for the large-deviation case insure that the quality of the approximation is still acceptable for values of θ far from $\hat{\theta}$.

Note that in the moderate-deviation case, the statistics in question may be expanded

$$\begin{aligned}[\ell_{;\hat{\theta}}(\theta) - \ell_{;\hat{\theta}}(\hat{\theta})]^T &= \ell_{\theta;\hat{\theta}}(\hat{\theta})(\theta - \hat{\theta}) + \tfrac{1}{2}(\theta - \hat{\theta})^T \ell_{\theta\theta;\hat{\theta}}(\hat{\theta})(\theta - \hat{\theta}) + O_p(n^{-1/2}) \\ &= \hat{\jmath}(\theta - \hat{\theta}) + \tfrac{1}{2}(\theta - \hat{\theta})^T \ell_{\theta\theta;\hat{\theta}}(\hat{\theta})(\theta - \hat{\theta}) + O_p(n^{-1/2});\end{aligned} \tag{6.10}$$

and

$$\ell_{\theta;\hat{\theta}}(\theta) = \ell_{\theta;\hat{\theta}}(\hat{\theta}) + \ell_{\theta\theta;\hat{\theta}}(\hat{\theta})(\theta - \hat{\theta}) + \cdots = \hat{\jmath} + \ell_{\theta\theta;\hat{\theta}}(\hat{\theta})(\theta - \hat{\theta}) + O_p(1). \tag{6.11}$$

These expansions will be useful in evaluating the moderate-deviation properties of the approximations.

6.7.2 APPROXIMATION BASED ON AN APPROXIMATELY ANCILLARY STATISTIC

Let f denote a generic s-dimensional function of the the parameter θ as well as the data y. We assume that f depends on the data only through the sufficient statistic $(\hat{\theta}, a)$ and, hence, we may write $f(\theta; y)$ or $f(\theta; \hat{\theta}, a)$; the symbol f will be used in both cases. We first consider the exact calculation of $f_{;\hat{\theta}}(\theta)$, the sample space derivative of a f with respect to $\hat{\theta}$. In the applications we will be considering, either $f(\theta) = \ell(\hat{\theta}) - \ell(\theta)$ or $f(\theta) = \ell_\theta(\theta)$. Assume that a is of dimension m, where $m \leq n - d$.

Consider the function $f(\theta)$ as a function of y, the vector of data values. By the chain rule, the sample space derivative

$$f_{;y}(\theta) = \frac{\partial f(\theta; y)}{\partial y}$$

may be written

$$f_{;y}(\theta) = (f_{;\hat{\theta}}(\theta) \quad f_{;a}(\theta)) \begin{pmatrix} \frac{\partial \hat{\theta}}{\partial y} \\ \frac{\partial a}{\partial y} \end{pmatrix};$$

hence

$$(f_{;\hat{\theta}}(\theta) \quad f_{;a}(\theta)) = f_{;y}(\theta) \begin{pmatrix} \frac{\partial \hat{\theta}}{\partial y} \\ \frac{\partial a}{\partial y} \end{pmatrix}^{-}$$

where M^{-} denotes any generalized inverse of M.

Write

$$(V_1 \quad V_2) = \begin{pmatrix} \frac{\partial \hat{\theta}}{\partial y} \\ \frac{\partial a}{\partial y} \end{pmatrix}^{-}$$

where V_1 is $n \times d$ and V_2 is $n \times m$. Then

$$f_{;\hat{\theta}}(\theta) = f_{;y}(\theta) V_1$$

where V_1 and V_2 satisfy

$$\frac{\partial \hat{\theta}}{\partial y} V_1 \frac{\partial \hat{\theta}}{\partial y} + \frac{\partial \hat{\theta}}{\partial y} V_2 \frac{\partial a}{\partial y} = \frac{\partial \hat{\theta}}{\partial y} \tag{6.12}$$

and

$$\frac{\partial a}{\partial y} V_1 \frac{\partial \hat{\theta}}{\partial y} + \frac{\partial a}{\partial y} V_2 \frac{\partial a}{\partial y} = \frac{\partial a}{\partial y}. \tag{6.13}$$

Suppose that V is a $n \times d$ matrix satisfying

$$\frac{\partial a}{\partial y} V = 0 \tag{6.14}$$

and

$$\left|\frac{\partial \hat{\theta}}{\partial y} V\right| \neq 0. \tag{6.15}$$

Then V_1 may be taken to be

$$V_1 = V\left(\frac{\partial \hat{\theta}}{\partial y} V\right)^{-1}$$

and V_2 may be taken to be a generalized inverse of $\partial a/\partial y$ satisfying

$$\frac{\partial \hat{\theta}}{\partial y} V_2 = 0;$$

it is straightforward to show that such a generalized inverse exists. Then

$$f_{;\hat{\theta}}(\theta) = f_{;y}(\theta) V \left(\frac{\partial \hat{\theta}}{\partial y} V\right)^{-1}. \tag{6.16}$$

Note that, by construction, the value of $f_{;\hat{\theta}}(\theta)$ does not depend on the choice of V provided that (6.24) and (6.25) are satisfied. If an explicit expression is available for the data vector y in terms of $(\hat{\theta}, a)$, we may take

$$V = \frac{\partial y}{\partial \hat{\theta}}.$$

Since $\hat{\theta}$ satisfies $\ell_\theta(\hat{\theta}) = 0$,

$$\ell_{\theta\theta}(\hat{\theta}) \frac{\partial \hat{\theta}}{\partial y} + \ell_{\theta;y}(\hat{\theta}) = 0$$

where $\hat{\jmath}$ denotes the observed information evaluated at $\hat{\theta}$. Hence,

$$\frac{\partial \hat{\theta}}{\partial y} = \hat{\jmath}^{-1} \ell_{\theta;y}(\hat{\theta})$$

and

$$f_{;\hat{\theta}}(\theta) = f_{;y}(\theta) V (\ell_{\theta;y}(\hat{\theta}) V)^{-1} \hat{\jmath}. \tag{6.17}$$

Note that (6.26) and (6.27) are exact expressions. Also note that these expressions do not require that a is an ancillary statistic.

Recall that we have assumed that a is an m-dimensional statistic such that $(\hat{\theta}, a)$ is sufficient and that $m \leq n - d$. Suppose that there exists an ancillary statistic b, of arbitrary dimension, such that $a = g(b)$ for some function g. If V satisfies

$$\frac{\partial b}{\partial y} V = 0$$

then, by the chain rule for differentiation, (6.14) holds. Hence, in (6.16) and (6.17) we may assume that a is of arbitrary dimension.

These result may be applied to obtain either $\ell_{;\hat{\theta}}(\theta)$ or $\ell_{\theta;\theta}(\theta)$. For instance,

$$\ell_{;\hat{\theta}}(\theta) = \ell_{;y}(\theta) V (\ell_{\theta;y}(\hat{\theta}) V)^{-1} \hat{\jmath}$$

and

$$\ell_{\theta;\hat{\theta}}(\theta) = \ell_{\theta;y}(\theta) V (\ell_{\theta;y}(\hat{\theta}) V)^{-1} \hat{\jmath}.$$

Since $\ell_{;y}(\theta)$ and $\ell_{\theta;y}(\theta)$ are typically straightforward to obtain, determination of the sample space derivatives requires only determination of the matrix V.

We now consider the construction of such a matrix V based on an approximately ancillary statistic. Let $\hat{a}$ denote an approximate ancillary statistic. Suppose that $\hat{a}$ is distribution constant, neglecting terms of order $O(n^{-j/2})$. We assume that, in this case, the exactly ancillary statistic a satisfies

$$\hat{a} = a + O_p(n^{-j/2}).$$

We may then use the matrix

$$\hat{V} = \frac{\partial y}{\partial \hat{\theta}},$$

where y is viewed as a function of $(\hat{\theta}, \hat{a})$, in calculating the sample space derivatives.

The resulting expressions,

$$\tilde{\ell}_{;\hat{\theta}}(\theta) = \ell_{;y}(\theta) \hat{V} (\ell_{\theta;y}(\hat{\theta}) \hat{V})^{-1} \hat{\jmath}$$

and

$$\tilde{\ell}_{\theta;\hat{\theta}}(\theta) = \ell_{\theta;y}(\theta) \hat{V} (\ell_{\theta;y}(\hat{\theta}) \hat{V})^{-1} \hat{\jmath},$$

are then approximations to the exact sample space derivatives.

The derivative $\tilde{\ell}_{;\hat{\theta}}(\theta)$ is the exact derivative of $\ell(\theta; \hat{\theta}, \hat{a})$ with respect to $\hat{\theta}$; note that here $\ell(\theta)$ is being viewed as a function of $\hat{\theta}, \hat{a}$. By the chain rule,

$$\tilde{\ell}_{;\hat{\theta}}(\theta) = \ell_{;\hat{\theta}}(\theta) + \ell_{;a}(\theta) \frac{\partial a(\hat{\theta}, \hat{a})}{\partial \hat{\theta}}.$$

For fixed θ, $\ell_{;\hat{\theta}}(\theta) = O(n)$. Since $a = \hat{a} + O(n^{-1})$,

$$\frac{\partial a(\hat{\theta}, \hat{a})}{\partial \hat{a}} = O(n^{-1}).$$

If a is of fixed dimension $\ell_{;a}(\theta) = O(n)$ and

$$\ell_{;a}(\theta)\frac{\partial a(\hat{\theta}, \hat{a})}{\partial \hat{\theta}} = O(1).$$

If the dimension of a is of order $O(n)$, then each component of $\ell_{;a}(\theta)$ is $O(1)$, but there n terms in the sum

$$\ell_{;a}(\theta)\frac{\partial a(\hat{\theta}, \hat{a})}{\partial \hat{\theta}},$$

so, again, this term is $O(1)$. In either case, the relative error of the approximation is $O(n^{-1})$.

A similar argument shows that, for fixed θ,

$$\tilde{\ell}_{\theta;\hat{\theta}}(\theta) = \ell_{\theta;\hat{\theta}}(\theta)[1 + O(n^{-1})].$$

Now consider the moderate-deviation case in which $\theta = \hat{\theta} + O(n^{-1/2})$. For notational simplicity, assume that θ is a scalar; the same result holds for the vector case. Using a Taylor's series expansion,

$$\ell_{;y}(\hat{\theta}) - \ell_{;y}(\theta) = \ell_{\theta;y}(\hat{\theta})(\hat{\theta} - \theta) - \tfrac{1}{2}\ell_{\theta\theta;y}(\hat{\theta})(\hat{\theta} - \theta)^2 + \tfrac{1}{6}\ell_{\theta\theta\theta;y}(\hat{\theta})(\hat{\theta} - \theta)^3 + \cdots .$$

Hence,

$$\begin{aligned}
&[\ell_{;y}(\hat{\theta}) - \ell_{;y}(\theta)]\hat{V}(\ell_{\theta;y}(\hat{\theta})\hat{V})^{-1}\hat{\jmath} \\
&= \hat{\jmath}(\hat{\theta} - \theta) - \tfrac{1}{2}[\ell_{\theta\theta;y}(\hat{\theta})\hat{V}(\ell_{\theta;y}(\hat{\theta})\hat{V})^{-1}\hat{\jmath}](\hat{\theta} - \theta)^2 \\
&\quad + \tfrac{1}{6}[\ell_{\theta\theta\theta;y}(\hat{\theta})\hat{V}(\ell_{\theta;y}(\hat{\theta})\hat{V})^{-1}\hat{\jmath}](\hat{\theta} - \theta)^3 + O_p(n^{-1}).
\end{aligned}$$

Applying the argument given previously in this section to $\ell_{\theta\theta}(\theta)$ and $\ell_{\theta\theta\theta}(\theta)$, we have that

$$\ell_{\theta\theta;y}(\hat{\theta})\hat{V}(\ell_{\theta;y}(\hat{\theta})\hat{V})^{-1}\hat{\jmath} = \ell_{\theta\theta;\hat{\theta}}(\hat{\theta}) + O(1)$$

and

$$\ell_{\theta\theta\theta;y}(\hat{\theta})\hat{V}(\ell_{\theta;y}(\hat{\theta})\hat{V})^{-1}\hat{\jmath} = \ell_{\theta\theta\theta;\hat{\theta}}(\hat{\theta}) + O(1).$$

Hence,

$$[\ell_{;y}(\hat{\theta}) - \ell_{;y}(\theta)]\hat{V}(\ell_{\theta;y}(\hat{\theta})\hat{V})^{-1}\hat{\jmath} = \hat{\jmath}(\hat{\theta} - \theta) - \tfrac{1}{2}\ell_{\theta\theta;\hat{\theta}}(\hat{\theta})(\hat{\theta} - \theta)^2 + \tfrac{1}{6}\ell_{\theta\theta\theta;\hat{\theta}}(\hat{\theta})(\hat{\theta} - \theta)^3 + O_p(n^{-1}).$$

Using an expansion of the form (6.10), but including an additional term, we have that

$$\ell_{;\hat{\theta}}(\hat{\theta}) - \ell_{;\hat{\theta}}(\theta) = \hat{\jmath}(\hat{\theta} - \theta) - \tfrac{1}{2}\ell_{\theta\theta;\hat{\theta}}(\hat{\theta})(\hat{\theta} - \theta)^2 + \tfrac{1}{6}\ell_{\theta\theta\theta;\hat{\theta}}(\hat{\theta})(\hat{\theta} - \theta)^3 + O_p(n^{-1}).$$

It follows that

$$\tilde{\ell}_{;\hat{\theta}}(\hat{\theta}) - \tilde{\ell}_{;\hat{\theta}}(\theta) = [\ell_{;\hat{\theta}}(\hat{\theta}) - \ell_{;\hat{\theta}}(\theta)][1 + O_p(n^{-3/2})].$$

A similar argument can be used to show that

$$\tilde{\ell}_{\theta;\hat{\theta}}(\hat{\theta}) - \tilde{\ell}_{\theta;\hat{\theta}}(\theta) = [\ell_{\theta;\hat{\theta}}(\hat{\theta}) - \ell_{\theta;\hat{\theta}}(\theta)][1 + O_p(n^{-3/2})].$$

The remaining issue is specification of the approximately ancillary statistic. Recall that, in forming $\hat{V}$, we may use an n-dimensional approximately ancillary statistic b such that there exists a function $\hat{a}$ of b such that $\hat{a}$ is approximately ancillary and $(\hat{\theta}, \hat{a})$ is sufficient.

Although many different statistics could, in principle, be used in this context, perhaps the simplest choice is the pivotal-based approximate ancillary described in Section 6.6.4. Let $F_j(\cdot; \theta)$ denote the distribution function of Y_j and let $b_j = F_j(y_j; \hat{\theta})$, $j = 1, \ldots, n$. Then $y_j = F_j^{-1}(b_j; \hat{\theta})$ so that

$$\frac{\partial y_j}{\partial \hat{\theta}} = -\frac{\partial F_j(y_j; \hat{\theta})/\partial \hat{\theta}}{p_j(y_j; \hat{\theta})}$$

where $p_j(\cdot; \theta)$ denotes the density function of Y_j. It follows that the matrix $\hat{V}$ may be taken to be

$$\hat{V} = \begin{pmatrix} -\dfrac{\partial F_1(y_1; \hat{\theta})/\partial \hat{\theta}}{p_1(y_j; \hat{\theta})} \\ \vdots \\ -\dfrac{\partial F_n(y_n; \hat{\theta})/\partial \hat{\theta}}{p_n(y_n; \hat{\theta})} \end{pmatrix}.$$

Example 6.30 Full-rank exponential family models
Suppose that the log-likelihood function for the model may be written

$$\ell(\theta) = ns(y)^T \eta(\theta) - nk(\eta(\theta))$$

where $s(y)$ is an $O_p(1)$ function of y and η is a one-to-one function of θ. Then we may write

$$\ell(\theta) = nk'(\eta(\hat{\theta}))\eta(\theta) - nk(\eta(\theta))$$

so that

$$\ell_{\theta;\hat{\theta}}(\theta) = n\eta'(\theta)^T k''(\eta(\hat{\theta}))\eta'(\hat{\theta}).$$

Since

$$\ell_\theta(\theta) = n\eta'(\theta)^T T(y) - n\eta'(\theta)^T k'(\eta(\theta))^T$$

and $\hat{j} = n\eta'(\hat{\theta})^T k''(\eta(\hat{\theta}))\eta'(\hat{\theta})$, it follows that the approximation given in this section is given by

$$n\eta'(\theta)^T s'(y)\hat{V}(n\eta'(\hat{\theta})^T s'(y)\hat{V})^{-1} n\eta'(\hat{\theta})^T k''(\eta(\hat{\theta}))\eta'(\hat{\theta}).$$

Provided that $\eta'(\hat{\theta})$ and $s'(y)\hat{V}$ are invertible, which is generally true, this approximation is exactly $\ell_{\theta;\hat{\theta}}(\theta)$. A similar result holds for $\ell_{;\hat{\theta}}(\hat{\theta}) - \ell_{;\hat{\theta}}(\theta)$. ∎

Example 6.31 Exponential hyperbola
Consider the exponential hyperbola model described in Example 6.21 and consider approximation of $\ell_{;\hat{\theta}}(\hat{\theta}) - \ell_{;\hat{\theta}}(\theta)$.

Here the vector of observations is given by $(y_1, \ldots, y_n, x_1, \ldots, x_n)$ where the Y_j are exponential random variables with mean θ and the X_j are exponential random variables with mean $1/\theta$. Hence, for this model, derivatives such as $\ell_{;y}(\theta)$ will refer to derivatives with respect to the vector $(y_1, \ldots, y_n, x_1, \ldots, x_n)$.

The log-likelihood function is given by

$$\ell(\theta) = -\sum y_j/\theta - \theta \sum x_j.$$

It follows that

$$\ell_{;y}(\theta) = -(\theta^{-1}, \ldots, \theta^{-1}, \theta, \ldots, \theta).$$

To determine the matrix $\hat{V}$, note that Y_j has distribution function $1 - \exp(-y/\theta)$ and X_j has distribution function $1 - \exp(-\theta x)$. It follows that

$$\hat{V} = \begin{pmatrix} y_1/\hat{\theta} \\ \vdots \\ y_n/\hat{\theta} \\ -x_1\hat{\theta} \\ \vdots \\ -x_n\hat{\theta} \end{pmatrix}.$$

Hence,

$$\ell_{;y}(\theta)\hat{V} = -n\left(\frac{\bar{y}}{\theta\hat{\theta}} - \frac{\bar{x}\theta}{\hat{\theta}}\right) = -na\left(\frac{1}{\theta} - \frac{\theta}{\hat{\theta}^2}\right).$$

A similar calculation shows that

$$\ell_{\theta;y}(\hat{\theta})\hat{V} = \hat{j} = \frac{2na}{\hat{\theta}^2}.$$

Hence,

$$\tilde{\ell}_{;\hat{\theta}}(\hat{\theta}) - \tilde{\ell}_{;\hat{\theta}}(\theta) = na\left(\frac{1}{\theta} - \frac{\theta}{\hat{\theta}^2}\right)$$

which is exactly equal to $\ell_{;\hat{\theta}}(\hat{\theta}) - \ell_{;\hat{\theta}}(\theta)$. ∎

6.7.3 APPROXIMATIONS BASED ON COVARIANCES

We begin by outlining the derivation of an approximation to

$$\ell_{;\hat{\theta}}(\theta; \theta_0, a) - \ell_{;\hat{\theta}}(\theta_0; \theta_0, a),$$

where θ_0 denotes a specific parameter point, for the case in which $\theta = \theta_0 + \delta/\sqrt{n}$. Then, by setting $\theta_0 = \hat{\theta}$, we obtain an approximation for $\ell_{;\hat{\theta}}(\theta) - \ell_{;\hat{\theta}}(\hat{\theta})$.

Note that, by expanding $\ell(\theta; \hat{\theta}, a)$ and $\ell(\theta_0; \hat{\theta}, a)$ around $\hat{\theta} = \theta_0$,

$$\begin{aligned}\ell(\theta; \hat{\theta}, a) - \ell(\theta_0; \hat{\theta}, a) &= \ell(\theta; \theta_0, a) - \ell(\theta_0; \theta_0, a) \\ &\quad + [\ell_{;\hat{\theta}}(\theta; \theta_0, a) - \ell_{;\hat{\theta}}(\theta_0; \theta_0, a)](\hat{\theta} - \theta_0) + \cdots\end{aligned}$$

and hence that

$$E\{\ell(\theta) - \ell(\theta_0)|a; \theta_0\} = \ell(\theta; \theta_0, a) - \ell(\theta_0; \theta_0, a) + O(n^{-1/2}).$$

It follows that

$$[\ell_{;\hat{\theta}}(\theta; \theta_0, a) - \ell_{;\hat{\theta}}(\theta_0; \theta_0, a)]^T = E[\ell_\theta(\theta_0)\{\ell(\theta) - \ell(\theta_0)\}|a; \theta_0] + O(n^{-1/2}),$$

conditionally on a. By expanding $\ell(\theta)$ around $\theta = \theta_0$ we have that

$$E[\ell_\theta(\theta_0)\{\ell(\theta) - \ell(\theta_0)\}|a; \theta_0] = E\{\ell_\theta(\theta_0)\ell_\theta(\theta_0)^T \delta|a; \theta_0\}n^{-1/2} + \cdots$$

and hence that

$$\begin{aligned}&E[\ell_\theta(\theta_0)\{\ell(\theta) - \ell(\theta_0)\}|a; \theta_0] - E[\ell_\theta(\theta_0)\{\ell(\theta) - \ell(\theta_0)\}; \theta_0] \\ &\quad = [E\{\ell_\theta(\theta_0)\ell_\theta(\theta_0)^T|a; \theta_0\} - E\{\ell_\theta(\theta_0)\ell_\theta(\theta_0)^T; \theta_0\}]\frac{\delta}{\sqrt{n}} + O(n^{-1/2}).\end{aligned}$$

We have seen that

$$E\{\ell_\theta(\theta_0)\ell_\theta(\theta_0)^T|a;\theta_0\} = -\ell_{\theta\theta}(\theta_0;\theta_0,a) + O(1).$$

It follows that

$$\begin{aligned}E[\ell_\theta(\theta_0)\{\ell(\theta)-\ell(\theta_0)\}|a;\theta_0] &= E[\ell_\theta(\theta_0)\{\ell(\theta)-\ell(\theta_0)\};\theta_0]\\ &\quad - \{\ell_{\theta\theta}(\theta_0;\theta_0,a)+i(\theta_0)\}\frac{\delta}{\sqrt{n}} + O(n^{-1/2})\end{aligned} \tag{6.18}$$

where $i(\theta_0)$ denotes the expected Fisher information matrix evaluated at $\theta = \theta_0$. Let

$$Q(\theta;\theta_0) = E\{\ell(\theta)\ell_\theta(\theta_0)^T;\theta_0\};$$

then (6.28) may be written

$$\begin{aligned}&E[\{\ell(\theta)-\ell(\theta_0)\}\ell_\theta(\theta_0)|a;\theta_0]\\ &\quad = [Q(\theta)-Q(\theta_0)]^T - \{\ell_{\theta\theta}(\theta_0;\theta_0,a)+i(\theta_0)\}\frac{\delta}{\sqrt{n}} + O(n^{-1/2}).\end{aligned} \tag{6.19}$$

Finally, note that $\delta/\sqrt{n} = \theta - \theta_0$ and that

$$[Q(\theta;\theta_0) - Q(\theta_0;\theta_0)]^T = i(\theta_0)(\theta-\theta_0) + O(1)$$

so that

$$\theta - \theta_0 = i(\theta_0)^{-1}[Q(\theta;\theta_0) - Q(\theta_0;\theta_0)]^T + O(n^{-1}).$$

Equation (6.19) can then be written

$$\begin{aligned}&E[\{\ell(\theta)-\ell(\theta_0)\}\ell_\theta(\theta_0)|a;\theta_0]\\ &\quad = -\ell_{\theta\theta}(\theta_0;\theta_0,a)i(\theta_0)^{-1}\{Q(\theta;\theta_0)-Q(\theta_0;\theta_0)\}^T + O(n^{-1/2})\end{aligned}$$

and hence

$$\begin{aligned}&\ell_{;\hat{\theta}}(\theta;\theta_0,a) - \ell_{;\hat{\theta}}(\theta_0;\theta_0,a)\\ &\quad = \{Q(\theta;\theta_0) - Q(\theta_0;\theta_0)\}i(\theta_0)^{-1}[-\ell_{\theta\theta}(\theta_0;\theta_0,a)] + O(n^{-1/2})\end{aligned} \tag{6.20}$$

conditionally on a.

Setting $\theta_0 = \hat{\theta}$, it follows that $\ell_{;\hat{\theta}}(\hat{\theta}) - \ell_{;\hat{\theta}}(\theta)$ can be approximated by

$$\bar{\ell}_{;\hat{\theta}}(\hat{\theta}) - \bar{\ell}_{;\hat{\theta}}(\theta) \equiv \{Q(\hat{\theta};\hat{\theta}) - Q(\theta;\hat{\theta})\}i(\hat{\theta})^{-1}\hat{j};$$

the relative error of this approximation is $O_p(n^{-1})$ for θ of the form $\hat{\theta} + O(n^{-1/2})$.

This result may be obtained more directly by using the expansion

$$[Q(\theta;\hat\theta)-Q(\hat\theta;\hat\theta)]^T = Q'(\hat\theta;\hat\theta)^T(\theta-\hat\theta)+\tfrac{1}{2}(\theta-\hat\theta)^T Q''(\hat\theta;\hat\theta)^T(\theta-\hat\theta) + O_p(n^{-1/2})$$

where

$$Q'(\theta;\theta_0)^T = \frac{\partial Q(\theta;\theta_0)^T}{\partial\theta}; \qquad Q''(\theta;\theta_0)^T = \frac{\partial^2 Q(\theta;\theta_0)^T}{\partial\theta\,\partial\theta^T}.$$

Note that $Q'(\hat\theta;\hat\theta) = i(\hat\theta)$ and

$$Q''(\hat\theta;\hat\theta)^T = \ell_{\theta\theta;\hat\theta}(\hat\theta) + O_p(\sqrt{n}).$$

Hence,

$$\begin{aligned}[\bar\ell_{;\hat\theta}(\theta)-\bar\ell_{;\hat\theta}(\hat\theta)]^T &= \hat\jmath\, i(\hat\theta)^{-1}[i(\hat\theta)(\theta-\hat\theta)+\tfrac{1}{2}(\theta-\hat\theta)^T\ell_{\theta\theta;\hat\theta}(\hat\theta)(\theta-\hat\theta)] \\ &\quad + O_p(n^{-1/2}) \\ &= \hat\jmath(\theta-\hat\theta)+\tfrac{1}{2}(\theta-\hat\theta)^T\ell_{\theta\theta;\hat\theta}(\hat\theta)(\theta-\hat\theta)+O_p(n^{-1/2}).\end{aligned}$$

Comparing this expansion to (6.10) yields the result.

For fixed θ, note that

$$E\{\ell(\theta)-\ell(\theta_0)|a;\theta_0\} = \ell(\theta;\theta_0,a)-\ell(\theta_0;\theta,a)+O(1)$$

and since

$$E\{\ell(\theta)-\ell(\theta_0)|a;\theta_0\} = E\{\ell(\theta)-\ell(\theta_0);\theta_0\}+O(\sqrt{n}),$$

it follows that

$$\ell(\theta;\theta_0,a)-\ell(\theta_0;\theta,a) = E\{\ell(\theta)-\ell(\theta_0);\theta_0\}+O(\sqrt{n}),$$

and hence that

$$\ell_{;\hat\theta}(\theta;\theta_0,a)-\ell_{;\hat\theta}(\theta_0;\theta,a) = Q(\theta;\theta_0)-Q(\theta_0;\theta_0)+O(\sqrt{n}).$$

Since $i(\hat\theta)^{-1}\hat\jmath = D + O_p(n^{-1/2})$, where D denotes an identity matrix, it follows that

$$\ell_{;\hat\theta}(\hat\theta)-\ell_{;\hat\theta}(\theta) = \{Q(\hat\theta;\hat\theta)-Q(\theta;\hat\theta)\}i(\hat\theta)^{-1}\hat\jmath + O_p(\sqrt{n});$$

for fixed θ, $\ell_{;\hat\theta}(\hat\theta)-\ell_{;\hat\theta}(\theta)$ is order $O_p(n)$, it follows that the relative error is $O(n^{-1/2})$.

A similar approximation can be developed for $\ell_{\theta;\hat{\theta}}(\theta; \theta_0, a)$. Let

$$I(\theta; \theta_0) = E[\ell_\theta(\theta)\ell_\theta(\theta_0)^T; \theta_0].$$

For $\theta = \theta_0 + \delta/\sqrt{n}$,

$$\ell_{\theta;\hat{\theta}}(\theta; \theta_0, a) - \ell_{\theta;\hat{\theta}}(\theta_0; \theta_0, a) = I(\theta; \theta_0) - I(\theta_0; \theta_0) + O(1)$$

and since

$$\begin{aligned} I(\theta; \theta_0) &+ \ell_{\theta;\hat{\theta}}(\theta_0; \theta_0, a) - I(\theta_0; \theta_0) \\ &= I(\theta; \theta_0) + j(\theta_0) - i(\theta_0) \\ &= I(\theta; \theta_0) + i(\theta_0)\{i(\theta_0)^{-1} j(\theta_0) - D\} \\ &= I(\theta; \theta_0) + \{I(\theta; \theta_0) + O(\sqrt{n})\}\{i(\theta_0)^{-1} j(\theta_0) - D\} \\ &= I(\theta; \theta_0) i(\theta_0)^{-1} j(\theta_0) + O(1), \end{aligned}$$

$\ell_{\theta;\hat{\theta}}(\theta; \theta_0, a)$ can be approximated by

$$I(\theta; \theta_0) i(\theta_0)^{-1} j(\theta_0).$$

Hence, an approximation to $\ell_{\theta;\hat{\theta}}(\theta)$ is given by

$$\bar{\ell}_{\theta;\hat{\theta}}(\theta) \equiv I(\theta; \hat{\theta}) i(\hat{\theta})^{-1} \hat{j}.$$

The relative error of this approximation is $O_p(n^{-1/2})$ for fixed θ. For $\theta = \hat{\theta} + O(n^{-1/2})$,

$$I(\theta; \hat{\theta}) i(\hat{\theta})^{-1} \hat{j} = I(\hat{\theta}; \hat{\theta}) i(\hat{\theta})^{-1} \hat{j} + I'(\hat{\theta}; \hat{\theta})(\theta - \hat{\theta}) i(\hat{\theta})^{-1} \hat{j} + O_p(1)$$

where

$$I'(\theta; \theta_0) = \frac{\partial}{\partial \theta} I(\theta; \theta_0).$$

Note that $I(\hat{\theta}; \hat{\theta}) = i(\hat{\theta})$; $I'(\hat{\theta}; \hat{\theta}) = \ell_{\theta\theta;\hat{\theta}}(\hat{\theta}) + O_p(\sqrt{n})$; $i(\hat{\theta}) = \hat{j} + O_p(\sqrt{n})$. Hence,

$$I(\theta; \hat{\theta}) i(\hat{\theta})^{-1} \hat{j} = \hat{j} + \ell_{\theta\theta;\hat{\theta}}(\hat{\theta})(\theta - \hat{\theta}) + O_p(1).$$

Comparing this to (6.11) shows that

$$I(\theta; \hat{\theta}) i(\hat{\theta})^{-1} \hat{j} = \ell_{\theta;\hat{\theta}}(\theta) + O(1);$$

since $\ell_{\theta;\hat{\theta}}(\theta)$ is of order $O_p(n)$, the relative error is $O(n^{-1})$.

Example 6.32 Exponential family model
As in Example 6.30, consider a model with log-likelihood function of the form

$$\ell(\theta) = ns(y)^T \eta(\theta) - nk(\eta(\theta)).$$

Then

$$\ell_\theta(\theta) = n\eta'(\theta)^T s(y) - n\eta'(\theta)^T k'(\eta(\theta))^T;$$

recall that $s(Y)$ has covariance matrix $k''(\eta(\theta))$ so that

$$I(\theta; \theta_0) = n\eta'(\theta)^T k''(\eta_0)\eta'(\theta_0).$$

Since

$$\hat{\jmath} = n\eta'(\hat{\theta})^T k''(\eta(\hat{\theta}))\eta'(\hat{\theta}) = I(\hat{\theta}; \hat{\theta}),$$

it follows that

$$\bar{\ell}_{\theta;\hat{\theta}}(\theta) = n\eta'(\theta)^T k''(\eta(\hat{\theta}))\eta'(\hat{\theta})$$

which is exactly equal to $\ell_{\theta;\hat{\theta}}(\theta)$.

A similar result holds for $\ell_{;\hat{\theta}}(\hat{\theta}) - \ell_{;\hat{\theta}}(\theta)$. ∎

Example 6.33 Exponential hyperbola
Consider the exponential hyperbola model considered in Example 6.31. Recall that the log-likelihood function is given by

$$\ell(\theta) = -\sum y_j/\theta - \theta \sum x_j.$$

Hence,

$$\ell_\theta(\theta) = \sum y_j/\theta^2 - \sum x_j$$

so that

$$Q(\theta; \theta_0) = n(1/\theta - \theta/\theta_0{}^2).$$

Since $i(\hat{\theta}) = 2n/\hat{\theta}^2$ and $\hat{\jmath} = 2na/\hat{\theta}^2$, it follows that

$$\bar{\ell}_{;\hat{\theta}}(\hat{\theta}) - \bar{\ell}_{;\hat{\theta}}(\theta) = na\left(\frac{1}{\theta} - \frac{\theta}{\hat{\theta}^2}\right)$$

which is exactly equal to $\ell_{;\hat{\theta}}(\hat{\theta}) - \ell_{;\hat{\theta}}(\theta)$. ∎

6.7.4 AN APPROXIMATION BASED ON EMPIRICAL COVARIANCES

In this section, we consider another level of approximation in which the expectations with respect to the distribution with parameter θ_0 are replaced by expectations with respect to the empirical distribution of the data.

Suppose that the data consist of n independent observations. Let $\ell^{(j)}(\theta)$ denote the log-likelihood function based on the jth observation alone, so that $\ell(\theta) = \sum \ell^{(j)}(\theta)$. Note that $Q(\theta; \theta_0)$ may be written

$$Q(\theta; \theta_0) = \mathrm{E}\Big\{\sum \ell^{(j)}(\theta)\ell_\theta^{(j)}(\theta_0)^T; \theta_0\Big\}.$$

When θ_0 is the true parameter value, or within $O(n^{-1/2})$ of the true parameter value, then

$$\frac{1}{n}\sum \ell^{(j)}(\theta)\ell_\theta^{(j)}(\theta_0)^T = \frac{1}{n}\mathrm{E}\Big\{\sum \ell^{(j)}(\theta)\ell_\theta^{(j)}(\theta_0)^T; \theta_0\Big\} + O_p(n^{-1/2}).$$

Hence, $Q(\theta; \theta_0)$ may be approximated by

$$\hat{Q}(\theta; \theta_0) = \sum \ell^{(j)}(\theta)\ell_\theta^{(j)}(\theta_0)^T;$$

similarly, $I(\theta; \theta_0)$ can be approximated by

$$\hat{I}(\theta; \theta_0) = \sum \ell_\theta^{(j)}(\theta)\ell_\theta^{(j)}(\theta_0)^T.$$

Let $\hat{\imath}(\hat{\theta}) = \hat{I}(\hat{\theta}; \hat{\theta})$ and let

$$\hat{\ell}_{;\hat{\theta}}(\theta) = \hat{Q}(\theta; \hat{\theta})\hat{\imath}(\hat{\theta})^{-1}\hat{\jmath}.$$

For fixed values of θ,

$$\hat{Q}(\theta; \hat{\theta}) = Q(\theta; \hat{\theta}) + O_p(\sqrt{n})$$

and

$$\hat{\imath}(\hat{\theta}) = i(\hat{\theta}) + O_p(\sqrt{n}).$$

Hence,

$$\hat{Q}(\theta; \hat{\theta})\hat{\imath}(\hat{\theta})^{-1}\hat{\jmath} = Q(\theta; \hat{\theta})i(\hat{\theta})^{-1}\hat{\jmath} + O_p(\sqrt{n});$$

that is,

$$\hat{\ell}_{;\hat{\theta}}(\theta) = \bar{\ell}_{;\hat{\theta}}(\theta) + O_p(\sqrt{n}).$$

Since the error of $\bar{\ell}_{;\hat{\theta}}(\theta)$ as an approximation to $\ell_{;\hat{\theta}}(\theta)$ is $O(\sqrt{n})$, $\hat{\ell}_{;\hat{\theta}}(\theta)$ also has error $O(\sqrt{n})$ and, hence, relative error $O(n^{-1/2})$.

Now consider the case in which $\theta = \hat{\theta} + O(n^{-1/2})$. Then

$$[\hat{\ell}_{;\hat{\theta}}(\theta) - \hat{\ell}_{;\hat{\theta}}(\hat{\theta})]^T$$
$$= \hat{j}\,\hat{\imath}(\hat{\theta})^{-1}\Big\{\hat{Q}'(\hat{\theta};\hat{\theta})^T(\theta-\hat{\theta}) + \tfrac{1}{2}(\theta-\hat{\theta})^T\hat{Q}''(\hat{\theta};\hat{\theta})^T(\theta-\hat{\theta}) + O_p(n^{-1/2})\Big\}$$

where

$$\hat{Q}'(\theta;\theta_0)^T = \frac{\partial \hat{Q}(\theta;\theta_0)^T}{\partial\theta}; \qquad \hat{Q}''(\theta;\theta_0)^T = \frac{\partial^2 \hat{Q}(\theta;\theta_0)^T}{\partial\theta\,\partial\theta^T}.$$

Note that $\hat{Q}'(\hat{\theta};\hat{\theta}) = \hat{I}(\hat{\theta};\hat{\theta}) = \hat{\imath}(\hat{\theta})$.

Hence,

$$\hat{j}\,\hat{\imath}(\hat{\theta})^{-1}\hat{Q}'(\hat{\theta};\hat{\theta})^T = \hat{j}i(\hat{\theta})^{-1}Q'(\hat{\theta})^T = \ell_{\theta;\hat{\theta}}(\hat{\theta})$$

exactly so that, using the fact that $\hat{Q}''(\hat{\theta};\hat{\theta}) = Q''(\hat{\theta};\hat{\theta}) + O_p(\sqrt{n})$, it follows that

$$\begin{aligned}[\hat{\ell}_{;\hat{\theta}}(\theta) - \hat{\ell}_{;\hat{\theta}}(\hat{\theta})]^T &= [\bar{\ell}_{;\hat{\theta}}(\theta) - \bar{\ell}_{;\hat{\theta}}(\hat{\theta})]^T \\ &\quad + \tfrac{1}{2}(\hat{\theta}_\psi - \hat{\theta})^T\{\hat{Q}''(\hat{\theta})^T - Q''(\hat{\theta})^T\}(\hat{\theta}_\psi - \hat{\theta}) + O_p(n^{-1/2}) \\ &= [\bar{\ell}_{;\hat{\theta}}(\theta) - \bar{\ell}_{;\hat{\theta}}(\theta)]^T + O_p(n^{-1/2}).\end{aligned}$$

Since

$$\bar{\ell}_{;\hat{\theta}}(\hat{\theta}) - \bar{\ell}_{;\hat{\theta}}(\theta) = \ell_{;\hat{\theta}}(\hat{\theta}) - \ell_{;\hat{\theta}}(\theta) + O_p(n^{-1/2})$$

and $\ell_{;\hat{\theta}}(\hat{\theta}) - \ell_{;\hat{\theta}}(\theta)$ is of order $O_p(\sqrt{n})$, it follows that the relative error of $\hat{\ell}_{;\hat{\theta}}(\hat{\theta}) - \hat{\ell}_{;\hat{\theta}}(\theta)$ as an approximation to $\ell_{;\hat{\theta}}(\hat{\theta}) - \ell_{;\hat{\theta}}(\theta)$ is of order $O(n^{-1})$.

Note that this results shows $\hat{\ell}_{;\hat{\theta}}(\hat{\theta}) - \hat{\ell}_{;\hat{\theta}}(\theta)$ and $\bar{\ell}_{;\hat{\theta}}(\hat{\theta}) - \bar{\ell}_{;\hat{\theta}}(\theta)$ agree to second order, rather than to first order, as might be expected. The reason for this is that the first terms in the expansions of these statistics are exactly equal.

Similarly, an approximation to $\ell_{\theta;\hat{\theta}}(\theta)$ is given by

$$\hat{\ell}_{\theta;\hat{\theta}}(\theta) \equiv \hat{I}(\theta;\hat{\theta})\hat{\imath}(\hat{\theta})^{-1}\hat{j}.$$

For fixed values of θ,

$$\hat{I}(\theta;\hat{\theta}) = I(\theta;\hat{\theta}) + O_p(\sqrt{n})$$

so that

$$\hat{\ell}_{\theta;\hat{\theta}}(\theta) = \bar{\ell}_{\theta;\hat{\theta}}(\theta) + O_p(\sqrt{n}).$$

It follows that

$$\hat{\ell}_{\theta;\hat{\theta}}(\theta) = \ell_{\theta;\hat{\theta}}(\theta) + O_p(\sqrt{n});$$

the relative error is $O(n^{-1/2})$.

For $\theta = \hat{\theta} + O(n^{-1/2})$, it can be shown that

$$\ell_{\theta;\hat{\theta}}(\theta) = \hat{\jmath} + \ell_{\theta\theta;\hat{\theta}}(\hat{\theta})(\theta - \hat{\theta}) + O_p(1)$$

and that

$$\hat{\ell}_{\theta;\hat{\theta}}(\theta) = \hat{\jmath} + \hat{I}'(\hat{\theta}; \hat{\theta})(\theta - \hat{\theta}) + O_p(1)$$

where $\hat{I}'(\theta; \theta_0) = \partial \hat{I}(\theta; \theta_0)/\partial\theta$. Since

$$\hat{I}'(\hat{\theta}; \hat{\theta}) = \ell_{\theta\theta;\hat{\theta}}(\hat{\theta}) + O_p(\sqrt{n}),$$

it follows that

$$\hat{\ell}_{\theta;\hat{\theta}}(\theta) = \ell_{\theta;\hat{\theta}}(\theta) + O_p(1);$$

the relative error of the approximation is $O(n^{-1})$.

It is straightforward to show that these results hold conditionally on a provided that

$$\hat{Q}(\theta; \theta_0) = Q(\theta; \theta_0) + O_p(\sqrt{n}) \tag{6.21}$$

and

$$\hat{I}(\theta; \theta_0) = I(\theta; \theta_0) + O_p(\sqrt{n}) \tag{6.22}$$

hold conditionally on a. Consider (6.21). Since

$$Q(\theta; \theta_0) = \mathrm{E}\Big\{\sum \ell^{(j)}(\theta)\ell_\theta^{(j)}(\theta_0)^T |a; \theta_0\Big\} + O_p(\sqrt{n})$$

holds conditionally on a, the desired result follows from the fact that

$$\frac{1}{n}\mathrm{E}\{\ell(\theta)\ell_\theta(\theta_0)^T |a; \theta_0\} = \frac{1}{n}\mathrm{E}\{\ell(\theta)\ell_\theta(\theta_0)^T; \theta_0\} + O(n^{-1/2}).$$

The result for (6.22) holds similarly. Hence, the asymptotic properties of $\hat{\ell}_{;\hat{\theta}}(\theta)$ and $\hat{\ell}_{\theta;\hat{\theta}}(\theta)$ hold conditionally on a, for a in the normal deviation range of $E(a) + O(n^{-1/2})$.

Note that, for the case of identically distributed observations, $\hat{Q}(\theta; \theta_0)$ and $\hat{I}(\theta; \theta_0)$ may be viewed as simply bootstrap estimates of $Q(\theta; \theta_0)$ and $I(\theta; \theta_0)$, respectively. However, given the simple forms of $Q(\theta; \theta_0)$ and $I(\theta; \theta_0)$, no random sampling is needed.

As with any type of estimate based on resampling, there is some arbitrariness in how the estimate is formed, particularly when there exists a sufficient statistic of dimension smaller than n. For instance, let (Y_{j1}, Y_{j2}), $j = 1, \ldots, n$, denote independent pairs of independent normally distributed random variables each with mean $\mu_j(\theta)$ and variance σ^2, where $\mu_j(\theta)$ is a nonlinear function of θ. Then the data could also be represented as $Y_1, \ldots, Y_n$ where $Y_j = (Y_{j1} + Y_{j2})/2$ has mean μ_j and variance $\sigma^2/2$. The approximations $\hat{Q}(\theta; \theta_0)$ and $\hat{I}(\theta; \theta_0)$ may be based on either representation.

In these cases, it is typically preferable to use the representation of the data that leads to the greatest number of independent components. It should be noted, however, that this issue arises almost exclusively with exponential family models and, for those models, the other approximation methods are generally preferable.

Example 6.34 Exponential family model
Let $Y_1, \ldots, Y_n$ denote independent random variables each distributed according to a full exponential family density of the form

$$\exp\{y^T \eta(\theta) - k(\eta(\theta)) + D(y)\};$$

here both y and η are vectors. Then

$$\hat{I}(\theta; \hat{\theta}) = \eta'(\theta)^T \sum (y_j - k'(\theta))(y_j - k'(\hat{\theta}))^T \eta'(\hat{\theta}).$$

Note that

$$\ell_\theta(\hat{\theta}) = \eta'(\hat{\theta})^T \sum (y_j - k'(\hat{\theta})) = 0,$$

so that

$$\hat{I}(\theta; \hat{\theta}) = \eta'(\theta)^T \sum (y_j - k'(\hat{\theta}))(y_j - k'(\hat{\theta}))^T \eta'(\hat{\theta}) = \eta'(\theta)^T [\eta'(\hat{\theta})^T]^{-1} \hat{\imath}(\hat{\theta}).$$

Since $\hat{\jmath} = n\eta'(\hat{\theta})^T k''(\hat{\theta})\eta'(\hat{\theta})$, it follows that

$$\hat{\ell}_{\theta;\hat{\theta}}(\theta) = \eta'(\theta)^T [\eta'(\hat{\theta})^T]^{-1} \hat{\imath}(\hat{\theta}) \hat{\imath}(\hat{\theta})^{-1} n\eta'(\hat{\theta})^T k''(\hat{\theta})\eta'(\hat{\theta}) = n\eta'(\theta)^T k''(\hat{\theta})\eta'(\hat{\theta})$$

which is exactly $\ell_{\theta;\hat{\theta}}(\theta)$.

A similar result holds for $\hat{\ell}_{;\hat{\theta}}(\hat{\theta}) - \hat{\ell}_{;\hat{\theta}}(\theta)$. ∎

Example 6.35 Exponential hyperbola
Consider the exponential hyperbola model considered in Examples 6.21 and 6.33. In this model, $\ell^{(j)}(\theta)$ is either $-y_j/\theta - \log(\theta)$ or $-x_j\theta + \log(\theta)$. Hence,

$$\begin{aligned} \hat{Q}(\theta; \theta_0) = &- \sum (y_j/\theta + \log\theta)(y_j/\theta_0{}^2 - 1/\theta_0) \\ &+ \sum (x_j\theta - \log\theta)(x_j - 1/\theta_0). \end{aligned}$$

Let $S_y = \sum (y_j - \bar{y})^2$ and $S_x = \sum (x_j - \bar{x})^2$. We may write

$$\hat{Q}(\theta; \hat{\theta}) = na\left(\frac{1}{\theta} - \frac{\theta}{\hat{\theta}^2}\right) + \theta S_x + n\theta\bar{x}^2 - \frac{1}{\theta\hat{\theta}^2} S_y - \frac{n}{\theta\hat{\theta}^2}\bar{y}^2 - \frac{2n(a-1)}{\hat{\theta}} \log\theta.$$

Similarly,

$$\hat{\imath}(\hat{\theta}) = \frac{1}{\hat{\theta}^4} S_y + S_x + n(a-1)^2 \frac{2}{\hat{\theta}^2}.$$

Using these results, together with the fact that $\hat{\jmath} = 2na/\hat{\theta}^2$, it is straightforward to construct $\hat{\ell}_{;\hat{\theta}}(\hat{\theta}) - \hat{\ell}_{;\hat{\theta}}(\theta)$.

Write

$$S_y = n\hat{\theta}^2 + \sqrt{n} Z_y; \qquad S_x = \frac{n}{\hat{\theta}^2} + \sqrt{n} Z_x$$

where Z_y and Z_x are $O_p(1)$ random variables. Then

$$\hat{\ell}_{;\hat{\theta}}(\hat{\theta}) - \hat{\ell}_{;\hat{\theta}}(\theta) = na\left(\frac{1}{\theta} - \frac{\theta}{\hat{\theta}^2}\right)[1 + O_p(n^{-1})] + na\left[(a^2 - a)\left(\frac{1}{\theta} - \frac{\theta}{\hat{\theta}^2}\right)\right.$$

$$\left. - \frac{2\log(\hat{\theta}/\theta)}{\hat{\theta}}(a-1) + \frac{(\hat{\theta}-\theta)^2}{2\theta}(Z_y/\hat{\theta}^4 - Z_x)\frac{1}{\sqrt{n}}\right].$$

In this expression, the $O_p(n^{-1})$ term does not depend on θ. From this expression, it follows that, for fixed θ, the relative error of the approximation is $O(n^{-1/2})$ while for θ of the form $\theta = \hat{\theta} + O(n^{-1/2})$, the relative error is $O(n^{-1})$. Both of these results require that $a = 1 + O(n^{-1/2})$. ∎

6.8 Discussion and references

The approach to conditional inference taken in this chapter is based on the work of Barndorff-Nielsen (1980, 1983), Cox (1980), Davison (1988), Efron and Hinkley (1978), Hinkley (1980b), McCullagh (1984), and Peers (1978). General discussions of conditional inference are given in Barndorff-Nielsen (1988), Barndorff-Nielsen and Cox (1994), Cox (1988), and McCullagh (1987).

The likelihood ratio approximation to the conditional distribution of the maximum likelihood estimate was first given by Barndorff-Nielsen (1980). This result is an extension of the exact result for transformation models (Fisher, 1934; Fraser, 1968; Hinkley, 1978) and approximations based on the saddlepoint method (Barndorff-Nielsen and Cox, 1979; Daniels, 1954, 1980; Durbin, 1980). Further discussion of the likelihood ratio approximation and related results are given by Barndorff-Nielsen (1983, 1984, 1988), Fraser and Reid (1988), Hinkley (1980b), and Skovgaard (1990). The derivation given here is based on Durbin (1980); alternative derivations are given by Barndorff-Nielsen and Cox (1994, Chapter 7), Fraser and Reid (1988), McCullagh (1984; 1987, Chapter 8), and Skovgaard (1990). The properties of the approximation in the lattice case are considered by Severini (2000b). The tail probability approximation based on p^* is based on the work of Barndorff-Nielsen (1990a), Daniels (1987), Fraser (1990, 1991), and Lugannani and Rice (1980). Further details are given in Barndorff-Nielsen and Cox (1994), and Jensen (1995). The problem of integrating p^* to approximate the distribution of some component of $\hat{\theta}$ has been studied by Barndorff-Nielsen (1990c), DiCiccio, Field, and Fraser (1990), and DiCiccio and Martin (1991).

Conditional inference in a location or a location-scale model is considered by Efron and Hinkley (1978), Hinkley (1980b), and DiCiccio (1986). DiCiccio (1988) considers conditional inference in linear regression models. General results for conditional inference in transformation models are given by Fraser (1968, 1979). Conditional inference about a normal mean with known coefficient of variation is studied in Hinkley (1977).

The concept of stable combinations of cumulants and stable inference is due to McCullagh (1987, Chapter 8); see Barndorff-Nielsen and Blæsild (1993) for a more general treatment. The use of the observed information in a conditional variance approximation for the maximum likelihood estimate was advanced by Efron and Hinkley (1978); see also Severini (1989) and Skovgaard (1985). The result in Section 6.4.4 on the stability of the likelihood ratio statistic is due to McCullagh (1984); see also McCullagh (1987, Chapter 8). The two measuring instruments example, originally due to Cox (1958), is also studied by Efron and Hinkley (1978).

The approximation of the conditional model given in Section 6.5 is based on McCullagh (1987, Chapter 8) and Skovgaard (1990). Local ancillarity was first considered by Cox (1980) and the use of a locally ancillary statistic in the likelihood ratio approximation was considered by McCullagh (1987, Chapter 8). The local method of constructing an approximate ancillary, discussed in Section 6.6.2, is due to Cox (1980); see also Skovgaard (1985) and Ryall (1981). McCullagh (1984) and Skovgaard (1985) considered the relationship between local ancillarity and the ancillary statistic of Efron and Hinkley (1978); see also Barndorff-Nielsen and Cox (1994, Section 7.2). The use of a local location model in constructing an approximate ancillary is due to Fraser (1964); further properties of these approximate ancillaries and their use in conditional inference are given in Fraser and Reid (1995). Fraser and Reid (1995, 1999) and Fraser, Reid, and Wu (1999) considered the use of pivotal quantities in deriving an approximately ancillary statistic. The likelihood ratio method of constructing an approximate ancillary was proposed by Barndorff-Nielsen (1980); see also Barndorff-Nielsen (1983, 1984), Barndorff-Nielsen and Cox (1994, Section 7.2), Jensen (1992), and Pedersen (1981). The possibility of adjusting a given approximately ancillary statistic to obtain a higher-order approximate ancillary was considered by Skovgaard (1986).

The method of approximating sample space derivatives based on an approximately ancillary statistic, discussed in Section 6.7.2, is due to Fraser and Reid (1995, 1999) and Fraser, Reid, and Wu (1999). The covariance method of approximation was used by Skovgaard (1996) and Severini (1998a); see also Barndorff-Nielsen (1995). The empirical covariance method of approximation is based on Severini (1999b). A review and comparison of the various methods for approximating sample space derivatives is given in Severini (2000c).

6.9 Exercises

6.1 Consider the p^* formula for the conditional density of the maximum likelihood estimate of a vector parameter θ. Consider a reparameterization of the model by $\phi = g(\theta)$ where g is a continuously differentiable function. Show that the p^* formula for the conditional density of $\hat{\phi}$ is identical to the result obtained by transforming the p^* formula for the conditional density of $\hat{\theta}$.

6.2 Consider the likelihood ratio approximation to the conditional density of the maximum likelihood estimate of a vector parameter θ for the case in which the underlying data have a lattice distribution. Consider a reparameterization of the model by $\phi = g(\theta)$ where g is a continuously differentiable function. Find a formula for the conditional density of $\hat{\phi}$ in terms of the density of $\hat{\theta}$.

6.3 Let $\mathcal{X}$ denote a lattice subset of $\Re^d$ and let S denote a subset of $\Re^d$. Then, in a certain sense, the Lebesgue measure of S is approximately proportional to the counting measure of $S \cap \mathcal{X}$.
Consider the measure m described in Section 6.3.3 and let S denote a subset of $\Re^d$. Show that, in this same sense,

$$\int_S m(s)\, ds$$

is approximately proportional to the counting measure of $S \cap \hat{\Theta}$.

6.4 Let $Y_1, \ldots, Y_n$ denote independent exponential random variables such that Y_j has mean λ_j^{-1} where

$$\log \lambda_j = \psi x_j$$

and $x_1, \ldots, x_n$ are fixed constants. Let

$$A_j = \exp\{\hat{\psi} x_j\} Y_j, \quad j = 1, \ldots, n.$$

Show that $A = (A_1, \ldots, A_n)$ is an ancillary statistic and use this fact to derive the p^* formula for the conditional density of $\hat{\psi}$ given a. Find an approximation to the probability $\Pr(\hat{\psi} \geq t)$ for the case $\psi = 0$.

6.5 Let Y and X denote independent Poisson random variables such that Y has mean $\psi\lambda$ and X has mean $(1 - \psi)\lambda$ where $0 < \psi < 1$ is an unknown parameter and $\lambda > 0$ is known. Find the p^* formula for the conditional density of $\hat{\psi}$ given an appropriately defined ancillary statistic a. Compare this approximation to the exact result.

6.6 Let $Y_1, \ldots, Y_n$ denote independent random variables each with a Gamma distribution with density of the form

$$\frac{\beta^\alpha}{\Gamma(\alpha)} y^{\alpha-1} \exp\{-\beta y\}, \quad y > 0$$

where $\alpha > 0$ and $\beta > 0$ are unknown parameters. Find the p^* formula for the approximate distribution of $(\hat{\alpha}, \hat{\beta})$.

6.7 Consider the same model considered in Exercise 6.6. Using the approximation for the distribution of $(\hat{\alpha}, \hat{\beta})$ found in Exercise 6.6, derive an approximation for the marginal density of $\hat{\alpha}$.

6.8 Using Laplace's method, show that

$$\int |\hat{\jmath}|^{1/2} \exp\{\ell(\theta) - \ell(\hat{\theta})\}\, d\hat{\theta} = (2\pi)^{d/2} + O(n^{-1})$$

where d denotes the dimension of θ.

6.9 For the case of continuously distributed data, find an approximation to the conditional density of the score function Z that is analogous to the p^* approximation to the conditional density of the maximum likelihood estimate. Assume that, given a, Z is a one-to-one function of $\hat{\theta}$, which is generally true for $\hat{\theta}$ near θ with probability rapidly approaching 1.

6.10 Consider a model parameterized by a scalar parameter θ and let

$$Z_2 - \frac{\nu_{11}}{\nu_2} Z$$

denote the statistic derived in Section 6.6.2. Show that the variance of this statistic is given by $\gamma^2(\theta)i(\theta)/n$ where $\gamma(\theta)$ denotes the statistical curvature of the model.

6.11 Consider a one-parameter exponential family model. Find the statistical curvature of the model.

6.12 For a model with a scalar parameter θ, derive approximations, with error $O(n^{-1})$, to the conditional cumulants $\nu_{02}(a)$, $\nu_{11}(a)$ that are analogous to the approximations to $\nu_2(a)$ and $\nu_3(a)$ derived in Section 6.2. Use these results to derive an approximation to the conditional curvature of the model, given by

$$\frac{\nu_{02}(a) - \nu_{11}(a)^2/\nu_2(a)}{\nu_2(a)^2}.$$

6.13 Consider a time-dependent Poisson process with intensity function $\lambda(t;\theta)$ where λ is a known function and θ is an unknown d-dimensional parameter. Suppose that the process is observed until n events have occurred and let $y_1, \ldots, y_n$ denote the event times. Find pivotals $P_1(y;\theta), \ldots, P_m(y;\theta)$ that could be used to construct an approximately ancillary statistic.

6.14 Consider the normal distribution with known coefficient of variation model considered in Example 6.11. Find $\tilde{\ell}_{;\theta}(\hat{\theta}) - \tilde{\ell}_{;\hat{\theta}}(\theta)$ compare it to the exact result.

6.15 Let $Y_1, \ldots, Y_n$ denote independent Bernoulli random variables such that Y_j has expected value $\pi_j(\theta)$, $j = 1, \ldots, n$, where π_j is a known function and θ is an unknown d-dimensional parameter. Calculate $\bar{\ell}_{\theta;\hat{\theta}}(\theta)$.

6.16 Let $Y_1, \ldots, Y_n$ denote independent random variables such that Y_j has a one-parameter exponential family distribution with natural parameter θx_j where $x_1, \ldots, x_n$ are known scalar constants. Find $\hat{\ell}_{\theta;\hat{\theta}}(\theta)$ and compare it to $\ell_{\theta;\hat{\theta}}(\theta)$.

6.17 Let $Y_1, \ldots, Y_n$ denote independent random variables, each distributed according to a distribution with a scalar parameter θ. For the approximation method discussed in Section 6.7.2, calculate the matrix V using the approximately ancillary statistic based on a local location model near $\theta = \hat{\theta}$ and compare it to the matrix $\hat{V}$.

7
The signed likelihood ratio statistic

7.1 Introduction

Consider a statistical model parameterized by a scalar parameter θ. For inference about θ, one approach is to use the signed likelihood ratio statistic,

$$R \equiv R(\theta) = \text{sgn}(\hat{\theta} - \theta)\,[2(\ell(\hat{\theta}) - \ell(\theta))]^{1/2}$$

where $\ell(\theta)$ denotes the log-likelihood function and $\hat{\theta}$ denotes the maximum likelihood estimator of θ. A test of the hypothesis $\theta = \theta_0$ may be based on the statistic $R(\theta_0)$ and a confidence region for θ is given by

$$\{\theta \in \Theta\colon R(\theta) \leq k\}$$

where k is a given constant.

For a model parameterized by $\theta = (\psi, \lambda)$, where ψ is a scalar parameter of interest and λ is a nuisance parameter, the signed likelihood ratio statistic is given by

$$R \equiv R(\psi) = \text{sgn}(\hat{\psi} - \psi)\,[2(\ell(\hat{\theta}) - \ell(\hat{\theta}_\psi))]^{1/2}.$$

Here $\hat{\theta}_\psi$ denotes the maximum likelihood estimator of θ with ψ held fixed; this may also be written $\hat{\theta}_\psi = (\psi, \hat{\lambda}_\psi)$. As in the scalar parameter case, R may be used to construct a test of the hypothesis $\psi = \psi_0$ or it may be used to construct a confidence region for ψ.

In order to use R for inference, we must compute, either exactly or approximately, its distribution. It has been shown in Chapter 4 that, under the distribution with parameter ψ, R is asymptotically distributed according to a standard normal distribution. Hence, tests and confidence regions with first-order accuracy are easily constructed using a normal approximation to the distribution of R. In many cases, however, the accuracy of this normal approximation is questionable and, hence, the goal of this chapter is to consider the construction of a modified version of the signed likelihood ratio statistic, denoted by R^*, that has a standard normal distribution to a high order of approximation, while still retaining the essential character of R.

The basic approach is to first approximate the conditional distribution of R given an ancillary statistic a; let $F(\cdot|a)$ denote the approximate conditional distribution function. Then $F(R|a)$ is distributed approximately as a uniform random variable on the interval $(0, 1)$; it follows that $\Phi^{-1}(F(R|a))$ is approximately distributed according to a standard normal distribution. The accuracy of the standard normal approximation to $\Phi^{-1}(F(R|a))$ depends on the accuracy of F as an approximation to the distribution function of R. Since R is approximately normally distributed,

$$\Phi^{-1}(F(R|a)) = R + O_p(n^{-1/2}).$$

7.2 Normalizing transformations

We begin by deriving a normalizing transformation for a random variable that has an approximately normal distribution. Let X denote a continuous $O_p(1)$ scalar random variable with a density of the form

$$\bar{c}h(x/\sqrt{n})\phi(x) \tag{7.1}$$

where h is $O(1)$, $h(0) = 1$, and $\bar{c}$ is a constant satisfying $\bar{c} = 1 + O(n^{-1})$.

An approximation to the distribution function of X is given by

$$\bar{c}\int_{-\infty}^{t} h(x/\sqrt{n})\phi(x)\,dx = 1 - \bar{c}\int_{t/\sqrt{n}}^{\infty} h(s)\sqrt{n}\phi(\sqrt{n}s)\,ds.$$

This integral may be approximated using the technique described in Section 2.5, leading to the following approximation to the distribution function of X:

$$F_n(t) = \Phi(t) - \phi(t)\frac{h(t/\sqrt{n}) - 1}{t}.$$

This approximation has relative error of order $O(n^{-3/2})$ for fixed values of t.

Hence, the transformed random variable

$$F_n(X) = \Phi(X) - \phi(X)\frac{h(X/\sqrt{n}) - 1}{X}$$

is uniformly distributed to order $O(n^{-3/2})$; note that $F_n(X) - \Phi(X) = O_p(n^{-1/2})$. To obtain a random variable that is approximately distributed according to a standard normal random variable we need to make a further transformation to $\Phi^{-1}[F_n(X)]$.

Using a Taylor's series expansion, the function Φ^{-1} evaluated at $t + c/\sqrt{n}$ may be expanded

$$\begin{aligned}\Phi^{-1}(t + c/\sqrt{n}) &= \Phi^{-1}(t) + \frac{1}{\phi(\Phi^{-1}(t))}\frac{c}{\sqrt{n}} + \frac{1}{2}\frac{\Phi^{-1}(t)}{\phi(\Phi^{-1}(t))^2}\frac{c^2}{n} + O(n^{-3/2}) \\ &= \Phi^{-1}(t) - \frac{1}{\Phi^{-1}(t)}\log\left[1 - \frac{\Phi^{-1}(t)}{\phi(\Phi^{-1}(t))}\frac{c}{\sqrt{n}}\right] + O(n^{-3/2}).\end{aligned}$$

Taking $t = \Phi(X)$ and

$$c = -\sqrt{n}\phi(X)\frac{h(X/\sqrt{n}) - 1}{X},$$

it follows that a normalizing transformation for X is given by

$$X^* = X - \frac{\log h(X/\sqrt{n})}{X}.$$

The random variable X^* has a standard normal distribution to order $O(n^{-3/2})$.

This transformation applies to a continuous random variable. We now consider the case of a random variable Y such that nY is a lattice random variable, with the minimal lattice taken to be the set of nonnegative integers. Suppose that

$$\Pr(Y = y) = \frac{1}{n} f'(y)h(y)\sqrt{n}\phi(\sqrt{n}f(y))[1 + O(n^{-3/2})] \tag{7.2}$$

for some $O(1)$ nonnegative functions h and f, where $h(0) = 1$ and $f'(y) \geq 0$ for all y; here ny is a nonnegative integer which we may write $y = k/n$.

Then $\Pr(Y \geq y)$ may be approximated by

$$\sum_{j=k}^{\infty} \frac{1}{n} f'(j/n)h(j/n)\sqrt{n}\phi(\sqrt{n}f(j/n)).$$

Using the approach described in Section 2.5, this sum may be approximated by the integral

$$\int_{k/n}^{\infty} \sqrt{n}\phi(\sqrt{n}f(y))h(y)\frac{f'(y)^2 f(y)}{1 - \exp\{-f'(y)f(y)\}}\, dy.$$

Hence, the distribution of Y may be approximated by the distribution of a continuous random variable $\tilde{Y}$ with density function

$$\sqrt{n}\phi(\sqrt{n}f(y))h(y)\frac{f'(y)^2 f(y)}{1 - \exp\{-f'(y)f(y)\}}.$$

Let $X = \sqrt{n}f(\tilde{Y})$; then X has density function

$$\phi(x)h(g(x/\sqrt{n}))\frac{f'(g(x/\sqrt{n}))x/\sqrt{n}}{1 - \exp\{-f'(g(x/\sqrt{n}))x/\sqrt{n}\}}$$

where $g = f^{-1}$. This is of the form $\tilde{h}(x/\sqrt{n})\phi(x)$ with

$$\tilde{h}(t) = h(g(t))\frac{tf'(g(t))}{1 - \exp\{-tf'(g(t))\}}.$$

Hence, the random variable

$$X - \frac{\log \tilde{h}(X/\sqrt{n})}{X}$$

is approximately distributed according to a standard normal distribution. Writing this in terms of the random variable Y we have that

$$Y^* = \sqrt{n}f(Y) - \frac{\log[h(Y)f'(Y)f(Y)/(1 - \exp\{-f'(Y)f(Y)\})]}{\sqrt{n}f(Y)}$$

has a standard normal distribution to order $O(n^{-3/2})$.

Some care is needed in using this result since, if Y has a lattice distribution,

$$\Pr(Y \leq y) \neq 1 - \Pr(Y \geq y).$$

The result given above applies to upper tail probabilities, that is,

$$\Pr(Y^* \geq y) = 1 - \Phi(y) + O(n^{-3/2}).$$

7.3 One-parameter models

In this section we consider the construction of an adjusted signed likelihood ratio statistic for the case in which there is only a single scalar parameter θ. In this case, the signed likelihood ratio statistic is given by

$$R = \text{sgn}(\hat{\theta} - \theta)\,[2(\ell(\hat{\theta}) - \ell(\theta))]^{1/2}.$$

We begin by deriving an approximation to the density of R based on the p^* approximation to the conditional density of $\hat{\theta}$.

The p^* approximation is given by

$$c\hat{j}^{1/2}\exp\{\ell(\theta) - \ell(\hat{\theta})\}$$

where $\ell(\theta) \equiv \ell(\theta; \hat{\theta}, a)$ is viewed as a function of $\hat{\theta}$ with the ancillary statistic a held fixed. Here c is a normalizing constant that may depend on θ and a; the constant c may be written $c = \bar{c}/\sqrt{(2\pi)}$ where $\bar{c} = 1 + O(n^{-1})$. The signed likelihood ratio statistic is a function of $\hat{\theta}$; we will assume that this function is one to one, although the final result holds provided only that the transformation is one to one on a neighborhood containing the value of $\hat{\theta}$ under consideration.

Note that

$$\tfrac{1}{2}R^2 = \ell(\hat{\theta}) - \ell(\theta) \tag{7.3}$$

so that the density of R may be approximated by

$$c\hat{j}^{1/2}\exp\Big\{-\frac{1}{2}r^2\Big\}\Big|\frac{\partial\hat{\theta}}{\partial r}\Big|$$

where $\hat{j}$ is viewed as a function of r. Using (7.3) it follows that

$$r\frac{\partial r}{\partial\hat{\theta}} = \ell_{;\hat{\theta}}(\hat{\theta}) - \ell_{;\hat{\theta}}(\theta);$$

hence,

$$\Big|\frac{\partial\hat{\theta}}{\partial r}\Big| = \frac{|r|}{|\ell_{;\hat{\theta}}(\hat{\theta}) - \ell_{;\hat{\theta}}(\theta)|}.$$

An approximation to the density of R is therefore given by

$$\bar{c}\frac{\hat{j}^{1/2}|r|}{|\ell_{;\hat{\theta}}(\hat{\theta}) - \ell_{;\hat{\theta}}(\theta)|}\phi(r);$$

in this expression $\hat{\theta}$ is to be interpreted as a function of r and $\bar{c} = 1 + O(n^{-1})$.

This approximation is of the form (7.1) with

$$h(r/\sqrt{n}) = \frac{\bar{c}\hat{j}^{1/2}|r|}{|\ell_{;\hat{\theta}}(\hat{\theta}) - \ell_{;\hat{\theta}}(\theta)|}.$$

Since $r = 0$ corresponds to $\hat{\theta} = \theta$, it is straightforward to show that $h(0) = 1$. It follows that a normalized version of R is given by

$$R^* = R + \frac{1}{R}\log(U/R)$$

where

$$U = \hat{j}^{-1/2}(\ell_{;\hat{\theta}}(\hat{\theta}) - \ell_{;\hat{\theta}}(\theta)).$$

The sign of U is taken to be the same as that of R so that U/R is never negative.

We now consider the asymptotic properties of R^*. Consider the probability $\Pr(R^* \leq t; \theta)$ where t is either $O(1)$ or $O(\sqrt{n})$. Viewing $R^* \equiv R^*(\hat{\theta})$ as a function of $\hat{\theta}$, we may write

$$\Pr(R^* \leq t; \theta) = \Pr(\hat{\theta} \leq c; \theta)$$

where c satisfies $R^*(c) = t$. Using the result in Section 6.3.1, we have that $\Pr(\hat{\theta} \leq c; \theta)$ may be approximated by

$$\Phi(r_c) + \phi(r_c)\left[\frac{1}{r_c} - \frac{1}{u_c}\right]$$

where r_c and u_c are the values of R and U, respectively, corresponding to $\hat{\theta} = c$. The error of this approximation is $O(n^{-3/2})$ for $c = \theta + O(n^{-1/2})$, corresponding to $t = O(1)$, and is $O(n^{-1})$ for $c = O(1)$, corresponding to $t = O(\sqrt{n})$.

Note that, in either case,

$$\Phi(r_c) + \phi(r_c)\left[\frac{1}{r_c} - \frac{1}{u_c}\right] = \Phi(r_c + \log(u_c/r_c)/r_c) + \phi(r_c)O(n^{-3/2}).$$

To prove this, we consider the cases $c = O(1)$ and $c = \theta + O(n^{-1/2})$ separately.

First suppose that $c = \theta + O(n^{-1/2})$; in this case, r_c and u_c are both $O(1)$ and $u_c/r_c = 1 + O(n^{-1/2})$. Using a Taylor's series expansion,

$$\Phi(r_c + \log(u_c/r_c)/r_c) = \Phi(r_c) - \phi(r_c)\log(r_c/u_c)\frac{1}{r_c} + \frac{1}{2}\phi'(r_c)\left[\log(r_c/u_c)\frac{1}{r_c}\right]^2 + \cdots.$$

Since $\phi'(x) = -x\phi(x)$, and using the fact that $\log(r_c/u_c) = O(n^{-1/2})$, this may be written

$$\Phi(r_c + \log(u_c/r_c)/r_c) =$$
$$\Phi(r_c) - \phi(r_c)\left[\log(r_c/u_c)\frac{1}{r_c} + \frac{1}{2}\phi(r_c)\log(r_c/u_c)^2\frac{1}{r_c} + O(n^{-3/2})\right].$$

Expanding $\log(r_c/u_c)$ in a Taylor's series expansion yields the result

$$\Phi(r_c + \log(u_c/r_c)/r_c) = \Phi(r_c) - \phi(r_c)\left[\left(\frac{1}{u_c} - \frac{1}{r_c}\right) + O(n^{-3/2})\right].$$

Now consider the case in which $c = O(1)$. In this case, r_c and u_c are both $O(\sqrt{n})$ and the ratio r_c/u_c is $O(1)$. Let $r_c^* = r_c + \log(u_c/r_c)/r_c$. Using integration by parts, it may be shown that

$$\begin{aligned}
\Phi(r_c) - \Phi(r_c^*) &= \int_{r_c^*}^{r_c} \phi(s)\,ds \\
&= -\frac{1}{s}\phi(s)\bigg|_{r_c^*}^{r_c} - \int_{r_c^*}^{r_c} \frac{1}{s^2}\phi(s)\,ds \\
&= \frac{1}{r_c^*}\phi(r_c^*) - \frac{1}{r_c}\phi(r_c) + \phi(r_c)O(n^{-3/2}).
\end{aligned}$$

Note that

$$\begin{aligned}
\frac{1}{r_c^*}\phi(r_c^*) - \frac{1}{r_c}\phi(r_c) &= \left(\frac{1}{r_c^*} - \frac{1}{r^*}\right)\phi(r_c^*) + \frac{1}{r_c}[\phi(r_c^*) - \phi(r_c)] \\
&= -\frac{\log(u_c/r_c)}{r_c^* r_c^2}\phi(r_c^*) + \frac{1}{r_c}\phi(r_c) \\
&\quad \times \left[\frac{r_c}{u_c}\exp\{-\log(u_c/r_c)^2/(2r_c^2)\} - 1\right] \\
&= O(n^{-3/2})\phi(r_c^*) + \frac{1}{r_c}\phi(r_c)\left\{\frac{r_c}{u_c}[1 + O(n^{-1})] - 1\right\} \\
&= \phi(r_c)\left[\frac{1}{u_c} - \frac{1}{r_c} + O(n^{-3/2})\right].
\end{aligned}$$

Hence,

$$\Phi(r_c) - \Phi(r_c^*) = \phi(r_c)\left[\frac{1}{u_c} - \frac{1}{r_c} + O(n^{-3/2})\right],$$

proving the result.

It follows that

$$\Phi(r_c) + \phi(r_c)\left[\frac{1}{r_c} - \frac{1}{u_c}\right]$$

may be approximated by $\Phi(r_c^*)$. For $c = \theta + O(n^{-1/2})$, $r_c = O(1)$ so that $\Phi(r_c)$ and $\phi(r_c)$ are both $O(1)$ so that the relative error of this approximation is $O(n^{-3/2})$. For $c = O(1)$, $r_c = O(\surd n)$ so that

$$\frac{\phi(r_c)}{\Phi(r_c)} = O(\surd n);$$

it follows that, in this case, the relative error of the approximation is $O(n^{-1})$.

Returning to the determination of $\Pr(R^* \leq t; \theta)$, note that, with relative error $O(n^{-3/2})$ for $t = O(1)$ and relative error $O(n^{-1})$ for $t = O(\surd n)$,

$$\begin{aligned}\Pr(R^* \leq t; \theta) = \Pr(\hat{\theta} \leq c; \theta) &= \Phi(r_c) + \phi(r_c)\left[\frac{1}{r_c} - \frac{1}{u_c}\right] \\ &= \Phi(r_c^*) = \Phi(t)\end{aligned}$$

since, by definition, $r_c^* = t$. Hence, in the moderate-deviation case, R^* is asymptotically normally distributed with error $O(n^{-3/2})$, while in the large-deviation case R^* is asymptotically normally distributed with error $O(n^{-1})$.

When $\theta = \hat{\theta}$, both U and R are 0 and, hence, the R^* formula breaks down. One way to deal with this is to consider the limiting value of the adjustment term $\log(U/R)/R$ as $\theta \to \hat{\theta}$. In practice however R^* is typically used to compute p-values or set confidence limits. For these uses, values of θ near $\hat{\theta}$ are typically not of much interest; that is, if θ is near $\hat{\theta}$ the p-value is clearly large and values of θ near $\hat{\theta}$ will be within the confidence limits for any commonly used value of the coverage probability. Hence, here we take the approach of simply setting R^* equal to R when R is close to 0, say $|R| \leq 0.10$.

Example 7.1 Exponential distribution
Let $Y_1, \ldots, Y_n$ denote independent identically distributed random variables each distributed according to an exponential distribution with rate parameter θ. The log-likelihood function is given by

$$\ell(\theta) = n \log \theta - n\theta \bar{y}$$

and

$$\begin{aligned}R &= \operatorname{sgn}(\hat{\theta} - \theta)\, \surd 2[n \log \hat{\theta} - n - n \log \theta + n\theta \bar{y}]^{1/2} \\ &= \operatorname{sgn}(1 - \theta \bar{y})\, \surd(2n)[\theta \bar{y} - 1 - \log(\theta \bar{y})]^{1/2}.\end{aligned}$$

Since $\hat{\theta} = 1/\bar{y}$, the log-likelihood function may be written

$$\ell(\theta) = n \log \theta - n\theta/\hat{\theta}.$$

Table 7.1 Confidence limits in Example 7.1

Probability	Confidence limit R^*	Confidence limit Exact	Coverage prob. for R^*
0.010	0.00362	0.00362	0.0100
0.025	0.00447	0.00447	0.0250
0.050	0.00530	0.00530	0.0500
0.100	0.00639	0.00639	0.1001
0.900	0.0188	0.0188	0.9001
0.950	0.0213	0.0213	0.9501
0.975	0.0237	0.0237	0.9750
0.990	0.0266	0.0266	0.9900

It follows that $\hat{j} = n/\hat{\theta}^2$ and

$$\ell_{;\hat{\theta}}(\theta) = \frac{n\theta}{\hat{\theta}^2};$$

hence, the statistic U is given by

$$U = \left(\frac{n}{\hat{\theta}^2}\right)^{1/2}\left(\frac{n}{\hat{\theta}} - \frac{n\theta}{\hat{\theta}^2}\right) = -\sqrt{n}\left(\frac{\theta}{\hat{\theta}} - 1\right) = \sqrt{n}(1 - \theta\bar{y}).$$

Exact confidence limits for θ may be based on the fact that $n\theta\bar{y}$ has a standard gamma distribution with index n. For the data in Table A.1, aircraft 1, Table 7.1 contains upper confidence limits for θ based both on R^* and this exact method. Note that lower confidence limits and confidence intervals are easily obtained from these results; for instance, a 95% lower confidence limit based on R^* is 0.00530 and a 95% confidence interval based on R^* is (0.00447, 0.0237). Clearly there is close agreement between the limits based on R^* and the exact limits. Table 7.1 also contains the exact coverage probabilities of the confidence limits based on R^*. ∎

Example 7.2 Two measuring instruments
Consider the two measuring instruments example considered in Example 6.5. The log-likelihood function for this model is given by

$$\ell(\theta) = \left(\theta\hat{\theta} - \frac{1}{2}\theta^2\right)\left[(n-a)\frac{1}{\sigma_0^2} + a\frac{1}{\sigma_1^2}\right]$$

and the signed likelihood ratio statistic is given by

$$R = \left[(n-a)\frac{1}{\sigma_0^2} + a\frac{1}{\sigma_1^2}\right](\hat{\theta} - \theta).$$

It follows that

$$\ell_{;\hat{\theta}}(\theta) = \theta\left[(n-a)\frac{1}{\sigma_0^2} + a\frac{1}{\sigma_1^2}\right]$$

and hence that

$$U = \left[(n-a)\frac{1}{\sigma_0^2} + a\frac{1}{\sigma_1^2}\right](\hat{\theta} - \theta) = R.$$

Therefore, in this example, $R^* = R$. We have seen in Example 6.5 that here R is exactly distributed according to a standard normal distribution, conditionally on a. ∎

Example 7.3 Normal distribution with known coefficient of variation
Consider the model in which $Y_1, \ldots, Y_n$ are independent identically distributed random variables each distributed according to a normal distribution with mean θ and standard deviation θ; see Example 6.11. Then

$$\ell(\theta) = -\frac{n}{2\theta^2}\left[\frac{2a^2\hat{\theta}^2}{1+2a^2-(1+4a^2)^{1/2}} - \frac{4\theta\hat{\theta}}{(1+4a^2)^{1/2}-1}\right]$$

where $a = \sqrt{n}[\sum y_j^2]^{1/2}/\sum y_j$. It follows that

$$\ell_{;\hat{\theta}}(\hat{\theta}) - \ell_{;\hat{\theta}}(\theta) = \frac{n(\hat{\theta}-\theta)}{\theta\hat{\theta}q}\left[\frac{(\hat{\theta}+\theta)}{\theta q} - \frac{1}{a}\right]$$

and

$$\hat{j} = \frac{2n}{\hat{\theta}^2}\left[1 + \frac{1}{2a^2q^2}\right]$$

where

$$q = \frac{(1+4a^2)^{1/2}-1}{2a}.$$

It follows that

$$U = \sqrt{n}\left(\frac{\hat{\theta}}{\theta} - 1\right)\frac{a(\hat{\theta}/\theta + 1)/q - 1}{[1+2a^2q^2]^{1/2}}.$$ ∎

The above discussion applies to the case in which R has a continuous distribution. When the distribution of R is not continuous an alternative approach is needed. Note that the distribution of R will not generally be a lattice distribution. Hence, when applying the results for the lattice case derived in Section 7.2, it is important to keep track of the lattice variables. The derivation of R^* for the lattice case closely follows the derivation of p^* for the lattice case given in Section 6.3.3.

Suppose there exists a one-to-one function g from the sample space of $\hat{\theta}$ to the set $\{0, 1, \dots\}$, as discussed in Section 6.3.3. For simplicity, assume that g is an increasing function. Consider the random variable $Y = \hat{\phi}/n$, where $\hat{\phi} = g(\hat{\theta})$. The density of Y with respect to counting measure may be approximated by

$$\frac{\bar{c}}{\sqrt{(2\pi)}} |\hat{j}|^{1/2} \exp\{\ell(\theta) - \ell(\hat{\theta})\} \left| \frac{\partial \hat{\theta}}{\partial \hat{\phi}} \right|$$

where $\hat{j} = -\ell_{\theta\theta}(\hat{\theta})$ and $\hat{\theta} = g^{-1}(ny)$. This approximation is of the form (7.2) with $f(y) = r/\sqrt{n}$ and

$$h(y) = \frac{|\hat{j}|^{1/2} r}{|\ell_{;\hat{\theta}}(\hat{\theta}) - \ell_{;\hat{\theta}}(\theta)|}.$$

It follows from the results of Section 7.2 that the modified signed likelihood ratio statistic is given by

$$R^* = R + \frac{1}{R} \log(U_L/R),$$

where

$$U_L = \frac{1 - \exp\{-[\ell_{;\hat{\theta}}(\hat{\theta}) - \ell_{;\hat{\theta}}(\theta)]\, m(\hat{\theta})\}}{|\hat{j}|^{1/2}\, m(\hat{\theta})}$$

and

$$m(\hat{\theta}) = \left| \frac{\partial g(t)}{\partial t} \Big|_{t=\hat{\theta}} \right|^{-1},$$

as defined in Section 6.3.3. As noted in Section 7.2, this form of R^* is for the calculation of upper tail probabilities.

Example 7.4 Binomial distribution

Let $Y_1, \dots, Y_n$ denote independent identically distributed random variables such that $\Pr(Y_j = 1) = 1 - \Pr(Y_j = 0) = \theta$. The log-likelihood function for this model is given by

$$n\hat{\theta} \log[\theta/(1-\theta)] + n \log(1-\theta).$$

Since $\hat{\theta}$ takes values in the set $\{0, 1/n, 2/n, \dots\}$, the function g is given by $g(\hat{\theta}) = n\hat{\theta}$ and $m(\hat{\theta}) = 1/n$.

Then

$$R = \operatorname{sgn}(\hat{\theta} - \theta) \left[2n\hat{\theta} \log \frac{\hat{\theta}}{1-\hat{\theta}} + 2n \log \frac{1-\hat{\theta}}{1-\theta} \right]^{1/2}.$$

It is straightforward to show that

$$\ell_{\hat{\theta}}(\hat{\theta}) - \ell_{;\hat{\theta}}(\theta) = n \log \frac{\hat{\theta}(1-\theta)}{\theta(1-\hat{\theta})}$$

and

$$-\ell_{\theta\theta}(\hat{\theta}) = \frac{n}{\hat{\theta}(1-\hat{\theta})}.$$

Table 7.2 p-values in Example 7.4

	p-value		
$n\hat{\theta}$	R^*	Exact	Normal
10	0.152	0.151	0.151
11	0.0596	0.0592	0.0607
12	0.0178	0.0176	0.0194
13	0.00377	0.00369	0.00491
14	0.000514	0.000488	0.000973

It follows that

$$U_L = [n\hat{\theta}(1-\hat{\theta})]^{1/2}\left[1 - \frac{1-\hat{\theta}}{\hat{\theta}}\frac{\theta}{1-\theta}\right].$$

Consider a test of the hypothesis $\theta = 1/2$ versus the alternative $\theta > 1/2$. Table 7.2 contains the p-value of the test based on R^* together with the p-value of the exact test for $n = 15$ and the p-value based on a normal approximation with a continuity correction, for various choices for $\hat{\theta}$. Clearly, the test based on R^* is in close agreement with the exact test. It should be noted however that, for the reasons described in Section 6.3.3, R^* cannot be used for the case $\hat{\theta} = 1$ since in that case $U = 0$ while R is finite, leading to an obviously incorrect p-value of 1. ∎

7.4 Scalar parameter of interest in the presence of a nuisance parameter

7.4.1 INTRODUCTION

We now consider a model parameterized by a scalar parameter of interest ψ together with a nuisance parameter λ which may be a vector. In this case, the signed likelihood ratio statistic is given by

$$R \equiv R(\psi) = \text{sgn}(\hat{\psi} - \psi)\,[2(\ell(\hat{\theta}) - \ell(\hat{\theta}_\psi))]^{1/2}$$

where $\hat{\theta}_\psi = (\psi, \hat{\lambda}_\psi)$ denotes the maximum likelihood estimator of θ with ψ held fixed.

To obtain the modified signed likelihood ratio statistic R^* we take the following approach. The conditional density of R given a may be written

$$p(r|a;\theta) = \frac{p(r, \hat{\lambda}_\psi|a;\theta)}{p(\hat{\lambda}_\psi|r, a;\theta)}.$$

Hence, we need to approximate the densities on the right-hand side of this expression.

To approximate $p(r, \hat{\lambda}_\psi|a;\theta)$ we begin with the likelihood ratio approximation to the conditional distribution of $\hat{\theta}$ given a, $p^*(\hat{\theta}|a;\theta)$. This approximation is then transformed to obtain a p^*-approximation to the conditional density of $(R, \hat{\lambda}_\psi)$ given a.

For the model in which ψ is held fixed, $(R, \hat{\lambda}_\psi, a)$ is sufficient and, according to the results of Section 6.6.5, R is approximately ancillary. Hence, we may use $p^*(\hat{\lambda}_\psi | r, a; \theta)$ to approximate the conditional distribution of $\hat{\lambda}_\psi$ given r, a. These approximations may then be used to construct the adjusted statistic R^*, using the same general approach as in Section 7.3. As in the one-parameter case, here

$$\Pr(R^* \leq t; \theta) = \Phi(t)[1 + O(n^{-3/2})]$$

for fixed values of t and

$$\Pr(R^* \leq t; \theta) = \Phi(t)[1 + O(n^{-1})]$$

for $t = O(\sqrt{n})$.

It is important to note, however, that since when ψ is fixed R is only a second-order locally ancillary statistic, the argument given in this section shows only that R^* is normally distributed with error $O(n^{-1})$. In order to establish that R^* is normally distributed with error $O(n^{-3/2})$, it is necessary to use a statistic that is ancillary to a higher order of approximation when approximating the conditional distribution of $\hat{\lambda}_\psi$; the statistic R^* itself can be used in place of R in this context. A more complete derivation of R^* which addresses this issue is given in Section 6.6 of Barndorff-Nielsen and Cox (1994).

The likelihood ratio approximation for the conditional distribution of $\hat{\theta}$ given an ancillary a is given by

$$p^*(\hat{\theta} | a; \theta) = \bar{c} \frac{|\hat{j}|^{1/2}}{(2\pi)^{d/2}} \exp\{\ell(\theta) - \ell(\hat{\theta})\}.$$

This may be transformed to an approximation to the density of $(R, \hat{\lambda}_\psi)$:

$$p^*(r, \hat{\lambda}_\psi | a; \theta) = \bar{c} \frac{|\hat{j}|^{1/2}}{(2\pi)^{d/2}} \exp\{\ell(\theta) - \ell(\hat{\theta}_\psi)\} \exp\left\{ -\frac{1}{2} r^2 \right\} |J|$$

where $|J|$ denotes the Jacobian of the transformation. However instead of calculating $J = \partial\hat{\theta}/\partial(r, \hat{\lambda}_\psi)$, we will calculate $J^{-1} = \partial(r, \hat{\lambda}_\psi)/\partial\hat{\theta}$ and then use the fact that $|J^{-1}| = |J|^{-1}$.

To obtain J^{-1}, first note that

$$r \frac{\partial r}{\partial \hat{\psi}} = \ell_{;\hat{\psi}}(\hat{\theta}) - \ell_{;\hat{\psi}}(\hat{\theta}_\psi)$$

and

$$r \frac{\partial r}{\partial \hat{\lambda}} = \ell_{;\hat{\lambda}}(\hat{\theta}) - \ell_{;\hat{\lambda}}(\hat{\theta}_\psi).$$

To complete the specification of J^{-1}, we need $\partial\hat{\lambda}_\psi/\partial\hat{\lambda}$ and $\partial\hat{\lambda}_\psi/\partial\hat{\psi}$. These may be obtained from the likelihood equations determining $\hat{\lambda}_\psi$:

$$\ell_\lambda(\psi, \hat{\lambda}_\psi; \hat{\psi}, \hat{\lambda}, a) = 0.$$

Differentiating this expression with respect to $\hat{\psi}$ and $\hat{\lambda}$ shows that

$$\ell_{\lambda\lambda}(\hat{\theta}_\psi)\frac{\partial\hat{\lambda}_\psi}{\partial\hat{\psi}} + \ell_{\lambda;\hat{\psi}}(\hat{\theta}_\psi) = 0$$

and

$$\ell_{\lambda\lambda}(\hat{\theta}_\psi)\frac{\partial\hat{\lambda}_\psi}{\partial\hat{\lambda}} + \ell_{\lambda;\hat{\lambda}}(\hat{\theta}_\psi) = 0.$$

Hence,

$$\frac{\partial\hat{\lambda}_\psi}{\partial\hat{\psi}} = j_{\lambda\lambda}(\hat{\theta}_\psi)^{-1}\ell_{\lambda;\hat{\psi}}(\hat{\theta}_\psi)$$

and

$$\frac{\partial\hat{\lambda}_\psi}{\partial\hat{\lambda}} = j_{\lambda\lambda}(\hat{\theta}_\psi)^{-1}\ell_{\lambda;\hat{\lambda}}(\hat{\theta}_\psi).$$

It follows that

$$J^{-1} = \begin{pmatrix} \dfrac{\partial r}{\partial\hat{\psi}} & \dfrac{\partial r}{\partial\hat{\lambda}} \\ \dfrac{\partial\hat{\lambda}_\psi}{\partial\hat{\psi}} & \dfrac{\partial\hat{\lambda}_\psi}{\partial\hat{\lambda}} \end{pmatrix}$$

$$= \begin{pmatrix} r^{-1}[\ell_{;\hat{\psi}}(\hat{\theta}) - \ell_{;\hat{\psi}}(\hat{\theta}_\psi)] & r^{-1}[\ell_{;\hat{\lambda}}(\hat{\theta}) - \ell_{;\hat{\lambda}}(\hat{\theta}_\psi)] \\ j_{\lambda\lambda}(\hat{\theta}_\psi)^{-1}\ell_{\lambda;\hat{\psi}}(\hat{\theta}_\psi) & j_{\lambda\lambda}(\hat{\theta}_\psi)^{-1}\ell_{\lambda;\hat{\lambda}}(\hat{\theta}_\psi) \end{pmatrix}$$

which may be written

$$\begin{pmatrix} r & 0 \\ 0 & j_{\lambda\lambda}(\hat{\theta}_\psi) \end{pmatrix}^{-1} \begin{pmatrix} \ell_{;\hat{\psi}}(\hat{\theta}) - \ell_{;\hat{\psi}}(\hat{\theta}_\psi) & \ell_{;\hat{\lambda}}(\hat{\theta}) - \ell_{;\hat{\lambda}}(\hat{\theta}_\psi) \\ \ell_{\lambda;\hat{\psi}}(\hat{\theta}_\psi) & \ell_{\lambda;\hat{\lambda}}(\hat{\theta}_\psi) \end{pmatrix}.$$

Hence,

$$|J| = r|j_{\lambda\lambda}(\hat{\theta}_\psi)| \Bigg/ \begin{vmatrix} \ell_{;\hat{\psi}}(\hat{\theta}) - \ell_{;\hat{\psi}}(\hat{\theta}_\psi) & \ell_{;\hat{\lambda}}(\hat{\theta}) - \ell_{;\hat{\lambda}}(\hat{\theta}_\psi) \\ \ell_{\lambda;\hat{\psi}}(\hat{\theta}_\psi) & \ell_{\lambda;\hat{\lambda}}(\hat{\theta}_\psi) \end{vmatrix}$$

$$\equiv r|j_{\lambda\lambda}(\hat{\theta}_\psi)| \Bigg/ \begin{vmatrix} \ell_{;\hat{\theta}}(\hat{\theta}) - \ell_{;\hat{\theta}}(\hat{\theta}_\psi) \\ \ell_{\lambda;\hat{\theta}}(\hat{\theta}_\psi) \end{vmatrix}.$$

Let

$$u = \begin{vmatrix} \ell_{;\hat{\theta}}(\hat{\theta}) - \ell_{;\hat{\theta}}(\hat{\theta}_\psi) \\ \ell_{\lambda;\hat{\theta}}(\hat{\theta}_\psi) \end{vmatrix} \Bigg/ (|j_{\lambda\lambda}(\hat{\theta}_\psi)|^{1/2}|\hat{j}|^{1/2}),$$

then

$$p^*(r, \hat{\lambda}_\psi|a;\theta) = \bar{c}\frac{1}{(2\pi)^{d/2}}\frac{r|j_{\lambda\lambda}(\hat{\theta}_\lambda)|^{1/2}}{u}\exp\{\ell(\theta) - \ell(\hat{\theta}_\psi)\}\exp\left\{-\frac{1}{2}r^2\right\}.$$

Using the model with ψ fixed, we may use the likelihood ratio approximation to approximate the conditional density of $\hat{\lambda}_\psi$ given r, a. It straightforward to show that

$$p^*(\hat{\lambda}_\psi | r, a; \theta) = \bar{c}_\psi \frac{|j_{\lambda\lambda}(\hat{\theta}_\psi)|^{1/2}}{(2\pi)^{(d-1)/2}} \exp\{\ell(\theta) - \ell(\hat{\theta}_\psi)\}$$

where $\bar{c}_\psi = 1 + O(n^{-1})$; as noted above, however, this approximation has error of order $O(n^{-1})$, rather than the usual $O(n^{-3/2})$.

These approximations lead to the following approximation for the conditional density of r given a:

$$\begin{aligned} p^*(r|a;\theta) &= \frac{p^*(r, \hat{\lambda}_\psi | a; \theta)}{p^*(\hat{\lambda}_\psi | r, a; \theta)} \\ &= \frac{\bar{c}}{\bar{c}_\psi} \frac{1}{\surd(2\pi)} \frac{r}{u} \exp\left\{ -\frac{1}{2} r^2 \right\}; \end{aligned} \tag{7.4}$$

this approximation has error of order $O(n^{-1})$.

Note that (7.4) is of the form (7.1) with the function h taken to be $h(r) = \surd n r/u$ where u is viewed as a function of r. Hence, the adjusted signed likelihood ratio statistic is given by

$$R^* = R + \frac{1}{R} \log(U/R)$$

where U is given by

$$U = \left| \begin{matrix} \ell_{;\hat{\theta}}(\hat{\theta}) - \ell_{;\hat{\theta}}(\hat{\theta}_\psi) \\ \ell_{\lambda;\hat{\theta}}(\hat{\theta}_\psi) \end{matrix} \right| \Big/ (|j_{\lambda\lambda}(\hat{\theta}_\psi)|^{1/2} |\hat{j}|^{1/2}).$$

Using the formula for the determinant of a partitioned matrix, an alternative form for U is

$$U = \frac{|\ell_{\lambda;\hat{\lambda}}(\hat{\theta}_\psi)|}{|j_{\lambda\lambda}(\hat{\theta}_\psi)|^{1/2} |\hat{j}_{\lambda\lambda}|^{1/2}} \times \frac{|\ell_{;\hat{\psi}}(\hat{\theta}) - \ell_{;\hat{\psi}}(\hat{\theta}_\psi) - \ell_{\lambda;\hat{\psi}}(\hat{\theta}_\psi)^T \ell_{\lambda;\hat{\lambda}}(\hat{\theta}_\psi)^{-1} [\ell_{;\hat{\lambda}}(\hat{\theta}) - \ell_{;\hat{\lambda}}(\hat{\theta}_\psi)]|}{|\hat{j}_{\psi\psi} - \hat{j}_{\psi\lambda} \hat{j}_{\lambda\lambda}^{-1} \hat{j}_{\lambda\psi}|^{1/2}}.$$

As in the scalar parameter case, the sign of U is taken to be the same as that of R.

The asymptotic properties of R^* are the same as those in the scalar parameter case: in the moderate-deviation case, R^* is asymptotically normally distributed with error $O(n^{-3/2})$, while in the large-deviation case R^* is asymptotically normally distributed with error $O(n^{-1})$.

Example 7.5 Linear regression model

Let $Y_1, \ldots, Y_n$ denote independent random variables of the form $Y_j = \psi + x_j \beta + \sigma \epsilon_j$ where $x_1, \ldots, x_n$ are known covariate vectors of length p satisfying $\sum x_j = 0$, ψ

is an unknown scalar parameter, β is an unknown parameter vector of length p, and the ϵ_j are independent unobservable standard normal random variables. Hence, the nuisance parameter of the model is $\lambda = (\beta^T, \sigma)^T$. Assume that the matrix x with columns $x_1, \ldots, x_n$ is of rank p.

Let $y = (y_1, \ldots, y_n)^T$ and $e = (1, \ldots, 1)^T$ denote column vectors each of length n. Note that we may write

$$\begin{aligned}\sum (y_j - \psi - x_j\beta)^2 &= (y - \psi e - x\beta)^T (y - \psi e - x\beta) \\ &= (y - \hat{\psi} e - x\hat{\beta})^T (y - \hat{\psi} e - x\hat{\beta}) \\ &\quad + n(\hat{\psi} - \psi)^2 + (\hat{\beta} - \beta)^T x^T x(\hat{\beta} - \beta) \\ &= n\hat{\sigma}^2 + n(\hat{\psi} - \psi)^2 + (\hat{\beta} - \beta)^T x^T x(\hat{\beta} - \beta)\end{aligned}$$

where $\hat{\psi}$, $\hat{\beta}$, and $\hat{\sigma}$ are the maximum likelihood estimates of ψ, β, and σ, respectively. Hence,

$$\begin{aligned}&\ell(\psi, \beta, \sigma; \hat{\psi}, \hat{\beta}, \hat{\sigma}) \\ &\quad = -n \log \sigma - \frac{1}{2\sigma^2}[n\hat{\sigma}^2 + n(\hat{\psi} - \psi)^2 + (\hat{\beta} - \beta)^T x^T x(\hat{\beta} - \beta)].\end{aligned}$$

It follows that

$$R = \operatorname{sgn}(\hat{\psi} - \psi)\left[2n \log \frac{\hat{\sigma}_\psi}{\hat{\sigma}}\right]^{1/2}.$$

To calculate the statistic U we use the facts that

$$\ell_{;\hat{\beta}}(\psi, \beta, \sigma) = -\frac{1}{\sigma^2} x^T x(\hat{\beta} - \beta); \qquad \ell_{;\hat{\sigma}}(\psi, \beta, \sigma) = -n\frac{\hat{\sigma}}{\sigma^2};$$

$$\ell_{;\hat{\psi}}(\psi, \beta, \sigma) = -\frac{n}{\sigma^2}(\hat{\psi} - \psi); \qquad \ell_{\beta;\hat{\beta}}(\psi, \beta, \sigma) = \frac{1}{\sigma^2} x^T x;$$

$$\ell_{\beta;\hat{\sigma}}(\psi, \beta, \sigma) = 0; \qquad \ell_{\sigma;\hat{\beta}}(\psi, \beta, \sigma) = \frac{2}{\sigma^3} x^T x(\hat{\beta} - \beta);$$

$$\ell_{\sigma;\hat{\sigma}}(\psi, \beta, \sigma) = 2n\frac{\hat{\sigma}}{\sigma^3}; \qquad \ell_{\beta;\hat{\psi}}(\psi, \beta, \sigma) = 0;$$

$$\ell_{\sigma;\hat{\psi}}(\psi, \beta, \sigma) = \frac{2n(\hat{\psi} - \psi)}{\sigma^3}.$$

It follows that

$$\begin{vmatrix} \ell_{;\hat{\psi}}(\hat{\theta}) - \ell_{;\hat{\psi}}(\hat{\theta}_\psi) & \ell_{;\hat{\lambda}}(\hat{\theta}) - \ell_{;\hat{\lambda}}(\hat{\theta}_\psi) \\ \ell_{\lambda;\hat{\psi}}(\hat{\theta}_\psi) & \ell_{\lambda;\hat{\lambda}}(\hat{\theta}_\psi) \end{vmatrix} = \frac{2n^2\hat{\sigma}}{\hat{\sigma}_\psi^{2p+3}} |x^T x| \, (\hat{\psi} - \psi)$$

where $\hat{\sigma}_\psi$ denotes the maximum likelihood estimator of σ with ψ held fixed.

It is straightforward to show that

$$|\hat{j}| = \frac{2n^2}{\hat{\sigma}^{2p+4}} |x^T x|$$

and that

$$|j_{\lambda\lambda}(\hat{\theta}_\psi)| = \frac{2n}{\hat{\sigma}_\psi^{2p+2}} |x^T x|$$

so that

$$U = \frac{\hat{\sigma}^{p+1}}{\hat{\sigma}_\psi^{p+2}} \sqrt{n}(\hat{\psi} - \psi).$$

∎

Example 7.6 Comparison of exponential distributions
Let $Y_1, \ldots, Y_n; X_1, \ldots, X_m$ denote independent exponential random variables such that the distribution of each Y_j has rate parameter λ_1 and the distribution of each X_j has rate parameter λ_2. Let $\psi = \lambda_1/\lambda_2$ denote the parameter of interest and take $\lambda = \lambda_2$ as the nuisance parameter.

The log-likelihood function for this model may be written

$$\ell(\theta) = n \log \psi + (n+m) \log \lambda - \frac{\lambda}{\hat{\lambda}} \left(n \frac{\psi}{\hat{\psi}} + m \right);$$

it follows that

$$R = \text{sgn}(\hat{\psi} - \psi) \, [2n \log(\hat{\psi}/\psi) + 2(n+m) \log(n(\psi/\hat{\psi}) + m) - 2(n+m) \log(n+m)]^{1/2}.$$

It is straightforward to show that

$$\ell_{;\hat{\theta}}(\theta) = \left(n \frac{\lambda}{\hat{\lambda}} \frac{\psi}{\hat{\psi}^2} \quad \frac{\lambda}{\hat{\lambda}^2} \left(n \frac{\psi}{\hat{\psi}} + m \right) \right);$$

$$\ell_{\lambda;\hat{\theta}}(\theta) = \left(n \frac{1}{\hat{\lambda}} \frac{\psi}{\hat{\psi}^2} \quad \frac{1}{\hat{\lambda}^2} \left(n \frac{\psi}{\hat{\psi}} + m \right) \right);$$

$$\hat{j} = \begin{pmatrix} \dfrac{n}{\hat{\psi}^2} & \dfrac{n}{\hat{\psi}\hat{\lambda}} \\ \dfrac{n}{\hat{\psi}\hat{\lambda}} & \dfrac{n+m}{\hat{\lambda}^2} \end{pmatrix};$$

$$j_{\lambda\lambda}(\hat{\theta}_\psi) = \frac{1}{\hat{\psi}^2 \hat{\lambda}^2} \frac{(n\psi + m\hat{\psi})^2}{n+m}.$$

It follows that

$$U = \sqrt{(nm)} \sqrt{(n+m)} \frac{\hat{\psi} - \psi}{n\psi + m\hat{\psi}}.$$

Table 7.3 Confidence limits in Example 7.6

Probability	R	R^*	Exact
0.010	0.131	0.128	0.128
0.025	0.156	0.153	0.153
0.050	0.182	0.178	0.178
0.100	0.215	0.211	0.211
0.900	0.671	0.669	0.669
0.950	0.787	0.786	0.786
0.975	0.904	0.905	0.905
0.990	1.065	1.068	1.068

Consider the data in data set A.6 of the Appendix. These data refer to the failure times, in mile–years, of two electric power transmission circuits in New Mexico. Of interest is a comparison of the failure rates of the two circuits. Table 7.3 gives upper confidence limits for ψ based on R and R^*, along with exact confidence limits based on the fact that $\psi \bar{Y}/\bar{X}$ has an F-distribution. All three sets of confidence limits are very close, with the limits based on R^* essentially equal to the exact limits. ∎

Example 7.7 Index of a Weibull regression model
Let $Y_1, \ldots, Y_n$ denote independent Weibull random variables such that Y_j has rate parameter $\exp\{x_j^T \lambda\}$ and index ψ, where x_j denotes a known vector of covariates and λ is an unknown parameter vector of dimension p. The density of Y_j is therefore given by

$$\exp\{x_j^T \lambda\} \psi y^{\psi - 1} \exp\{-\exp\{x_j^T \lambda\} y^{\psi}\}, \quad y > 0.$$

Each observation Y_j may be written

$$\log Y_j = \frac{1}{\psi}[Z_j - x_j^T \lambda],$$

where Z_j has a known distribution. Hence, this is a transformation model and there exists an exactly ancillary statistic $A = (A_1, \ldots, A_n)$ where

$$A_j = \exp\{x_j^T \hat{\lambda}\} Y_j^{\hat{\psi}}$$

and $(\hat{\psi}, \hat{\lambda})$ denotes the maximum likelihood estimate of (ψ, λ).

The log-likelihood function for the model may be written

$$\ell(\psi, \lambda) = \sum x_j^T \lambda + n \log \psi + \frac{\psi}{\hat{\psi}} \sum \log a_j - \frac{\psi}{\hat{\psi}} \sum x_j^T \hat{\lambda}$$
$$- \sum \exp\left\{x_j^T \lambda + \frac{\psi}{\hat{\psi}}(\log a_j - x_j^T \hat{\lambda})\right\}.$$

It follows that

$$\ell_{;\hat{\psi}}(\psi, \lambda) = \frac{\psi}{\hat{\psi}} \sum (\log y_j)(y_j^{\psi} \exp\{x_j^T \lambda\} - 1);$$

$$\ell_{;\hat{\lambda}}(\psi, \lambda) = \frac{\psi}{\hat{\psi}} \sum (y_j \exp\{x_j^T \lambda\} - 1) x_j^T;$$

$$\ell_{\lambda;\hat{\psi}}(\psi, \lambda) = \frac{\psi}{\hat{\psi}} \sum y_j^{\psi} \exp\{x_j^T \lambda\} \log(y_j) x_j;$$

$$\ell_{\lambda;\hat{\lambda}}(\psi, \lambda) = \frac{\psi}{\hat{\psi}} \sum y_j^{\psi} \exp\{x_j^T \lambda\} x_j x_j^T.$$

Also,

$$j_{\lambda\lambda}(\hat{\theta}_\psi) = \sum y_j \psi \exp\{x_j^T \hat{\lambda}_\psi\} x_j x_j^T$$

and

$$\hat{j} = \begin{pmatrix} \sum y_j^{\hat{\psi}} (\log y_j)^2 \exp\{x_j^T \hat{\lambda}\} + n/\hat{\psi}^2 & \sum y_j^{\hat{\psi}} \log(y_j) \exp\{x_j^T \hat{\lambda}\} x_j \\ \sum y_j^{\hat{\psi}} \log(y_j) \exp\{x_j^T \hat{\lambda}\} x_j^T & \sum y_j^{\hat{\psi}} \exp\{x_j^T \hat{\lambda}\} x_j x_j^T \end{pmatrix}.$$

Using these results, it is a simple matter to calculate R^* for a given set of data, although the algebraic expression for R^* is quite complicated.

Consider the data in Table A.7 of the Appendix. In these data, Y_j represents the survival time (from diagnosis) of a leukaemia patient and $x_j = (1, w_j)$ where w_j represents the log of the white blood cell count of the patient. Table 7.4 contains confidence limits for ψ based on R and R^*. Table 7.4 also contains the results of a small simulation study performed to evaluate the frequency properties of confidence limits based on R and R^*. Random variates were drawn from Weibull distributions with parameters $\hat{\lambda}_j$, $j = 1, \ldots, 17$, and index ψ where $\hat{\lambda}_j$ is the maximum likelihood estimate of λ_j based on the data described above. A Monte Carlo sample size of

Table 7.4 Confidence limits in Example 7.7

	Confidence limit		Coverage probability	
Probability	R	R^*	R	R^*
0.010	0.610	0.532	0.004	0.010
0.025	0.667	0.585	0.009	0.025
0.050	0.717	0.634	0.019	0.050
0.100	0.779	0.692	0.042	0.101
0.900	1.308	1.119	0.779	0.899
0.950	1.396	1.285	0.870	0.947
0.975	1.478	1.363	0.924	0.973
0.990	1.576	1.457	0.962	0.989

10 000 was used. Several values of ψ were considered; since the results are essentially independent of ψ however only the values for $\psi = 1$ are reported here.

These results indicate that, for this set of data, confidence limits based on R^* differ considerably from those based on R. Also the simulation study suggests that frequency properties of limits based on R^* are much better than those of the limits based on R. ∎

Example 7.8 Ratio of normal means
Let (X_j, Y_j), $j = 1, \ldots, n$, denote independent identically distributed bivariate normal random vectors with mean $(\mu, \psi\mu)$ and covariance matrix

$$\Sigma = \begin{pmatrix} \sigma_1^2 & \rho\sigma_1\sigma_2 \\ \rho\sigma_1\sigma_2 & \sigma_2^2 \end{pmatrix}.$$

The parameter of interest ψ is therefore the ratio of the mean of Y_j to the mean of X_j with $(\mu, \sigma_1, \sigma_2, \rho)$ as the nuisance parameter. The log-likelihood function for this model is given by

$$\ell(\theta) = -\frac{n}{2}\log(\sigma_1^2\sigma_2^2(1-\rho^2)) \\ -\frac{1}{2}\frac{1}{1-\rho^2}\sum\left[\frac{(x_j-\mu)^2}{\sigma_1^2} + \frac{(y_j-\psi\mu)^2}{\sigma_2^2} - 2\rho\frac{(x_j-\mu)(y_j-\psi\mu)}{\sigma_1\sigma_2}\right];$$

in terms of the maximum likelihood estimates we may write

$$\ell(\theta) = -\frac{n}{2}\log(\sigma_1^2\sigma_2^2(1-\rho^2)) - \frac{n}{2}\left[\frac{\hat\sigma_1^2}{\sigma_1^2} + \frac{\hat\sigma_2^2}{\sigma_2^2} + \frac{(\hat\mu-\mu)^2}{\sigma_1^2} + \frac{(\hat\psi\hat\mu-\psi\mu)^2}{\sigma_2^2} \\ - 2\rho\hat\rho\frac{\hat\sigma_1\hat\sigma_2}{\sigma_1\sigma_2} - 2\rho\frac{(\hat\mu-\mu)(\hat\psi\hat\mu-\psi\mu)}{\sigma_1\sigma_2}\right].$$

Using this expression, it is straightforward, but tedious, to calculate R^*.

Consider the data in Table A.8 of the Appendix. These data represent the results of a small bioequivalence study. In that study, each of $n = 8$ subjects received three treatments, a placebo, an 'approved' treatment, and a 'new' treatment. The response is the blood level of a particular hormone. Let X_j denote the difference between the approved and placebo results and let Y_j denote the difference between the new and approved results. Of interest is a confidence interval for ψ, which should contain the value 0 if the approved and new treatments are 'equivalent'. Note that for this example, the effect of the nuisance parameters is potentially important, as there are 5 parameters and only 8 observations.

Exact confidence limits for ψ may be based on Fieller's method which uses the fact that

$$\frac{\sqrt{n}(\bar y - \psi\bar x)}{[\psi\hat\Sigma_{11} - 2\psi\hat\Sigma_{12} + \hat\Sigma_{22}]^{1/2}}$$

has a rescaled t-distribution with $n-1$ degrees of freedom. Table 7.5 gives confidence limits for ψ based on R, R^* and Fieller's method. There is close agreement between the confidence limits based on R^* and those based on Fieller's method.

Table 7.5 Confidence limits in Example 7.8

Probability	R	R^*	Fieller
0.010	−0.310	−0.343	−0.345
0.025	−0.265	−0.289	−0.290
0.050	−0.230	−0.248	−0.249
0.100	−0.193	−0.206	−0.207
0.900	0.077	0.097	0.098
0.950	0.136	0.169	0.170
0.975	0.201	0.250	0.252
0.990	0.300	0.386	0.390

∎

7.4.2 INFERENCE WITHOUT AN EXPLICIT NUISANCE PARAMETER

The expression for R^* given in this section is based on a parameterization of the model of the form $\theta = (\psi, \lambda)$ where ψ is the parameter of interest and λ is a nuisance parameter. We now show that it is possible to calculate R^* without choosing an explicit form for the nuisance parameter. That is, consider a model with parameter θ and let $\psi \equiv \psi(\theta)$ denote the real-valued parameter of interest. We may calculate R^* based on this structure alone, without explicitly selecting a form for the nuisance parameter λ. This is entirely a matter of convenience; the resulting form for R^* is identical to the one that would be obtained by choosing an explicit form for λ. Hence, this result also serves to verify that R^* is invariant under interest-respecting reparameterizations.

As noted in Section 4.3, calculation of R itself does not require an explicit form for the nuisance parameter. Hence, in the expression for R^*, an explicit form for λ was only used in calculation of U, specifically in the calculation of the derivatives $\ell_{\lambda;\hat{\theta}}(\theta)$, and $\ell_{\lambda\lambda}(\theta)$.

Let $f(\theta)$ be an arbitrary function of θ and consider calculation of $f_\lambda(\theta) = \partial f(\theta)/\partial\lambda$. By the chain rule,

$$f_\lambda(\theta) = f_\theta(\theta)\frac{\partial\theta}{\partial\lambda};$$

hence, it is sufficient to obtain an expression for $\partial\theta/\partial\lambda$. Note that

$$\begin{pmatrix} \dfrac{\partial\theta}{\partial\psi} & \dfrac{\partial\theta}{\partial\lambda} \end{pmatrix} = \begin{pmatrix} \dfrac{\partial\psi}{\partial\theta} \\ \dfrac{\partial\lambda}{\partial\theta} \end{pmatrix}^{-1}$$

and, hence, $\partial\theta/\partial\lambda$ satisfies

$$\frac{\partial\lambda}{\partial\theta}\frac{\partial\theta}{\partial\lambda} = D; \qquad \frac{\partial\psi}{\partial\theta}\frac{\partial\theta}{\partial\lambda} = 0;$$

here D represents the identity matrix with rank $d - 1$ where $d = \dim(\theta)$.

Let $B \equiv B(\theta)$ denote a $d \times (d-1)$ matrix such that for all θ,

$$\frac{\partial \psi}{\partial \theta} B = 0$$

and

$$\frac{\partial \lambda}{\partial \theta} B$$

is nonsingular. Then

$$\frac{\partial \theta}{\partial \lambda} = B\Big(\frac{\partial \lambda}{\partial \theta} B\Big)^{-1}.$$

Hence,

$$\begin{pmatrix} \ell_{;\hat{\theta}}(\hat{\theta}) - \ell_{;\hat{\theta}}(\hat{\theta}_\psi) \\ \ell_{\lambda;\hat{\theta}}(\hat{\theta}_\psi) \end{pmatrix} = \begin{pmatrix} D & 0 \\ 0 & \Big(\frac{\partial \lambda}{\partial \theta}(\hat{\theta}_\psi) B(\hat{\theta}_\psi)\Big)^{-1} \end{pmatrix}^T \begin{pmatrix} \ell_{;\hat{\theta}}(\hat{\theta}) - \ell_{;\hat{\theta}}(\hat{\theta}_\psi) \\ B(\hat{\theta}_\psi)^T \ell_{\theta;\hat{\theta}}(\hat{\theta}_\psi) \end{pmatrix}$$

and

$$\ell_{\lambda\lambda}(\hat{\theta}_\psi) = \Big[\Big(\frac{\partial \lambda}{\partial \theta}(\hat{\theta}_\psi) B(\hat{\theta}_\psi)\Big)^{-1}\Big]^T B(\hat{\theta}_\psi)^T \ell_{\theta\theta}(\hat{\theta}_\psi) B(\hat{\theta}_\psi) \Big(\frac{\partial \lambda}{\partial \theta}(\hat{\theta}_\psi) B(\hat{\theta}_\psi)\Big)^{-1}.$$

It follows that

$$U = \frac{\begin{vmatrix} \ell_{;\hat{\theta}}(\hat{\theta}) - \ell_{;\hat{\theta}}(\hat{\theta}_\psi) \\ \ell_{\lambda;\hat{\theta}}(\hat{\theta}_\psi) \end{vmatrix}}{|j_{\lambda\lambda}(\hat{\theta}_\psi)|^{1/2} |\hat{j}|^{1/2}} = \frac{\begin{vmatrix} \ell_{;\hat{\theta}}(\hat{\theta}) - \ell_{;\hat{\theta}}(\hat{\theta}_\psi) \\ B(\hat{\theta}_\psi)^T \ell_{\theta;\hat{\theta}}(\hat{\theta}_\psi) \end{vmatrix}}{|B(\hat{\theta}_\psi)^T j(\hat{\theta}_\psi) B(\hat{\theta}_\psi)|^{1/2} |\hat{j}|^{1/2}}.$$

In this expression, $B(\theta)$ is any matrix such that

$$\frac{\partial \psi}{\partial \theta} B = 0 \quad \text{and} \quad \Big|\frac{\partial \lambda}{\partial \theta} B\Big| \neq 0$$

for some nuisance parameter λ.

For a given set of data and a given value of ψ, $B(\hat{\theta}_\psi)$ may be calculated numerically. For instance, let e_j denote the $d \times 1$ vector consisting of the jth column of the identity matrix of rank d and let

$$\hat{e}_j = e_j - \frac{(\partial \psi/\partial \theta)(\hat{\theta}_\psi) e_j}{(\partial \psi/\partial \theta)(\hat{\theta}_\psi)\,(\partial \psi/\partial \theta)(\hat{\theta}_\psi)^T} \frac{\partial \psi}{\partial \theta}(\hat{\theta}_\psi)^T.$$

Then

$$\frac{\partial \psi}{\partial \theta}(\hat{\theta}_\psi) \hat{e}_j = 0.$$

The set $\{\hat{e}_1, \ldots, \hat{e}_d\}$ has only $d-1$ linearly independent elements. Let $\hat{e}_{i_1}, \ldots, \hat{e}_{i_{d-1}}$ denote a set of $d-1$ linearly independent vectors and let $\hat{B}$ be the matrix with the jth column taken to be $\hat{e}_{i_j}$. Clearly,

$$\frac{\partial \psi}{\partial \theta}(\hat{\theta}_\psi) \hat{B} = 0.$$

Consider a nuisance parameter $\lambda = (\lambda_1, \ldots, \lambda_{d-1})$ such that

$$\frac{\partial \lambda_j}{\partial \theta}(\hat{\theta}_\psi) = \hat{e}_{i_j}^T.$$

Then, for this choice of λ,

$$\frac{\partial \lambda}{\partial \theta}(\hat{\theta}_\psi)\hat{B} = \hat{B}^T \hat{B}$$

which is nonsingular. It follows that for calculating R_ψ^*, we may use $B = \hat{B}$.

Example 7.9 Normal-theory linear model with known variance
Let $Y_1, \ldots, Y_n$ denote independent normally distributed random variables such that Y_j has mean $x_j\theta$ where $x_1, \ldots, x_n$ are known $1 \times d$ vectors and θ is an unknown parameter vector. Assume that the variance of Y_j is known; for simplicity, we take the variance to be 1. For numerical work, the case in which the variance is an unknown parameter can be handled using the same general approach; the algebraic expression for R^* however is much more complicated.

The log-likelihood function is given by

$$\ell(\theta) = -\tfrac{1}{2}(y - x\theta)^T(y - x\theta)$$

where $y = (y_1, \ldots, y_n)^T$ and x denotes the $n \times d$ matrix with jth row x_j. Note that $\ell(\theta)$ may be written

$$\ell(\theta) = -\tfrac{1}{2}(\hat{\theta} - \theta)^T x^T x(\hat{\theta} - \theta).$$

Let $\psi(\theta) = c^T\theta$ denote the parameter of interest, where c is a known vector.

It is straightforward to show that $\ell_{\theta\theta}(\theta) = -x^T x$,

$$\hat{\theta}_\psi = \hat{\theta} + \frac{(x^T x)^{-1}c}{c^T(x^T x)^{-1}c}(\psi - \hat{\psi});$$

$$\ell_{;\hat{\theta}}(\theta) = -(x^T x)(\hat{\theta} - \theta); \qquad \ell_{\theta;\hat{\theta}}(\theta) = x^T x.$$

Hence,

$$U = |x^T x|^{1/2} \begin{vmatrix} (\hat{\theta} - \hat{\theta}_\psi)^T \\ B(\hat{\theta}_\psi)^T \end{vmatrix} \Big/ |B(\hat{\theta}_\psi)^T x^T x B(\hat{\theta}_\psi)|$$

where $B(\theta)$ satisfies $c^T B(\theta) = 0$.

Note that

$$|x^T x|^{1/2} \begin{vmatrix} (\hat{\theta} - \hat{\theta}_\psi)^T \\ B(\hat{\theta}_\psi)^T \end{vmatrix} = \left| \begin{pmatrix} (\hat{\theta} - \hat{\theta}_\psi)^T \\ B(\hat{\theta}_\psi)^T \end{pmatrix} x^T x \begin{pmatrix} (\hat{\theta} - \hat{\theta}_\psi)^T \\ B(\hat{\theta}_\psi)^T \end{pmatrix}^T \right|^{1/2}$$
$$= |B(\hat{\theta}_\psi)^T x^T x B(\hat{\theta}_\psi)|^{1/2} |(\hat{\theta} - \hat{\theta}_\psi)^T x^T x(\hat{\theta} - \hat{\theta}_\psi)$$
$$- M^T[B(\hat{\theta}_\psi)^T x^T x B(\hat{\theta}_\psi)]^{-1} M|^{1/2}$$

where

$$M = B(\hat{\theta}_\psi)^T (x^T x)(\hat{\theta} - \hat{\theta}_\psi).$$

Using the expression for $\hat{\theta}_\psi$ given above, along with the fact that $c^T B(\theta) = 0$ for all θ, it follows that

$$B(\theta)^T (x^T x)(\hat{\theta} - \hat{\theta}_\psi) = 0$$

and hence that

$$U = \frac{\hat{\psi} - \psi}{c^T x^T x c}.$$

It is straightforward to show that $R = U$ so that $R^* = R$. This expected since R itself is exactly distributed according to a standard normal distribution. ∎

Example 7.10 Comparison of exponential distributions

Consider the comparison of exponential distributions example considered in Example 7.6. For this model, $Y_1, \ldots, Y_n; X_1, \ldots, X_m$ are independent exponential random variables such that the distribution of each Y_j has rate parameter λ_1 and the distribution of each X_j has rate parameter λ_2. Here we take $\theta_1 = \lambda_1$ and $\theta_2 = \lambda_2$; the parameter of interest is given by $\psi(\theta) = \theta_1/\theta_2$.

The log-likelihood function for this model is given by

$$\ell(\theta) = n \log \theta_1 + m \log \theta_2 - n\theta_1/\hat{\theta}_1 - m\theta_2/\hat{\theta}_2.$$

It is straightforward to show that

$$\hat{\theta}_\psi = \left(\frac{n+m}{n + m\hat{\psi}/\psi} \hat{\theta}_1, \frac{n+m}{m + n\psi/\hat{\psi}} \hat{\theta}_2 \right); \qquad \ell_{;\hat{\theta}}(\theta) = (n\theta_1/\hat{\theta}_1^2 \quad m\theta_2/\hat{\theta}_2^2);$$

$$\ell_{\theta;\hat{\theta}}(\theta) = \begin{pmatrix} n/\hat{\theta}_1^2 & 0 \\ 0 & m/\hat{\theta}_2^2 \end{pmatrix}; \qquad -\ell_{\theta\theta}(\theta) = \begin{pmatrix} n/\theta_1^2 & 0 \\ 0 & m/\theta_2^2 \end{pmatrix}.$$

Let $B(\theta) = (b_1, b_2)^T$. A tedious, but elementary, calculation shows that

$$U = \frac{\sqrt{(nm)(n+m)}(\hat{\psi} - \psi)\psi}{(n\psi + m\hat{\psi})^2} \frac{m\hat{\psi} b_2 + n b_1}{[m\psi^2 b_2^2 + n b_1^2]^{1/2}}. \tag{7.5}$$

To determine b_1, b_2, first note that

$$\frac{\partial \psi}{\partial \theta}(\theta) = (1/\theta_2 \quad -\theta_1/\theta_2^2).$$

Let $e_1 = (1, 0)^T$ and $e_2 = (0, 1)^T$. Then

$$\hat{e}_1 = \begin{pmatrix} \dfrac{\psi^2}{1+\psi^2} \\ \dfrac{\psi}{1+\psi^2} \end{pmatrix}; \qquad \hat{e}_2 = \begin{pmatrix} \dfrac{\psi}{1+\psi^2} \\ \dfrac{1}{1+\psi^2} \end{pmatrix}.$$

We may take B to be either $\hat{e}_1$ or $\hat{e}_2$. It is straightforward to show that either choice leads to

$$U = \surd(nm)\surd(n+m)\frac{\hat{\psi} - \psi}{n\psi + m\hat{\psi}}, \tag{7.6}$$

the same result obtained in Example 7.6 using $\lambda = \theta_2$ as the nuisance parameter.

Alternatively, we could note that the condition that

$$\frac{\partial \psi}{\partial \theta}(\theta) B = 0$$

implies that $b_1 = \psi b_2$. Substituting ψb_2 for b_1 in (7.5) also yields (7.6). ∎

7.5 Approximations to R^*

7.5.1 INTRODUCTION

Although R^* is, in general, an improvement over the unadjusted likelihood ratio statistic, it has the drawback that the statistic U required for its calculation is often difficult to determine. Computation of U requires determination of the sample space derivatives $\ell_{\theta;\hat{\theta}}(\theta)$ and $\ell_{;\hat{\theta}}(\theta)$. As discussed in Section 6.7, exact computation of these derivatives generally requires that we be able to write the sufficient statistic of the model as $(\hat{\theta}, a)$ where a denotes an ancillary statistic. Hence, exact computation is, with few exceptions, limited to exponential family models and transformation models.

In this section, we consider approximations to R^*. Several different methods of approximation are considered. In Section 7.5.2, we consider a method based on a nuisance parameter that is orthogonal to ψ. The methods described in Sections 7.5.3–7.5.5 correspond to the approximations to sample space derivatives presented in Sections 6.7.2–6.7.4, respectively.

7.5.2 AN APPROXIMATION BASED ON ORTHOGONAL PARAMETERS

The first approach to approximating R^* is based on the use of orthogonal parameters. Suppose that ψ and λ are orthogonal. Recall that this implies that

$$\frac{\partial}{\partial \psi}\hat{\lambda}_\psi \bigg|_{\psi=\hat{\psi}} = O_p(n^{-1/2})$$

so that for $\psi = \hat{\psi} + O(n^{-1/2})$,

$$\hat{\lambda}_\psi = \hat{\lambda} + O_p(n^{-1}).$$

It follows that

$$\frac{\partial \hat{\lambda}_\psi}{\partial \hat{\lambda}} = D + O_p(n^{-1})$$

and

$$\frac{\partial \hat{\lambda}_\psi}{\partial \hat{\psi}} = O_p(n^{-1});$$

here D denotes an identity matrix. Since

$$\frac{\partial \hat{\lambda}_\psi}{\partial \hat{\lambda}} = \hat{j}_{\lambda\lambda}(\hat{\theta}_\psi)^{-1} \ell_{\lambda;\hat{\lambda}}(\psi, \hat{\lambda}_\psi)$$

and

$$\frac{\partial \hat{\lambda}_\psi}{\partial \hat{\psi}} = \hat{j}_{\lambda\lambda}(\hat{\theta}_\psi)^{-1} \ell_{\lambda;\hat{\psi}}(\psi, \hat{\lambda}_\psi),$$

it follows that

$$\ell_{\lambda;\hat{\lambda}}(\hat{\theta}_\psi) = \hat{j}_{\lambda\lambda}(\hat{\theta}_\psi) + O(1)$$

and

$$\ell_{\lambda;\hat{\psi}}(\hat{\theta}_\psi) = O(1).$$

Using these results, we have that

$$U = \frac{|\ell_{;\hat{\psi}}(\hat{\theta}) - \ell_{;\hat{\psi}}(\hat{\theta}_\psi)| \, |\hat{j}_{\lambda\lambda}(\hat{\theta}_\psi)|^{1/2}}{|\hat{j}|^{1/2}} + O_p(n^{-1}).$$

Hence, to approximate U we need to approximate

$$\ell_{;\hat{\psi}}(\hat{\theta}) - \ell_{;\hat{\psi}}(\hat{\theta}_\psi) \equiv \ell_{;\hat{\psi}}(\hat{\psi}, \hat{\lambda}) - \ell_{;\hat{\psi}}(\psi, \hat{\lambda}_\psi).$$

Using a Taylor's series expansion along with the fact that $\hat{\lambda}_\psi = \hat{\lambda} + O_p(n^{-1})$, we have that

$$\ell_{;\hat{\psi}}(\psi, \hat{\lambda}_\psi) - \ell_{;\hat{\psi}}(\hat{\psi}, \hat{\lambda}) = \ell_{\psi;\hat{\psi}}(\hat{\theta})(\psi - \hat{\psi}) + \tfrac{1}{2}\ell_{\psi\psi;\hat{\psi}}(\hat{\theta})(\psi - \hat{\psi})^2 + O_p(n^{-1/2}).$$

Let $n\nu_{ijk}(\theta)$ denote the (i, j, k)th cumulant of $(\ell_\psi(\theta), \ell_{\psi\psi}(\theta), \ell_{\psi\psi\psi}(\theta))$ and recall the results of Section 6.4.2 relating sample space derivatives to quantities of the form ν_{ijk}. Using these results, it is straightforward to show that

$$\ell_{\psi\psi;\hat{\psi}}(\hat{\theta}) = n\nu_{11}(\hat{\theta}) + O_p(n^{1/2}).$$

This result, together with the fact that $\ell_{\psi;\hat{\psi}}(\hat{\theta}) = \hat{j}_{\psi\psi}$, shows that

$$\ell_{;\hat{\psi}}(\psi, \hat{\lambda}_\psi) - \ell_{;\hat{\psi}}(\hat{\psi}, \hat{\lambda}) = \hat{j}_{\psi\psi}(\psi - \hat{\psi}) + \frac{n}{2}\nu_{11}(\hat{\theta})(\psi - \hat{\psi})^2 + O_p(n^{-1/2}).$$

A similar expansion exists for the score function for ψ:

$$\begin{aligned}\ell_\psi(\psi, \hat{\lambda}_\psi) - \ell_\psi(\hat{\psi}, \hat{\lambda}) &= \ell_{\psi\psi}(\hat{\psi}, \hat{\lambda})(\psi - \hat{\psi}) + \frac{1}{2}\ell_{\psi\psi\psi}(\hat{\psi}, \hat{\lambda})(\psi - \hat{\psi})^2 \\ &\quad + O_p(n^{-1/2}) \\ &= -\hat{j}_{\psi\psi}(\psi - \hat{\psi}) + \frac{n}{2}\nu_{001}(\hat{\theta})(\psi - \hat{\psi})^2 + O_p(n^{-1/2}).\end{aligned}$$

Hence,

$$\ell_{;\hat{\psi}}(\psi, \hat{\lambda}_\psi) - \ell_{;\hat{\psi}}(\hat{\psi}, \hat{\lambda}) = -\ell_\psi(\psi, \hat{\lambda}_\psi) + \frac{n}{2}[\nu_{11}(\hat{\theta}) + \nu_{001}(\hat{\theta})](\psi - \hat{\psi})^2 + O_p(n^{-1/2}).$$

Note that

$$n[\nu_{11}(\theta) + \nu_{001}(\theta)] = -\frac{\partial}{\partial\psi} i_{\psi\psi}(\psi, \lambda).$$

Hence,

$$n[\nu_{11}(\hat{\theta}) + \nu_{001}(\hat{\theta})](\psi - \hat{\psi}) = i_{\psi\psi}(\psi, \hat{\lambda}_\psi) - i_{\psi\psi}(\hat{\psi}, \hat{\lambda}) + O_p(1).$$

It follows that

$$\begin{aligned} &\ell_{;\hat{\psi}}(\psi, \hat{\lambda}_\psi) - \ell_{;\hat{\psi}}(\hat{\psi}, \hat{\lambda}) \\ &\quad = -\ell_\psi(\psi, \hat{\lambda}_\psi) + \tfrac{1}{2}[i_{\psi\psi}(\psi, \hat{\lambda}_\psi) - i_{\psi\psi}(\hat{\psi}, \hat{\lambda})](\psi - \hat{\psi}) + O_p(n^{-1/2}) \end{aligned}$$

and using the fact that

$$\psi - \hat{\psi} = -\frac{\ell_\psi(\psi, \hat{\lambda}_\psi)}{\hat{\jmath}_{\psi\psi}} + O_p(n^{-1}) = -\frac{\ell_\psi(\psi, \hat{\lambda}_\psi)}{i_{\psi\psi}(\hat{\theta})} + O_p(n^{-1}),$$

we have that

$$\begin{aligned} \ell_{;\hat{\psi}}(\psi, \hat{\lambda}_\psi) - \ell_{;\hat{\psi}}(\hat{\psi}, \hat{\lambda}) &= -\ell_\psi(\psi, \hat{\lambda}_\psi)\left\{1 - \frac{1}{2}\left[\frac{i_{\psi\psi}(\psi, \hat{\lambda}_\psi)}{i_{\psi\psi}(\hat{\psi}, \hat{\lambda})} - 1\right]\right\} + O_p(n^{-1/2}) \\ &= -\frac{i_{\psi\psi}(\hat{\psi}, \hat{\lambda})^{1/2}}{i_{\psi\psi}(\psi, \hat{\lambda}_\psi)^{1/2}} \ell_\psi(\psi, \hat{\lambda}_\psi) + O_p(n^{-1/2}). \end{aligned}$$

Hence, U may be approximated by

$$U_O = \ell_\psi(\psi, \hat{\lambda}_\psi) \frac{|\hat{\jmath}_{\lambda\lambda}(\hat{\theta}_\psi)|^{1/2} i_{\psi\psi}(\hat{\theta})^{1/2}}{|\hat{\jmath}|^{1/2} i_{\psi\psi}(\hat{\theta}_\psi)^{1/2}}$$

and R^* may be approximated by

$$R_O^* = R + \frac{1}{R}\log(U_O/R).$$

The argument above shows that

$$R_O^* = R^* + O_p(n^{-1})$$

for ψ of the form $\hat{\psi} + O(n^{-1/2})$.

Example 7.11 Linear regression model
Consider the linear regression model considered in Example 7.5. The log-likelihood function for this model is given by

$$\ell(\psi, \beta, \sigma) = -n \log \sigma - \frac{1}{2\sigma^2}[n\hat{\sigma}^2 + n(\hat{\psi} - \psi)^2 + (\hat{\beta} - \beta)^T z^T z(\hat{\beta} - \beta)];$$

it is straightforward to verify that ψ and (β, σ) are orthogonal.

Here

$$-\ell_{\psi\psi}(\theta) = \frac{n}{\sigma^2}$$

so that

$$i_{\psi\psi}(\hat{\theta}) = \frac{n}{\hat{\sigma}^2}$$

and

$$i_{\psi\psi}(\hat{\theta}_\psi) = \frac{n}{\hat{\sigma}_\psi^2}.$$

Using the facts that

$$\ell_\psi(\hat{\theta}_\psi) = \frac{n}{\hat{\sigma}_\psi^2}(\hat{\psi} - \psi);$$

$$|\hat{j}| = \frac{2n}{\hat{\sigma}^{2p+4}}|z^T z|; \qquad |j_{\lambda\lambda}(\hat{\theta}_\psi)| = \frac{2n}{\hat{\sigma}_\psi^{2p+2}}|z^T z|,$$

it follows that

$$U_O = \frac{\hat{\sigma}^{p+1}}{\hat{\sigma}_\psi^{p+2}}\sqrt{n}(\hat{\psi} - \psi)$$

which is exactly equal to U. ∎

There are several drawbacks to U_O and, hence, to R_O^*. The most obvious is that it requires an orthogonal parameterization of the model. Although an orthogonal parameterization is always available when ψ is a scalar, it is often a nontrivial exercise to solve the necessary differential equations. Also, U_O depends on the orthogonal parameterization used, although the difference in R_O^* corresponding to two choice of parameterization is of order $O_p(n^{-1})$. Finally, R_O^* only approximates R^* for ψ in the moderate-deviation range $\hat{\psi} + O(n^{-1/2})$; for fixed ψ, $R_O^* = R^* + O_p(1)$.

Example 7.12 Linear regression model
Consider the linear regression model considered in Example 7.10. The model may be parameterized by (ψ, β, ϕ) where $\phi = \sigma^\alpha$ and α is a known constant. It is

straightforward to show that, in this case,

$$U_O = \frac{\hat{\sigma}^{p+\alpha}}{\hat{\sigma}_{\psi}^{p+\alpha+1}} \sqrt{n}(\hat{\psi} - \psi).$$

It follows that

$$R_O^* = R^* + (\alpha - 3)\frac{1}{R}\log(\hat{\sigma}/\hat{\sigma}_{\psi}) = R^* - (\alpha - 3)\frac{R}{2n}.$$

In general, the properties of R_O^* depend on the value of α, that is, they depend on the parameterization used.

For instance, suppose that $n = 10$ and $p = 4$ and consider the probability that $R_O^* \leq \Phi^{-1}(0.95)$. When the model is parameterized by σ, in which case $R_O^* = R^*$, this probability is 0.9496, when the model is parameterized by σ^2, this probability is 0.9618 while when the model is parameterized by the precision, $1/\sigma$, this probability is 0.9235. Although the quality of the normal approximation to the distribution of R_O^* does depend on the parameterization used, the use of any of these parameterizations is significantly better than using the unadjusted statistic R, for which the analogous probability is 0.8661. ∎

7.5.3 AN APPROXIMATION BASED ON AN APPROXIMATELY ANCILLARY STATISTIC

Let $Y_1, \ldots, Y_n$ denote independent scalar random variables such that Y_j has distribution function $F_j(y; \theta)$ and density function $p_j(y; \theta)$. Let $\tilde{\ell}_{;\hat{\theta}}(\theta)$ and $\tilde{\ell}_{\theta;\hat{\theta}}(\theta)$ denote the approximations to $\ell_{;\hat{\theta}}(\theta)$ and $\ell_{\theta;\hat{\theta}}(\hat{\theta})$, respectively, derived in Section 6.7.2. Recall that

$$\tilde{\ell}_{;\hat{\theta}}(\theta) = \ell_{;y}(\theta)\hat{V}\left(\ell_{\theta;y}(\hat{\theta})\hat{V}\right)^{-1}\hat{\jmath}$$

and

$$\tilde{\ell}_{\theta;\hat{\theta}}(\theta) = \ell_{\theta;y}(\theta)\hat{V}\left(\ell_{\theta;y}(\hat{\theta})\hat{V}\right)^{-1}\hat{\jmath}$$

where

$$\hat{V} = \begin{pmatrix} -\dfrac{\partial F_1(y_1; \hat{\theta})/\partial\hat{\theta}}{p_1(y_j; \hat{\theta})} \\ \vdots \\ -\dfrac{\partial F_n(y_n; \hat{\theta})/\partial\hat{\theta}}{p_n(y_n; \hat{\theta})} \end{pmatrix}.$$

Then $\tilde{\ell}_{;\hat{\theta}}(\hat{\theta}) - \tilde{\ell}_{;\hat{\theta}}(\hat{\theta}_{\psi})$ and $\tilde{\ell}_{\theta;\hat{\theta}}(\hat{\theta}_{\psi})$ have relative error $O(n^{-1})$ for fixed values of ψ and relative error $O(n^{-3/2})$ for ψ of the form $\psi = \hat{\psi} + O(n^{-1/2})$.

Let $\tilde{U}$ denote the statistic U calculated using these approximations and let

$$\tilde{R}^* = R^* + \frac{1}{R}\log(\tilde{U}/R).$$

Note that either the expression for U based on an explicitly defined nuisance parameter or the expression for U derived in Section 7.4.3 may be used in this context. Then

$$\tilde{R}^*_\psi = R^*_\psi + O_p(n^{-3/2})$$

for $\psi = \hat{\psi} + O(n^{-1/2})$ and

$$\tilde{R}^*_\psi = R^*_\psi + O_p(n^{-1})$$

for fixed ψ. Hence, under the distribution with parameter ψ, $\tilde{R}^*$ is asymptotically distributed according to a standard normal distribution with error $O(n^{-3/2})$.

Example 7.13 Index of a Weibull distribution
Consider the Weibull distribution example considered in Example 7.7. In this example, Y_j has a Weibull distribution with index ψ and rate parameter $x_j^T\lambda$. Hence,

$$F_j(y_j;\theta) = 1 - \exp\{-\exp\{x_j^T\lambda\}y_j^\psi\}$$

and

$$p_j(y_j;\theta) = \psi\exp\{x_j^T\lambda\}y_j^{\psi-1}\exp\{-\exp\{-x_j^T\lambda\}y_j^\psi\}.$$

It follows that

$$\hat{V} = -\frac{1}{\hat{\psi}}\begin{pmatrix} y_1\log y_1 & y_1x_1^T \\ \vdots & \vdots \\ y_n\log y_n & y_nx_n^T \end{pmatrix}.$$

The log-likelihood function for this model is given by

$$\ell(\theta) = n\log\psi + \psi\sum\log y_j + \sum x_j^T\lambda - \sum\exp\{x_j^T\lambda\}\,y_j^\psi.$$

Hence,

$$\ell_{;y}(\theta) = \left(\frac{\psi}{y_1}(1 - y_1^\psi\exp\{x_1^T\lambda\})\cdots\frac{\psi}{y_n}(1 - y_n^\psi\exp\{x_n^T\lambda\})\right)$$

and

$$\ell_{;y}(\theta)\hat{V} = \left(\frac{\psi}{\hat{\psi}}\sum\log y_j(y_j^\psi\exp\{x_j^T\lambda\} - 1) \quad \frac{\psi}{\hat{\psi}}\sum(y_j^\psi\exp\{x_j^T\lambda\} - 1)x_j^T\right).$$

Comparing this to the result in Example 7.7, we see that

$$\ell_{;y}(\theta)\hat{V} = \ell_{;\hat{\theta}}(\theta).$$

Similarly,

$$\ell_{\lambda;y}(\theta) = (\psi \exp\{x_1^T \lambda\}\, y_1^{\psi-1} x_1 \cdots \psi \exp\{x_n^T \lambda\}\, y_n^{\psi-1} x_n)$$

so that

$$\ell_{\lambda;y}(\theta)\hat{V} = \ell_{\lambda;\hat{\theta}}(\theta).$$

Also,

$$\ell_{\theta;y}(\hat{\theta})\hat{V} = \ell_{\theta\theta}(\hat{\theta}).$$

It follows that $\tilde{R}^* = R^*$. ∎

Example 7.14 Logistic growth model with t-distributed errors
Consider independent random variables $Y_1, \ldots, Y_n$ each of the form

$$Y_j = \mu_j + \sigma\epsilon_j, \quad j = 1, \ldots, n$$

where $\epsilon_1, \ldots, \epsilon_n$ are independent random variables each distributed according to a t-distribution with 3 degrees of freedom, and

$$\mu_j = \frac{\beta}{1 + \exp(\gamma - \psi x_j)}, \quad j = 1, \ldots, n$$

where $x_1, \ldots, x_n$ are fixed covariates. The nuisance parameter is therefore $\lambda = (\beta, \gamma, \sigma)$ where $\sigma > 0$.

Here

$$F_j(y; \theta) = H((y - \mu_j)/\sigma)$$

where $H(\cdot)$ denotes the distribution function of the t-distribution with 3 degrees of freedom and

$$\mu_j \equiv \mu_j(\theta) = \frac{\beta}{1 + \exp\{\gamma - \psi x_j\}}.$$

It follows that

$$\hat{V} = \begin{pmatrix} \frac{\partial \mu_1}{\partial \psi}(\hat{\theta}) & \frac{\partial \mu_1}{\partial \beta}(\hat{\theta}) & \frac{\partial \mu_1}{\partial \gamma}(\hat{\theta}) & \frac{y_1 - \mu_1(\hat{\theta})}{\hat{\sigma}} \\ \vdots & \vdots & \vdots & \vdots \\ \frac{\partial \mu_n}{\partial \psi}(\hat{\theta}) & \frac{\partial \mu_n}{\partial \beta}(\hat{\theta}) & \frac{\partial \mu_n}{\partial \gamma}(\hat{\theta}) & \frac{y_n - \mu_n(\hat{\theta})}{\hat{\sigma}} \end{pmatrix}.$$

The log-likelihood function for this model is given by

$$\ell(\theta) = -2 \sum \log\left(1 + \frac{(y_j - \mu)^2}{3\sigma^2}\right) - n \log \sigma;$$

determination of $\ell_{;y}(\theta)$ and $\ell_{\theta;y}(\theta)$ is straightforward.

Table 7.6 Confidence limits in Example 7.14

	Confidence limit		Coverage probability	
Probability	R	$\tilde{R}^*$	R	$\tilde{R}^*$
0.010	0.597	0.580	0.022	0.010
0.025	0.609	0.596	0.046	0.024
0.050	0.619	0.609	0.079	0.051
0.100	0.630	0.623	0.135	0.105
0.900	0.704	0.718	0.862	0.898
0.950	0.717	0.737	0.921	0.949
0.975	0.731	0.757	0.954	0.974
0.990	0.749	0.783	0.979	0.989

Consider the data in Table A.10 of the Appendix. These data give the weight of onions as a function of growing time. Table 7.6 gives confidence limits for ψ based on R and $\tilde{R}^*$. To investigate the frequency properties of $\tilde{R}^*$ in this setting a small simulation study was conducted. Data were simulated from this model with the parameter values taken to be the maximum likelihood estimates based on the data in Table A.10. A Monte Carlo sample size of 10 000 was used. The estimated coverage probabilities of confidence limits based on R and $\tilde{R}^*$ are also given in Table 7.6. ∎

Example 7.15 Normal distributions with common mean
Let $Y_{jk}, k = 1, \ldots, n_j, j = 1, \ldots, m$ denote independent normal random variables such that Y_{jk} has mean μ and standard deviation σ_j. Take the common mean μ as the parameter of interest and take $(\sigma_1, \ldots, \sigma_m)$ as the nuisance parameter.

It follows that the first column of $\hat{V}$ is a column of ones. The second column is given by

$$\left(\frac{y_{11} - \hat{\mu}}{\hat{\sigma}_1}, \ldots, \frac{y_{1n_1} - \hat{\mu}}{\hat{\sigma}_1}, 0, \ldots, 0\right)^T;$$

the third column is of the form

$$\left(0, \ldots, 0, \frac{y_{21} - \hat{\mu}}{\hat{\sigma}_2}, \ldots, \frac{y_{2n_2} - \hat{\mu}}{\hat{\sigma}_2}, 0, \ldots, 0\right)^T$$

and so on.

Consider the data in Table A.9 of the Appendix. These data represent measurements of the strength of six samples of cotton yarn; four measurements were taken of each sample. Table 7.7 contains confidence limits for μ based on $\tilde{R}^*$ along with those based on R. Table 7.7 also contains the results of a small simulation study designed to assess the coverage probabilities of confidence limits based on R and $\tilde{R}^*$. The sample sizes used in the simulation study were taken from the data in Table A.9 and the parameter values used were the maximum likelihood estimates based on these data.

Table 7.7 Confidence limits in Example 7.15

	Confidence limit		Coverage probability	
Probability	R	$\tilde{R}^*$	R	$\tilde{R}^*$
0.010	15.05	14.83	0.036	0.012
0.025	15.15	15.00	0.070	0.025
0.050	15.22	15.11	0.108	0.052
0.100	15.29	15.22	0.172	0.104
0.900	15.72	15.79	0.824	0.890
0.950	15.80	15.91	0.885	0.945
0.975	15.87	16.03	0.926	0.973
0.990	15.97	16.21	0.962	0.989

The results are based on 10 000 Monte Carlo samples. The results indicate that the coverage probabilities of the limits based on $\tilde{R}^*$ are a great improvement over those based on R. ∎

7.5.4 AN APPROXIMATION BASED ON COVARIANCES

In Section 6.7.3 it was shown that $\ell_{;\hat{\theta}}(\hat{\theta}) - \ell_{;\hat{\theta}}(\theta)$ and $\ell_{\theta;\hat{\theta}}(\theta)$ may be approximated by

$$\bar{\ell}_{;\hat{\theta}}(\hat{\theta}) - \bar{\ell}_{;\hat{\theta}}(\theta) = \{Q(\hat{\theta};\hat{\theta}) - Q(\theta;\hat{\theta})\} i(\hat{\theta})^{-1}\hat{\jmath}$$

and

$$\bar{\ell}_{\theta;\hat{\theta}}(\theta) = I(\theta;\hat{\theta}) i(\hat{\theta})^{-1}\hat{\jmath},$$

respectively, where

$$I(\theta;\theta_0) = E[\ell_\theta(\theta)\ell_\theta(\theta_0)^T;\theta_0],$$

$$Q(\theta;\theta_0) = E[\ell(\theta)\ell_\theta(\theta_0)^T;\theta_0];$$

and $i(\hat{\theta}) = I(\hat{\theta};\hat{\theta})$.

Let $\bar{U}$ denote the statistic U based on these approximations and let

$$\bar{R}^*_\psi \equiv \bar{R}^* = R + \frac{1}{R}\log(\bar{U}/R).$$

Note that $\bar{U}$ may be calculated using either an explicit parameterization for the nuisance parameter or the implicit nuisance parameter approach described in Section 7.4.3.

The results of Section 6.7.3 show that

$$\bar{R}^*_\psi = R^*_\psi + O_p(n^{-1})$$

for ψ of the form $\hat{\psi} + O(n^{-1/2})$. Hence, under the distribution with parameter value ψ, $\bar{R}^*_\psi$ is distributed according to a standard normal distribution, with error $O(n^{-1})$, conditionally on a.

For fixed ψ, $\bar{U}/R$ and U/R are both of order $O_p(1)$ and $\bar{U}/R = U/R + O_p(n^{-1/2})$; since $R = O_p(\sqrt{n})$, again we have $\bar{R}^*_\psi = R^*_\psi + O_p(n^{-1})$. In this case however it is more accurate to think of $\bar{R}^*_\psi$ as a first-order approximation to R^*_ψ since the adjustment term in $\bar{R}^*$ is only a first-order approximation.

Example 7.16 Normal distributions with common mean
Let Y_{jk}, $k = 1, \ldots, n_j$, $j = 1, \ldots, m$ denote independent normal random variables such that Y_{jk} has mean μ and standard deviation σ_j; see Example 7.15. Let

$$Y_j = \frac{1}{n_j} \sum Y_{jk}$$

and

$$S_j = \sum (Y_{jk} - Y_j)^2;$$

by sufficiency, the analysis may be based on $(Y_1, S_1), \ldots, (Y_m, S_m)$.

The log-likelihood function is given by

$$\ell(\theta) = -\sum n_j \log \sigma_j - \frac{1}{2} \sum \frac{n_j}{\sigma_j^2}(y_j - \mu)^2 - \frac{1}{2} \sum s_j/\sigma_j^2.$$

It follows that

$$\ell_\mu(\theta) = \sum \frac{n_j}{\sigma_j^2}(y_j - \mu)$$

and

$$\ell_{\sigma_j}(\theta) = -\frac{n_j}{\sigma_j} + \frac{n_j}{\sigma_j^3} + \frac{s_j}{\sigma_j^3}.$$

Hence,

$$E[\ell_\mu(\theta_0)\ell_\mu(\theta); \theta_0] = \sum \frac{n_j}{\sigma_j^2}; \qquad E[\ell_\mu(\theta_0)\ell_{\sigma_j}(\theta); \theta_0] = \frac{2n_j(\mu_0 - \mu)}{\sigma_j^3};$$

$$E[\ell_{\sigma_j}(\theta_0)\ell_\mu(\theta); \theta_0] = 0; \qquad E[\ell_{\sigma_j}(\theta_0)\ell_{\sigma_k}(\theta); \theta_0] = \begin{cases} 0 & \text{if } j \neq k \\ \dfrac{2n_j\sigma_{j0}}{\sigma_j^3} & \text{if } j = k \end{cases};$$

$$E[\ell(\theta)\ell_\mu(\theta_0); \theta_0] = \sum \frac{n_j}{\sigma_j^2}(\mu - \mu_0); \qquad E[\ell(\theta)\ell_{\sigma_j}(\theta_0); \theta_0] = -n_j \frac{\sigma_{0j}}{\sigma_j^2}.$$

Calculation of $Q(\theta; \theta_0)$ and $I(\theta; \theta_0)$ follows easily from these results. Although the analytical form of $\bar{U}$ is complicated, it is straightforward to determine the value of $\bar{U}$ and hence $\bar{R}^*$ numerically for a given set of data.

Table 7.8 contains confidence limits for μ based on $\bar{R}^*$ for the data described in Example 7.15, along with the confidence limits based on R and $\tilde{R}^*$. Table 7.8 also contains the coverage probabilities of these confidence limits based on a small Monte Carlo study as described in Example 7.15.

Table 7.8 Confidence limits in Example 7.16

	Confidence limit			Coverage probability		
Probability	R	$\bar{R}^*$	$\tilde{R}^*$	R	$\bar{R}^*$	$\tilde{R}^*$
0.010	15.05	14.85	14.83	0.036	0.015	0.012
0.025	15.15	15.01	15.00	0.070	0.033	0.025
0.050	15.22	15.12	15.11	0.108	0.064	0.052
0.100	15.29	15.23	15.22	0.172	0.118	0.104
0.900	15.72	15.78	15.79	0.824	0.876	0.890
0.950	15.80	15.90	15.91	0.885	0.933	0.945
0.975	15.87	16.01	16.03	0.926	0.965	0.973
0.990	15.97	16.17	16.21	0.962	0.985	0.989

∎

Example 7.17 Nonlinear regression model
Let $Y_1, \ldots, Y_n$ denote independent normal random variables such that Y_j has mean $f_j(\psi, \phi)$ and standard deviation σ. Here f_j is a known function, ψ is the parameter of interest and $\lambda = (\phi, \sigma)$ is the nuisance parameter.

The log-likelihood function for this model is given by

$$\ell(\theta) = -\frac{1}{2\sigma^2} \sum (y_j - f_j(\psi, \phi))^2 - n \log \sigma;$$

the log-likelihood derivatives needed for the computation of $I(\theta; \theta_0)$ and $Q(\theta; \theta_0)$ are given by

$$\ell_\psi(\theta) = \frac{1}{\sigma^2} \sum (y_j - f_j(\psi, \phi)) \frac{\partial f_j}{\partial \psi}(\psi, \phi);$$

$$\ell_\phi(\theta) = \frac{1}{\sigma^2} \sum (y_j - f_j(\psi, \phi)) \frac{\partial f_j}{\partial \phi}(\psi, \phi);$$

$$\ell_\sigma(\theta) = \frac{1}{\sigma^3} \sum (y_j - f_j(\psi, \phi))^2 - \frac{n}{\sigma}.$$

It follows that

$$I(\theta; \theta_0) = \frac{1}{\sigma^2} \begin{pmatrix} z(\psi, \phi)^T z(\psi_0, \phi_0) & z(\psi, \phi)^T x(\psi_0, \phi_0) & 0 \\ x(\psi, \phi)^T z(\psi_0, \phi_0) & x(\psi, \phi)^T x(\psi_0, \phi_0) & 0 \\ 2\delta^T z(\psi_0, \phi_0)/\sigma & 2\delta^T x(\psi_0, \phi_0)/\sigma & 2/(\sigma\sigma_0) \end{pmatrix}$$

where $z(\psi, \phi)$ denotes an $n \times 1$ vector with jth element given by $\partial f_j(\psi, \phi)/\partial \psi$, $x(\psi, \phi)$ denotes an $n \times p$ matrix with (i, j) element given by

$$x_{ij}(\psi, \phi) = \frac{\partial f_i}{\partial \phi_j}(\psi, \phi)$$

and δ denotes an $n \times 1$ vector with jth element $f_j(\psi_0, \phi_0) - f_j(\psi, \phi)$.

Table 7.9 Confidence limits in Example 7.17

Probability	R	$\bar{R}^*$
0.010	194.6	186.3
0.025	197.4	187.6
0.050	199.7	194.6
0.100	202.1	199.0
0.900	218.4	221.5
0.950	221.1	226.5
0.975	223.6	235.1
0.990	226.8	236.7

It is straightforward to show that

$$Q(\theta;\theta_0) = \left(-\frac{1}{\sigma^2}\delta^T z(\psi_0,\phi_0) \quad -\frac{1}{\sigma^2}\delta^T x(\psi_0,\phi_0) \quad -n\frac{\sigma_0}{\sigma^2} \right).$$

Calculation of $\bar{U}$ and $\bar{R}^*$ follows easily from these results.

Consider the model

$$Y_j = \frac{\psi x_j}{\phi + x_j} + \sigma\epsilon_j.$$

Table 7.9 contains confidence limits for ψ based on $\bar{R}^*$ along with those based on R for the data in Table A.4. ∎

7.5.5 AN APPROXIMATION BASED ON EMPIRICAL COVARIANCES

In Section 6.7.4, it was shown that the covariances used to form $\bar{R}^*$ may be replaced by empirical covariances without changing the order of the approximation. Let

$$\hat{Q}(\theta;\theta_0) = \sum \ell^{(j)}(\theta)\ell_\theta^{(j)}(\theta_0)^T,$$

$$\hat{I}(\theta;\theta_0) = \sum \ell_\theta^{(j)}(\theta)\ell_\theta^{(j)}(\theta_0)^T,$$

and $\hat{\imath}(\hat{\theta}) = \hat{I}(\hat{\theta};\hat{\theta})$. Then $\ell_{;\hat{\theta}}(\hat{\theta}) - \ell_{;\hat{\theta}}(\theta)$ and $\ell_{\theta;\hat{\theta}}(\theta)$ may be approximated by

$$\hat{\ell}_{;\hat{\theta}}(\hat{\theta}) - \hat{\ell}_{;\hat{\theta}}(\theta) = \{\hat{Q}(\hat{\theta};\hat{\theta}) - \hat{Q}(\theta;\hat{\theta})\}\hat{\imath}(\hat{\theta})^{-1}\hat{\jmath}$$

and

$$\hat{\ell}_{\theta;\hat{\theta}}(\theta) = \hat{I}(\theta;\hat{\theta})\hat{\imath}(\hat{\theta})^{-1}\hat{\jmath},$$

respectively.

Table 7.10 Confidence limits in Example 7.18

	Confidence limit			Coverage probability		
Probability	R	R^*	$\hat{R}^*$	R	R^*	$\hat{R}^*$
0.010	0.610	0.532	0.546	0.004	0.010	0.008
0.025	0.667	0.585	0.602	0.009	0.025	0.021
0.050	0.717	0.634	0.652	0.019	0.050	0.042
0.100	0.779	0.692	0.712	0.042	0.101	0.089
0.900	1.308	1.119	1.236	0.779	0.899	0.888
0.950	1.396	1.285	1.325	0.870	0.947	0.942
0.975	1.478	1.363	1.405	0.924	0.973	0.970
0.990	1.576	1.457	1.502	0.962	0.989	0.988

Let $\hat{U}$ denote the statistic U based on these approximations and let

$$\hat{R}^* = R + \frac{1}{R}\log(\hat{U}/R).$$

The results in Section 6.7.4 show that $\hat{R}^*_\psi = R^*_\psi + O_p(n^{-1})$ for ψ of the form $\hat{\psi} + O(n^{-1/2})$. Hence, under the distribution with parameter ψ, $\hat{R}^*_\psi$ is distributed according to a standard normal distribution, with error $O(n^{-1})$, conditionally on a. As was the case for $\bar{U}$, for fixed ψ, $\hat{U}/R$ and U/R are both of order $O_p(1)$ and $\hat{U}/R = U/R + O_p(n^{-1/2})$; hence, $\hat{R}^*_\psi = R^*_\psi + O_p(n^{-1})$.

Example 7.18 Index of a Weibull distribution
Consider the Weibull distribution example considered in Examples 7.7 and 7.13. Numerical calculation of $\hat{R}^*$ for this model is straightforward, although the analytic expression for $\hat{R}^*$ is complicated.

Consider the data Table A.7 that was analyzed in Example 7.7. Table 7.10 contains confidence limits for ψ based on R, R^*, and $\hat{R}^*$. The results indicate that confidence limits based on $\hat{R}^*$ closely approximate those based on R^*.

The frequency properties of confidence limits based on $\hat{R}^*$ were also considered, using the same type of study used in Example 7.7 to study the frequency properties of confidence limits based on R and R^*. Those results are also presented in Table 7.10. ∎

Example 7.19 Logistic growth model with t-distributed errors
Consider the logistic growth model considered in Example 7.14. Table 7.11 contains confidence limits for ψ based on $\hat{R}^*$ for the onion growth data in Table A.10 , along with the coverage probability of those limits based on Monte Carlo simulation as described in Example 7.14. For comparison, the confidence limits based on R and $\tilde{R}^*$ are included as well. The coverage probabilities of confidence limits based on $\hat{R}^*$ are very close to those based on $\tilde{R}^*$; there are small differences in the values of the limits themselves for the data under consideration.

Table 7.11 Confidence limits in Example 7.19

	Confidence limit			Coverage probability		
Probability	R	$\hat{R}^*$	$\tilde{R}^*$	R	$\hat{R}^*$	$\tilde{R}^*$
0.010	0.597	0.582	0.580	0.022	0.010	0.010
0.025	0.609	0.597	0.596	0.046	0.025	0.024
0.050	0.619	0.609	0.609	0.079	0.050	0.051
0.100	0.630	0.621	0.623	0.135	0.103	0.105
0.900	0.704	0.703	0.718	0.862	0.900	0.898
0.950	0.717	0.719	0.737	0.921	0.952	0.949
0.975	0.731	0.734	0.757	0.954	0.978	0.974
0.990	0.749	0.756	0.783	0.979	0.992	0.989

∎

Table 7.12 Confidence limits in Example 7.20

Probability	R	$\hat{R}^*$
0.010	0.462	0.444
0.025	0.494	0.478
0.050	0.522	0.508
0.100	0.554	0.543
0.900	0.720	0.719
0.950	0.735	0.735
0.975	0.748	0.748
0.990	0.762	0.762

Example 7.20 Mixture of normal distributions
Let $Y_1, \ldots, Y_n$ denote independent identically distributed random variables such that Y_j has a normal distribution with mean μ_1 and standard deviation σ_1 with probability ψ and a normal distribution with mean μ_2 and standard deviation σ_2 with probability $1 - \psi$. Assume $\mu_1 > \mu_2$; under this assumption, the parameters of the model are identifiable.

Consider the data in Table A.2 of the appendix. These data represent the lengths in micrometers of a sample of 1000 trypanosome protozoa. Two different strains of protozoan were included in the sample. Table 7.12 contains confidence limits for ψ based on $\hat{R}^*$ along with those based on $\hat{R}$. Even with the large sample size, $\hat{R}^*$ still provides a slight adjustment to the confidence limits based on R. ∎

7.6 Discussion and references

The signed likelihood ratio statistic has a long history. Lawley (1956) considers adjustments to R designed to improve the accuracy of the normal approximation; further results along these lines are given by DiCiccio (1984), McCullagh (1984), and Jensen

(1986, 1987). The modified signed likelihood ratio statistic considered in this chapter was first proposed by Barndorff-Nielsen (1986); see also Fraser (1990), Barndorff-Nielsen (1991b) and Jensen (1992, 1993, 1997). The material in Section 7.3 draws heavily from Jensen (1995, Chapter 5).

General discussions of the derivation of R^* and its properties are given in Barndorff-Nielsen and Cox (1994), Jensen (1995), Reid (1996), and Skovgaard (2000). The asymptotic properties of R^* have been studied by a number of authors. A proof that R^* follows a standard normal distribution to order $O(n^{-3/2})$ is given by Barndorff-Nielsen (1986); see also Barndorff-Nielsen (1990b, 1991b), Jensen (1995), Sweeting (1995), and Skovgaard (1996, 2000). The large deviation properties of R^* have been considered by Jensen (1992, 1995), Skovgaard (1996), and Barndorff-Nielsen and Wood (1998). It should be noted that, although the discussion in this section is for general models, rigorous proofs of the large-deviation properties of R^* are generally restricted to exponential family models.

A number of numerical examples are available which indicate the high accuracy of these methods in applications. See, for example, Barndorff-Nielsen (1990a, 1991b), Fraser (1990), Barndorff-Nielsen and Chamberlin (1991), Jensen (1992, 1995), Fraser and Reid (1993), Skovgaard (1996), and Severini (1999b). In particular, Pierce and Peters (1992) give a detailed discussion of the application of these methods to models for discrete data and consider several examples.

The approximation to R^* based on orthogonal parameters, discussed in Section 7.5.2, is due to DiCiccio and Martin (1993) who proposed a class of adjusted signed likelihood ratio statistics based on a connection with Bayesian inference. The approximation to R^* based on covariances, $\bar{R}^*$, is due to Skovgaard (1996) who gives a detailed study of its properties; in particular, the large-deviation properties of $\bar{R}^*$ are considered. The empirical version $\hat{R}^*$ is due to Severini (1999b).

The adjusted signed likelihood ratio statistic, derived in Section 7.5.3 and denoted by $\tilde{R}^*$, is due to Fraser, Reid, and Wu (1999). Although here $\tilde{R}^*$ is presented as a version of R^*, it was originally derived using an alternative approach in which the given model is approximated by a model of exponential family form. This approach is developed by Fraser (1988, 1990), Fraser and Reid (1993, 1995, 1999) and Cakmak, Fraser, and Reid (1994).

Another approximation to R^*, not discussed in this chapter, has been proposed by Barndorff-Nielsen and Chamberlin (1994); see also Barndorff-Nielsen (1994, 1995). A detailed discussion of this alternative approximation is available in Barndorff-Nielsen and Cox (1994, Section 8.4).

7.7 Exercises

7.1 Let X denote a normally distributed random variable with mean $\mu/\sqrt{n}$ and standard deviation 1. Construct X^*, the transformed version of X derived in Section 7.2. Find the distribution of X^*.

7.2 Consider a model for a scalar parameter θ and suppose that the underlying data have a lattice distribution. Find a version of the modified signed likelihood ratio statistic that may be used for the calculation of lower tail probabilities.

7.3 Let $Y_1, \ldots, Y_n$ denote independent identically distributed random variables each distributed according to the density

$$\frac{\theta}{\Gamma(\theta)} y^{\theta-1} \exp\{-\theta y\}, \quad y > 0$$

where $\theta > 0$. Find R^*.

7.4 Let $Y_1, \ldots, Y_n$ denote independent Poisson random variables each with mean θ. Find R^*.

7.5 Consider a curved exponential family model with log-likelihood function of the form

$$y^T \phi(\theta) - k(\phi(\theta))$$

where the model with log-likelihood function

$$y^T \phi - k(\phi)$$

is a full-rank d-parameter exponential family model and θ is a scalar parameter. Using the approximation $\bar{\ell}_{;\hat{\theta}}(\theta)$ for the sample space derivative $\ell_{;\hat{\theta}}(\theta)$, construct an approximation to R^*.

7.6 For the exponential hyperbola model, described in Example 6.21, construct the approximation to R^* derived in Exercise 7.5.

7.7 For a one-parameter model, based on a continuously distributed data, show that

$$\tilde{R}^* = R + \frac{1}{R} \log(\tilde{U}/R)$$

where

$$\tilde{U} = \sqrt{\hat{\jmath}} \frac{[\ell_{;y}(\hat{\theta}) - \ell_{;y}(\theta)]\hat{V}}{\ell_{\theta;y}(\hat{\theta})\hat{V}}.$$

Compute this approximation for exponential distribution example considered in Example 7.1 and compare the result to the exact value.

7.8 Consider a model with parameters (ψ, λ) where ψ is a scalar parameter of interest and λ is a vector nuisance parameter. Suppose that $\hat{\lambda}_\psi$, the maximum likelihood estimate of λ for fixed ψ, does not depend on ψ. Give a simplified form for R^* in this case.

7.9 Let $Y_1, \ldots, Y_n$ denote independent random variables each distributed according to the density

$$\frac{\lambda^\psi}{\Gamma(\psi)} y^{\psi-1} \exp\{-\lambda y\}, \quad y > 0$$

where $\lambda > 0$ and $\psi > 0$ are unknown parameters. Find R^* for testing the hypothesis $\psi = 1$.

7.10 Let $Y_1, \ldots, Y_n$ denote independent d-dimensional multivariate normal random variables with unknown mean vector μ and known covariance matrix Σ. Find R^* for a scalar parameter of interest of the form $\psi(\mu)$. Show that if Σ is the identity matrix, then $R^* = R$.

7.11 Show directly that $U/R = 1 + O(|\hat{\psi} - \psi|)$ and, hence, that $R^* = R + O_p(n^{-1/2})$.

7.12 Let $Y_1, \ldots, Y_n$ denote independent normally distributed random variables such that Y_j has mean $\mu_j(\lambda)$ and standard deviation σ. Here $\mu_j(\lambda)$ is a known function of an unknown p-dimensional parameter λ. Find R^*_σ. Compare the result in the general case to the result in the case in which each $\mu_j(\lambda)$ is a linear function of λ.

7.13 Show that we may write

$$\tilde{U} = \begin{vmatrix} [\ell_{;y}(\hat{\theta}) - \ell_{;y}(\hat{\theta}_\psi)]\hat{V} \\ \ell_{\lambda;y}(\hat{\theta}_\psi)\hat{V} \end{vmatrix} |\hat{j}|^{1/2} \Big/ (|j_{\lambda\lambda}(\hat{\theta}_\psi)|^{1/2} |\ell_{\theta;y}(\hat{\theta})\hat{V}|^{-1})$$

and

$$\bar{U} = \begin{vmatrix} Q(\hat{\theta}; \hat{\theta}) - Q(\hat{\theta}_\psi; \hat{\theta}) \\ I_{\lambda;\hat{\theta}}(\hat{\theta}_\psi; \hat{\theta}) \end{vmatrix} |\hat{j}|^{1/2} \Big/ (|j_{\lambda\lambda}(\hat{\theta}_\psi)|^{1/2} |i(\hat{\theta})|)$$

where

$$I_{\lambda;\hat{\theta}}(\theta; \theta_0) = E[\ell_\lambda(\theta)\ell_\theta(\theta_0)^T; \theta_0].$$

8 Likelihood functions for a parameter of interest

8.1 Introduction

Consider a model parameterized by a parameter θ which may be written $\theta = (\psi, \lambda)$ where ψ is the parameter of interest and λ is a nuisance parameter. In general, both ψ and λ may be vectors, although often ψ is a scalar. For models without a nuisance parameter, inference may be based directly on the likelihood function $L(\psi)$; for instance, a test of $\psi = \psi_1$ versus $\psi = \psi_2$ may be based on the likelihood ratio $L(\psi_1)/L(\psi_2)$. When a nuisance parameter is present however it is not possible to use the likelihood function to directly compare different values of ψ. For instance, the value of $L(\psi_1, \lambda)/L(\psi_2, \lambda)$ will, in general, depend on the value of λ. The greater the dimension of the nuisance parameter, the greater is its potential effect on the conclusions regarding the parameter of interest.

Example 8.1 Normal-theory linear regression model
Let $Y_1, \ldots, Y_n$ denote independent random variables of the form

$$Y_j = x_j\beta + \sigma\epsilon_j, \quad j = 1, \ldots, n$$

where $\epsilon_1, \ldots, \epsilon_n$ are independent standard normal random variables, $x_1, \ldots, x_n$ are known $1 \times p$ vectors of constants, β is an unknown $p \times 1$ parameter vector, and $\sigma > 0$ is an unknown scalar parameter. Take σ to be the parameter of interest, with β as the nuisance parameter.

The log-likelihood function is given by

$$\ell(\sigma, \beta) = -n \log \sigma - \frac{1}{2\sigma} \sum (y_j - x_j\beta)^2.$$

Clearly, in order to draw conclusions regarding σ some method of handling β is needed. ∎

Hence, in order to draw inferences regarding the parameter of interest in a given model, we must deal in some way with the nuisance parameters. Ideally, we would like to construct a likelihood function for ψ alone. Perhaps the simplest method of doing this is to construct a likelihood function based on a statistic T such that the distribution of T depends only on ψ. In this case, we may form a genuine likelihood function for ψ based on the density function of T; this is called a *marginal* likelihood function since it is based on the marginal distribution of T. The main drawback of this approach is that we may not be using all of the available information about ψ; also, it may be difficult to find such a statistic.

Another approach is available whenever there exists a statistic S such that the conditional distribution of the data Y given $S = s$ depends only on ψ. In this case, we may form a genuine likelihood function for ψ based on the conditional density function of Y given $S = s$; this is called a *conditional* likelihood function. Again, the main drawback of this approach is that we are discarding the part of the likelihood function based on the marginal distribution of S and this may contain useful information about ψ.

Conditional and marginal likelihood functions are examples of what are called *pseudo-likelihood* functions. A pseudo-likelihood function for a parameter ψ is a function of the data and ψ that may be used as a likelihood function for ψ. In this chapter, we consider conditional and marginal likelihood functions as well as integrated likelihood functions in which the nuisance parameter is eliminated by integration with respect to some weighting function.

This chapter focuses on exact conditional and marginal likelihoods. In Chapter 9, we consider the construction of pseudo-likelihood functions based on the approximation of a conditional or marginal likelihood.

8.2 Conditional likelihood functions

8.2.1 INTRODUCTION

One approach to constructing a pseudo-likelihood function for ψ is to use conditioning to eliminate λ from the likelihood function. For instance, suppose that the minimal sufficient statistic model may be written (t, s) such that

$$p(t, s; \psi, \lambda) = p(t|s; \psi)p(s; \psi, \lambda).$$

The statistic s is a sufficient statistic in the model with ψ held fixed. A likelihood function for ψ may be based on $p(t|s; \psi)$ which does not depend on λ; the resulting conditional likelihood function for ψ is a genuine likelihood function. Note however that this conditional distribution is not needed. The conditional log-likelihood function is given by

$$\ell(\theta) - \ell(\theta; s)$$

where $\ell(\theta; s)$ denotes the log-likelihood function based on the marginal distribution of S.

In using this approach, we are making two assumptions regarding S. The first is that S is not sufficient in the general model with parameters (ψ, λ), for if it was, the conditional likelihood would not depend on either ψ or λ. The other assumption is that S, the sufficient statistic when ψ is fixed, is the same for all ψ, that is, S does not depend on ψ. This assumption will be relaxed in Section 8.2.3.

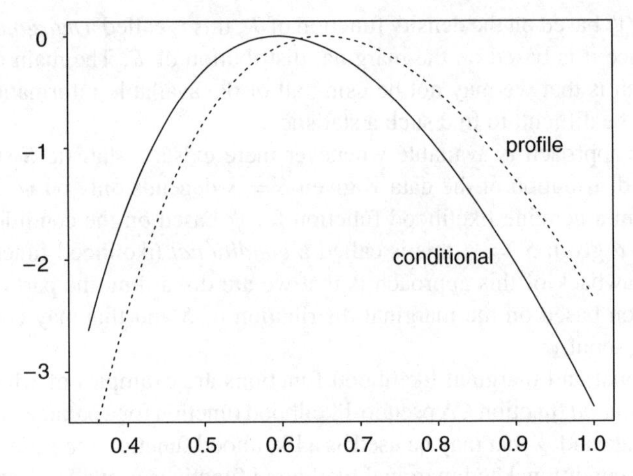

Fig. 8.1 Likelihood functions for Example 8.2.

Example 8.2 Gamma distributions with common index
For each $j = 1, \ldots, m$, let $Y_{j1}, \ldots, Y_{jn_j}$ denote independent gamma random variables with rate parameter λ_j and index ψ and assume that Y_{jk} and $Y_{i\ell}$ are also independent. Then the likelihood function for $(\psi, \lambda_1, \ldots, \lambda_n)$ is given by

$$\frac{(\lambda_1^{n_1} \cdots \lambda_m^{n_m})^{\psi}}{\Gamma(\psi)^{\sum n_j}} (t_1 \cdots t_m)^{\psi - 1} \exp\left\{-\sum \lambda_j s_j\right\}$$

where $t_j = y_{j1} \cdots y_{jn_j}$ and $s_j = \sum y_{jk}$.

For fixed values of ψ, $(s_1, \ldots, s_m)$ is sufficient so that conditional distribution of the data given $s_1, \ldots, s_m$ does not depend on $\lambda_1, \ldots, \lambda_m$. Since each s_j has a gamma distribution with index $n_j\psi$ and rate parameter λ_j, it is straightforward to show that the conditional log-likelihood function is given by

$$\sum \log \Gamma(n_j\psi) - \log \Gamma(\psi) \sum n_j + \psi \sum \log t_j - \psi \sum n_j \log s_j.$$

Consider the data in Table A.1 of the Appendix. These data refer to the hours between failures of air conditioning equipment in certain aircraft. Here we model the results for the four aircraft as independent samples from gamma distributions with common index ψ and rate parameter λ_j depending on the aircraft. Figure 8.1 contains a plot of the conditional log-likelihood function for ψ together with the profile log-likelihood function. ∎

Example 8.3 Poisson regression
Let $Y_1, \ldots, Y_n$ denote independent random variables such that y_j has a Poisson distribution with mean $\exp\{\lambda + \psi x_j\}$ where $x_1, \ldots, x_n$ are known constants. The conditional distribution of $y_1, \ldots, y_n$ given $s = \sum y_j$ does not depend on λ. Since s has a Poisson distribution with mean

$$\sum \exp\{\lambda + \psi x_j\},$$

it is straightforward to show that the conditional log-likelihood function is given by

$$\psi \sum x_j y_j - s \log\left[\sum \exp\{\psi x_j\}\right].$$ ∎

Example 8.4 A capture–recapture model
Consider drawing two independent samples from a closed population of ψ individuals. Let λ_1 denote the probability that a given individual is in the first sample and let λ_2 denote the probability that a given individual is in the second sample. Let y_1 denote the number of individuals in the first sample only, y_2 denote the number of individuals in the second sample only, and let y_3 denote the number of individuals in both samples. The likelihood function based on y_1, y_2, y_3 is given by

$$\binom{\psi}{y_1, y_2, y_3} \lambda_1^{y_1+y_3} \lambda_2^{y_2+y_3} (1-\lambda_1)^{\psi-y_1-y_3} (1-\lambda_2)^{\psi-y_2-y_3}.$$

Let $s_1 = y_1 + y_2 + 2y_3$; $s_2 = y_1 + y_3$; $s = (s_1, s_2)$. For fixed ψ, s is sufficient, so that a likelihood function for ψ may be based on the conditional distribution of y_1, y_2, y_3 given s; this conditional likelihood function is given by

$$\binom{\psi}{y_1, y_2, y_3} \Big/ \left[\binom{\psi}{s_1 - s_2}\binom{\psi}{s_2}\right] = \binom{\psi}{y_1, y_2, y_3} \Big/ \left[\binom{\psi}{y_2 + y_3}\binom{\psi}{y_1 + y_3}\right].$$

Note that conditional inference given s is equivalent to treating the number of individuals captured and the number of individuals recaptured as fixed constants. ∎

8.2.2 ANCILLARITY IN THE PRESENCE OF A NUISANCE PARAMETER

It has been noted that if s is also sufficient in the model with parameter (ψ, λ), then the resulting conditional likelihood is useless for inference about ψ. More generally, one possible drawback of a conditional likelihood function is that the conditioning variable s may contain information about ψ and this information is lost when s is taken to be fixed. The following example illustrates this possibility.

Example 8.5 Normal-theory linear regression with repetition
Let (Y_{j1}, Y_{j2}), $j = 1, \ldots, n$ denote independent pairs of independent random variables such that Y_{jk} is normally distributed with mean $x_j\lambda$ and variance ψ; here $x_1, \ldots, x_n$ are fixed $1 \times p$ vectors and λ is a $p \times 1$ parameter vector. For each j, the conditional distribution of (y_{j1}, y_{j2}) given $s_j = y_{j1} + y_{j2}$ depends only on ψ and, hence, we may base a conditional likelihood function on the conditional distribution of $y_1, \ldots, y_n$ given $s_1, \ldots, s_n$, where $y_j = y_{j1} - y_{j2}$. Since y_j and s_j are independent and y_j is normally distributed with mean 0 and variance 2ψ, this conditional log-likelihood function is given by

$$-\frac{n}{2}\log\psi - \frac{1}{4\psi}\sum y_j^2.$$

By conditioning on $s_1, \ldots, s_n$ we have lost the information regarding ψ in the distribution of $(s_1, \ldots, s_n)$. Note that the s_j are independent normally distributed random variables and that s_j has mean $2x_j\lambda$ and variance 2ψ. ∎

Hence, it is appropriate to require that the distribution of the conditioning statistic s not contain any information regarding ψ; in this case s is said to be *ancillary for ψ in the presence of λ*. It is important to note that ancillarity in the presence of a nuisance parameter is qualitatively different than the type of general ancillarity described in Section 1.6 and used extensively in Chapter 6. Unfortunately, it is surprisingly difficult to make precise the idea of 'no information for ψ' and, therefore, many different definitions of ancillarity in the presence of a nuisance parameter have been proposed, differing in the way we interpret this idea.

First suppose that the distribution of S depends only on λ; then clearly S contains no information about ψ. Hence, we might define S to be ancillary for ψ provided that the distribution of S depends only on λ. This definition of ancillarity in the presence of a nuisance parameter is unsuitable since it depends on how the nuisance parameter of the model is chosen. This idea however is used as the basis of S-ancillarity.

A statistic s is said to be S-ancillary for ψ in the presence of λ if the family of density functions $\{p(s; \psi, \lambda) \colon \lambda \in \Lambda\}$ is the same for each ψ; here Λ denotes the space of possible λ. That is, for each ψ_0, ψ_1, and λ_0 there exists a λ_1 such that

$$p(s; \psi_0, \lambda_0) = p(s; \psi_1, \lambda_1);$$

we may write $\lambda_1 = h(\psi_0, \psi_1, \lambda_0)$. Fix ψ_1 and let $h_1(\psi, \lambda) = h(\psi, \psi_1, \lambda)$; then for any ψ and λ,

$$p(s; \psi, \lambda) = p_1(s; h_1(\psi, \lambda))$$

where $p_1(s; h_1(\psi, \lambda)) = p(s; \psi_1, h_1(\psi, \lambda))$.

Therefore s is S-ancillary if and only if there exists an interest-respecting parameterization of the model (ψ, ϕ) such that the distribution of s depends only on ϕ.

Example 8.6 Poisson regression

Consider the Poisson regression model considered in Example 8.3. Since s has a Poisson distribution with mean

$$\phi \equiv \sum \exp\{\lambda + \psi x_j\},$$

s is S-ancillary. ∎

Example 8.7 Gamma distribution

Consider the gamma distributions considered in Example 8.2 for the case $m = 1$ so that we observe a single sample of gamma random variables with index ψ and rate parameter λ. Here $s = \sum y_j$ has a gamma distribution with rate parameter λ and index $n\psi$. It is easy to show that s is not S-ancillary. For instance, suppose that $\psi = 1$. Then the coefficient of variation of s is $1/\sqrt{n}$ for all values of λ; if $\psi = 4$ then the coefficient of variation of s is $2/\sqrt{n}$ for all λ. ∎

Another approach to defining ancillarity in the presence of a nuisance parameter may be based on the partial information for ψ. As in Section 3.6.3, let $i_\psi(\theta)$ denote the partial information for ψ based on the distribution of Y and let $i_\psi(\theta; S)$ denote the partial information for ψ based on the distribution of S alone. Then, since

$$\ell_{\psi\psi}(\theta) = \ell_{\psi\psi}(\theta; s) + \ell_{\psi\psi}(\theta; y|s);$$
$$\ell_{\psi\lambda}(\theta) = \ell_{\psi\lambda}(\theta; s); \qquad \ell_{\lambda\lambda}(\theta) = \ell_{\lambda\lambda}(\theta; s),$$

it follows that $i_\psi(\theta) - i_\psi(\theta; S)$ is equal to the partial information for ψ based on the conditional distribution of the data given S.

The statistic S is said to be P-ancillary if $i_\psi(\theta; S) = 0$ for all θ. In this case, the partial information for ψ based on the conditional distribution of Y given $S = s$ is equal to the partial information for ψ based on the full distribution of Y.

Example 8.8 Poisson regression
Consider the Poisson regression model considered in Examples 8.3 and 8.6. The log-likelihood function based on the distribution of S is given by

$$\ell(\psi, \lambda; s) = s \log \sum \exp\{\lambda + \psi x_j\} - \sum \exp\{\lambda + \psi x_j\}.$$

Then

$$i_{\psi\psi}(\theta; S) = \frac{\left[\sum x_j \exp\{\lambda + \psi x_j\}\right]^2}{\sum \exp\{\lambda + \psi x_j\}};$$

$$i_{\psi\lambda}(\theta; S) = \sum x_j \exp\{\lambda + \psi x_j\}; \qquad i_{\lambda\lambda}(\theta; S) = \sum \exp\{\lambda + \psi x_j\}.$$

It follows that $i_\psi(\theta; S) = 0$ so that S is P-ancillary. ∎

Example 8.9 Gamma distribution
Consider the gamma distribution considered in Example 8.7. The log-likelihood function based on the distribution of s is given by

$$\ell(\psi, \lambda; s) = n\psi \log(n\lambda) - \log\Gamma(n\psi) + n\psi \log(s) - \lambda s.$$

It is straightforward to show that

$$i_\psi(\theta; S) = n^2\Psi'(n\psi) - \frac{n}{\psi} > 0.$$

Hence, S is not P-ancillary. ∎

As these examples suggest, P-ancillarity is closely related to S-ancillarity. First note that the condition $i_\psi(\theta; S) = 0$ is invariant under interest-respecting reparameterizations. Hence, without loss of generality, we may assume that the distribution of an S-ancillary statistic depends only on the nuisance parameter λ. It follows easily that $i_\psi(\theta; S) = 0$.

Now suppose that $i_\psi(\theta; S) = 0$. First note that

$$\ell_\psi(\theta; s) - i_{\psi\lambda}(\theta; S)i_{\lambda\lambda}(\theta; S)^{-1}\ell_\lambda(\theta; s)$$

has variance $i_\psi(\theta; S)$; here $i_{\psi\lambda}(\theta; S)$ and $i_{\lambda\lambda}(\theta; S)$ denote components of the expected information matrix based on the distribution of S. Hence, if $i_\psi(\theta; S) = 0$, then for each θ,

$$\ell_\psi(\theta; s) = i_{\psi\lambda}(\theta; S)i_{\lambda\lambda}(\theta; S)^{-1}\ell_\lambda(\theta; s)$$

with probability 1. Consider an interest-respecting parameterization of the model (ψ, ϕ) where $\phi = h(\psi, \lambda)$ and $\lambda = g(\psi, \phi)$ and let $\tilde{\ell}(\psi, \phi; s)$ denote the log-likelihood function based on s in the (ψ, ϕ)-parameterization. Suppose that the function g satisfies the differential equation

$$\frac{\partial g}{\partial \psi}(\psi, \phi) = i_{\lambda\lambda}(\psi, g(\psi, \phi); S)^{-1} i_{\lambda\psi}(\psi, g(\psi, \phi); S);$$

note that a solution to this equation generally exists. Then

$$\begin{aligned}\tilde{\ell}_\psi(\psi, \phi; s) &= \ell_\psi(\psi, g(\psi, \phi); s) + \ell_\lambda(\psi, g(\psi, \phi); s)^T \frac{\partial g}{\partial \psi}(\psi, \phi) \\ &= \ell_\psi(\psi, \lambda; s) - i_{\psi\lambda}(\psi, \lambda; S)i_{\lambda\lambda}(\psi, \lambda; S)^{-1}\ell_\lambda(\psi, \lambda; s).\end{aligned}$$

It follows that $\tilde{\ell}_\psi(\psi, \phi; s) = 0$ with probability 1 so that the distribution of S depends only on ϕ. If the set of possible ϕ is the same for each ψ, then S is S-ancillary as well; in a sense, P-ancillarity is a local form of S-ancillarity. As the following example shows however these two definitions of ancillarity are not equivalent.

Example 8.10 Normal-theory linear regression with known variance
Let $Y_1, \ldots, Y_n$ denote independent normally distributed random variables such that Y_j has mean $\lambda + \psi x_j$ and variance 1 where $\lambda > 0$ and $-\infty < \psi < \infty$; here $x_1, \ldots, x_n$ are known scalar constants not all equal. For fixed ψ, $s = \bar{y}$ is sufficient and, hence, inference about ψ may be based on the conditional likelihood function given s. The statistic s is normally distributed with mean $n(\lambda + \psi\bar{x})$ and variance n; for simplicity, assume $\bar{x} > 0$.

The log-likelihood function based on s is given by

$$\ell(\psi, \lambda; s) = -\frac{1}{2n}[s - n(\lambda + \psi\bar{x})]^2$$

and it is straightforward to show that, for each θ, $i_\psi(\theta; S) = 0$ so that S is P-ancillary.

However, the set of possible distributions of S depends on ψ. For instance, suppose $\psi = 0$. Then S is normally distributed with mean $n\lambda$ so that the mean of S is positive. If $\psi = -1$, then S is normally distributed with mean $n\lambda - \bar{x}$ so that the mean of S is any real number greater than $-\bar{x}$. It follows that S is not S-ancillary. ∎

Both S-ancillarity and P-ancillarity are strong conditions that are not often satisfied. However both conditions are designed to determine when no information is lost by conditioning on s; in practice, some information loss may be considered an acceptable price to pay for eliminating the nuisance parameter from the model. In these cases, a large-sample version of ancillarity in the presence of a nuisance parameter may be appropriate. The simplest type of asymptotic ancillarity is based on P-ancillarity. We will say that S is asymptotically ancillary for ψ in the presence of λ if, for each θ,

$$\frac{i_\psi(\theta; S)}{i_\psi(\theta)} \to 0 \quad \text{as } i_\psi(\theta) \to \infty$$

so that the information in the distribution of S relative to that in the entire sample goes to 0. Since, in general, $i_\psi(\theta) = O(n)$, S is asymptotically ancillary provided that

$$\frac{i_\psi(\theta; S)}{n} \to 0 \quad \text{as } n \to \infty$$

Example 8.11 Gamma distribution
Consider the gamma distribution considered in Example 8.9. There it was shown that

$$i_\psi(\theta; S) = n^2 \Psi'(n\psi) - \frac{n}{\psi}.$$

Using the standard asymptotic expansion for $\Psi'(n\psi)$, it may be shown that

$$i_\psi(\theta; S) = \frac{1}{2}\frac{1}{\psi^2} + O(n^{-1})$$

so that S is asymptotically ancillary for ψ. ∎

Example 8.12 Normal-theory linear regression with repetition
Consider the normal-theory linear regression model considered in Example 8.5. In this example, $S = (S_1, \ldots, S_n)$ where S_j is normally distributed with mean $2x_j\lambda$ and variance 2ψ. It is straightforward to show that

$$i_\psi(\theta; S) = \frac{n}{2}\frac{1}{\psi^2};$$

it follows that S is not asymptotically ancillary for ψ.

Suppose that we condition on the $p \times 1$ vector $S_0 = \sum x_j^T S_j$ instead of on S. Note that S_0 has a multivariate normal distribution with mean vector $2\sum x_j^T x_j \lambda$ and covariance matrix $2\psi \sum x_j^T x_j$. It follows that the conditional log-likelihood function for ψ is given by

$$-(n - \frac{p}{2}) \log \psi - \frac{1}{2\psi} Q,$$

where Q denotes the sum of squared least-squares residuals and that

$$i_\psi(\theta; S_0) = \frac{p}{2}\frac{1}{\psi^2};$$

hence, S_0 is asymptotically ancillary for ψ. ∎

8.2.3 MODELS WITH MANY NUISANCE PARAMETERS

There is an important class of models in which the dimension of the nuisance parameter is of the same order as the sample size. For instance, we may have independent observations $Y_1, \ldots, Y_n$, which may be vectors, such that the distribution of Y_j depends on the parameter of interest ψ, which is common to all Y_j, as well as the nuisance parameter λ_j which is specific to Y_j. The nuisance parameters $\lambda_1, \ldots, \lambda_n$ are sometimes described as incidental nuisance parameters. The parameter space for such a model is effectively infinite-dimensional and thus does not fall into the general framework described in Chapter 3. Hence, rather than present a general theory, we only consider some examples which illustrate some of the issues involved.

Suppose that, for each j, there exists a statistic S_j such that the conditional distribution of Y_j given s_j depends only on ψ. Then a conditional likelihood function for ψ may be constructed from the conditional distribution of $Y_1, \ldots, Y_n$ given $s_1, \ldots, s_n$.

Example 8.13 Normal distributions with common standard deviation
Let (X_j, Y_j), $j = 1, \ldots, n$, denote independent pairs of independent normal random variables such that X_j and Y_j have mean μ_j and standard deviation σ, which is the parameter of interest. The log-likelihood function is

$$\ell(\sigma, \mu_1, \ldots, \mu_n) = -2n \log \sigma - \frac{1}{2\sigma^2} \sum [(x_j - \mu_j)^2 + (y_j - \mu_j)^2].$$

For fixed σ, $s_j = x_j + y_j$ is sufficient for μ_j. It is straightforward to show that the conditional log-likelihood function given $s_1, \ldots, s_n$ is given by

$$-n \log \sigma - \frac{1}{4\sigma^2} \sum (x_j - y_j)^2. \tag{8.1}$$

Inference based on this conditional log-likelihood differs considerably from standard methods of inference based on the log-likelihood function for the full data. For instance, the maximum likelihood estimate of σ is given by

$$\hat{\sigma} = \left[\frac{1}{4n} \sum (x_j - y_j)^2\right]^{1/2},$$

which is not consistent as $n \to \infty$. The conditional maximum likelihood estimate is given by

$$\left[\frac{1}{2n} \sum (x_j - y_j)^2\right]^{1/2}. \qquad \blacksquare$$

Example 8.14 Time-dependent Poisson processes with common rate parameter
For each j, let $Y_{j1}, Y_{j2}, \ldots, Y_{jn_j}$ denote the event times of a time-dependent Poisson process observed during the interval $(0, t_j)$ and having rate function $\lambda_j \exp\{\psi t\}$. Suppose that m such processes are observed; assuming the individual processes are

independent, the log-likelihood function for this model is given by

$$\sum n_j \log \lambda_j + \psi \sum y_j - \frac{1}{\psi} \sum \lambda_j (\exp\{\psi t_j\} - 1)$$

where $y_j = \sum y_{jk}$. Note that here $n_1, \dots, n_m$ are random variables.

For fixed values of $\psi, n_1, \dots, n_m$ are sufficient. Recall that the n_j are independent Poisson random variables such that n_j has mean

$$\frac{\lambda_j}{\psi}(\exp\{\psi t_j\} - 1).$$

It follows that the conditional log-likelihood function is given by

$$\sum n_j \log \psi + \psi \sum y_j - \sum n_j \log(\exp\{\psi t_j\} - 1).$$

For this model, this is identical to the profile log-likelihood function so that, for example, the conditional maximum likelihood estimate is identical to the usual maximum likelihood estimate. ∎

An alternative approach to analyzing models with incidental parameters is to model the incidental parameters as independent identically distributed random variables with density function $f(\cdot; \lambda)$, where λ is an unknown parameter. This may be viewed as a 'random effects' version of the model. The resulting model for the data $y_1, \dots, y_n$ is mixture model obtained by integrating out $\lambda_1, \dots, \lambda_n$; this model has parameters ψ, λ. Inference regarding ψ can then proceed using standard methods for parametric models.

If a conditional likelihood function is available in the model with incidental parameters $\lambda_1, \dots, \lambda_n$ then a conditional likelihood function is available in the model with parameters ψ, λ and the two conditional likelihood functions are identical. Let S_j denote a statistic such that the conditional distribution of Y_j given s_j depends only on ψ. Note that the density of Y_j under the model with parameters (ψ, λ) is given by

$$\bar{p}(y_j; \psi, \lambda) = \int p(y_j; \psi, \lambda_j) f(\lambda_j; \lambda)\, d\lambda_j;$$

a similar expression is available for $\bar{p}(s_j; \psi, \lambda)$, the density of S_j under the model with parameters (ψ, λ). Now suppose that

$$\frac{p(y_j; \psi, \lambda_j)}{p(s_j; \psi, \lambda_j)} \equiv p(y_j | s_j; \psi)$$

does not depend on λ_j so that a conditional likelihood function exists in the model with parameters $(\psi, \lambda_1, \dots, \lambda_n)$. Then

$$\begin{aligned} \bar{p}(y_j; \psi, \lambda) &= \int \frac{p(y_j; \psi, \lambda_j)}{p(s_j; \psi, \lambda_j)} p(s_j; \psi, \lambda_j) f(\lambda_j; \lambda)\, d\lambda_j \\ &= p(y_j | s_j; \psi) \bar{p}(s_j; \psi, \lambda). \end{aligned}$$

It follows that in the model with parameters (ψ, λ) the conditional likelihood function given s_j may be formed from

$$\frac{\bar{p}(y_j; \psi, \lambda)}{\bar{p}(s_j; \psi, \lambda)} = p(y_j | s_j; \psi)$$

so that this conditional likelihood function is identical to the one based on the model with parameters $(\psi, \lambda_1, \ldots, \lambda_n)$.

Example 8.15 Normal distributions with common standard deviation
Consider the model considered in Example 8.13. In the random effects version of this model, $\mu_1, \ldots, \mu_n$ are modeled as independent identically distributed random variables. Suppose that the μ_j are taken to be normally distributed with mean μ and variance τ. Under this model, the pair (X_j, Y_j) has a bivariate normal distribution such that X_j and Y_j both have mean μ and standard deviation $(\sigma^2 + \tau)^{1/2}$ and that X_j and Y_j have covariance τ.

The resulting log-likelihood function is given by

$$-\frac{n}{2}\log(\tau + \sigma^2/2) - n\log\sigma - \frac{1}{4(2\tau + \sigma^2)}\sum(s_j - 2\mu)^2$$
$$-\frac{1}{4\sigma^2}\sum(x_j - y_j)^2;$$

recall that $s_j = x_j + y_j$. For fixed σ, $s_1, \ldots, s_n$ is sufficient and the resulting conditional likelihood function is identical to (8.1) obtained in Example 8.13. ∎

Although the fixed effects and random effects versions of the model lead to the same conditional likelihood function, the type of model used does have an effect on the assessment of the information available for ψ in the conditioning statistic $s_1, \ldots, s_n$. This is illustrated on the following example.

Example 8.16 Normal distributions with common standard deviation
Consider the example considered in Examples 8.13 and 8.15. First consider the fixed effects version of the model with parameters $\mu_1, \ldots, \mu_n, \psi$. The log-likelihood function based on the distribution of the statistic $s = (s_1, \ldots, s_n)$ is given by

$$-n\log\sigma - \frac{1}{4\sigma^2}\sum(s_j - 2\mu_j)^2.$$

It follows that σ and μ_j are orthogonal parameters and, hence, the partial information for σ based on the distribution of S is given by

$$i_\sigma(\theta; S) = \frac{2n}{\sigma^2}.$$

The partial information available in the entire sample is given by $i_\sigma(\theta) = 4n/\sigma^2$. Hence, s is neither P-ancillary nor asymptotically P-ancillary. This suggests that conditioning on s leads to a substantial loss of information regarding σ.

Note, however, that these evaluations of the available information appear to be overly optimistic. For instance, the conditional maximum likelihood estimate has asymptotic variance $\sigma^2/2$ which exceeds the bound provided by the inverse of the average partial information, $\sigma^2/4$.

Now consider the random effects version of the model. Under this model, s_j has a normal distribution with mean 2μ and variance $4\tau + 2\sigma^2$. It follows that the log-likelihood function based on the distribution of s is given by

$$-\frac{n}{2}\log(2\tau+\sigma^2) - \frac{1}{4(2\tau+\sigma^2)}\sum(s_j - 2\mu)^2$$

and, hence, that $i_\sigma(\theta; S) = 0$ so that s is P-ancillary. The total partial information in the sample is easily calculated to be $2n/\sigma^2$. Hence, according to this model, conditioning on s does not result in any information loss.

In this example, it seems clear that in the fixed effects version of the model the partial information in the distribution of s overstates the amount of information actually available for inference about σ since, in order to use that information, consistent estimation of $\mu_1, \ldots, \mu_n$ is needed, which is not possible in this model. ∎

8.2.4 EXPONENTIAL FAMILY MODELS

One common situation in which conditional inference is often used is in the case of inference about the canonical parameter of an exponential family model. Suppose that T is a scalar statistic and S is a vector statistic with the same dimension as λ such that the density of (T, S) is a full-rank exponential family density of the form

$$\exp\{n\psi t + n\lambda^T s - nd(\psi, \lambda) + Q(t, s)\}.$$

Clearly s is a complete sufficient statistic for fixed ψ so that the conditional distribution given s depends only on ψ.

We now consider an approximation to the conditional likelihood given s. The log-likelihood function based on the full data is

$$n\psi t + n\lambda^T s - nd(\psi, \lambda)$$

and the conditional log-likelihood function is the full log-likelihood minus the log-likelihood function based on the marginal distribution of s. To approximate this marginal log-likelihood we can use a saddlepoint approximation for the density of S, evaluated at the observed value s of S.

The cumulant-generating function of S is given by

$$K(z) = n[d(\psi, \lambda + z/n) - d(\psi, \lambda)].$$

The saddlepoint equation is therefore given by

$$d_\lambda(\psi, \lambda + \hat{z}) = s;$$

when s is taken to be the observed value of S, this is exactly the likelihood equation for the model with ψ held fixed. Hence, $\lambda + \hat{z} = \hat{\lambda}_\psi$ where $\hat{\lambda}_\psi$ denotes the maximum likelihood estimator of λ for fixed ψ. We may therefore approximate the marginal likelihood function based on S by

$$|d_{\lambda\lambda}(\psi, \hat{\lambda}_\psi)|^{-1/2} \exp\{n[d(\psi, \hat{\lambda}_\psi) - d(\psi, \lambda)] - (\hat{\lambda}_\psi - \lambda)^T s\};$$

the resulting approximation to the conditional log-likelihood function is given by

$$\begin{aligned} n\psi t + n\hat{\lambda}_\psi^T s - n d(\psi, \hat{\lambda}_\psi) + \tfrac{1}{2} \log |d_{\lambda\lambda}(\psi, \hat{\lambda}_\psi)| \\ \equiv \ell(\psi, \hat{\lambda}_\psi) + \tfrac{1}{2} \log |d_{\lambda\lambda}(\psi, \hat{\lambda}_\psi)|. \end{aligned} \quad (8.3)$$

Example 8.17 Poisson regression
Consider the Poisson regression model considered in Example 8.3. The log-likelihood function is given by

$$\ell(\psi, \lambda) = \lambda \sum y_j + \psi \sum x_j y_j - \sum \exp\{\lambda + \psi x_j\}.$$

It follows that

$$\hat{\lambda}_\psi = \log\left(\frac{\sum y_j}{\sum \exp\{\psi x_j\}}\right).$$

The function $d(\psi, \lambda)$ is given by

$$d(\psi, \lambda) = \frac{1}{n} \sum \exp\{\lambda + \psi x_j\}.$$

It follows that the approximate conditional log-likelihood function given $s = \sum y_j$ is given by

$$\begin{aligned} &\hat{\lambda}_\psi \sum y_j + \psi \sum x_j y_j - \sum \exp\{\hat{\lambda}_\psi + \psi x_j\} + \frac{1}{2} \log \sum \exp\{\hat{\lambda}_\psi + \psi x_j\} \\ &= \psi \sum x_j y_j - s \log \sum \exp\{\psi x_j\}, \end{aligned}$$

neglecting terms not depending on ψ.

Comparing this result to the exact conditional likelihood given in Example 8.3 shows that for this example the approximation is exact. ∎

Example 8.18 Gamma distribution
Consider the gamma distribution model considered in Example 8.7. The log-likelihood function for the model may be written

$$\ell(\psi, \lambda) = \psi \sum \log y_j - \lambda s + n\psi \log \lambda - n \log \Gamma(\psi).$$

It follows that

$$\hat{\lambda}_\psi = \frac{n\psi}{s}$$

and

$$d(\psi, \lambda) = \log \Gamma(\psi) - \psi \log \lambda.$$

The approximate conditional log-likelihood function is given by

$$\psi \sum \log y_j - \hat{\lambda}_\psi s + n\psi \log \hat{\lambda}_\psi - n \log \Gamma(\psi) + \tfrac{1}{2} \log(\psi/\hat{\lambda}_\psi^2)$$
$$= \psi \sum \log y_j - n\psi \log s - n\psi + n\psi \log(n\psi) - n \log \Gamma(\psi) - \tfrac{1}{2} \log(n\psi).$$

The exact conditional log-likelihood function is given by

$$\psi \sum \log y_j - n\psi \log s + \log \Gamma(n\psi) - n \log \Gamma(\psi).$$

The difference between the approximate and exact conditional log-likelihood functions is

$$n\psi \log(n\psi) - n\psi - \tfrac{1}{2} \log(n\psi) - \log \Gamma(n\psi).$$

Using the fact that

$$\log \Gamma(x) = (\log(x) - 1)x - \tfrac{1}{2} \log(x) + \tfrac{1}{2} \log(2\pi) + O(x^{-1})$$

as $x \to \infty$, it follows that the difference between the approximate and exact conditional log-likelihood functions is of order $O((n\psi)^{-1})$ as $n \to \infty$. ∎

To investigate the asymptotic ancillarity of S in this context, we need to consider the partial information for ψ based on S. The log-likelihood function based on the marginal distribution of S is given by

$$\ell(\psi, \lambda; s) = n\lambda^T s - nd(\psi, \lambda) + \log \int \exp\{n\psi t + Q(t, s)\}\, dt$$

so that the score functions are

$$\ell_\lambda(\psi, \lambda; s) = ns - nd_\lambda(\psi, \lambda)$$

and

$$\ell_\psi(\psi, \lambda; s) = \frac{n \int t \exp\{n\psi t + Q(t, s)\}\, dt}{\int \exp\{n\psi t + Q(t, s)\}\, dt} - nd_\psi(\psi, \lambda).$$

Note that the conditional density of T given $S = s$ is given by

$$\frac{\exp\{n\psi t + Q(t, s)\}}{\int \exp\{n\psi t + Q(t, s)\}\, dt}$$

so that

$$\mathrm{E}[T|S = s; \psi] = \frac{\int t \exp\{n\psi t + Q(t, s)\}\, dt}{\int \exp\{n\psi t + Q(t, s)\}\, dt}.$$

Hence, $\ell_\psi(\psi, \lambda; s)$ may be written as

$$n\mathrm{E}[T|S = s; \psi] - nd_\psi(\psi, \lambda).$$

Using the mixed approximation for the conditional distribution of T given $S = s$ discussed in Section 2.10.4, we have that

$$\mathrm{E}[T|S = s; \psi] = d_\psi(\psi, \tilde{\lambda}_\psi) + O_p(n^{-1})$$

where $\tilde{\lambda}_\psi$ is the maximum likelihood estimate of λ for fixed ψ based on the observation of $S = s$. Hence,

$$\begin{aligned}\ell_\psi(\psi, \lambda) &= n[d_\psi(\psi, \tilde{\lambda}_\psi) - d_\psi(\psi, \lambda)] + O_p(1)\\ &= nd_{\psi\lambda}(\psi, \lambda)(\tilde{\lambda}_\psi - \lambda) + O_p(1).\end{aligned}$$

Using a Taylor's series expansion for the likelihood equation determining $\tilde{\lambda}_\psi$ we have that

$$\sqrt{n}(\tilde{\lambda}_\psi - \lambda) = d_{\lambda\lambda}(\psi, \lambda)^{-1}(s - d_\lambda(\psi, \lambda)) + O_p(n^{-1/2}).$$

Hence,

$$\ell_\psi(\psi, \lambda; s) = nd_{\psi\lambda}(\psi, \lambda)d_{\lambda\lambda}(\psi, \lambda)^{-1}(s - d_\lambda(\psi, \lambda)) + O_p(1).$$

It follows that the variance of $\ell_\psi(\psi, \lambda; s)$ is given by

$$nd_{\psi\lambda}(\psi, \lambda)d_{\lambda\lambda}(\psi, \lambda)^{-1}d_{\lambda\psi}(\psi, \lambda) + O(1).$$

The variance of $\ell_\lambda(\psi, \lambda; s)$ is $nd_{\lambda\lambda}(\psi, \lambda)$. Using the properties of conditional expectation, it is straightforward to show that the covariance of $\mathrm{E}[T|S; \psi]$ and S is identical to the covariance of T and S and is therefore given by $d_{\psi\lambda}(\psi, \lambda)/n$. Hence, the partial information for ψ based on the marginal distribution of S is given by

$$\begin{aligned}&nd_{\psi\lambda}(\psi, \lambda)d_{\lambda\lambda}(\psi, \lambda)^{-1}d_{\lambda\psi}(\psi, \lambda)\\ &\quad + O(1) - nd_{\psi\lambda}(\psi, \lambda)d_{\lambda\lambda}(\psi, \lambda)^{-1}d_{\lambda\psi}(\psi, \lambda) = O(1).\end{aligned}$$

It follows that S is asymptotically ancillary.

8.2.5 CONDITIONING ON A PARAMETER-DEPENDENT FUNCTION

In some cases, the sufficient statistic for the model with ψ fixed depends on the value of ψ. Let S_ψ denote such a statistic. Then the conditional distribution of the data given s_ψ depends only on ψ and, hence, this conditional distribution can, in principle, be used to form a conditional likelihood function. It is important to note, however, that it is not possible to take the conditional log-likelihood function to be simply

$$\ell(\theta) - \ell(\theta; s_\psi),$$

as it is in the case in which s_ψ does not depend on ψ. The following example illustrates the problems that occur in this setting.

Example 8.19 Ratio of normal means
Consider two independent samples of size n from normal distributions with means μ_1 and μ_2, respectively and each with standard deviation 1. Let X and Y denote the sample means and write $\psi = \mu_1/\mu_2$ and $\lambda = \mu_2$; assume that $\psi > 0$. Then the log-likelihood function is given by

$$\ell(\theta) = -\frac{n}{2}[(x - \psi\lambda)^2 + (y - \lambda)^2]$$
$$= n\lambda(\psi x + y) - \frac{n}{2}(\psi^2 + 1)\lambda^2.$$

Consider the model with ψ held fixed; the maximum likelihood estimator of λ in this model is $\hat{\lambda}_\psi = (\psi x + y)/(\psi^2 + 1)$ which is also the sufficient statistic in this model. The marginal distribution of $\hat{\lambda}_\psi$ is normal with mean λ and variance $1/[n(\psi^2+1)]$. Hence, the log-likelihood function based on this marginal distribution, written in terms of x and y, is given by

$$\ell(\theta; \hat{\lambda}_\psi)$$
$$= -\frac{n}{2}\left[\frac{(\psi x + y)^2}{\psi^2 + 1} + (\psi^2 + 1)\lambda^2 - 2\lambda(\psi x + y)\right] + \frac{1}{2}\log(\psi^2 + 1).$$

It follows that

$$\ell(\theta) - \ell(\theta; \hat{\lambda}_\psi) = \frac{n}{2}\frac{(\psi x + y)^2}{\psi^2 + 1} - \frac{1}{2}\log(\psi^2 + 1). \tag{8.3}$$

Another approach to constructing a conditional log-likelihood function given $\hat{\lambda}_\psi$ is to find a statistic t such that $(t, \hat{\lambda}_\psi)$ is sufficient in the full model with parameters (ψ, λ), determine the conditional distribution of t given $\hat{\lambda}_\psi$, and then use this conditional distribution as the basis for the conditional log-likelihood function. For instance, take $t = x$. Then

$$p(x|\hat{\lambda}_\psi; \psi) = \frac{p(x, \hat{\lambda}_\psi; \theta)}{p(\hat{\lambda}_\psi; \theta)}.$$

The density $p(x, \hat{\lambda}_\psi; \theta)$ may be obtained from $p(x, y; \theta)$ using the usual change-of-variable formula. The resulting conditional log-likelihood function is

$$\frac{n}{2}\frac{(\psi x + y)^2}{\psi^2 + 1} + \frac{1}{2}\log(\psi^2 + 1). \tag{8.4}$$

The same approach may be used with $t = y$, leading to the conditional log-likelihood function

$$\frac{n}{2}\frac{(\psi x + y)^2}{\psi^2 + 1} + \frac{1}{2}\log(\psi^2 + 1) - \log\psi \tag{8.5}$$

or with $t = \hat{\psi}$, leading to the conditional log-likelihood function

$$\frac{n}{2}\frac{(\psi x + y)^2}{\psi^2 + 1} + \frac{1}{2}\log(\psi^2 + 1) - \log|\psi x + y|. \tag{8.6}$$

Hence, there are at least four different conditional likelihood functions, depending on the way in which the calculation is performed. ∎

This example shows that the conditional log-likelihood function given a function s_ψ is not well defined whenever s_ψ depends on ψ. Consider the calculation of the conditional likelihood given s_ψ using a complementary statistic t. The conditional density of t given s_ψ is given by

$$p(t|s_\psi; \psi) = \frac{p(t, s_\psi; \theta)}{p(s_\psi; \theta)}.$$

Associated with each density function is a differential element, i.e., the density function $p(x)$ for a random variable X is perhaps more accurately written as $p(x)\,dx$. Roughly speaking, this quantity represents the probability that X lies in an interval $(x, x + dx)$.

A conditional density such as $p(t|s_\psi; \psi)$ may be written

$$\frac{p(t, s_\psi; \theta)\,d(t, s_\psi)}{p(s_\psi; \theta)\,ds_\psi}.$$

If S_ψ does not depend on ψ then the differential elements $d(t, s_\psi)$ and ds_ψ do not depend on ψ and, hence, they may be ignored when forming the conditional likelihood function. Whenever s_ψ does depend on ψ however these differential elements also depend on ψ and, hence, this dependence must be accounted for. Stated another way, the density $p(s_\psi; \theta)$ represents the probability that S_ψ lies in an interval $(s_\psi, s_\psi + ds_\psi)$; the length of this interval depends on ψ and, hence, must be accounted for. Similar considerations apply to the density of (T, S_ψ).

One approach to dealing with this issue is to express all the density functions with respect to a metric that is independent of ψ. Let z denote a variable of the same dimension as (t, s_ψ) such that (t, s_ψ) is a one-to-one function of z and that the Euclidean metric is thought to be appropriate for z. Then the density of (T, S_ψ) expressed in terms of this metric is given by

$$p(t, s_\psi; \theta)\left|\frac{\partial(t, s_\psi)}{\partial z}\right|.$$

The situation is not as simple for $p(s_\psi; \theta)$ since s_ψ and z are not of the same dimension; hence, $\partial s_\psi/\partial z$ is a $p \times q$ matrix where p is the dimension of s_ψ and q is the dimension of z, $q > p$. It may be shown that the correct Jacobian term in this case is

$$\left|\frac{\partial s_\psi}{\partial z}\frac{\partial s_\psi}{\partial z}^T\right|^{1/2}.$$

Hence, the marginal density of S_ψ is given by

$$p(s_\psi;\theta)\left|\frac{\partial s_\psi}{\partial z}\frac{\partial s_\psi}{\partial z}^T\right|^{1/2}.$$

It follows that the conditional density of t given s_ψ, expressed in terms of a Euclidean metric for z, is given by

$$\frac{p(t,s_\psi;\theta)}{p(s_\psi;\theta)}\frac{|\partial(t,s_\psi)/\partial z|}{|(\partial s_\psi/\partial z)\,(\partial s_\psi/\partial z)^T|^{1/2}}.$$

This form of the conditional density may then be used to construct a conditional likelihood function.

Example 8.20 Ratio of normal means
Recall the model considered in Example 8.19. Consider the conditional log-likelihood function based on the conditional distribution of X given $\hat\lambda_\psi$. To express the conditional density in terms of a ψ-independent metric, we first need to specify the variable z. For this model, $z=(x,y)$ seems appropriate. The next step is the calculation of

$$\frac{\partial(x,\hat\lambda_\psi)}{\partial z} \quad\text{and}\quad \frac{\partial\hat\lambda_\psi}{\partial z}.$$

It is straightforward to show that

$$\left|\frac{\partial(x,\hat\lambda_\psi)}{\partial z}\right| = \begin{vmatrix}1 & 0\\ \psi/(\psi^2+1) & 1/(\psi^2+1)\end{vmatrix} = \frac{1}{\psi^2+1}$$

and that

$$\frac{\partial\hat\lambda_\psi}{\partial z} = (\psi/(\psi^2+1)\quad 1/(\psi^2+1))$$

so that

$$\left|\frac{\partial s_\psi}{\partial z}\frac{\partial s_\psi}{\partial z}^T\right|^{1/2} = \frac{1}{(\psi^2+1)^{1/2}}.$$

It follows that the conditional log-likelihood function is (8.4) minus $[\log(\psi^2+1)]/2$, which yields

$$\frac{n}{2}\frac{(\psi x+y)^2}{\psi^2+1}. \tag{8.7}$$

Alternatively, we can take $t=y$. Then

$$\left|\frac{\partial(y,\hat\lambda_\psi)}{\partial z}\right| = \begin{vmatrix}0 & 1\\ \psi/(\psi^2+1) & 1/(\psi^2+1)\end{vmatrix} = \frac{\psi}{\psi^2+1}.$$

It is straightforward to show that this also leads to the log-likelihood function (8.7). Using $t=\hat\psi$ leads to (8.7) as well. ∎

Often the simplest approach to determining the conditional likelihood function in these cases is to use

$$\frac{p(y;\theta)}{p(s_\psi;\theta)} = \exp\{\ell(\theta) - \ell(\theta; s_\psi)\}$$

where y represents a sufficient statistic for the model, which may be taken to be the full set of data. When dz is taken to have metric $|(dz)^T dz|^{1/2}$, the resulting conditional log-likelihood function is given by

$$\ell(\theta) - \ell(\theta; s_\psi) + \frac{1}{2}\log\left|\frac{\partial y}{\partial z}\frac{\partial y}{\partial z}^T\right| - \frac{1}{2}\log\left|\frac{\partial s_\psi}{\partial z}\frac{\partial s_\psi}{\partial z}^T\right|$$

$$= \ell(\theta) - \ell(\theta; s_\psi) - \frac{1}{2}\log\left|\frac{\partial s_\psi}{\partial z}\frac{\partial s_\psi}{\partial z}^T\right|.$$

Example 8.21 Weibull distribution
Let $Y_1, \ldots, Y_n$ denote independent random variables each distributed according to a Weibull distribution with rate parameter λ and index ψ. The log-likelihood function for this model is given by

$$\ell(\theta) = n\log\psi + n\log\lambda + \psi\sum\log y_j - \lambda\sum y_j^\psi.$$

When ψ is known, $s_\psi = \sum y_j^\psi$ is a sufficient statistic and, hence, inference regarding ψ may be based on the conditional likelihood function given s_ψ. The marginal distribution of s_ψ is a gamma distribution with rate parameter λ and index n so that

$$\ell(\theta; s_\psi) = n\log\lambda + (n-1)\log\sum y_j^\psi - \lambda\sum y_j^\psi.$$

It follows that

$$\ell(\theta) - \ell(\theta; s_\psi) = n\log\psi + \psi\sum\log y_j - (n-1)\log\sum y_j^\psi.$$

To compute the conditional likelihood function, we need to specify the variable z which will be taken to have a Euclidean metric. First consider $z = (y_1, \ldots, y_n)$. Then, since

$$\frac{\partial s_\psi}{\partial y_j} = \psi y_j^{\psi-1}, \qquad \left|\frac{\partial s_\psi}{\partial z}\frac{\partial s_\psi}{\partial z}^T\right| = \psi^2\sum y_j^{2(\psi-1)}.$$

It follows that the conditional log-likelihood function is given by

$$n\log\psi + \psi\sum\log y_j - (n-1)\log\sum y_j^\psi - \log\psi - \frac{1}{2}\log\sum y_j^{2(\psi-1)}. \tag{8.8}$$

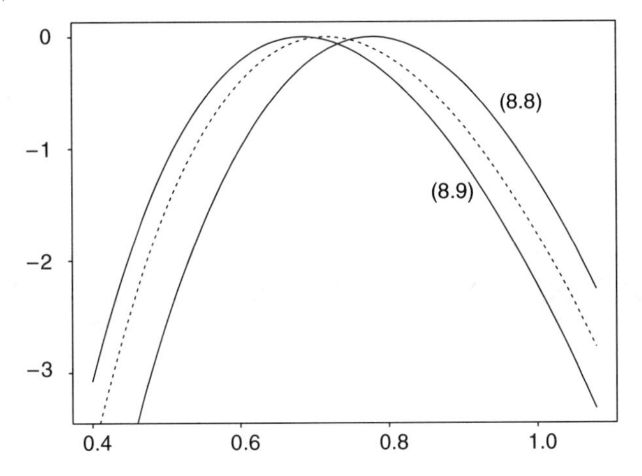

Fig. 8.2 Likelihood functions for Example 8.21.

Another choice for z is $z = (\log y_1, \dots, \log y_n)$; note that the model for $\log y_j$ is a location-scale model with scale parameter ψ. Since $s_\psi = \sum \exp\{\psi \log y_j\}$,

$$\frac{\partial s_\psi}{\partial \log y_j} = \psi \exp\{\psi \log y_j\} = \psi y_j^\psi;$$

hence,

$$\left| \frac{\partial s_\psi}{\partial z} \frac{\partial s_\psi}{\partial z}^T \right| = \psi^2 \sum y_j^{2\psi}.$$

It follows that the conditional log-likelihood function is given by

$$n \log \psi + \psi \sum \log y_j - (n-1) \log \sum y_j^\psi - \log \psi - \frac{1}{2} \log \sum y_j^{2\psi}. \quad (8.9)$$

Consider the data in Table A.3 of the Appendix. These data represent the failure times in hours of 20 pressure vessels subjected to a certain constant pressure. Figure 8.2 contains plots of the conditional log-likelihood functions (8.8) and (8.9), along with the profile log-likelihood. These results indicate that, at least in some cases, the choice of the variable z has a substantial effect on the resulting conditional likelihood.

∎

Example 8.22 Linear regression model
Consider the linear regression model considered in Example 7.5. In this example, $Y_1, \dots, Y_n$ denote independent random variables of the form $Y_j = \psi + x_j\beta + \sigma\epsilon_j$ where $x_1, \dots, x_n$ are known covariate vectors of length p satisfying $\sum x_j = 0$, ψ is an unknown scalar parameter, β is an unknown parameter vector of length p, and the ϵ_j are independent unobservable standard normal random variables. The log-likelihood

function is given by

$$\ell(\psi, \beta, \sigma) = -n \log \sigma - \frac{1}{2\sigma^2}[n\hat{\sigma}^2 + n(\hat{\psi} - \psi)^2 + (\hat{\beta} - \beta)^T x^T x(\hat{\beta} - \beta)].$$

For fixed ψ, the sufficient statistic for the model is given by $(\hat{\beta}, \hat{\sigma}^2_\psi)$ where

$$\hat{\sigma}^2_\psi = \hat{\sigma}^2 + (\hat{\psi} - \psi)^2$$

denotes the maximum likelihood estimator of σ for fixed ψ; hence, we may take $S = (\hat{\beta}, \hat{\sigma}^2_\psi)$. Using the usual results regarding the distribution of the maximum likelihood estimates of parameters in a normal-theory linear model, it follows that

$$\ell(\psi, \beta, \sigma; s) = -n \log \sigma - \frac{1}{2\sigma^2}[n\hat{\sigma}^2_\psi + (\hat{\beta} - \beta)^T x^T x(\hat{\beta} - \beta)]$$
$$+ (n - p - 2) \log \hat{\sigma}_\psi$$

and hence that

$$\ell(\psi, \beta, \sigma) - \ell(\psi, \beta, \sigma; s) = -(n - p - 2) \log \hat{\sigma}_\psi.$$

To complete specification of the conditional likelihood function, we need to compute the differential element. A reasonable choice for the variable z is $z = y$ and

$$\frac{\partial(\hat{\beta}, \hat{\sigma}^2_\psi)}{\partial y} = \begin{pmatrix} (x^T x)^{-1} x^T \\ -2/n(y - \psi e - x\hat{\beta})^T \end{pmatrix};$$

here e denotes a vector of all ones. Hence, neglecting constant terms,

$$\left| \frac{\partial s}{\partial z} \frac{\partial s}{\partial z}^T \right| = \hat{\sigma}^2_\psi.$$

It follows that the conditional log-likelihood function is given by

$$-(n - p - 1) \log \hat{\sigma}_\psi.$$

■

8.3 Marginal likelihood functions

8.3.1 INTRODUCTION

Another approach to constructing a pseudo-likelihood function for ψ is to base a likelihood function on the distribution of a statistic with a marginal distribution that depends only on ψ. Clearly, the resulting marginal likelihood function is a genuine likelihood function for ψ.

Suppose that there exists a statistic T such that the density of the data Y may be written

$$p(y; \psi, \lambda) = p(t; \psi) p(y|t; \psi, \lambda).$$

In this case, inference about ψ may be based on the marginal distribution of T which does not depend on λ. In particular, the marginal likelihood function based on t is given by

$$L(\psi; t) = p(t; \psi).$$

Example 8.23 Standard deviation of a normal distribution
Let $Y_1, \ldots, Y_n$ denote independent identically distributed normal random variables each with mean λ and standard deviation ψ. This example was discussed in Examples 5.4 and 5.6. Inference about ψ may be based on the marginal distribution of $T = \sum(Y_j - \bar{Y})^2$. Since $\sum(Y_j - \bar{Y})^2/\psi^2$ has a chi-square distribution with $n-1$ degrees of freedom, the resulting marginal log-likelihood function is given by

$$\ell(\psi; t) = -(n-1)\log\psi - \frac{1}{2}\frac{\sum(y_j - \bar{y})^2}{\psi^2}.$$

∎

Example 8.24 Inverse Gaussian distribution
Let $Y_1, \ldots, Y_n$ denote independent observations each distributed according to an inverse Gaussian distribution with density of the form

$$[\psi/(2\pi y^3)]^{1/2}\exp\left\{-\frac{\psi}{2\lambda^2 y}(y-\lambda)^2\right\}, \quad y > 0 \tag{8.10}$$

where $\psi > 0$ and $\lambda > 0$. The parameter of interest ψ is a scale parameter for the distribution.

Let $t = \sum(y_j^{-1} - \bar{y}^{-1})$. It may be shown that ψT has a chi-square distribution with $n-1$ degrees of freedom. Hence, the marginal log-likelihood function based on t is given by

$$\frac{n-1}{2}\log\psi - \frac{\psi}{2}t.$$

∎

Example 8.25 Two Poisson distributions
Let $Y_1, \ldots, Y_n, X_1, \ldots, X_m$ denote independent Poisson random variables such that y_j has mean λ_1 and X_j has mean λ_2; assume that $m \geq n$. Let $\psi = \lambda_1 + \lambda_2$ denote the parameter of interest and take $\lambda = \lambda_2$ as the nuisance parameter.

Let

$$Y = \sum_1^n Y_j; \qquad X = \sum_1^n X_j.$$

Then $T = X + Y$ has a Poisson distribution with mean $n\psi$. The marginal log-likelihood function based on t is given by

$$t\log\psi - n\psi.$$

∎

Example 8.26 Multinomial distribution
Consider a multinomial distribution of order d with index n. The probability of a given observation $(y_1, \ldots, y_d)$ is given by

$$\frac{n!}{y_1!\cdots y_{d+1}!}\pi_1^{y_1}\cdots\pi_d^{y_d}\pi_{d+1}^{y_{d+1}}$$

where $y_{d+1} = n - y_1 - \cdots - y_d$ and $\pi_{d+1} = 1 - \pi_1 - \cdots - \pi_d$. Suppose that the parameter of interest is $\psi = \pi_1$ and take $\pi_2, \ldots, \pi_d$ as nuisance parameters.

The marginal distribution of Y_1 is binomial with index n and success probability ψ. Hence,

$$\ell(\psi; y_1) = y_1 \log \psi + (n - y_1) \log(1 - \psi).$$ ∎

8.3.2 SUFFICIENCY IN THE PRESENCE OF A NUISANCE PARAMETER

As is the case with conditional likelihood functions, there is generally some loss of information involved in the use of a marginal likelihood function, specifically the information in the conditional distribution of Y given $T = t$ is not available for inference about ψ. Hence, it is reasonable to require that this conditional distribution does not contain any information about ψ; the statistic T is then said to be sufficient for ψ in the presence of λ. As with ancillarity in the presence of a nuisance parameter, there have been many attempts to define sufficiency in the presence of a nuisance parameter and many of these definitions are analogous to definitions of ancillarity.

For instance, if the conditional distribution of Y given $T = t$ depends only on λ, then T is sufficient for ψ in some sense. More generally, T is said to be S-sufficient for ψ in the presence of λ if the set of conditional density functions

$$\{p(y|t; \psi, \lambda) \colon \lambda \in \Lambda\}$$

is the same for each ψ. Let $i_\psi(\theta; T)$ denote the marginal information for ψ based on the marginal distribution of T. If $i_\psi(\theta; T) = i_\psi(\theta)$ then T is said to be P-sufficient for ψ. A large-sample version of P-sufficiency is also easily formulated. If, as $n \to \infty$,

$$\frac{i_\psi(\theta) - i_\psi(\theta; T)}{n} \to 0,$$

then T is said to be asymptotically sufficient for ψ.

Example 8.27 Inverse Gaussian distribution

Consider the inverse Gaussian distribution considered in Example 8.24. As noted previously, $(\bar{Y}, T)$ is minimal sufficient and $\bar{Y}$ and T are independent. Hence, in determining the information loss in using a likelihood based on the marginal distribution of T we may consider the conditional distribution of $\bar{Y}$ given T or, equivalently, the marginal distribution of $\bar{Y}$. The density function of $\bar{Y}$ is also of the form (8.10) with ψ replaced by $n\psi$. Hence, it is straightforward to show that

$$i_\psi(\theta; \bar{Y}) = \frac{1}{2} \frac{1}{\psi^2}.$$

It follows that T is neither S- nor P-sufficient for ψ. However, since $i_\psi(\theta; \bar{Y})/n = o(1)$, T is asymptotically sufficient for ψ. ∎

Example 8.28 Multinomial distribution

Consider the multinomial distribution considered in Example 8.26. The conditional distribution of $y_2, \ldots, y_d$ given y_1 is also multinomial with index $n - y_1$

and probabilities $\pi_j/(1-\psi)$, $j = 2, \ldots, d$. Since for any fixed ψ, the vector $(\pi_2/(1-\psi), \ldots, \pi_q/(1-\psi))$ may take any value in the set

$$\{(x_1, \ldots, x_{d-1}) \in [0, 1]^{d-1}\colon x_1 + \cdots + x_{d-1} \leq 1\},$$

it follows that Y_1 is S-sufficient. ∎

Example 8.29 Matched exponential samples
Consider independent pairs of independent random variables, $(X_1, Y_1), \ldots, (X_n, Y_n)$, such that X_j and Y_j are gamma random variables with index m and rate parameters λ_{1j} and λ_{2j}, respectively; here m is a known positive integer and λ_{1j} and λ_{2j} are unknown parameters. The observations X_j and Y_j may each be considered to be the sums of independent exponential random variables; by sufficiency, the individual observations are not needed. Let $\psi = \lambda_{1j}/\lambda_{2j}$ denote the parameter of interest, which is assumed to be the same for all j, and take $\lambda_j = \lambda_{2j}$ as the nuisance parameter. The log-likelihood function is given by

$$mn \log \psi + 2 \sum m \log \lambda_j - \sum \lambda_j(\psi x_j + y_j).$$

Let $T_j = X_j/Y_j$; the marginal distribution of $T = (T_1, \ldots, T_n)$ depends only on ψ and, hence, may be used to form a marginal likelihood function. The marginal density of T is easily obtained from the density of $(X_1, Y_1), \ldots, (X_n, Y_n)$ leading to the marginal log-likelihood function

$$\ell(\psi; T) = mn \log \psi - 2m \sum \log(\psi x_j + y_j).$$

To assess the information lost by using the marginal likelihood based on T, consider the partial information for ψ. For the full data, the partial information is given by

$$i_\psi(\theta) = \frac{mn}{2} \frac{1}{\psi^2};$$

for the marginal likelihood based on T, the partial information is given by

$$i_\psi(\theta; T) = \frac{mn}{2} \frac{2m}{2m+1} \frac{1}{\psi^2}.$$

Hence, the information loss is given by

$$\frac{mn}{2} \frac{1}{2m+1} \frac{1}{\psi^2};$$

as a proportion of the total partial information for ψ, the information loss is $1/(2m+1)$.

For the asymptotic scenario in which n, the number of pairs (X_j, Y_j), stays fixed while m, the number of observations making up each X_j, Y_j increases indefinitely, the information loss is negligible in large samples. On the other hand, for the asymptotic scenario in which n increases while m stays fixed, the information loss does not decrease in large samples.

An alternative formulation of this model involves modeling $\lambda_1, \ldots, \lambda_n$ as random variables with a distribution depending on a parameter λ; see Section 8.2.3. Inference for ψ may then be based on the model with parameter (ψ, λ).

For instance, suppose that the λ_j are independent exponential random variables with rate parameter λ. Under this model, the pairs (X_j, Y_j) are independent with density function

$$\frac{\lambda\psi^m\Gamma(2m+1)}{\Gamma(m)^2}\frac{x^{m-1}y^{m-1}}{(\psi x+y+\lambda)^{2m+1}}, \quad x>0, \ y>0$$

and the log-likelihood function is given by

$$n\log\lambda + mn\log\psi - (2m+1)\sum\log(\psi x_j + y_j + \lambda).$$

As in the fixed effects version of the model, the marginal distribution of T depends only on ψ and leads to the log-likelihood function

$$m\log\psi - 2m\sum\log(\psi x_j + y_j).$$

The partial information for ψ in the total data may be shown to be

$$i_\psi(\theta) = \frac{mn}{2}\frac{2m+1}{2m+2}\frac{1}{\psi^2};$$

the partial information for ψ is, as above, given by

$$i_\psi(\theta; T) = \frac{mn}{2}\frac{2m}{2m+1}\frac{1}{\psi^2}.$$

Hence, the random effects formulation of the model leads to the same general conclusions regarding the approximation sufficiency of T as does the fixed effects formulation; in particular, for the scenario in which n increases while m stays fixed, there is some loss of information in using the marginal likelihood function based on T. ∎

8.3.3 COMPOSITE TRANSFORMATION MODELS

The most common class of models for which a marginal likelihood function exists is composite transformation models which were discussed in Section 1.3. Suppose that the family of distributions of Y forms a composite transformation model with index parameter ψ and group parameter λ. Then a marginal likelihood function may be based on the maximal invariant statistic T. In this case, the conditional distribution of the data given T also forms a composite transformation model; the statistic T is said to be G-sufficient for ψ.

Example 8.30 Exponential regression

Let $Y_1, \ldots, Y_n$ denote independent exponential random variables such that y_j has mean $[\lambda\exp\{\psi x_j\}]^{-1}$ where $x_1, \ldots, x_n$ are fixed constants. This model may be viewed as a composite transformation model with group parameter λ. The marginal likelihood based on the maximal invariant statistic $(t_1, \ldots, t_n), t_j = y_j/y_1$, is given by

$$\frac{\exp\{\psi \sum x_j\}}{[\sum y_j \exp\{\psi x_j\}]^n}.$$

The conditional distribution of Y_1 given $T = t$ is gamma distribution with rate parameter $\lambda \sum \exp\{\psi x_j\} t_j$ and index n. It follows that T is S-sufficient as well as G-sufficient. ∎

Example 8.31 Uniform distribution
Let $Y_1, \ldots, Y_n$ denote independent random variables each uniformly distributed on the interval $(\lambda, \lambda + \psi)$, where $\psi > 0$. Clearly, the minimum and maximum of the sample, $Y_{(1)}$ and $Y_{(n)}$, are sufficient; the joint density of $(Y_{(1)}, Y_{(n)})$ is given by

$$\frac{n(n-1)}{\psi^n}[y_{(n)} - y_{(1)}]^{n-2}, \quad \lambda < y_{(1)} < y_{(n)} < \lambda + \psi.$$

For each fixed value of ψ this is a transformation model with group parameter λ; the maximal invariant statistic is $y_{(n)} - y_{(1)}$. It is straightforward to show that the marginal likelihood based on this statistic is

$$\psi^{-(n-1)}, \quad \psi > y_{(n)} - y_{(1)}.$$ ∎

8.3.4 MARGINAL LIKELIHOOD BASED ON A PARAMETER-DEPENDENT FUNCTION

A marginal likelihood may be based on a function of the data and ψ that has a distribution depending only on ψ. However, as in the case in which a conditional likelihood is based on conditioning on a parameter-dependent function, some care is needed. The issues involved are essentially the same as those encountered when constructing a conditional likelihood given a parameter-dependent function.

Example 8.32 Ratio of normal means
Consider the ratio of normal means example considered in Example 8.19. In this example, X and Y are independent normal random variables with means $\psi\lambda$ and λ, respectively and variance n^{-1}, where $\psi > 0$.

The function $X - \psi Y$ is normally distributed with mean 0 and variance $(1+\psi^2)/n$. Hence, the marginal log-likelihood function based on the distribution of $X - \psi Y$ is given by

$$-\frac{n}{2}\frac{(x - \psi y)^2}{1 + \psi^2} - \frac{1}{2}\log(1 + \psi^2).$$

Alternatively, a marginal likelihood could be based on the function

$$\frac{\sqrt{n}(X - \psi Y)}{[1 + \psi^2]^{1/2}}$$

which has a standard normal distribution. The resulting marginal log-likelihood function is given by

$$-\frac{n}{2}\frac{(x-\psi y)^2}{1+\psi^2}.$$

A third possibility is to consider the function

$$\frac{n(X-\psi Y)^2}{1+\psi^2}$$

which has a chi-square distribution with 1 degree of freedom. The resulting marginal log-likelihood function is given by

$$-\frac{n}{2}\frac{(x-\psi y)^2}{1+\psi^2}+\frac{1}{2}\log(1+\psi^2)-\log|x-\psi y|.$$

Hence, the form of the marginal likelihood depends on the exact form of the function used for its derivation. ∎

As in the case of a conditional likelihood, the solution to this problem is to include the Jacobian term in the specification of the marginal likelihood. Let T_ψ denote a function of the data and ψ with density $p(t_\psi;\psi)$ depending only on ψ and let z denote a variable such that a Euclidean metric for z is considered appropriate. Then the marginal likelihood based on T_ψ is given by

$$p(t_\psi;\psi)\left|\frac{\partial t_\psi}{\partial z}\frac{\partial t_\psi}{\partial z}\right|^{1/2}.$$

Example 8.33 Ratio of normal means

Consider the example considered in Example 8.32. A reasonable choice for the variable z is $z=(x,y)$. First consider $t_\psi = x-\psi y$. Then

$$\frac{\partial t_\psi}{\partial z}=(1 \quad -\psi)$$

and

$$\left|\frac{\partial t_\psi}{\partial z}\frac{\partial t_\psi}{\partial z}^T\right|^{1/2}=[1+\psi^2]^{1/2};$$

the resulting marginal log-likelihood is

$$-\frac{n}{2}\frac{(x-\psi y)^2}{1+\psi^2}. \tag{8.11}$$

Now take

$$t_\psi=\frac{\sqrt{n}(x-\psi y)}{[1+\psi^2]^{1/2}}.$$

Then

$$\frac{\partial t_\psi}{\partial z} = \frac{\sqrt{n}}{[1+\psi^2]^{1/2}}(1 \quad -\psi)$$

and

$$\left|\frac{\partial t_\psi}{\partial z}\frac{\partial t_\psi}{\partial z}^T\right|^{1/2} = \sqrt{n};$$

the resulting marginal log-likelihood is also given by (8.11).

Finally, consider

$$t_\psi = \frac{n(x-\psi y)^2}{1+\psi^2}.$$

Then

$$\frac{\partial t_\psi}{\partial z} = \frac{n}{1+\psi^2}(2(x-\psi y) - \psi 2(x-\psi y))$$

and

$$\left|\frac{\partial t_\psi}{\partial z}\frac{\partial t_\psi}{\partial z}^T\right|^{1/2} = \frac{2n}{(1+\psi^2)^{1/2}}|x-\psi y|;$$

the resulting marginal log-likelihood again is given by (8.11). ∎

Example 8.34 Linear regression model

Consider the linear regression model considered in Example 8.22. In this model,

$$T_\psi \equiv \frac{\hat{\psi}-\psi}{S/\sqrt{n}}$$

has a t-distribution with $n-(p+1)$ degrees of freedom, where

$$S^2 = \frac{n}{n-(p+1)}\hat{\sigma}^2.$$

The log-likelihood function based on T_ψ is given by

$$-\frac{n-p}{2}\log\left(1+\frac{t_\psi^2}{n-p-1}\right).$$

To complete specification of the marginal likelihood, calculation of $\partial t_\psi/\partial z$, for some variable z, is needed. A reasonable choice for z is $z=(y_1,\ldots,y_n)$. Then it is straightforward to show that

$$\left|\frac{\partial t_\psi}{\partial z}\frac{\partial t_\psi}{\partial z}^T\right| = \frac{n-p-1}{n\hat{\sigma}^2}\left[1+\frac{t_\psi^2}{(n-p-1)^2}\right].$$

The resulting marginal likelihood for ψ is given by

$$-\frac{n-p-1}{2}\log\left(1+\frac{t_\psi^2}{n-p-1}\right).$$

∎

8.4 Integrated likelihood functions

8.4.1 INTRODUCTION

The methods described in the previous two sections, based on conditioning and marginalization, respectively, can only be applied in certain circumstances. For instance, to use a conditional likelihood function it is necessary to be able to construct a statistic s such that the conditional distribution of the data given s depends only on ψ, the parameter of interest. Another approach, that can be applied generally, is to average the likelihood function $L(\psi, \lambda)$ with respect to a weight function on λ.

Let $\pi(\lambda)$ denote a weight function defined on Λ, the space of possible λ. Then an *integrated likelihood function* for ψ is given by

$$L(\psi; \pi) = \int_{\Lambda} L(\psi, \lambda)\pi(\lambda)\, d\lambda.$$

Inference about ψ may then be based on $L(\psi; \pi)$ or, equivalently, on the integrated log-likelihood function $\ell(\psi; \pi) = \log L(\psi; \pi)$.

Integrated likelihood functions have the advantage that, unlike conditional or marginal likelihoods, they are always available and, in principle, are relatively easy to determine, although sophisticated computational methods may be needed to evaluate the integrals that arise.

8.4.2 UNIFORM INTEGRATED LIKELIHOOD FUNCTION

The form of the integrated likelihood function depends on the weight function π used. Here we will consider the use of a uniform density so that $\pi(\lambda) = 1$; the resulting integrated likelihood function will be denoted $L_U(\psi; \lambda)$ to emphasize that the uniform weight function applies to λ.

Example 8.35 Index of a negative binomial distribution
Let $Y_1, \ldots, Y_n$ denote independent negative binomial random variables each with density function

$$\frac{\Gamma(\psi + y)}{\Gamma(y+1)\Gamma(\psi)}\lambda^y(1-\lambda)^\psi, \quad y = 0, 1, 2, \ldots,$$

where $\psi > 0$ and $0 < \lambda < 1$. The integrated likelihood function corresponding to a uniform weight function for λ is given by

$$\begin{aligned} L_U(\psi; \lambda) &= \frac{\prod \Gamma(y_j + \psi)}{\Gamma(\psi)^n} \int_0^1 \lambda^{n\bar{y}}(1-\lambda)^{n\psi}\, d\lambda \\ &= \frac{\prod \Gamma(y_j + \psi)}{\Gamma(\psi)^n} \frac{\Gamma(n\psi + 1)\Gamma(n\bar{y} + 1)}{\Gamma(n\bar{y} + n\psi + 2)}. \end{aligned}$$

∎

Example 8.36 Normal mean
Let $Y_1, \ldots, Y_n$ denote independent normally distributed random variables each with mean μ and standard deviation σ. Take μ as the parameter of interest so that in the

general notation $\psi = \mu$ and $\lambda = \sigma$. The likelihood function for (μ, σ) is given by

$$L(\mu, \sigma) = \sigma^{-n} \exp\left\{-\frac{1}{2\sigma^2}\sum(y_j - \mu)^2\right\}$$

so that the integrated likelihood function is given by

$$\begin{aligned} L_U(\mu; \sigma) &= \int_0^\infty \sigma^{-n} \exp\left\{-\frac{1}{2\sigma^2}\sum(y_j - \mu)^2\right\} d\sigma \\ &= \left[1 + \frac{(\mu - \bar{y})^2}{\hat{\sigma}^2}\right]^{-(n-1)/2} \end{aligned}$$

where $\hat{\sigma}^2 = \sum(y_j - \bar{y})^2/n$.

Suppose that the model is parameterized in terms of $\log\sigma$ instead of σ. A uniform weight function for $\log\sigma$ is equivalent to a weight function of the form σ^{-1} for σ. The resulting integrated likelihood function is given by

$$L_U(\mu; \log(\sigma)) = \left[1 + \frac{(\mu - \bar{y})^2}{\hat{\sigma}^2}\right]^{-n/2}.$$

If the model is parameterized in terms of the variance σ^2, the uniform integrated likelihood function is given by

$$L_U(\mu; \sigma^2) = \left[1 + \frac{(\mu - \bar{y})^2}{\hat{\sigma}^2}\right]^{-(n-2)/2}.$$ ∎

This example illustrates an important property of the uniform integrated likelihood function: the resulting likelihood function depends on the parameterization used. For many models, there is a natural parameterization for which a uniform weight function for λ is likely to be reasonable. However, invariance under interest-respecting reparameterizations is certainly preferable.

8.4.3 RELATIONSHIP WITH CONDITIONAL AND MARGINAL LIKELIHOOD FUNCTIONS

In many cases in which a conditional or marginal likelihood function is available, that likelihood function corresponds to an integrated likelihood function with respect to a specific weighting function.

Example 8.37 A capture–recapture model
Consider the capture–recapture model considered in Example 8.4. The likelihood function for the model is given by

$$\binom{\psi}{y_1, y_2, y_3} \lambda_1^{y_1+y_3} \lambda_2^{y_2+y_3} (1 - \lambda_1)^{\psi - y_1 - y_3} (1 - \lambda_2)^{\psi - y_2 - y_3}.$$

Let π denote the weighting function given by $\pi(\lambda_1, \lambda_2) = \lambda_1^{-1}\lambda_2^{-1}$. Then

$$\ell(\psi; \pi) = \binom{\psi}{y_1, y_2, y_3} \Big/ \left[\binom{\psi}{y_2 + y_3}\binom{\psi}{y_1 + y_3}\right]$$

which is exactly the conditional likelihood function given previously. ∎

Suppose that a conditional likelihood function is based on the conditional distribution of the data given a statistic s and suppose that s is S-ancillary. Then there is an integrated likelihood function that is identical to the conditional likelihood function $L(\psi; y|s)$. As noted in Section 8.2, if s is S-ancillary then there exists a function $h(\lambda; \psi_0, \psi_1)$, taking values in Λ, such that for any ψ_0, ψ_1, λ,

$$p(s; \psi_0, \lambda) = p(s; \psi_1, h(\lambda; \psi_0, \psi_1));$$

hence, the distribution of s depends only on the parameter $\phi \equiv h(\lambda; \psi_0, \psi_1)$ which takes values in the set Λ. It follows that the integrated likelihood function based on any weight function for the parameter ϕ of the form $\pi(\phi)$ is identical to the conditional likelihood function given S.

Example 8.38 Poisson regression

Consider the Poisson regression model considered in Examples 8.3 and 8.6; the likelihood function for the model is given by

$$L(\psi, \lambda) = \exp\left\{\lambda \sum y_j + \psi \sum x_j y_j\right\} \exp\left\{-\sum \exp\{\lambda + \psi x_j\}\right\}.$$

The conditioning statistic $S = \sum Y_j$ is S-ancillary and has a Poisson distribution with mean

$$\phi \equiv \sum \exp\{\lambda + \psi x_j\};$$

the likelihood function in terms of (ψ, ϕ) is given by

$$\frac{\phi^{\sum y_j}}{[\sum \exp\{\psi x_j\}]^{\sum y_j}} \exp\left\{-\phi + \psi \sum x_j y_j\right\}.$$

An integrated likelihood function with respect to a weight function $\pi(\phi)$ is given by

$$\frac{\exp\{\psi \sum x_j y_j\}}{[\sum \exp\{\psi x_j\}]^{\sum y_j}} \int_0^\infty \phi^{\sum y_j} \exp\{-\phi\}\pi(\phi)\, d\phi$$

which is equivalent to the conditional likelihood function given S. ∎

Consider a composite transformation model with index parameter ψ and group parameter λ. Then, it may be shown that the marginal likelihood function based on the maximal invariant statistic is equivalent to an integrated likelihood function with respect to a weight function taken to be the density of the right-invariant measure on $\mathcal{G}$, the group of transformations.

Example 8.39 Exponential regression
Consider the exponential regression model considered in Example 8.30. As noted in Section 8.3.3, this model is a composite transformation model with λ as the group parameter; the right-invariant measure has density λ^{-1}. It is straightforward to show that the integrated likelihood function with respect to this density is exactly the marginal density based on the maximal invariant. ∎

8.5 Inference based on a pseudo-likelihood function

8.5.1 INTRODUCTION

A marginal, conditional or integrated likelihood function depends only on the parameter of interest ψ and, hence, simplifies the statistical analysis. In this section, we consider the properties of statistical procedures based on one of these likelihoods.

8.5.2 MARGINAL LIKELIHOOD FUNCTIONS

In many respects, the simplest case is that of a marginal likelihood function. Consider a marginal likelihood based on the distribution of a statistic T that does not depend on ψ. Provided that the marginal log-likelihood satisfies the regularity conditions required of a log-likelihood, the properties of procedures based on $\ell(\psi; T)$ discussed in Chapters 4 through 7 will hold.

Example 8.40 Standard deviation of a normal distribution
Let $Y_1, \ldots, Y_n$ denote independent identically distributed normal random variables each with mean λ and standard deviation ψ; see Example 8.23. The marginal likelihood function is given by

$$\ell(\psi; t) = -(n-1)\log\psi - \frac{1}{2}\frac{\sum(y_j - \bar{y})^2}{\psi^2}$$

where $T = \sum(Y_j - \bar{Y})^2$. This marginal log-likelihood is maximized by

$$\hat{\psi}^2 = \frac{\sum(y_j - \bar{y})^2}{n-1}$$

which differs from the maximum likelihood estimate of ψ.

The formal cumulants of $\sqrt{n}(\hat{\psi} - \psi)$ may be obtained using the results presented in Section 5.3, with the quantities ν_2, ν_{11}, and so on based on the log-likelihood function $\ell(\psi; t)$. Since

$$\ell_\psi(\psi; t) = -\frac{n-1}{\psi} + \frac{\sum(y_j - \bar{y})^2}{\psi^3};$$

$$\ell_{\psi\psi}(\psi; t) = \frac{n-1}{\psi^2} - 3\frac{\sum(y_j - \bar{y})^2}{\psi^4};$$

$$\ell_{\psi\psi\psi}(\psi; t) = -2\frac{n-1}{\psi^3} + 12\frac{\sum(y_j - \bar{y})^2}{\psi^5},$$

it follows that

$$\nu_2(\psi) = \frac{2(n-1)}{n}\frac{1}{\psi^2}; \quad \nu_{11}(\psi) = -\frac{6(n-1)}{n}\frac{1}{\psi^3};$$
$$\nu_{30}(\psi) = \frac{8(n-1)}{n}\frac{1}{\psi^3}; \quad \nu_{001}(\psi) = \frac{10(n-1)}{n}\frac{1}{\psi^3}.$$

Using the results of Section 5.3, it follows that the first three formal cumulants of $\sqrt{(n\nu_2(\psi))}(\hat{\psi} - \psi)$ are given by

$$-\frac{1}{2\sqrt{2}}\frac{1}{\sqrt{(n-1)}}; \quad 1; \quad \frac{1}{\sqrt{2}}\frac{1}{\sqrt{(n-1)}}.$$

These expressions have error $O(n^{-3/2})$ or smaller. ∎

Example 8.41 Inverse Gaussian distribution
Consider the inverse Gaussian distribution considered in Example 8.24. The marginal log-likelihood function is given by

$$\ell(\psi; t) = \frac{n-1}{2}\log\psi - \frac{\psi}{2}t$$

where $t = \sum(y_j^{-1} - \bar{y}^{-1})$. Note that $\hat{\psi} = (n-1)/t$ so that we may write the marginal log-likelihood function as

$$\frac{n-1}{2}\log\psi - \frac{n-1}{2}\psi/\hat{\psi}.$$

The likelihood ratio approximation to the density of $\hat{\psi}$ is given by

$$p^*(\hat{\psi}; \psi) = c\frac{\psi^{n-1/2}}{\hat{\psi}^{n+1/2}}\exp\left\{-\frac{n-1}{2}\psi/\hat{\psi}\right\}.$$

Using the fact that $(n-1)\psi/\hat{\psi}$ has a chi-square distribution with $n-1$ degrees of freedom, it is straightforward to show that this approximation is exact. ∎

Another approach to constructing a signed likelihood ratio statistic for a scalar parameter of interest when there is a nuisance parameter is to first eliminate that nuisance parameter using either a marginal or conditional likelihood and then use the formula for R^* for a single parameter model.

Example 8.42 Standard deviation of a linear model
Consider a normal-theory linear model in which $Y_1, \ldots, Y_n$ are independent normally distributed random variables such that Y_j has mean $x_j^T\beta$ and standard deviation σ

where x_j is a known vector of covariates and β is an unknown parameter vector of length p. Take σ to be the parameter of interest with β as a nuisance parameter. The log-likelihood function for this model may be written

$$\ell(\beta, \sigma) = -n \log \sigma - \frac{1}{2\sigma^2}[n\hat{\sigma}^2 + (\hat{\beta} - \beta)^T x^T x(\hat{\beta} - \beta)]$$

where $(\hat{\sigma}, \hat{\beta})$ denotes the maximum likelihood estimator of (σ, β) and x denotes the $n \times p$ matrix with j row given by x_j^T.

It is straightforward to show that

$$R = \sqrt{n}\, \text{sgn}(\hat{\sigma} - \sigma)\, [(\hat{\sigma}^2/\sigma^2 - 1) - \log(\hat{\sigma}^2/\sigma^2)]^{1/2}$$

and that

$$U = (n/2)^{1/2}(\hat{\sigma}/\sigma)^p\left(\frac{\hat{\sigma}^2}{\sigma^2} - 1\right).$$

Another approach to inference in this model is to use a marginal likelihood function based on the distribution of $T = (Y - x\hat{\beta})^T(Y - x\hat{\beta})$, which depends only on σ. Using the fact that T/σ^2 has a chi-square distribution with $n - p$ degrees of freedom, this marginal log-likelihood function is given by

$$\ell(\sigma; t) = -(n - p) \log \sigma - \frac{1}{2}\frac{t}{\sigma^2}.$$

This marginal likelihood function is maximized by $\tilde{\sigma}$ where

$$\tilde{\sigma}^2 = \frac{t}{n - p} = \frac{n}{n - p}\hat{\sigma}.$$

Hence, the marginal log-likelihood function may be written

$$\ell(\sigma; t) = -(n - p)\left[\log \sigma + \frac{1}{2}\frac{\tilde{\sigma}^2}{\sigma^2}\right].$$

This log-likelihood function may be used to construct a version of the signed likelihood ratio statistic and the modified signed likelihood ratio statistic, which will be denoted by R_T and R_T^*, respectively. It is straightforward to show that

$$R_T = (n - p)^{1/2}\text{sgn}(\tilde{\sigma} - \sigma)\, [(\tilde{\sigma}^2/\sigma^2 - 1) - \log(\tilde{\sigma}^2/\sigma^2)]^{1/2}$$

and that the statistic U based on $\ell(\sigma; t)$ is given by

$$U_T = [(n - p)/2]^{1/2}(\tilde{\sigma}^2/\sigma^2 - 1).$$

Both R^* and R_T^* are distributed according to a standard normal distribution to a high degree of approximation. To compare these two approximations, $\Pr[R^* \le z]$ and $\Pr[R_T^* \le z]$ where calculated for several choices of z for the case $n = 10$ and $p = 5$; the results are presented in Table 8.1. These results show that, at least in this case, R_T^* is preferable.

Table 8.1 Coverage probabilities of confidence limits in Example 8.42

	Likelihood function	
Probability	Marginal	Full data
0.010	0.0100	0.0199
0.025	0.0250	0.0444
0.050	0.0500	0.0811
0.100	0.1001	0.1447
0.900	0.9004	0.9268
0.950	0.9502	0.9647
0.975	0.9751	0.9829
0.990	0.9901	0.9934

∎

The situation is much different whenever the statistic T depends on ψ. The main difficulty is that in this case the Bartlett identities no longer hold, in general, for the marginal likelihood function. Recall that the Bartlett identities follow from the fact that

$$E[\exp\{\ell(\theta) - \ell(\theta_0)\}; \theta_0] = 1$$

for all θ, θ_0. For a marginal log-likelihood function $\ell(\psi; t)$ based on a statistic T that does not depend on ψ, $\exp\{\ell(\psi; t) - \ell(\psi_0; t)\}$ depends on the data only through t and, hence,

$$E[\exp\{\ell(\psi; T) - \ell(\psi_0; T)\}; \psi_0] = 1$$

for all ψ, ψ_0.

Now suppose that $h(\psi; t_\psi)$ is a log-likelihood function based on a function T_ψ. Then, since the statistics t_ψ and t_{ψ_0} are not the same, $\exp\{h(\psi; t_\psi) - h(\psi_0; t_{\psi_0})\}$ depends, in general, on the entire underlying random variable. Hence, the identity upon which the Bartlett identities are based does not necessarily hold.

Example 8.43 Ratio of normal means
Consider the ratio of normal means example considered in Examples 8.32 and 8.33. The marginal likelihood function based on the function $(x - \psi y)$ is given by

$$-\frac{n}{2}\frac{(x - \psi y)^2}{(1 + \psi^2)}.$$

Let $\theta_0 = (1, \lambda_0)$ and consider the case $n = 1$. It may be shown that

$$E\left[\frac{\exp\{-(n/2)((x - \psi y)^2/(1 + \psi^2))\}}{\exp\{-(n/2)((x - y)^2/2)\}}; \theta_0\right] = \frac{\sqrt{2}(1 + \psi^2)^{1/2}}{1 + \psi}.$$

It follows that the Bartlett identities do not hold. ∎

8.5.3 CONDITIONAL LIKELIHOOD FUNCTIONS

The situation becomes more complicated when inference is based on a conditional likelihood function. First suppose that the conditional likelihood is formed by conditioning on a statistic s that does not depend on ψ. The resulting conditional likelihood function is a genuine likelihood function and, hence, first-order asymptotic theory as well as the higher-order asymptotic theory discussed in Chapter 5 holds, provided that the conditional log-likelihood function satisfies the usual regularity conditions.

Example 8.44 Index of a gamma distribution
Let $Y_1, \ldots, Y_n$ denote independent identically distributed gamma random variables each with index ψ and rate parameter λ; see Examples 5.5, 5.8, and 8.18. The statistic $s = \sum y_j$ is sufficient when ψ is fixed and the resulting conditional log-likelihood function is given by

$$\log \Gamma(n\psi) - n \log \Gamma(\psi) + \psi \sum \log y_j - n\psi \log \sum y_j.$$

Note that $\tilde{\psi}$, the maximizer of this conditional log-likelihood function, satisfies

$$n\Psi(n\tilde{\psi}) - n\Psi(\tilde{\psi}) = n \log \sum y_j - \sum \log y_j$$

and, hence, the conditional maximum likelihood estimate is different than the unconditional maximum likelihood estimate which satisfies

$$n \log \hat{\psi} - n\Psi(\hat{\psi}) = n \log \sum y_j - \sum \log y_j - n \log n.$$

In comparing these estimates note that, for fixed $\tilde{\psi}$,

$$n\Psi(n\tilde{\psi}) = n \log \tilde{\psi} + n \log n + O(1);$$

hence, we expect that the conditional and unconditional maximum likelihood estimates will be close in large samples. A more formal analysis shows that the difference between the two estimates is of order $O_p(n^{-1})$.

It is straightforward to show that the ν_{ijk} based on the conditional log-likelihood function are given by

$$\nu_2 = \Psi'(\psi) - n\Psi'(n\psi); \qquad \nu_{11} = 0;$$
$$\nu_{001} = -\nu_{30} = n^2\Psi''(n\psi) - \Psi''(\psi).$$

Hence, $\sqrt{n}(\tilde{\psi} - \psi)$ is asymptotically normally distributed with asymptotic variance

$$\frac{1}{\Psi'(\psi) - n\Psi'(n\psi)}.$$

Recall that the asymptotic variance of the maximum likelihood estimate is

$$\frac{1}{\Psi'(\psi) - 1/\psi}.$$

Note that

$$\frac{1}{\Psi'(\psi) - n\Psi'(n\psi)} - \frac{1}{\Psi'(\psi) - 1/\psi} = \frac{1}{2\psi^2} \frac{1}{[\Psi'(\psi) - 1/\psi]^2} \frac{1}{n} + O(n^{-2}).$$

The first formal cumulant of $\sqrt{n}(\tilde{\psi} - \psi)$ is given by

$$\frac{1}{2} \frac{n^2\Psi''(n\psi) - \Psi''(\psi)}{[\Psi'(\psi) - n\Psi'(n\psi)]^2} \frac{1}{\sqrt{n}} + O(n^{-3/2})$$

$$= -\frac{1}{2} \frac{\Psi''(\psi) + 1/\psi^2}{[\Psi'(\psi) - 1/\psi^2]^2} \frac{1}{\sqrt{n}} + O(n^{-3/2})$$

while the first formal cumulant of $\sqrt{n}(\hat{\psi} - \psi)$ is given by

$$\frac{1}{2} \frac{\Psi'(\psi)/\psi - \Psi''(\psi) - 2/\psi^2}{[\Psi'(\psi) - 1/\psi]^2} \frac{1}{\sqrt{n}} + O(n^{-3/2}).$$

∎

An important difference between inference based on a conditional likelihood and inference based on a marginal likelihood is due to the fact that all functions of a marginal likelihood have distributions depending only on ψ. The same is not true for a conditional likelihood since it depends on the conditioning statistic s, which, in general, has a distribution depending on both ψ and λ.

Example 8.45 Comparison of Poisson distributions
Let X and Y denote independent Poisson random variables with means $n\psi\lambda$ and $n\lambda$, respectively; X and Y may be viewed as the sufficient statistics based on n observations from each distribution. The parameter of interest ψ is the ratio of the means of the two distributions.

The log-likelihood function is given by

$$x \log \psi + (x + y) \log \lambda - n(1 + \psi)\lambda.$$

For fixed ψ, $x + y$ is sufficient; the conditional log-likelihood is given by

$$x \log \psi - (x + y) \log(1 + \psi).$$

The second derivative of the conditional log-likelihood is

$$-\frac{x}{\psi^2} + \frac{x + y}{(1 + \psi)^2};$$

it follows that

$$\nu_2 = -\nu_{01} = \frac{\lambda}{\psi(1 + \psi)}.$$

Hence, the asymptotic variance of the conditional maximum likelihood estimate depends on the value of the nuisance parameter λ.

∎

Since the conditional likelihood function depends only on ψ, it is clear from this result that the p^* formula cannot be applied to the conditional likelihood to obtain an approximation to the distribution of the conditional maximum likelihood estimate. The p^* formula can be used to obtain an approximation to the conditional distribution of the conditional maximum likelihood estimate given the conditioning statistic s, as well as any ancillary statistic in the model. This follows from the fact that

$$\begin{aligned} p(\tilde{\psi}|s, a; \psi) &= \frac{p(\tilde{\psi}|s, a; \psi)}{p(\tilde{\psi}|s, a; \psi_0)} p(\tilde{\psi}|s, a; \psi_0) \\ &= \exp\{[\ell(\psi) - \ell(\psi; s)] - [\ell(\psi_0) - \ell(\psi_0; s)]\}\, p(\tilde{\psi}|s, a; \psi_0) \end{aligned}$$

for any ψ_0; here $\tilde{\psi}$ denotes the conditional maximum likelihood estimate. The derivation of p^* now follows in the usual manner, except that the conditional log-likelihood is used in place of the usual log-likelihood function.

Example 8.46 Comparison of Poisson distributions
Consider the model considered in Example 8.45. The conditional maximum likelihood estimate of ψ is given by $\tilde{\psi} = x/y$. The conditional log-likelihood function is given by

$$x \log \psi - s \log(1 + \psi) = \left[\frac{\tilde{\psi}}{1 + \tilde{\psi}} \log \psi - \log(1 + \psi)\right] s$$

where $s = x + y$ and is considered fixed; hence, no ancillary statistic is needed.

The likelihood ratio approximation to the conditional distribution of $\tilde{\psi}$ is given by

$$p^*(\tilde{\psi}|s; \psi) = c \frac{(1 + \tilde{\psi})^{s-1}}{\tilde{\psi}^{s\tilde{\psi}/(1+\tilde{\psi})+1/2}} \frac{\psi^{s\tilde{\psi}/(1+\tilde{\psi})}}{(1 + \psi)^s}.$$

Note that $\tilde{\psi} = x/(s - x)$ is not a lattice variable. Since $s\tilde{\psi}/(1 + \tilde{\psi})$ is a lattice variable the dominating measure for p^* has density

$$m(\tilde{\psi}) = (\tilde{\psi} + 1)^2/s.$$

It follows that probabilities regarding $\tilde{\psi}$ may be approximated by

$$p^*(\tilde{\psi}|s; \tilde{\psi}) m(\tilde{\psi}) = c \frac{(1 + \tilde{\psi})^{s+1}}{\tilde{\psi}^{s\tilde{\psi}/(1+\tilde{\psi})+1/2}} \frac{\psi^{s\tilde{\psi}/(1+\tilde{\psi})}}{(1 + \psi)^s}$$

for some constant c.

It is straightforward to show that the conditional distribution of x given s is a binomial distribution with index s and parameter $\psi/(1 + \psi)$. It follows that the exact density of $\tilde{\psi}$ is given by

$$\binom{s}{\frac{\tilde{\psi}}{\tilde{\psi}+1}s} \psi^{(s\tilde{\psi}/(1+\tilde{\psi}))/(1+\psi)^s}.$$

The likelihood ratio approximation may be obtained from this exact expression by approximating the terms in the binomial coefficient using the usual asymptotic expansion for the gamma function. ∎

When using a conditional likelihood function, the properties of R^* hold conditionally on the conditioning statistic used, as well as an ancillary statistic for the model.

Example 8.47 Comparison of binomial distributions
Let Y and X denote independent binomial random variables such that Y has index n and parameter π_1 and X has index m and parameter π_2. Let

$$\psi = \log \frac{\pi_1}{1-\pi_1} - \log \frac{\pi_2}{1-\pi_2}$$

denote the parameter of interest and take

$$\lambda = \log \frac{\pi_2}{1-\pi_2}$$

as the nuisance parameter. The log-likelihood function for the model is given by

$$y\psi + (x+y)\lambda - n\log(1+\exp\{\psi+\lambda\}) - m\log(1+\exp\{\lambda\}).$$

For fixed values of ψ, $x+y$ is sufficient and it is straightforward to show that the conditional likelihood function given $s = x+y$ is given by

$$\psi y - \log \sum \binom{n}{j}\binom{m}{s-j} \exp\{j\psi\} = \psi y - \log K(\psi)$$

where

$$K(\psi) = \sum \binom{n}{j}\binom{m}{s-j} \exp\{j\psi\}.$$

and the sum is over all j from $\max(0, s-m)$ to $\min(n, s)$.

It follows that, based on the conditional log-likelihood,

$$\hat{j} = \frac{K''(\hat{\psi})}{K(\hat{\psi})} - \left(\frac{K'(\hat{\psi})}{K(\hat{\psi})}\right)^2.$$

Table 8.2 Confidence limits in Example 8.47

				Method		
n	m	s	y	R	R^*	Exact
12	8	11	8	0.105	0.203	0.205
12	8	11	9	0.0146	0.0395	0.0399
12	8	11	10	0.000775	0.00324	0.00322
10	10	9	7	0.0126	0.0346	0.0349
10	10	9	8	0.000653	0.00276	0.00274
8	12	8	5	0.0506	0.112	0.113
8	12	8	6	0.00484	0.0152	0.0154
8	12	8	7	0.000155	0.000758	0.000770

The maximum likelihood estimate $\hat{\psi}$ satisfies

$$y = \frac{K'(\hat{\psi})}{K(\hat{\psi})};$$

since y takes values in the integer lattice, $m(\hat{\psi}) = 1/\hat{\jmath}$. Also,

$$\ell_{;\hat{\psi}}(\hat{\psi}) - \ell_{;\hat{\psi}}(\psi) = (\hat{\psi} - \psi)\frac{\partial y}{\partial \hat{\psi}} = (\hat{\psi} - \psi)/m(\hat{\psi}).$$

Hence,

$$U_L = \left[\frac{K''(\hat{\psi})}{K(\hat{\psi})} - \left(\frac{K'(\hat{\psi})}{K(\hat{\psi})}\right)^2\right]^{1/2} [1 - \exp\{-(\hat{\psi} - \psi)\}].$$

Consider a test of the null hypothesis $\psi = 0$ versus the alternative $\psi > 0$. The test based on the conditional distribution of Y given $X + Y$ is Fisher's exact test. The test based on R^* rejects the null hypothesis for large values of R^*. Table 8.2 contains the p-value of the test based on R^* along with the p-value of Fisher's exact test and the p-value of the test based on R for several values of n, m, s, and y. There is close agreement between the p-value for the exact test and the p-value for the test based on R^*. ∎

As in the case of a marginal likelihood function, the situation is much different whenever the conditioning statistic depends on ψ. Again, the main difficulty is that in this case the Bartlett identities no longer hold, in general, for the conditional likelihood function.

Example 8.48 Ratio of normal means
Recall the ratio of normal means example considered in Example 8.43. The marginal likelihood considered in Example 8.43 was also derived as a conditional likelihood function in Example 8.20. Hence, the results of Example 8.43 also show that a conditional likelihood based on a parameter-dependent function does not necessarily obey the Bartlett identities. ∎

8.5.4 INTEGRATED LIKELIHOOD FUNCTIONS

There is an important difference between inference based on an integrated likelihood function and inference based on a marginal or conditional likelihood function: an integrated likelihood function is not a genuine likelihood function. That is, an integrated likelihood function does not, in general, correspond to either the conditional or marginal density of some function of the random variable under consideration.

For instance, let $\ell_U(\psi; \lambda) = \log L_U(\psi; \lambda)$. Then, if $\ell_U(\psi; \lambda)$ is a genuine log-likelihood function

$$\mathrm{E}[\ell'_U(\psi; \lambda); \theta] = 0.$$

However, in general,

$$\mathrm{E}[\ell'_U(\psi; \lambda); \theta] = O(1) \quad \text{as } n \to \infty.$$

Example 8.49 Index of a negative binomial distribution
Consider the negative binomial distribution considered in Example 8.35. The log-likelihood function corresponding to the uniform integrated log-likelihood function is given by

$$\ell_U(\psi; \lambda) = \sum \log \Gamma(\psi + y_j) - n \log \Gamma(\psi) + \log \Gamma(n\psi + 1) - \log \Gamma(n\psi + n\bar{y} + 2)$$

so that

$$\ell'_U(\psi; \lambda) = \sum \Psi(\psi + y_j) - n\Psi(\psi) + n\Psi(n\psi + 1) - n\Psi(n\psi + n\bar{y} + 2).$$

The log-likelihood function corresponding to a single observation Y is given by

$$\ell(\psi, \lambda; y) = \log \Gamma(\psi + y) - \log \Gamma(\psi) + y \log \lambda + \psi \log(1 - \lambda)$$

so that

$$\ell_\psi(\psi, \lambda; y) = \Psi(\psi + y) - \Psi(\psi) + \log(1 - \lambda).$$

Since $\mathrm{E}[\ell_\psi(\psi, \lambda; Y); \psi, \lambda] = 0$, it follows that

$$\mathrm{E}[\Psi(\psi + Y); \psi, \lambda] = \Psi(\psi) - \log(1 - \lambda).$$

Using this fact, together with the fact that $\Psi(x+1) = x^{-1} + \Psi(x)$, it is straightforward to show that

$$\mathrm{E}[\ell'_U(\psi; \lambda); \psi, \lambda] = 1 - 2\left[\frac{1-\lambda}{\psi}\right] + O(n^{-1}).$$ ∎

Although an integrated likelihood function is not a genuine likelihood function, it is closely related to the profile likelihood function. The uniform integrated likelihood function is given by

$$L_U(\psi; \lambda) = \int L(\psi, \lambda)\, d\lambda = \int \exp\{\ell(\psi, \lambda)\}\, d\lambda.$$

Applying Laplace's method and noting that, for fixed ψ, $\ell(\psi, \lambda)$ is maximized by $\hat{\lambda}_\psi$, shows that

$$L_U(\psi; \lambda) = \exp\{\ell(\psi, \hat{\lambda}_\psi)\} \,|-\ell_{\lambda\lambda}(\psi, \hat{\lambda}_\psi)|^{-1/2}[1 + O(n^{-1})]$$

and hence that

$$\ell_U(\psi; \lambda) = \ell_p(\psi) - \tfrac{1}{2}\log|-\ell_{\lambda\lambda}(\psi, \hat{\lambda}_\psi)| + O(n^{-1}).$$

It follows that for ψ of the form $\psi = \hat{\psi} + O(n^{-1/2})$,

$$\ell_U(\psi; \lambda) - \ell_U(\hat{\psi}; \lambda) = \ell_p(\psi) - \ell_p(\hat{\psi}) + O(n^{-1/2}).$$

Using this result, it may be shown that the first-order properties of procedures based on the uniform integrated likelihood function are the same as the first-order properties of procedures based on the profile likelihood.

8.6 Discussion and references

The problem of eliminating nuisance parameters occurs in virtually all statistical applications. A good general discussion of likelihood-based inference in the presence of nuisance parameters and the construction of pseudo-likelihood functions is given by Pace and Salvan (1997, Chapter 4). A critical discussion of nonBayesian approaches to the elimination of nuisance parameters is given by Basu (1977, 1978); see also Dawid (1975).

Marginal and conditional likelihood methods are considered by Andersen (1970, 1980), Kalbfleisch and Sprott (1970, 1973), and Fraser (1979). S-ancillarity was proposed by Sandved (1965) and Sverdrup (1966); see also Fraser (1956), Andersen (1970), Sandved (1972), Barndorff-Nielsen and Blæsild (1975), and Sprott (1975). The definition of ancillarity and sufficiency in the presence of a nuisance parameter used here, based on the partial information, is based on Bhapkar (1991), Liang (1983, 1984), and Zhu and Reid (1994). A definition of approximate ancillarity based on S-ancillarity was proposed by Severini (1993a); see also Severini (1994). Other definitions of ancillarity or sufficiency in the presence of a nuisance parameter have been given by Cox (1958), Barndorff-Nielsen (1973), Godambe (1976, 1980), Remon (1984), and Jørgensen (1993). A survey and comparison of several definitions of ancillarity and sufficiency is given by Barndorff-Nielsen (1976).

The problems associated with models with many nuisance parameters were first considered by Neyman and Scott (1948). The use of conditional inference in these models was discussed by Andersen (1970, 1971, 1973, 1980) and Kalbfleisch and Sprott (1973), and van der Vaart (1988). Methods of inference based on modeling the nuisance parameters as random variables are considered by Kiefer and

Wolfowitz (1956) and Lindsay (1980, 1983a, 1983b, 1995). Other useful references on these topics include Haberman (1977), Kumon and Amari (1984), and Pfanzagl (1993).

The issues involved in basing a conditional or marginal likelihood function on a parameter-dependent function were considered by Kalbfleisch and Sprott (1970, 1973) and Fraser and Reid (1989). A good reference for the determination of the differential elements that arise in this context is Tjur (1980).

The partial likelihood provides a generalization of marginal and conditional likelihoods; see Cox (1975) and Wong (1986).

The use of an integrated likelihood function to eliminate nuisance parameters arises naturally in a Bayesian approach to statistical inference. An integrated likelihood function however may also be useful in nonBayesian inference. A comprehensive discussion of integrated likelihood methods is given by Berger, Liseo, and Wolpert (1999); see also Aitkin and Stasinopoulos (1989), Liseo (1993), and Osborne and Severini (2000). The problem of selecting a prior distribution is discussed in detail in Kass and Wasserman (1996). The relationship between integrated likelihood functions and conditional likelihood functions given an S-ancillary statistic is considered in Dawid (1980). The relationship between integrated likelihood functions and the marginal likelihood function in a composite transformation model is considered in Barndorff-Nielsen (1983) and Barndorff-Nielsen, Blæsild, and Eriksen (1989). The relationship between Bayesian and nonBayesian methods of constructing a pseudo-likelihood function is considered in Severini (1999a).

The general problem of constructing mean and variance adjustments to signed likelihood ratio statistics based on a pseudo-likelihood function is considered in DiCiccio and Stern (1994). In particular, this work shows that it is often preferable to modify the likelihood ratio statistic based on a pseudo-likelihood function rather than modify the usual signed likelihood ratio statistic R, in agreement with the conclusion of Example 8.42.

Tweedie (1957) gives a detailed discussion of the properties of the inverse Gaussian distribution that are used in Examples 8.24 and 8.27. The use of conditional inference in the capture–recapture model considered in Example 8.4 is discussed in Viveros and Sprott (1986).

8.7 Exercises

8.1 Let $Y_1, \ldots, Y_n$ denote independent observations each with density function

$$\frac{\Gamma(\psi + y)}{\Gamma(y+1)\Gamma(\psi)} \frac{\lambda^y \psi^\psi}{(\lambda + \psi)^{y+\psi}}, \quad y = 0, 1, \ldots.$$

Find a conditional likelihood function for ψ.

8.2 Suppose that (X, Y) has density function

$$\binom{n}{x, y, z} \theta_1^x \theta_2^y (1 - \theta_1 - \theta_2)^{(n-x-y)},$$

$$x = 0, 1, \ldots, n, \quad y = 0, 1, \ldots, n, \quad x + y \leq n;$$

here $0 \leq \theta_1 \leq 1$, $0 \leq \theta_2 \leq 1$, and $\theta_1 + \theta_2 \leq 1$. Let $\psi = \theta_1/\theta_2$. Find a conditional likelihood function for ψ.

8.3 In Exercise 8.2, is the conditioning statistic used to form the conditional likelihood S-ancillary?

8.4 Let $Y_1, \ldots, Y_n$ denote independent random variables such that

$$Y_j = \alpha + \beta X_j + \sigma\epsilon_j, \quad j = 1, \ldots, n$$

where $\epsilon_1, \ldots, \epsilon_n$ are independent standard normal random variables, α and β are unknown parameters and σ is a known constant. Let c denote a known constant and let ψ denote the solution in x of

$$\alpha + \beta x = c;$$

i.e., $\psi = (c - \alpha)/\beta$. Find a conditional likelihood function for ψ.

8.5 Let $Y_1, \ldots, Y_n$ denote independent random variables each distributed according to a two-parameter exponential distribution with density of the form

$$\psi \exp\{-\psi(y - \mu)\}, \quad y > \mu$$

where $\mu > 0$ and $\psi > 0$ are unknown parameters. Find a conditional likelihood function for ψ.

8.6 Let $Y_1, \ldots, Y_n; X_1, \ldots, X_m$ denote independent exponential random variables such that the distribution of each Y_j has rate parameter $\psi\lambda$ and the distribution of each X_j has rate parameter λ. Hence, the parameter of interest is the ratio of the rate parameters. Find a statistic S_ψ that is sufficient for the model with ψ fixed and calculate the corresponding conditional likelihood function.

8.7 Let $Y_1, \ldots, Y_n; X_1, \ldots, X_m$ denote independent normal random variables such that the Y_j have mean μ_1 and standard deviation σ_1 and the X_j have mean μ_2 and standard deviation σ_2. Let $\psi = \sigma_1/\sigma_2$ denote the parameter of interest. Find a marginal likelihood function for ψ.

8.8 Is the statistic used in Exercise 8.7 to form the marginal likelihood function P-sufficient? Are there conditions under which the statistic is asymptotically sufficient?

8.9 Suppose X_1, X_2, X_3 have a multinomial distribution with index n and expected values $n[\phi^2 + 2\phi(1 - \phi - \lambda)]$, $n[\lambda^2 + 2\lambda(1 - \phi - \lambda)]$, and $2n\phi\lambda$, respectively. Here ϕ and λ are nonnegative parameters satisfying $0 < \phi + \lambda < 1$. This model arises in the analysis of blood-type data; see Sham (1998, Example 2.18). Find marginal likelihood functions for ϕ and λ.

8.10 Consider the two Poisson distributions example considered in Example 8.25. Find conditions on n, m such that (a) T is P-sufficient and (b) T is asymptotically sufficient.

8.11 In Exercise 8.2, find the uniform integrated likelihood function, first taking $\lambda = \theta_2$ as the nuisance parameter and then taking $\lambda = \theta_1$ as the nuisance parameter. Compare these results to the conditional likelihood obtained in Exercise 8.2.

8.12 For the exponential distributions model considered in Exercise 8.6, find the integrated likelihood function based on a uniform prior for λ.

8.13 Suppose that T is a scalar statistic and S is a vector statistic with the same dimension as λ such that the density of (T, S) is a full-rank exponential family density of the form

$$\exp\{n\psi t + n\lambda^T s - nd(\psi, \lambda) + Q(t, s)\}.$$

We have seen that an approximation to the conditional log-likelihood function given S is given by

$$\ell(\psi, \hat{\lambda}_\psi) + \frac{1}{2} \log |d_{\lambda\lambda}(\psi, \hat{\lambda}_\psi)|.$$

Find a weight function $\pi(\lambda|\psi)$ such that the resulting integrated likelihood function is approximately equal to the conditional likelihood function given S.

8.14 For the exponential distributions model considered in Exercise 8.6, find the marginal likelihood function based on the distribution of $\bar{Y}/\bar{X}$. Using that marginal likelihood, determine the likelihood ratio approximation to the density of $\hat{\psi}$, the maximum likelihood estimate of ψ. Compare the approximation to the exact density.

8.15 For the exponential distributions model considered in Exercise 8.14, compute the modified signed likelihood ratio statistic R^* based on the marginal likelihood function. Compare this version of R^* to the one based on the full likelihood; see Example 7.6.

9 The modified profile likelihood function

9.1 Introduction

In Chapter 8 it was shown that, in some cases, inference about a parameter of interest ψ may be based on a marginal or conditional likelihood function. Those methods are only available however when the model has a particular structure. Furthermore, even when a marginal or conditional likelihood function exists, calculation of the likelihood function is often difficult.

In this chapter, we consider the modified profile likelihood, a pseudo-likelihood function that is available for general models. The modified profile likelihood may be derived as an approximation to either a marginal or conditional likelihood when either of those likelihoods exists. Furthermore, the calculation of the modified profile likelihood function does not require the existence of a marginal or conditional likelihood and, hence, it has been adopted for general use.

As the name suggests, the modified profile likelihood function may also be viewed as a modification to the usual profile likelihood function. A number of examples suggests that the inferential properties of the modified profile likelihood function are, in general, superior to those of the profile likelihood function. The main focus of this chapter is the derivation of the modified profile likelihood and its application to a wide range of examples.

One drawback of the modified profile likelihood function is that it is often difficult to calculate. This has led to several proposals for approximations to the modified profile likelihood function, a number of which are presented in this chapter.

9.2 Profile likelihood

One simple method of eliminating λ from the likelihood function is to replace it by the partial maximum likelihood estimator $\hat{\lambda}_\psi$, the maximum likelihood estimator of λ keeping ψ fixed; we will often write $\hat{\theta}_\psi = (\psi, \hat{\lambda}_\psi)$. The resulting likelihood function $L_p(\psi) = L(\hat{\theta}_\psi)$ is called the profile likelihood function and $\ell_p(\psi) = \log L_p(\psi)$ is called the profile log-likelihood function. As discussed in Section 4.6, many commonly used procedures may be described in terms of $\ell_p(\psi)$.

It is important to realize that, in general, $L_p(\psi)$ is not a genuine likelihood function, that is, it is not based on the density function of any random variable. In large samples, replacing λ by its maximum likelihood estimate has a relatively minor effect on inferences regarding ψ. In small samples however replacing λ by an estimator may have a large impact on inferences, particularly when the dimension of λ is large. Consider the following example which, although artificial, illustrates the potential problems that can arise in using the profile likelihood function.

Example 9.1 Special normal distribution
Let $Y_1, \ldots, Y_n$ denote independent normally distributed random variables each with mean ψ and variance 1. Let Z denote a normally distributed random variable, independent of $(Y_1, \ldots, Y_n)$, such that Z has mean λ and variance $\exp\{-n\psi^2\}$. The likelihood function for ψ based on the entire data is given by

$$L(\psi, \lambda) = \exp\left\{\frac{n}{2}\psi^2\right\} \exp\left\{-\frac{1}{2}\sum(y_j - \psi)^2\right\}$$
$$\times \exp\left\{-\frac{1}{2}\exp\{n\psi^2\}(z - \lambda)^2\right\}.$$

The partial maximum likelihood estimator of λ is simply z, so that the profile likelihood function is given by

$$L_p(\psi) = \exp\left\{\frac{n}{2}\psi^2\right\} \exp\left\{-\frac{1}{2}\sum(y_j - \psi)^2\right\} = \exp\left\{\psi \sum y_j\right\}.$$

Hence, the profile likelihood function is either strictly increasing or strictly decreasing depending on the sign of $\sum y_j$.

Note that a marginal log-likelihood for ψ may be based on the distribution of $Y_1, \ldots, Y_n$ and is given by $-\sum(y_j - \psi)^2/2$. ∎

The following example is more representative of the problems that occur in practice when using the profile likelihood function.

Example 9.2 Normal-theory linear regression model
Consider the normal-theory linear model considered in Example 8.1. Let $Y_1, \ldots, Y_n$ denote independent random variables of the form

$$Y_j = x_j\beta + \sigma\epsilon_j, \quad j = 1, \ldots, n$$

where $\epsilon_1, \ldots, \epsilon_n$ are independent standard normal random variables, $x_1, \ldots, x_n$ are known $1 \times p$ vectors of constants, β is an unknown $p \times 1$ parameter vector, and $\sigma > 0$ is an unknown scalar parameter. Take σ to be the parameter of interest, with β as the nuisance parameter.

For fixed σ, the maximum likelihood estimate of β is given by

$$\hat{\beta}_\sigma \equiv \hat{\beta} = (x^T x)^{-1} x^T y$$

where x denotes the $n \times p$ matrix with jth row x_j and y denotes the vector $(y_1, \ldots, y_n)^T$. Hence, the profile log-likelihood for σ is given by

$$\ell_p(\sigma) = -n\log(\sigma) - \frac{1}{2}\frac{(y - x\hat{\beta})^T(y - x\hat{\beta})}{\sigma^2}.$$

The profile log-likelihood function takes its maximum value at $\sigma = \hat{\sigma}$ where

$$\hat{\sigma}^2 = \frac{(y - x\hat{\beta})^T (y - x\hat{\beta})}{n}.$$

When p is large relative to n, $\hat{\sigma}$ tends to be significantly less than σ. For instance, if $n = 10$ and $p = 4$, the probability that $\hat{\sigma} < \sigma$ is about 0.875 and the probability that $\hat{\sigma} < 0.75\sigma$ is greater than 0.5.

Let $z = y - x\hat{\beta}$ denote the residuals from the model. Then, since

$$z = y - x(x^T x)^{-1} x^T y = (D_n - x(x^T x)^{-1} x^T)y \equiv By$$

where D_n denotes the identity matrix of rank n, z has a multivariate normal distribution with mean vector 0 and covariance matrix $\sigma^2 B$. Note, however, that B has rank $n - p$ and, hence, the distribution of z is degenerate and the density of z does not exist in the usual sense. Let $t = Hz$ where H is any $(n - p) \times n$ matrix of full rank; then $\Sigma = HBH^T$, which is proportional to the covariance matrix of t, is invertible. It follows that t has a multivariate normal distribution with mean vector 0 and covariance matrix $\sigma^2 \Sigma$.

The marginal log-likelihood function based on the distribution of t is given by

$$\ell(\sigma; t) = -(n - p) \log \sigma - \frac{1}{2\sigma^2} t^T \Sigma^{-1} t,$$

discarding factors depending only on the data. Note that

$$[H^T (HBH^T)^{-1} H] B [H^T (HBH^T)^{-1} H] = H^T (HBH^T)^{-1} H$$

so that B is a generalized inverse of $H^T (HBH^T)^{-1} H$; since the rank of $H^T (HBH^T)^{-1} H$ is equal to the rank of B, it follows that $H^T (HBH^T)^{-1} H$ is a generalized inverse of B. Hence,

$$BH^T (HBH^T)^{-1} HB = B.$$

It follows that

$$\begin{aligned} t^T \Sigma t &= (Hz)^T (HBH^T)^{-1} (Hz) = z^T (HBH^T)^{-1} Hz \\ &= y^T B^T H^T (HBH^T)^{-1} HBy = y^T By = (y - x\hat{\beta})^T (y - x\hat{\beta}). \end{aligned}$$

Hence

$$\ell_T(\sigma) = -(n - p) \log \sigma - \frac{1}{2\sigma^2} (y - x\hat{\beta})^T (y - x\hat{\beta}).$$

This marginal log-likelihood function is of the same form as $\ell_p(\sigma)$ except that there is now an adjustment for the degrees of freedom.

A conditional approach could also be used. The log-likelihood function may be written

$$\ell(\sigma, \beta) = -n \log \sigma - \frac{1}{2\sigma^2}[(y - x\hat{\beta})^T (y - x\hat{\beta}) + (\hat{\beta} - \beta)^T x^T x(\hat{\beta} - \beta)].$$

Hence, for fixed σ, $\hat{\beta}$ is a sufficient statistic and the conditional distribution of Y given $\hat{\beta}$ does not depend on β. It is straightforward to show that $\hat{\beta}$ has a multivariate normal distribution with mean β and covariance matrix $\sigma^2(x^T x)^{-1}$; it follows that the marginal log-likelihood function based on $\hat{\beta}$ is given by

$$\ell(\sigma, \beta; \hat{\beta}) = -p \log \sigma - \frac{1}{2\sigma^2}[(\hat{\beta} - \beta)^T x^T x(\hat{\beta} - \beta)].$$

Hence, the conditional log-likelihood function given $\hat{\beta}$ is given by

$$\ell(\sigma; y|\hat{\beta}) = \ell(\sigma, \beta) - \ell_{\hat{\beta}}(\sigma, \beta)$$
$$= -(n - p) \log \sigma - \frac{1}{2\sigma^2}(y - x\hat{\beta})^T (y - x\hat{\beta}). \quad \blacksquare$$

Although, in general, the profile likelihood function is not a genuine likelihood function, in certain cases it does coincide with a marginal or conditional likelihood. For instance, consider a conditional likelihood function based on the conditional distribution of the data given a statistic S that is S-ancillary. Then the profile log-likelihood function is given by

$$\ell_p(\psi) = \sup_{\lambda} \ell(\psi, \lambda; S) + [\ell(\psi, \lambda) - \ell(\psi, \lambda; S)];$$

note that the second term on the right-hand side of this expression does not depend on λ. Since S is S-ancillary, there exists a parameterization (ψ, ϕ) of the model such that the distribution of S depends only on ϕ. It follows that

$$\sup_{\lambda} \ell(\psi, \lambda; S)$$

does not depend on ψ and, hence, neglecting terms not depending on ψ,

$$\ell_p(\psi) = \ell(\psi, \lambda) - \ell(\psi, \lambda; S),$$

which is identical to the conditional likelihood function.

A similar result holds for a marginal likelihood function based on a statistic T that is S-sufficient.

Example 9.3 Poisson regression

Consider the Poisson regression model considered in Examples 8.3, 8.6, and 8.8. In this model $Y_1, \ldots, Y_n$ are independent Poisson random variables such that Y_j has mean $\exp\{\lambda + \psi x_j\}$; the log-likelihood function is given by

$$\lambda \sum y_j + \psi \sum x_j y_j - \sum \exp\{\lambda + \psi x_j\}.$$

The conditional distribution of the data given $S = \sum Y_j$ depends only on ψ and the conditional log-likelihood function is given by

$$\psi \sum x_j y_j - s \log\left[\sum \exp\{\psi x_j\}\right];$$

furthermore, it was shown in Example 8.6 that S is S-ancillary.

The maximum likelihood estimate of λ for fixed ψ is given by

$$\hat{\lambda}_\psi = \log\left(\frac{\sum y_j}{\sum \exp\{\psi x_j\}}\right)$$

and it is straightforward to show that $\ell_p(\psi)$ is identical to the conditional log-likelihood. ∎

Example 9.4 Comparison of exponential distributions
Consider a model for random variables $\bar{Y}$ and $\bar{X}$. Assume that $n\bar{Y}$ and $m\bar{X}$ are independent gamma random variables with indices n and m and rate parameters $\psi\lambda$ and λ respectively; see Example 7.6. The log-likelihood function for this model is given by

$$\ell(\psi, \lambda) = n \log \psi + (n+m) \log \lambda - \lambda(n\psi\bar{y} + m\bar{x}).$$

A marginal log-likelihood may be based on the distribution of $T = n\bar{Y}/(m\bar{X})$ and is given by

$$n \log \psi - (m+n) \log(n\psi\bar{y} + m\bar{x});$$

it may be shown that T is S-sufficient.

The maximum likelihood estimate of λ for fixed ψ is given by

$$\hat{\lambda}_\psi = \frac{n\psi\bar{y} + m\bar{x}}{n+m}$$

and it is straightforward to show that $\ell_p(\psi)$ is identical to the marginal log-likelihood function given above. ∎

9.3 Modified profile likelihood

9.3.1 INTRODUCTION

The poor performance of the profile likelihood function in some cases may be partially explained by the fact that it does not attempt to approximate a genuine conditional or marginal likelihood function. In this section, we derive a modification to the profile likelihood based on approximating a marginal or conditional likelihood function.

First note that, since a given likelihood function is a function of ψ we must specify the values of ψ for which a given approximation holds; this same issue arose in Section 7.5 where approximation of the modified signed likelihood ratio statistic R^* was considered.

If an approximation $h(\psi)$ to a given conditional log-likelihood function $\ell_C(\psi)$ satisfies

$$h(\psi) = \ell_C(\psi) + O(n^{-1})$$

for each fixed ψ we say that h approximates ℓ_C with error $O(n^{-1})$ in the *large-deviation sense*. Here *large deviation* refers to the fact that the accuracy of the approximation is not restricted to values of ψ near any particular value.

In contrast, suppose that $h(\psi)$ satisfies

$$h(\psi) = \ell_C(\psi) + O(n^{-1})$$

for ψ of the form $\psi = \hat{\psi} + O(n^{-1/2})$. In this case we say that h approximates ℓ_C with error $O(n^{-1})$ in the *moderate-deviation sense* or for ψ in the *moderate-deviation range*. It is important to remember that in verifying these conditions we may freely add terms depending only on the data, e.g., we may always assume that $h(\hat{\psi}) = \ell_C(\hat{\psi})$.

Since $\hat{\psi} = \psi_0 + O_p(n^{-1/2})$ where ψ_0 denotes the true value of ψ, in practice we are most often interested in values of ψ close to $\hat{\psi}$, i.e., in the moderate-deviation range. Hence, the moderate-deviation results are typically of most interest. However, it is also desirable that a given approximation is accurate in the large-deviation sense since this suggests that the quality of the approximation does not deteriorate as ψ moves away from $\hat{\psi}$.

Consider the model with ψ held fixed; this model has parameter λ. We assume that this model satisfies the conditional model discussed in Section 6.1. That is, there exists a statistic a_0 such that $(\hat{\lambda}_\psi, a_0)$ is sufficient when ψ is held fixed and a_0 is ancillary when ψ is held fixed; that is, the distribution of a_0 may depend on ψ, but not on λ. In general, a_0, like $\hat{\lambda}_\psi$, may depend on ψ.

The modified profile likelihood function is derived under either of the following assumptions regarding the model. First suppose that in the restricted model the sufficient statistic is given by $(\hat{\lambda}, a, \hat{\psi})$ where a is the ancillary statistic from the unrestricted model and $\hat{\psi}$ is ancillary in the restricted model. Then a marginal likelihood function may be based on the distribution of $\hat{\psi}$ which does not depend on λ.

Alternatively, suppose that $(\hat{\lambda}, a)$ is sufficient in the restricted model with ψ held fixed. Then a conditional likelihood function may be based on the conditional distribution of $(\hat{\psi}, \hat{\lambda}, a)$ given $(\hat{\lambda}, a)$ which does not depend on λ. Either of these two approaches leads to the modified profile likelihood function.

The modified profile likelihood is given by

$$L_M(\psi) = \left| \frac{\partial \hat{\lambda}_\psi}{\partial \hat{\lambda}} \right|^{-1} |\hat{j}_{\lambda\lambda}(\psi, \hat{\lambda}_\psi)|^{-1/2} L_p(\psi);$$

$\ell_M(\psi) = \log L_M(\psi)$ is called the modified profile log-likelihood.

The Jacobian term

$$\left| \frac{\partial \hat{\lambda}_\psi}{\partial \hat{\lambda}} \right|$$

appearing in $L_M(\psi)$ may be expressed in terms of a sample space derivative of the log-likelihood function. Note that $\hat{\lambda}_\psi$ satisfies

$$\ell_\lambda(\psi, \hat{\lambda}_\psi; \hat{\psi}, \hat{\lambda}, a) = 0.$$

Differentiating this expression with respect to $\hat{\lambda}$ yields the identity

$$\ell_{\lambda\lambda}(\psi, \hat{\lambda}_\psi)\frac{\partial \hat{\lambda}_\psi}{\partial \hat{\lambda}} + \ell_{\lambda;\hat{\lambda}}(\psi, \hat{\lambda}_\psi) = 0.$$

Hence,

$$\frac{\partial \hat{\lambda}_\psi}{\partial \hat{\lambda}} = j_{\lambda\lambda}(\hat{\theta}_\psi)^{-1}\ell_{\lambda;\hat{\lambda}}(\hat{\theta}_\psi).$$

The modified profile likelihood function may therefore be written

$$\begin{aligned} L_M(\psi) &= \left|j_{\lambda\lambda}(\hat{\theta}_\psi)^{-1}\ell_{\lambda;\hat{\lambda}}(\hat{\theta}_\psi)\right|^{-1}|j_{\lambda\lambda}(\hat{\theta}_\psi)|^{-1/2}L_p(\psi) \\ &= \frac{|j_{\lambda\lambda}(\hat{\theta}_\psi)|^{1/2}}{|\ell_{\lambda;\hat{\lambda}}(\hat{\theta}_\psi)|}L_p(\psi). \end{aligned}$$

9.3.2 DERIVATION AS AN APPROXIMATION TO A MARGINAL LIKELIHOOD

First consider the case in which $(\hat{\lambda}, a, \hat{\psi})$ is sufficient and $\hat{\psi}$ is ancillary when ψ is held fixed. Note that the conditional distribution of $\hat{\psi}$ given a may be written

$$p(\hat{\psi}|a; \psi) = \frac{p(\hat{\psi}, \hat{\lambda}|a; \psi, \lambda)}{p(\hat{\lambda}|\hat{\psi}, a; \psi, \lambda)}. \tag{9.1}$$

We may derive a p^*-type approximation to $p(\hat{\psi}|a; \psi)$ by using the p^*-formula to approximate the two densities on the right-hand side of (9.1).

First consider approximation of $p(\hat{\lambda}, \hat{\psi}|a; \psi, \lambda)$. This density may be approximated directly by the p^*-formula. Recall that

$$p^*(\hat{\psi}, \hat{\lambda}|a; \psi, \lambda) = c_1(\psi, \lambda, a)|\hat{j}|^{1/2}\exp\{\ell(\psi, \lambda) - \ell(\hat{\psi}, \hat{\lambda})\}$$

where $c_1(\psi, \lambda, a) = (2\pi)^{d/2}[1 + O(n^{-1})]$ and d denotes the dimension of (ψ, λ).

The second component needed to approximate the conditional density of $\hat{\psi}$ given a is the conditional density of $\hat{\lambda}$ given $\hat{\psi}, a$. This may be approximated by using p^* to approximate the conditional density of $\hat{\lambda}_\psi$ given $(a, \hat{\psi})$ in the model with ψ held fixed and then transforming that approximation to an approximation for the conditional density of $\hat{\lambda}$.

Applying the p^* formula to the model with ψ fixed, we have that

$$p^*(\hat{\lambda}_\psi|\hat{\psi}, a; \psi, \lambda) = c_2(\psi, \lambda, a, b)|j_{\lambda\lambda}(\psi, \hat{\lambda}_\psi)|^{1/2}\exp\{\ell(\psi, \lambda) - \ell(\psi, \hat{\lambda}_\psi)\}$$

where $c_2(\psi, \lambda, a, b) = (2\pi)^{-(d-p)/2}[1 + O(n^{-1})]$; here p denotes the dimension of ψ. Hence,

$$p^*(\hat{\lambda}|\hat{\psi}, a; \psi, \lambda) = c_2(\psi, \lambda, a, b)|j_{\lambda\lambda}(\psi, \hat{\lambda}_\psi)|^{1/2} \times \exp\{\ell(\psi, \lambda) - \ell(\psi, \hat{\lambda}_\psi)\}\left|\frac{\partial \hat{\lambda}_\psi}{\partial \hat{\lambda}}\right|.$$

Hence, using (9.1), an approximation to the conditional density of $\hat{\psi}$ given a is given by

$$\frac{c_1(\psi, \lambda, a)|\hat{j}|^{1/2} \exp\{\ell(\psi, \lambda) - \ell(\hat{\psi}, \hat{\lambda})\}}{c_2(\psi, \lambda, a, b)|j_{\lambda\lambda}(\psi, \hat{\lambda}_\psi)|^{1/2} \exp\{\ell(\psi, \lambda) - \ell(\psi, \hat{\lambda}_\psi)\}|\partial\hat{\lambda}_\psi/\partial\hat{\lambda}|}.$$

Recall that

$$c_1(\psi, \lambda, a) = (2\pi)^{d/2}[1 + O(n^{-1})].$$

It follows that there exists a constant c such that

$$\frac{c_1(\psi, \lambda, a)}{c_2(\psi, \lambda, a, b)} = c[1 + O(n^{-1})]$$

for fixed ψ, λ. For $(\psi, \lambda) = (\hat{\psi}, \hat{\lambda}) + O_p(n^{-1/2})$,

$$\frac{c_1(\psi, \lambda, a)}{c_2(\psi, \lambda, a, b)} = c[1 + O(n^{-3/2})].$$

It follows that the approximate marginal likelihood for ψ based on $\hat{\psi}$ is given by

$$L_M(\psi) = \left|\frac{\partial \hat{\lambda}_\psi}{\partial \hat{\lambda}}\right|^{-1} |j_{\lambda\lambda}(\psi, \hat{\lambda}_\psi)|^{-1/2} L_p(\psi). \tag{9.2}$$

The argument given above shows that whenever the distribution of $\hat{\psi}$ does not depend on λ, then $\ell_M(\psi)$ approximates a genuine marginal log-likelihood function with error $O(n^{-1})$ for fixed ψ and with error $O(n^{-3/2})$ for ψ of the form $\psi = \hat{\psi} + O_p(n^{-1/2})$.

9.3.3 DERIVATION AS AN APPROXIMATION TO A CONDITIONAL LIKELIHOOD

The modified profile likelihood function may also be derived as an approximation to a conditional likelihood function. Consider the case in which $(\hat{\lambda}, a)$ is sufficient in the model with ψ held fixed. A conditional likelihood function may be based on the conditional density of $\hat{\psi}$ given $(\hat{\lambda}, a)$ which is given by

$$p(\hat{\psi}|\hat{\lambda}, a; \theta) = \frac{p(\hat{\psi}, \hat{\lambda}|a; \theta)}{p(\hat{\lambda}|a; \theta)}. \tag{9.4}$$

An approximation to this conditional density may be obtained by approximating the densities on the right-hand side of this expression using the p^* approximation. As might be expected, the derivation closely follows the one for the marginal likelihood function given above.

First consider $p(\hat{\lambda}|a;\theta)$. This density may be approximated by approximating the conditional density of $\hat{\lambda}_\psi$ and then using the usual change-of-variable formula. The p^*-formula for the conditional density of $\hat{\lambda}_\psi$ given a is given by

$$c_1(\psi,\lambda,a)|j_{\lambda\lambda}(\hat{\theta}_\psi)|^{1/2}\exp\{\ell(\theta)-\ell(\hat{\theta}_\psi)\}.$$

Hence, the conditional density of $\hat{\lambda}$ given a may be approximated by

$$c_1(\psi,\lambda,a)|j_{\lambda\lambda}(\hat{\theta}_\psi)|^{1/2}\exp\{\ell(\theta)-\ell(\hat{\theta}_\psi)\}\left|\frac{\partial\hat{\lambda}_\psi}{\partial\hat{\lambda}}\right|.$$

Using the p^* approximation

$$c_2(\psi,\lambda,a)|\hat{j}|^{1/2}\exp\{\ell(\theta)-\ell(\hat{\theta})\}$$

for the conditional density of $(\hat{\psi},\hat{\lambda})$ given a, using the properties of c_1 and c_2 described above, and ignoring terms not depending on the parameters, leads to $L_M(\psi)$ as an approximation to the conditional likelihood function.

For the case in which $(\hat{\lambda},a)$ is sufficient for fixed ψ, $\ell_M(\psi)$ approximates the conditional likelihood given $\hat{\lambda}$ to order $O(n^{-1})$ in the large-deviation case and to order $O(n^{-3/2})$ in the moderate-deviation case.

9.3.4 EXAMPLES

Example 9.5 Normal-theory linear regression model
Let $Y_1,\ldots,Y_n$ denote independent random variables of the form $Y_j=\psi+x_j\beta+\sigma\epsilon_j$ where $x_1,\ldots,x_n$ are known covariate vectors of length p satisfying $\sum x_j=0$, ψ is an unknown scalar parameter, β is an unknown parameter vector of length p, and the ϵ_j are independent unobservable standard normal random variables. Hence, the nuisance parameter of the model is $\lambda=(\beta,\sigma)^T$. Assume that the matrix x with columns $x_1,\ldots,x_n$ is of rank p.

Let $y=(y_1,\ldots,y_n)^T$ and $e=(1,\ldots,1)^T$ denote column vectors each of length n. Note that we may write

$$\begin{aligned}\sum(y_j-\psi-x_j\beta)^2&=(y-\psi e-x\beta)^T(y-\psi e-x\beta)\\&=(y-\hat{\psi}e-x\hat{\beta})^T(y-\hat{\psi}e-x\hat{\beta})+n(\hat{\psi}-\psi)^2+(\hat{\beta}-\beta)^Tx^Tx(\hat{\beta}-\beta)\\&=n\hat{\sigma}^2+n(\hat{\psi}-\psi)^2+(\hat{\beta}-\beta)^Tx^Tx(\hat{\beta}-\beta).\end{aligned}$$

Hence,

$$\begin{aligned}&\ell(\psi,\beta,\sigma;\hat{\psi},\hat{\beta},\hat{\sigma})\\&\quad=-n\log\sigma-\frac{1}{2\sigma^2}[n\hat{\sigma}^2+n(\hat{\psi}-\psi)^2+(\hat{\beta}-\beta)^Tx^Tx(\hat{\beta}-\beta)].\end{aligned}$$

It is straightforward to show that

$$\ell_{\beta}(\psi, \beta, \sigma) = \frac{1}{\sigma^2} x^T x(\hat{\beta} - \beta)$$

and

$$\ell_{\sigma}(\psi, \beta, \sigma) = -\frac{n}{\sigma} + \frac{1}{\sigma^3}[n\hat{\sigma}^2 + n(\hat{\psi} - \psi)^2 + (\hat{\beta} - \beta)^T x^T x(\hat{\beta} - \beta)].$$

It follows that $\hat{\beta}_{\psi} = \hat{\beta}$,

$$\hat{\sigma}_{\psi}^2 = \hat{\sigma}^2 + (\hat{\psi} - \psi)^2$$

and $\ell_p(\psi) = -n \log \hat{\sigma}_{\psi}$.

Also

$$\hat{j}_{\lambda\lambda}(\psi, \hat{\lambda}_{\psi}) = \begin{pmatrix} x^T x/\hat{\sigma}_{\psi}^2 & 0 \\ 0 & 2/\hat{\sigma}_{\psi}^2 \end{pmatrix}$$

and

$$\ell_{\lambda;\hat{\lambda}}(\psi, \hat{\lambda}_{\psi}) = \begin{pmatrix} x^T x/\hat{\sigma}_{\psi}^2 & 0 \\ 0 & 2n\hat{\sigma}/\hat{\sigma}_{\psi}^3 \end{pmatrix}.$$

It follows that

$$|j_{\lambda\lambda}(\psi, \hat{\lambda}_{\psi})| = 2n|x^T x|\hat{\sigma}_{\psi}^{-2(p+1)}$$

and

$$|\ell_{\lambda;\hat{\lambda}}(\psi, \hat{\lambda}_{\psi})| = 2n|x^T x|\hat{\sigma}\hat{\sigma}_{\psi}^{-[2(p+1)+1]}$$

so that, neglecting terms that do not depend on ψ,

$$\ell_M(\psi) = (p + 2) \log \hat{\sigma}_{\psi} - n \log \hat{\sigma}_{\psi} = -(n - p - 2) \log \hat{\sigma}_{\psi}.$$

Hence, the modified profile likelihood provides a 'degrees-of-freedom' correction to the profile likelihood. ∎

Example 9.6 Normal distributions with common standard deviation
Consider the model for pairs of random variables from several normal distributions with differing means and common standard deviation; see Example 8.13. In this example, (X_j, Y_j), $j = 1, \ldots, n$, denote independent pairs of independent normal random variables such that X_j and Y_j have mean μ_j and standard deviation σ, which is the parameter of interest. The log-likelihood function is

$$\ell(\theta) = -2n \log \sigma - \frac{1}{2\sigma^2} \sum [(x_j - \mu_j)^2 + (y_j - \mu_j)^2]$$

where $\theta = (\sigma, \mu_1, \ldots, \mu_n)$. In terms of the maximum likelihood estimates we may write

$$\ell(\theta) = -2n \log \sigma - \frac{1}{\sigma^2}\left[n\hat{\sigma}^2 + \sum(\hat{\mu}_j - \mu_j)^2\right].$$

It follows that

$$\ell_{\mu_j;\hat{\mu}_k}(\hat{\theta}_\psi) = \begin{cases} 0 & \text{if } j \neq k \\ \dfrac{2}{\sigma^2} & \text{if } j = k \end{cases}$$

and that

$$-\ell_{\mu_j\mu_j}(\hat{\theta}_\psi) = \begin{cases} 0 & \text{if } j \neq k \\ \dfrac{2}{\sigma^2} & \text{if } j = k. \end{cases}$$

Hence, neglecting terms that do not depend on ψ,

$$\frac{|j_{\lambda\lambda}(\hat{\theta}_\psi)|^{1/2}}{|\ell_{\lambda;\hat{\lambda}}(\hat{\theta}_\psi)|} = \sigma^n.$$

The profile log-likelihood function is given by

$$\ell_p(\sigma) = -2n \log \sigma - n\frac{\hat{\sigma}^2}{\sigma^2}$$

so that

$$\ell_M(\sigma) = -n \log \sigma - n\frac{\hat{\sigma}^2}{\sigma^2};$$

since

$$\hat{\sigma}^2 = \frac{1}{4n}\sum(x_j - y_j)^2,$$

this is identical to the conditional log-likelihood function derived in Example 8.13. ∎

Example 9.7 Exponential regression

Let $Y_1, \ldots, Y_n$ denote independent exponential random variables such that Y_j has mean $\lambda \exp(\psi x_j)$ where $x_1, \ldots, x_n$ are scalar constants and ψ and λ are unknown parameters. The log-likelihood function may be written

$$\ell(\theta) = -n \log \lambda - \psi \sum x_j - \frac{1}{\lambda}\sum \exp\{-\psi x_j\} y_j.$$

In this model the maximum likelihood estimates are not sufficient and an ancillary statistic is needed. Let

$$a_j = \log y_j - \log \hat{\lambda} - \hat{\psi} y_j, \quad j = 1, \ldots n$$

and take $a = (a_1, \ldots, a_n)$. The log-likelihood function may be written

$$\ell(\theta) = -n \log \lambda - \psi \sum x_j - \frac{\hat{\lambda}}{\lambda}\sum \exp\{(\hat{\psi} - \psi)x_j + a_j\}.$$

It follows that

$$\ell_{\lambda;\hat{\lambda}}(\hat{\theta}_\psi) = \frac{n}{\hat{\lambda}_\psi \hat{\lambda}}$$

and that

$$-\ell_{\lambda\lambda}(\hat{\theta}_\psi) = \frac{n}{\hat{\lambda}_\psi^2};$$

here

$$\hat{\lambda}_\psi = \hat{\lambda}\frac{\sum \exp\{(\hat{\psi} - \psi)x_j\}a_j}{n} = \frac{\sum \exp\{-\psi x_j\}y_j}{n}.$$

Hence, the factor

$$\frac{|j_{\lambda\lambda}(\hat{\theta}_\psi)|^{1/2}}{|\ell_{\lambda;\hat{\lambda}}(\hat{\theta}_\psi)|}$$

depends only on the data and

$$\ell_M(\psi) = \ell_p(\psi) = -n \log \hat{\lambda}_\psi - \psi \sum x_j.$$ ∎

Example 9.8 Logistic regression
Let $Y_1, \ldots, Y_n$ denote independent binary random variables with

$$\Pr(Y_j = 1) = 1 - \Pr(Y_j = 0) = \pi_j$$

where

$$\log\left(\frac{\pi_j}{1 - \pi_j}\right) = \alpha + \beta x_j + \psi z_j, \quad j = 1, \ldots, n$$

where $x_1, \ldots, x_n; z_1, \ldots, z_n$ are fixed scalar constants and $\theta = (\psi, \alpha, \beta)$. The log-likelihood function for this model is given by

$$\ell(\theta) = \alpha \sum y_j + \beta \sum x_j y_j + \psi \sum z_j y_j \\ - \sum \log(1 + \exp\{\alpha + \beta x_j + \psi z_j\}).$$

It follows that

$$|\hat{j}_{\lambda\lambda}(\hat{\theta}_\psi)| = \sum x_j^2 w_j(\psi) \sum w_j(\psi) - \left[\sum x_j w_j(\psi)\right]^2$$

where

$$w_j(\psi) = \frac{\exp\{\hat{\alpha}_\psi + \hat{\beta}_\psi x_j + \psi z_j\}}{[1 + \exp\{\hat{\alpha}_\psi + \hat{\beta}_\psi x_j + \psi z_j\}]^2}.$$

Note that

$$\ell_\alpha(\theta) = \sum y_j - \sum \frac{\exp\{\alpha + \beta x_j + \psi z_j\}}{1 + \exp\{\alpha + \beta x_j + \psi z_j\}}$$

so that $\ell_{\alpha;\hat{\alpha}}(\hat{\theta}_\psi)$ and $\ell_{\alpha;\hat{\beta}}(\hat{\theta}_\psi)$ both will depend only on the data; the same is true of $\ell_{\beta;\hat{\alpha}}(\hat{\theta}_\psi)$ and $\ell_{\beta;\hat{\beta}}(\hat{\theta}_\psi)$. It follows that

$$L_M(\psi) = |j_{\lambda\lambda}(\hat{\theta}_\psi)|^{1/2} L_p(\psi).$$

∎

Example 9.9 Inverse Gaussian distribution
Consider the inverse Gaussian distribution considered in Examples 8.24 and 8.27. The log-likelihood function for this model is given by

$$\ell(\psi, \lambda) = \frac{n}{2}\log\psi - \frac{\psi}{2\lambda^2}\sum(y_j - \lambda)^2/y_j.$$

It follows that

$$\ell_\lambda(\psi, \lambda) = \frac{\psi}{\lambda^3}\sum(y_j - \lambda)^2/y_j + \frac{\psi}{\lambda^2}\sum(y_j - \lambda)/y_j = \frac{\psi}{\lambda^3}\left[\sum(y_j - \lambda)\right]$$

and, hence, that $\hat{\lambda}_\psi = \bar{y}$. The profile log-likelihood function is therefore given by

$$\ell_p(\psi) = \frac{n}{2}\log\psi - \frac{\psi}{2\bar{y}^2}\sum(y_j - \bar{y})^2/y_j.$$

Note that since $\hat{\lambda}_\psi$ does not depend on ψ,

$$L_M(\psi) = |\hat{j}_{\lambda\lambda}(\psi, \hat{\lambda}_\psi)|^{-1/2} L_p(\psi).$$

It is straightforward to show that

$$\ell_{\lambda\lambda}(\psi, \lambda) = -3\frac{\psi}{\lambda^4}\sum(y_j - \lambda) - n\frac{\psi}{\lambda^3}$$

so that

$$j_{\lambda\lambda}(\psi, \lambda) = n\psi/\bar{y}^3$$

and

$$\ell_M(\psi) = \frac{(n-1)}{2}\log\psi - \frac{\psi}{2\bar{y}^2}\sum(y_j - \bar{y})^2/y_j.$$

∎

Example 9.10 A capture–recapture model
Consider the capture–recapture model considered in Example 8.4. The log-likelihood function is given by

$$\ell(\psi, \lambda) = \log\binom{\psi}{y_1, y_2, y_3} + (y_1 + y_3)\log\lambda_1 + (y_2 + y_3)\log\lambda_2$$
$$+ (\psi - y_1 - y_3)\log(1 - \lambda_1) + (\psi - y_2 - y_3)\log(1 - \lambda_2).$$

For fixed ψ, $y_1 + y_3$ and $y_2 + y_3$ are independent binomial random variables each with index ψ and means $\psi\lambda_1$ and $\psi\lambda_2$, respectively. It follows that

$$\hat{\lambda}_\psi = ((y_1 + y_3)/\psi, (y_2 + y_3)/\psi)$$

and that the profile log-likelihood function for ψ is given by

$$\ell_p(\psi) = \log \binom{\psi}{y_1, y_2, y_3} + (\psi - y_1 - y_3)\log(\psi - y_1 - y_3)$$
$$+ (\psi - y_2 - y_3)\log(\psi - y_2 - y_3) - 2\psi\log(\psi),$$
$$\psi \geq \max(y_1 + y_3, y_2 + y_3).$$

It is straightforward to show that

$$\ell_{\lambda_1}(\psi, \lambda) = \frac{y_1 + y_3}{\lambda_1} - \frac{\psi - y_1 - y_3}{1 - \lambda_1}$$

and

$$\ell_{\lambda_2}(\psi, \lambda) = \frac{y_2 + y_3}{\lambda_2} - \frac{\psi - y_2 - y_3}{1 - \lambda_2}.$$

Note that

$$y_1 + y_3 = \hat{\lambda}_1 \hat{\psi}; \qquad y_2 + y_3 = \hat{\lambda}_2 \hat{\psi}.$$

It follows that

$$\ell_{\lambda_1;\hat{\lambda}_1}(\psi, \lambda) = \frac{\hat{\psi}}{\lambda_1} - \frac{\hat{\psi}}{1 - \lambda_1} = \frac{\hat{\psi}}{\lambda_1(1 - \lambda_1)},$$

$$\ell_{\lambda_2;\hat{\lambda}_2}(\psi, \lambda) = \frac{\hat{\psi}}{\lambda_1(1 - \lambda_1)},$$

and

$$\ell_{\lambda_1;\hat{\lambda}_2}(\psi, \lambda) = \ell_{\lambda_2;\hat{\lambda}_1}(\psi, \lambda) = 0.$$

Hence,

$$|\ell_{\lambda;\hat{\lambda}}(\psi, \hat{\lambda}_\psi)| = \frac{\hat{\psi}^2 \psi^4}{(y_1 + y_3)(y_2 + y_3)(\psi - y_1 - y_3)(\psi - y_2 - y_3)}.$$

Since

$$\ell_{\lambda\lambda}(\psi, \lambda) = \begin{pmatrix} \frac{y_1 + y_3}{\lambda_1^2} + \frac{\psi - y_1 - y_3}{(1 - \lambda_1)^2} & 0 \\ 0 & \frac{y_2 + y_3}{\lambda_2^2} + \frac{\psi - y_2 - y_3}{(1 - \lambda_2)^2} \end{pmatrix},$$

it follows that

$$|j_{\lambda\lambda}(\psi, \hat{\lambda}_\psi)| = \psi^6 \left[\frac{1}{(y_1 + y_3)(y_2 + y_3)(\psi - y_1 - y_3)(\psi - y_2 - y_3)} \right]$$

so that, ignoring terms not depending on ψ,

$$\frac{|j_{\lambda\lambda}(\psi, \lambda)|^{1/2}}{|\ell_{\lambda;\hat{\lambda}}(\psi, \hat{\lambda}_\psi)|} = [(\psi - y_1 - y_3)(\psi - y_2 - y_3)]^{1/2} / \psi.$$

The modified profile log-likelihood function is therefore given by

$$\ell_M(\psi) = \log\binom{\psi}{y_1, y_2, y_3} + (\psi - y_1 - y_3 + 1/2)\log(\psi - y_1 - y_3)$$
$$+ (\psi - y_2 - y_3 + 1/2)\log(\psi - y_2 - y_3) - (2\psi + 1)\log(\psi)$$

for $\psi \geq \max(y_1 + y_3, y_2 + y_3)$.

Recall that the conditional log-likelihood function is given by

$$\ell_C(\psi) \equiv \ell(\psi; y|s_1, s_2) = \log\binom{\psi}{y_1, y_2, y_3} - \log\binom{\psi}{s_1 - s_2} - \log\binom{\psi}{s_2},$$
$$\psi \geq \max(s_2, s_1 - s_2).$$

Both $\ell_p(\psi)$ and $\ell_M(\psi)$ may be obtained from $\ell_C(\psi)$ by using the usual asymptotic approximation for the gamma function:

$$\log\Gamma(n+1) = \{\log(n) - 1\}n + \tfrac{1}{2}\log(n) + \tfrac{1}{2}\log(2\pi) + O(n^{-1}).$$

The profile likelihood function corresponds to using the first term in the approximation while the modified profile likelihood corresponds to using the first two terms. ∎

9.4 Calculation of the modified profile likelihood without an explicit nuisance parameter

Recall that in Section 7.4.2 it was shown that the likelihood derivatives needed to calculate the modified signed likelihood ratio statistic R^* can be calculated without explicitly specifying the nuisance parameter λ. Consider a model with parameter θ and let $\psi = \psi(\theta)$ denote the parameter of interest. As in Section 7.4.2, let $B \equiv B(\theta)$ denote a $d \times (d-1)$ matrix such that for all θ,

$$\frac{\partial\psi}{\partial\theta}B = 0$$

and

$$\frac{\partial\lambda}{\partial\theta}B$$

is nonsingular for some nuisance parameter λ. Then

$$\ell_{\lambda;\hat{\lambda}}(\hat{\theta}_\psi) = \left[\left(\frac{\partial\lambda}{\partial\theta}(\hat{\theta}_\psi)B(\hat{\theta}_\psi)\right)^{-1}\right]^T B(\hat{\theta}_\psi)^T \ell_{\theta;\hat{\theta}}(\hat{\theta}_\psi)B(\hat{\theta})\left(\frac{\partial\lambda}{\partial\theta}(\hat{\theta})B(\hat{\theta})\right)^{-1}$$

and

$$\ell_{\lambda\lambda}(\hat{\theta}_\psi) = \left[\left(\frac{\partial\lambda}{\partial\theta}(\hat{\theta}_\psi)B(\hat{\theta}_\psi)\right)^{-1}\right]^T B(\hat{\theta}_\psi)^T \ell_{\theta\theta}(\hat{\theta}_\psi)B(\hat{\theta}_\psi)$$
$$\times\left(\frac{\partial\lambda}{\partial\theta}(\hat{\theta}_\psi)B(\hat{\theta}_\psi)\right)^{-1}.$$

Further details on the construction of the matrix B are given in Section 7.4.2.

It follows that, omitting terms depending only on the data,

$$\frac{|j_{\lambda\lambda}(\hat{\theta}_\psi)|^{1/2}}{|\ell_{\lambda;\hat{\lambda}}(\hat{\theta}_\psi)|} = \frac{|B(\hat{\theta}_\psi)^T j(\hat{\theta}_\psi) B(\hat{\theta}_\psi)|^{1/2}}{|B(\hat{\theta}_\psi)^T \ell_{\theta;\hat{\theta}}(\hat{\theta}_\psi) B(\hat{\theta})|}.$$

Example 9.11 Normal-theory linear regression model
Let $Y_1, \ldots, Y_n$ denote independent random variables of the form

$$Y_j = x_j\beta + \sigma\epsilon_j, \quad j = 1, \ldots, n$$

where $x_1, \ldots, x_n$ are fixed $1 \times (p+1)$ vectors, β is an unknown $(p+1) \times 1$ parameter vector, $\sigma > 0$ is an unknown scalar parameter, and $\epsilon_1, \ldots, \epsilon_n$ are unobservable independent standard normal random variables. The dimension of the x_j is taken to be $p + 1$ to facilitate comparison with the results of Example 9.5.

The parameter of the model is $\theta = (\beta, \sigma)$ and take as the parameter of interest a function $\psi \equiv \psi(\theta) = c^T \beta$ where c is a $(p + 1) \times 1$ vector of constants.

The log-likelihood function for this model is given by

$$\ell(\theta) = -n \log \sigma - \frac{n}{2}\frac{\hat{\sigma}^2}{\sigma^2} - \frac{1}{2\sigma^2}(\hat{\beta} - \beta)^T x^T x(\hat{\beta} - \beta)$$

where x denotes the $n \times (p + 1)$ matrix with jth row x_j. It follows that

$$\ell_{\theta;\hat{\theta}}(\hat{\theta}_\psi) = \frac{1}{\hat{\sigma}_\psi^2}\begin{pmatrix} x^T x & 0 \\ 2(\hat{\beta} - \hat{\beta}_\psi)^T x^T x/\hat{\sigma}_\psi & 2n\hat{\sigma}/\hat{\sigma}_\psi \end{pmatrix}$$

and

$$j(\hat{\theta}_\psi) = \frac{1}{\hat{\sigma}_\psi^2}\begin{pmatrix} x^T x & 2x^T x(\hat{\beta} - \hat{\beta}_\psi)/\hat{\sigma}_\psi \\ 2(\hat{\beta} - \hat{\beta}_\psi)^T x^T x/\hat{\sigma}_\psi & 2n \end{pmatrix}.$$

It is straightforward to show that $\hat{\beta}_\psi$ satisfies

$$x^T x(\hat{\beta} - \hat{\beta}_\psi) = (\hat{\psi} - \psi)\frac{c}{c^T(x^T x)^{-1}c}.$$

Hence, we may write

$$\ell_{\theta;\hat{\theta}}(\hat{\theta}_\psi) = \frac{1}{\hat{\sigma}_\psi^2}\begin{pmatrix} x^T x & 0 \\ 2c^T(\hat{\psi} - \psi)/[\hat{\sigma}_\psi c^T(x^T x)^T c] & 2n\hat{\sigma}/\hat{\sigma}_\psi \end{pmatrix}$$

and

$$j(\hat{\theta}_\psi) = \frac{1}{\hat{\sigma}_\psi^2}\begin{pmatrix} x^T x & 2c(\hat{\psi} - \psi)/[\hat{\sigma}_\psi c^T(x^T x)^{-1}c] \\ 2c^T(\hat{\psi} - \psi)/[\hat{\sigma}_\psi c^T(x^T x)^{-1}c] & 2n \end{pmatrix}.$$

Let $B = (B_0 \quad B_1)^T$ where B_0 is a $p \times p$ matrix and B_1 is a $1 \times p$ vector. The condition that $\psi'(\theta)B = 0$ may be written as $c^T B_0 = 0$. Hence, this condition does not put any requirements on B_1; note, however, that the matrix B must be of full rank. Since $c^T B_0 = 0$ the rows of B_0 are linearly dependent. It follows that B_1 may not be in the row space of B_0. The simplest approach is to take as B_1 a vector orthogonal to the rows of B_0; such a vector must be proportional to c^T. Hence, we take $B_0 = c^T$. Note that B does not depend on θ.

It follows that

$$|B^T \ell_{\theta;\hat{\theta}}(\hat{\theta}_\psi) B| = \frac{|B_0^T x^T x B_0 + B_1^T B_1 2n\hat{\sigma}/\hat{\sigma}_\psi|}{\hat{\sigma}_\psi^{2p+2}}$$

and

$$|B^T j(\hat{\theta}_\psi) B| = \frac{|B_0^T x^T x B_0 + 2n B_1^T B_1|}{\hat{\sigma}^{2p+2}}.$$

Note that $B_0^T x^T x B_0$ has rank $p-1$ and hence,

$$|B_0^T x^T x B_0 + B_1^T B_1 2n\hat{\sigma}/\hat{\sigma}_\psi|$$

is proportional to $\hat{\sigma}_\psi^{-1}$. This follows from writing

$$|B_0^T x^T x B_0 + B_1^T B_1 2n\hat{\sigma}/\hat{\sigma}_\psi| = \begin{vmatrix} B_0^T x^T x B_0 & -B_1^T \\ B_1 2n\hat{\sigma}/\hat{\sigma}_\psi & 1 \end{vmatrix}.$$

Hence, neglecting terms not depending on ψ,

$$\frac{|B(\hat{\theta}_\psi)^T j(\hat{\theta}_\psi) B(\hat{\theta}_\psi)|^{1/2}}{|B(\hat{\theta}_\psi)^T \ell_{\theta;\hat{\theta}}(\hat{\theta}_\psi) B(\hat{\theta})|} = \hat{\sigma}_\psi^{p+2}$$

and

$$L_M(\psi) = \hat{\sigma}^{-(n-p-2)},$$

in agreement with the results of Example 9.5. ∎

Example 9.12 Inference for the effect size

Let $X_1, \ldots, X_n; Y_1, \ldots, Y_m$ denote independent normally distributed random variables such that each X_j has mean μ_1 and standard deviation σ and each Y_j has mean μ_2 and standard deviation σ. Here $\theta = (\mu_1, \mu_2, \sigma)$ and we take the 'effect size',

$$\psi = \frac{\mu_1 - \mu_2}{\sigma},$$

as the parameter of interest.

The log-likelihood function for this model may be written

$$\ell(\theta) = -(n+m)\log\sigma - \frac{1}{2\sigma^2}[(n+m)\hat{\sigma}^2 + n(\hat{\mu}_1 - \mu_1)^2 + m(\hat{\mu}_2 - \mu_2)^2].$$

It follows that

$$\ell_{\theta;\hat{\theta}}(\hat{\theta}_\psi) = \frac{1}{\hat{\sigma}_\psi^2}\begin{pmatrix} n & 0 & 0 \\ 0 & m & 0 \\ 2n(\hat{\mu}_1 - \hat{\mu}_{1\psi})/\hat{\sigma}_\psi & 2m(\hat{\mu}_2 - \hat{\mu}_{2\psi})/\hat{\sigma}_\psi & 2(n+m)\hat{\sigma}/\hat{\sigma}_\psi \end{pmatrix}$$

and

$$j(\hat{\theta}_\psi) = \frac{1}{\hat{\sigma}_\psi^2}\begin{pmatrix} n & 0 & 2n(\hat{\mu}_1 - \hat{\mu}_{1\psi})/\hat{\sigma}_\psi \\ 0 & m & 2m(\hat{\mu}_2 - \hat{\mu}_{2\psi})/\hat{\sigma}_\psi \\ 2n(\hat{\mu}_1 - \hat{\mu}_{1\psi})/\hat{\sigma}_\psi & 2m(\hat{\mu}_2 - \hat{\mu}_{2\psi})/\hat{\sigma}_\psi & 2(n+m)\hat{\sigma}/\hat{\sigma}_\psi \end{pmatrix}.$$

Since

$$\psi'(\theta) = \begin{pmatrix} \frac{1}{\sigma} & \frac{1}{\sigma} & -\frac{\mu_1 - \mu_2}{\sigma^2} \end{pmatrix} = \frac{1}{\sigma}(1 \quad -1 \quad -\psi),$$

the 3×2 matrix B must satisfy

$$(1 \quad -1 \quad -\psi)B = 0.$$

For instance, we may take

$$B = \begin{pmatrix} 1 & \psi \\ 1 & 0 \\ 0 & 1 \end{pmatrix}.$$

Although the algebraic expression for $L_M(\psi)$ is complicated, $L_M(\psi)$ is easily calculated numerically for a given set of data. ∎

9.5 Approximations to the modified profile likelihood

9.5.1 INTRODUCTION

Like the modified likelihood ratio statistic R^*, he modified profile likelihood function has the drawback that it is often difficult to determine. In this section we consider a number of methods of approximating these functions. The methods of approximation used here are analogous to the methods used in Section 7.5 in approximating R^*.

9.5.2 AN APPROXIMATION BASED ON ORTHOGONAL PARAMETERS

Recall that the modified profile likelihood function may be written

$$L_M(\psi) = \left|\frac{\partial\hat{\lambda}_\psi}{\partial\hat{\lambda}}\right|^{-1} |j_{\lambda\lambda}(\psi, \hat{\lambda}_\psi)|^{-1/2} L_p(\psi).$$

If $\hat{\lambda}_\psi$ does not depend on ψ then $\hat{\lambda}_\psi = \hat{\lambda}$ and

$$L_M(\psi) = |j_{\lambda\lambda}(\psi, \hat{\lambda}_\psi)|^{-1/2} L_p(\psi)$$

which is easily calculated for virtually any model.

It is not generally possible to parameterize a model so that $\hat{\lambda}_\psi = \hat{\lambda}$; however it is often possible to parameterize a model so that $\hat{\lambda}_\psi \doteq \hat{\lambda}$. Suppose that ψ and λ are orthogonal parameters. It was shown in Section 3.6.4, that for ψ of the form $\psi = \hat{\psi} + O_p(n^{-1/2})$,

$$\hat{\lambda}_\psi = \hat{\lambda} + O_p(n^{-1}).$$

In this case, we may use the approximation

$$\left|\frac{\partial \hat{\lambda}_\psi}{\partial \hat{\lambda}}\right| = 1 + O_p(n^{-1}).$$

Hence, suppose that λ is orthogonal to ψ. Define the *adjusted profile likelihood* function by

$$L_A(\psi) = |j_{\lambda\lambda}(\psi, \hat{\lambda}_\psi)|^{-1/2} L_p(\psi)$$

and let $\ell_A(\psi) = \log L_A(\psi)$. As noted above, for ψ of the form $\hat{\psi} + O_p(n^{-1/2})$, i.e., ψ in the moderate deviation range,

$$\ell_M(\psi) - \ell_M(\hat{\psi}) = \ell_A(\psi) - \ell_A(\hat{\psi}) + O_p(n^{-1}).$$

There are two drawbacks to the adjusted profile likelihood function. One is that it requires an orthogonal parameterization. Construction of an orthogonal nuisance parameter was discussed in Section 3.6.4; recall that such a construction is possible, in principle, whenever ψ is a scalar, but in general an orthogonal parameterization does not exist.

The other drawback of the adjusted profile log-likelihood is that it is not invariant under interest-respecting reparameterizations. This fact has two consequences. One is that to calculate $\ell_A(\psi)$ it is necessary to determine an orthogonal parameterization, or at least determine $\partial\lambda/\partial\psi$ where λ is an orthogonal nuisance parameter. The other consequence is that $\ell_A(\psi)$ depends on the orthogonal parameterization used. Since if λ is orthogonal to ψ then so is $g(\lambda)$ for any smooth function $g(\cdot)$, $\ell_A(\psi)$ is only determined up to a factor $\log|g_\lambda(\hat{\lambda}_\psi)|$. Since ψ and λ are orthogonal, $\hat{\lambda}_\psi = \hat{\lambda} + O(n^{-1})$ for $\psi = \hat{\psi} + \delta/\sqrt{n}$ and, hence, any two versions of $\ell_A(\psi)$ agree to order $O_p(n^{-1})$ for ψ of the form $\psi = \hat{\psi} + \delta/\sqrt{n}$.

Example 9.13 Linear regression model

Consider the linear regression model considered in Example 9.5. It is straightforward to show that ψ is orthogonal to the nuisance parameter (β, σ). Recall that $\ell_p(\psi) = -n\log\hat{\sigma}_\psi$ and $|\hat{j}_{\lambda\lambda}(\hat{\theta}_\psi)|$ is proportional to $\hat{\sigma}_\psi^{-2(p+1)}$; it follows that

$$\ell_A(\psi) = -(n-p-1)\log\hat{\sigma}_\psi.$$

If instead the model is parameterized by the variance, then

$$\ell_A(\psi) = \ell_M(\psi) = -(n-p-2)\log\hat{\sigma}_\psi,$$

while if the model is parameterized by the precision, $1/\sigma^2$, then

$$\ell_A(\psi) = -(n - p + 2)\log\hat{\sigma}_\psi.$$

In fact, any function of the form $-(n - q)\log\hat{\sigma}_\psi$ for fixed q is a valid version of $\ell_A(\psi)$. ∎

Example 9.14 Inverse Gaussian distribution
Consider the inverse Gaussian distribution considered in Example 9.9. For this model, the log-likelihood function is given by

$$\ell(\theta) = \frac{n}{2}\log\psi - \frac{\psi}{2\lambda^2}\sum(y_j - \lambda)^2/y_j.$$

Recall that

$$\ell_\lambda(\theta) = \frac{\psi}{\lambda^3}\sum(y_j - \lambda)$$

which implies that $\mathrm{E}(y_j; \theta) = \lambda$. Hence, $i_{\psi\lambda}(\theta) = 0$ so that ψ and λ are orthogonal parameters. Furthermore, $\hat{\lambda}_\psi = \hat{\lambda}$ so that

$$L_M(\psi) = |j_{\lambda\lambda}(\hat{\theta}_\psi)|^{-1/2}L_p(\psi) = L_A(\psi).$$

Note that, since $\hat{\lambda}_\psi$ does not depend on ψ, $L_A(\psi)$ is invariant under interest-respecting reparameterizations for this model. ∎

9.5.3 AN APPROXIMATION BASED ON AN APPROXIMATELY ANCILLARY STATISTIC

The sample space derivative needed to calculate the modified profile likelihood function may be approximated using the method described in Section 6.7.2. Let

$$\tilde{\ell}_{\theta;\hat{\theta}}(\theta) = \ell_{\theta;y}(\theta)\hat{V}\left(\ell_{\theta;y}(\theta)\hat{V}\right)^{-1}\hat{\jmath}$$

denote the approximation to $\ell_{\theta;\hat{\theta}}(\theta)$ derived in Section 6.7.2. Here

$$\hat{V} = \begin{pmatrix} -\dfrac{\partial F_1(y_1;\hat{\theta})/\partial\hat{\theta}}{p_1(y_j;\hat{\theta})} \\ \vdots \\ -\dfrac{\partial F_n(y_n;\hat{\theta})/\partial\hat{\theta}}{p_n(y_n;\hat{\theta})} \end{pmatrix}$$

where F_j and p_j denote the distribution function and density function, respectively, of Y_j.

If an explicit nuisance parameter is not available, then an approximation to $\ell_M(\psi)$, $\tilde{\ell}_M(\psi)$, may be obtained by replacing $\ell_{\theta;\hat{\theta}}(\theta)$ by $\tilde{\ell}_{\theta;\hat{\theta}}(\theta)$ in the expression for the adjustment factor given in Section 9.4. However, when an explicit nuisance parameter is available, a simpler procedure may be used.

To calculate $\ell_{\lambda;\hat{\lambda}}(\theta)$, we may use the argument in Section 6.7.2, but considering only the derivative with respect to $\hat{\lambda}$ and replacing a by $(\hat{\psi}, a)$. Hence,

$$\ell_{\lambda;\hat{\lambda}}(\theta) = \ell_{\lambda;y}(\theta) V_\lambda \left(\frac{\partial \hat{\lambda}}{\partial y} V_\lambda \right)^{-1}$$

where V_λ is an $n \times (d-1)$ matrix satisfying

$$\frac{\partial(\hat{\psi}, a)}{\partial y} V_\lambda = 0$$

and

$$\left| \frac{\partial \hat{\lambda}}{\partial y} V_\lambda \right| \neq 0.$$

Using the approximate ancillary statistic described in Section 6.7.2, it is straightforward to show that $\ell_{\lambda;\hat{\lambda}}(\theta)$ may be approximated by taking $V_\lambda = \hat{V}_\lambda$ where

$$\hat{V}_\lambda = \begin{pmatrix} -\dfrac{\partial F_1(y_1; \hat{\theta})/\partial \hat{\lambda}}{p_1(y_j; \hat{\theta})} \\ \vdots \\ -\dfrac{\partial F_n(y_n; \hat{\theta})/\partial \hat{\lambda}}{p_n(y_n; \hat{\theta})} \end{pmatrix}.$$

Hence, neglecting terms not depending on ψ, an approximation to $\ell_M(\psi)$ is given by

$$\tilde{\ell}_M(\psi) = \ell_p(\psi) + \log \frac{|j_{\lambda\lambda}(\hat{\theta}_\psi)|^{1/2}}{|\ell_{\lambda;y}(\hat{\theta}_\psi)\hat{V}_\lambda|}.$$

Example 9.15 Exponential regression

Consider the exponential regression model considered in Example 9.7. The log-likelihood function for this model is given by

$$\ell(\theta) = -n \log \lambda - \psi \sum x_j - \frac{1}{\lambda} \sum \exp\{-\psi x_j\} y_j$$

where $x_1, \ldots, x_n$ are known scalar constants and $Y_1, \ldots, Y_n$ are independent exponential random variables such that Y_j has mean $\lambda \exp\{\psi x_j\}$. Hence, Y_j has density function

$$\lambda^{-1} \exp\{-\psi x_j\} \exp\{-\exp(-\psi x_j)\, y_j/\lambda\}$$

and distribution function

$$1 - \exp\{-\exp(-\psi x_j) y_j/\lambda\}.$$

It follows that the jth row of $\hat{V}_\lambda$ is given by $y_j/\hat{\lambda}$. Also,

$$\ell_{\lambda;y_j}(\theta) = \lambda^{-2} \exp\{-\psi x_j\}.$$

It is straightforward to show that

$$\ell_{\lambda;y}(\hat{\theta}_\psi)\hat{V}_\lambda = \sum y_j \exp\{-\psi x_j\}/(\hat{\lambda}\hat{\lambda}_\psi^2);$$

it follows that

$$\ell_{\lambda;y}(\hat{\theta}_\psi)\hat{V}_\lambda = \frac{\sum y_j \exp\{-\psi x_j\}}{\hat{\lambda}\hat{\lambda}_\psi^2},$$

which is exactly equal to $\ell_{\lambda;\hat{\lambda}}(\hat{\theta}_\psi)$. Hence, $\tilde{\ell}_M(\psi) = \ell_M(\psi)$. ∎

Example 9.16 Normal-theory nonlinear regression

Let $Y_1, \ldots, Y_n$ denote independent random variables of the form

$$Y_j = \mu_j(\phi) + \sigma\epsilon_j, \quad j = 1, \ldots, n$$

where $\mu_1, \ldots, \mu_n$ are known functions of an unknown p-dimensional parameter ϕ, $\sigma > 0$ is an unknown scalar parameter and $\epsilon_1, \ldots, \epsilon_n$ are unobservable independent standard normal random variables. Hence, the parameter of the model is $\theta = (\phi, \sigma)$.

Since Y_j has distribution function

$$F_j(y_j; \theta) = \Phi\left(\frac{y_j - \mu_j(\phi)}{\sigma}\right),$$

where $\Phi(\cdot)$ denotes the standard normal distribution function, the matrix $\hat{V}$ is given by

$$\hat{V} = \frac{1}{\hat{\sigma}}\begin{pmatrix} x(\hat{\phi}) & e(\hat{\theta}) \end{pmatrix}$$

where $x(\phi)$ denotes the $n \times p$ matrix with jth row given by

$$\mu_j'(\phi) = \frac{\partial}{\partial\phi}\mu_j(\phi)$$

and $e(\theta) = (e_1(\theta), \ldots, e_n(\theta))^T$ where $e_j(\theta) = (y_j - \mu_j(\phi))/\sigma$, $j = 1, \ldots, n$.

It is straightforward to show that

$$\ell_{\theta;y}(\theta) = \frac{1}{\sigma^2}\begin{pmatrix} X(\phi)^T \\ 2e(\theta)^T \end{pmatrix}.$$

It follows that

$$\ell_{\theta;y}(\theta)\hat{V} = \frac{1}{\sigma^2\hat{\sigma}}\begin{pmatrix} x(\phi)^T x(\hat{\phi}) & x(\phi)^T e(\hat{\theta}) \\ 2e(\theta)^T x(\hat{\phi}) & e(\hat{\theta})^T e(\theta) \end{pmatrix}.$$

Hence,

$$\tilde{\ell}_{\theta;\hat{\theta}}(\theta) = \frac{\hat{\sigma}^2}{\sigma^2}\begin{pmatrix} x(\phi)^T x(\hat{\phi}) & x(\phi)^T e(\hat{\theta}) \\ 2e(\theta)x(\hat{\phi}) & 2e(\hat{\theta})^T e(\theta) \end{pmatrix}\begin{pmatrix} x(\hat{\phi})^T x(\hat{\phi}) & 0 \\ 0 & 2 \end{pmatrix}^{-1}\hat{\jmath}.$$

This result may be used to determine $\tilde{\ell}_M(\psi)$ for any given parameter of interest $\psi \equiv \psi(\theta)$. ∎

9.5.4 AN APPROXIMATION BASED ON COVARIANCES OF LOG-LIKELIHOOD DERIVATIVES

In Section 6.7.3 it was shown that the sample space derivatives needed to calculate $\ell_M(\psi)$ may be approximated by certain covariances of the log-likelihood function and its derivatives.

Recall that $\ell_{\theta;\hat{\theta}}(\hat{\theta}_\psi)$ may be approximated by

$$\bar{\ell}_{\theta;\hat{\theta}}(\theta) = I(\hat{\theta}_\psi; \hat{\theta}) i(\hat{\theta})^{-1} \hat{\jmath}$$

where

$$I(\theta; \theta_0) = \mathrm{E}\{\ell_\theta(\theta)\ell_\theta(\theta_0)^T; \theta_0\}.$$

If an explicit nuisance parameter is not available, then an approximation to $\ell_M(\psi)$, $\bar{\ell}_M(\psi)$, may be obtained by replacing $\ell_{\theta;\hat{\theta}}(\theta)$ by $\bar{\ell}_{\theta;\hat{\theta}}(\theta)$ in the expression for the adjustment factor given in Section 9.4. However, as was the case with the approximation described in Section 9.5.3, when an explicit nuisance parameter is available, a simpler procedure may be used.

Using the same general argument as in Section 6.7.3, it may be shown that, for approximating $\ell_M(\psi)$, $\ell_{\lambda;\hat{\lambda}}(\hat{\theta}_\psi)$ may be approximated by $I(\hat{\theta}_\psi; \hat{\theta})$ where

$$I_{\lambda;\lambda}(\theta; \theta_0) = E\{\ell_\lambda(\theta)\ell_\lambda(\theta_0)^T; \theta_0\}.$$

Hence, we may approximate $\ell_M(\psi)$ by

$$\bar{\ell}_M(\psi) = \ell_p(\psi) + \log \frac{|j_{\lambda\lambda}(\hat{\theta}_\psi)|^{1/2}}{|I_{\lambda;\lambda}(\hat{\theta}_\psi; \hat{\theta})|}.$$

The argument given in Section 6.7.3 shows that, in general, $\bar{\ell}_M(\psi)$ is accurate to $O(n^{-1})$ in the moderate-deviation sense and to $O(n^{-1/2})$ in the large-deviation sense.

Example 9.17 Exponential regression
Consider the exponential regression model considered in Example 9.7. The log-likelihood function is given by

$$\ell(\theta) = -n \log \lambda - \psi \sum x_j - \frac{1}{\lambda} \sum \exp\{-\psi x_j\} y_j.$$

Hence,

$$\ell_\psi(\theta) = -\sum x_j + \frac{1}{\lambda} \sum \exp\{-\psi x_j\} x_j y_j$$

and

$$\ell_\lambda(\theta) = -\frac{n}{\lambda} + \frac{1}{\lambda^2} \sum \exp\{-\psi x_j\} y_j.$$

It follows that

$$I_{\lambda;\lambda}(\theta;\theta_0) = \frac{1}{\lambda^2}\exp\{(\psi_0 - \psi)x_j\}$$

and hence

$$I_{\lambda;\lambda}(\hat{\theta}_\psi;\hat{\theta}) = \frac{1}{\hat{\lambda}_\psi^2}\sum\exp\{(\hat{\psi} - \psi)x_j\}.$$

Recall that the in this example we may determine $\ell_{\lambda;\hat{\lambda}}(\hat{\theta}_\psi)$ exactly and it is given by

$$\ell_{\lambda;\hat{\lambda}}(\hat{\theta}_\psi) = \frac{n}{\hat{\lambda}\hat{\lambda}_\psi}.$$

It is straightforward to show that

$$\frac{I_{\lambda;\lambda}(\hat{\theta}_\psi)}{\ell_{\lambda;\hat{\lambda}}(\hat{\theta}_\psi)} = \frac{\sum\exp\{(\hat{\psi} - \psi)x_j\}}{\sum\exp\{(\hat{\psi} - \psi)x_j + a_j\}}$$

where $a_j = \log y_j - \log\hat{\lambda} - \hat{\psi}y_j$, $j = 1, \ldots, n$. Expanding the exponential terms in the numerator and denominator of this expression, we may write

$$\frac{I_{\lambda;\lambda}(\hat{\theta}_\psi)}{\ell_{\lambda;\hat{\lambda}}(\hat{\theta}_\psi)} = \frac{n + (\hat{\psi} - \psi)\sum x_j + O(1)}{\sum\exp\{a_j\} + (\hat{\psi} - \psi)\sum x_j\exp\{a_j\} + O(1)}.$$

Using the likelihood equations for $\hat{\psi}$, $\hat{\lambda}$, it follows that

$$\sum\exp\{a_j\} = n; \qquad \sum x_j\exp\{a_j\} = \sum x_j.$$

Then

$$\frac{I_{\lambda;\lambda}(\hat{\theta}_\psi)}{\ell_{\lambda;\hat{\lambda}}(\hat{\theta}_\psi)} = 1 + O(n^{-1}),$$

in accordance with the general theory of this section.

The approximation to $\ell_M(\psi)$ is given by

$$\bar{\ell}_M(\psi) = \ell_p(\psi) + \log\frac{\hat{\lambda}_\psi}{\sum\exp\{(\hat{\psi} - \psi)x_j\}}.$$

Consider the data in Table A.7 in the Appendix. In these data, y_j represents the survival time of a leukaemia patient, in weeks after diagnosis, and x_j represents the log of the white blood count of the patient. Figure 9.1 contains a plot of the modified profile likelihood along with the approximation derived above. Clearly, the approximation is extremely accurate in this example.

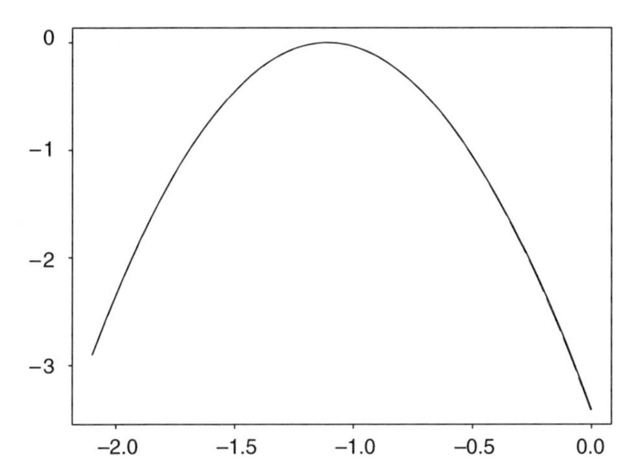

Fig. 9.1 Likelihood functions for Example 9.17.

∎

Example 9.18 Normal distributions with common mean
Let $Y_{jk}, k = 1, \ldots, n_j, j = 1, \ldots, m$ denote independent normal random variables such that Y_{jk} has mean μ and standard deviation σ_j. Take the common mean μ as the parameter of interest and take $(\sigma_1, \ldots, \sigma_m)$ as the nuisance parameter. Let

$$Y_j = \frac{1}{n_j}\sum Y_{jk}; \qquad S_j = \sum (Y_{jk} - Y_j)^2;$$

by sufficiency, the analysis may be based on $(Y_1, S_1), \ldots, (Y_m, S_m)$.

The log-likelihood function is given by

$$\ell(\theta) = -\sum n_j \log \sigma_j - \frac{1}{2}\sum \frac{n_j}{\sigma_j^2}(y_j - \mu)^2 - \frac{1}{2}\sum s_j/\sigma_j^2.$$

It follows that

$$\ell_\mu(\theta) = \sum \frac{n_j}{\sigma_j^2}(y_j - \mu)$$

and

$$\ell_{\sigma_j}(\theta) = -\frac{n_j}{\sigma_j} + \frac{n_j}{\sigma_j^3}(y_j - \mu)^2 + \frac{s_j}{\sigma_j^3}.$$

Hence,

$$E[\ell_\mu(\theta_0)\ell_\mu(\theta); \theta_0] = \sum \frac{n_j}{\sigma_j^2},$$

$$E[\ell_\mu(\theta_0)\ell_{\sigma_j}(\theta); \theta_0] = \frac{2n_j(\mu_0 - \mu)}{\sigma_j^3},$$

$$E[\ell_{\sigma_j}(\theta_0)\ell_\mu(\theta); \theta_0] = 0,$$

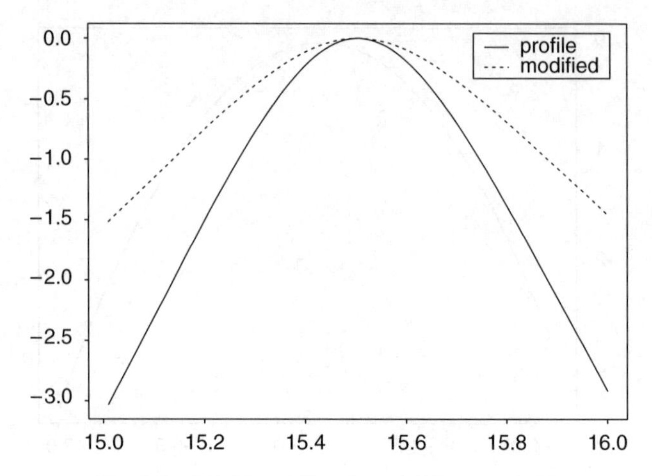

Fig. 9.2 Likelihood functions for Example 9.18.

and

$$E[\ell_{\sigma_j}(\theta_0)\ell_{\sigma_k}(\theta); \theta_0] = \begin{cases} 0 & \text{if } j \neq k \\ (2n_j\sigma_{j0})/(\sigma_j^3) & \text{if } j = k. \end{cases}$$

It follows that, neglecting terms not depending on ψ,

$$|I_{\lambda;\lambda}(\hat{\theta}_\psi; \hat{\theta})| = (\hat{\sigma}_{1\psi} \cdots \hat{\sigma}_{m\psi})^{-3}.$$

Since,

$$-\ell_{\sigma_j\sigma_k}(\hat{\theta}_\psi) = \begin{cases} 0 & \text{if } j \neq k \\ (2n_j)/(\hat{\sigma}_{j\psi}^2) & \text{if } j = k, \end{cases}$$

$$|j_{\lambda\lambda}(\hat{\theta}_\psi)| = \frac{2^m n_1 \cdots n_m}{[\hat{\sigma}_{1\psi} \cdots \hat{\sigma}_{m\psi}]^2}$$

and

$$\bar{\ell}_M(\mu) = \ell_p(\mu) + \sum \log \hat{\sigma}_{j\psi}^2 = -\sum (n_j - 2) \log \hat{\sigma}_{j\psi}.$$

Consider the data in Table A.9 of the Appendix and analyzed in Example 7.15. Figure 9.2 contains plots of $\bar{\ell}_M$ and ℓ_p. ∎

Example 9.19 Poisson regression
Let $Y_1, \ldots, Y_n$ denote independent Poisson random variables such that Y_j has mean $\exp\{x_j\theta\}$, $j = 1, \ldots, n$. Here $x_1, \ldots, x_n$ are known $1 \times d$ vectors and θ is an unknown $d \times 1$ parameter vector. Take $\psi = c^T\theta$, where c is a known vector not depending on θ, as the parameter of interest.

The log-likelihood function for this model is given by

$$\ell(\theta) = \sum x_j \theta y_j - \sum \exp\{x_j \theta\}$$

so that

$$\ell_\theta(\theta) = \sum y_j x_j^T - \sum \exp\{x_j \theta\} x_j^T .$$

It follows that $I(\theta; \theta_0)$ depends only on θ_0 and, hence,

$$\bar{\ell}_M(\psi) = \ell_p(\psi) + \frac{1}{2} \log \left| B(\hat{\theta}_\psi)^T \left[\sum \exp\{x_j \theta\} x_j^T x_j \right] B(\hat{\theta}_\psi) \right|$$

where $B(\theta)$ is a $d \times (d-1)$ matrix of full rank satisfying $c^T B(\theta) = 0$.

Note that

$$\sum \exp\{x_j \theta\} x_j^T x_j$$

may be written $x^T W_\theta x$ where x denotes the $n \times d$ matrix with jth row x_j and W_θ denotes an $n \times n$ diagonal matrix with jth diagonal element $\exp\{x_j \theta\}$. Let M denote an arbitrary $d \times (d-1)$ matrix of full rank; then we may take

$$B = (D_d - c(c^T c)^{-1} c^T) M$$

where D_d denotes the identity matrix of rank d. Hence,

$$\begin{aligned} &|B(\hat{\theta}_\psi)^T x^T W_\theta x B(\hat{\theta}_\psi)| \\ &\quad = |M^T (D_d - c(c^T c)^{-1} c^T) x^T W_\theta x (D_d - c(c^T c)^{-1} c^T) M|; \end{aligned}$$

the expression for L_M does not depend on the choice of M. Alternatively,

$$|B(\hat{\theta}_\psi)^T x^T W_\theta x B(\hat{\theta}_\psi)|$$

may be calculated as the product of the nonzero eigenvalues of

$$|(D_d - c(c^T c)^{-1} c^T) x^T W_\theta x|. \qquad \blacksquare$$

Example 9.20 Dispersion parameter

Let Y denote a multivariate normal random variable, of dimension n, with mean $x\lambda$ and covariance matrix $\Sigma(\psi)$. Here x is a known $n \times p$ matrix of constants, λ is an unknown $p \times 1$ parameter vector, and Σ is an $n \times n$ positive definite matrix that depends on an unknown vector parameter ψ, which is taken as the parameter of interest.

The log-likelihood function is given by

$$\ell(\theta) = -\tfrac{1}{2}\log|\Sigma(\psi)| - \tfrac{1}{2}(y - x\lambda)^T \Sigma(\psi)^{-1}(y - x\lambda).$$

Hence,

$$\ell_\lambda(\theta) = x^T \Sigma(\psi)^{-1}(y - x\lambda)$$

and

$$E[\ell_\lambda(\theta)\ell_\lambda(\theta_0)^T; \theta_0] = x^T \Sigma(\psi)^{-1} x.$$

It follows that

$$\begin{aligned}\bar{\ell}_M(\psi) = &- \tfrac{1}{2}\log|\Sigma(\psi)| - \tfrac{1}{2}\log|x^T \Sigma(\psi)^{-1} x| \\ &- \tfrac{1}{2}(y - x\hat{\lambda}_\psi)^T \Sigma(\psi)^{-1}(y - x\hat{\lambda}_\psi).\end{aligned}$$ ∎

9.5.5 AN APPROXIMATION BASED ON EMPIRICAL COVARIANCES

In Section 6.7.4 it was shown that, provided the data consist of n independent observations, the covariances used in forming $\bar{\ell}_M(\psi)$ can be replaced by empirical covariances without changing the order of the error of the approximation.

Let $\ell^{(j)}(\theta)$ denote the log-likelihood function based on the jth observation alone so that $\ell(\theta) = \sum \ell^{(j)}(\theta)$. Note that $I(\theta; \theta_0)$ may be written

$$I(\theta; \theta_0) = E_0\Big\{\sum \ell_\theta^{(j)}(\theta_0)\ell_\theta^{(j)}(\theta)^T\Big\}.$$

Hence, $I(\theta; \theta_0)$ may be approximated by

$$\hat{I}(\theta; \theta_0) = \sum \ell_\theta^{(j)}(\theta_0)\ell_\theta^{(j)}(\theta)^T.$$

An approximation to $\ell_M(\psi)$ may be constructed using $\hat{I}(\theta; \theta_0)$ in place of $I(\theta; \theta_0)$ in the approximation $\bar{\ell}_M(\psi)$. For the case in which an explicit nuisance parameter is available, this approximation is given by

$$\hat{\ell}_M(\psi) = \ell_p(\psi) + \log \frac{|j_{\lambda\lambda}(\hat{\theta}_\psi)|^{1/2}}{|\hat{I}_{\lambda;\lambda}(\hat{\theta}_\psi; \hat{\theta})|}$$

where

$$\hat{I}_{\lambda;\lambda}(\hat{\theta}_\psi; \hat{\theta}) = \sum \ell_\lambda^{(j)}(\hat{\theta})\ell_\lambda^{(j)}(\hat{\theta}_\psi)^T.$$

The argument given in Section 6.7.4 shows that

$$\hat{\ell}_M(\psi) = \ell_M(\psi) + O_p(n^{-1/2})$$

for fixed values of ψ and for ψ of the form $\psi = \hat{\psi} + O(n^{-1/2})$,

$$\hat{\ell}_M(\psi) = \ell_M(\psi) + O(n^{-1}).$$

Example 9.21 Exponential regression

Consider the exponential regression model considered in Examples 9.7 and 9.17. The log-likelihood function for this model is given by

$$\ell(\theta) = -n \log \lambda - \psi \sum x_j - \frac{1}{\lambda} \sum \exp\{-\psi x_j\} y_j$$

and hence

$$\ell_{\lambda}^{(j)}(\theta) = -\frac{1}{\lambda} + \frac{1}{\lambda^2} \exp\{-\psi x_j\} y_j.$$

It follows that

$$\hat{I}_{\lambda;\lambda}(\hat{\theta}_{\psi}; \hat{\theta}) = \frac{1}{\hat{\lambda}^2 \hat{\lambda}^2} \sum [\exp\{-\psi x_j\} y_j - \hat{\lambda}_{\psi}][\exp\{-\hat{\psi} x_j\} y_j - \hat{\lambda}].$$

Alternatively, we may write

$$\hat{I}_{\lambda;\lambda}(\hat{\theta}_{\psi}; \hat{\theta}) = \hat{\lambda}^2 \sum \exp\{(\hat{\psi} - \psi) x_j + a_j\} (\exp\{a_j\} - 1)$$

where $(a_1, \ldots, a_n)$ is the ancillary statistic described in Example 9.17.

Recall that

$$\ell_{\lambda;\hat{\lambda}}(\hat{\theta}_{\psi}) = \frac{n^2}{\hat{\lambda}^2} \frac{1}{\sum \exp\{(\psi - \hat{\psi}) x_j + a_j\}}.$$

Hence, there exists a constant c such that

$$\frac{\hat{I}_{\lambda;\lambda}(\hat{\theta}_{\psi})}{\ell_{\lambda;\hat{\lambda}}(\hat{\theta}_{\psi})} = c\left[1 + \frac{\sum (x_j - \bar{x}) \exp\{a_j\}(\exp\{a_j\} - 1)}{\sum \exp\{a_j\}(\exp\{a_j\} - 1)} \right.$$
$$\left. \times (\hat{\psi} - \psi) + O(|\hat{\psi} - \psi|^2)\right].$$

It is straightforward to show that

$$\frac{\sum \exp\{a_j\} (\exp\{a_j\} - 1)}{n} = 1 + O_p(n^{-1/2})$$

and that

$$\frac{\sum \exp\{a_j\} (\exp\{a_j\} - 1)(x_j - \bar{x})}{n} = \frac{\sum (x_j - \bar{x})}{n} + O_p(n^{-1/2})$$
$$= O_p(n^{-1/2}).$$

Hence,

$$\frac{\sum (x_j - \bar{x}) \exp\{a_j\} (\exp\{a_j\} - 1)}{\sum \exp\{a_j\} (\exp\{a_j\} - 1)} = O(n^{-1/2}),$$

so that

$$\frac{\hat{I}_{\lambda;\lambda}(\hat{\theta}_{\psi})}{\ell_{\lambda;\hat{\lambda}}(\hat{\theta}_{\psi})} = c + O(n^{-1}),$$

in accordance with the general results of this section. ∎

9.6 Discussion and references

The modified profile likelihood function was proposed by Barndorff-Nielsen (1983); see also Barndorff-Nielsen (1985, 1991a, 1994) and Barndorff-Nielsen and McCullagh (1993). A detailed discussion of the modified profile likelihood function is given in Barndorff-Nielsen and Cox (1994, Chapter 8). Alternative versions of the modified profile likelihood function are presented in Fraser and Reid (1989) and in Severini (2000a).

The adjusted profile likelihood function is due to Cox and Reid (1987); see also Cox and Reid (1992, 1993). The approximation to the modified profile likelihood based on approximating sample space derivatives by covariances was originally proposed by Barndorff-Nielsen (1995) for curved exponential family models and was extended to general models by Severini (1998a). This approach is also closely related to that used by DiCiccio and Stern (1993). Another method of approximating the modified profile likelihood function is given by Barndorff-Nielsen (1994) and is discussed in detail in Barndorff-Nielsen and Cox (1994, Chapter 8). McCullagh and Tibshirani (1990) proposed a different method of adjusting the profile likelihood function; their method involves modifying the profile likelihood so that it satisfies the Bartlett identities to a given order of approximation. Efron (1993) considers the construction of a pseudo-likelihood based on a method of constructing confidence intervals. The empirical approximation to the modified profile likelihood function given in Section 9.5.4 is based on Severini (1999b).

In general, the modified profile likelihood has properties closer to those of a genuine likelihood function. For instance, Ferguson *et al.* (1991) show that the score function based on the modified profile likelihood function has mean approximately zero. Mukerjee (1992) shows that the likelihood ratio test based on the modified profile likelihood has, to second order, the same power as the likelihood ratio test for the case in which λ is known; the approximations to the modified profile likelihood function discussed in this chapter also have this property. The extent to which the modified profile likelihood and related pseudo-likelihood functions satisfy the second Bartlett identity is considered in DiCiccio *et al.* (1996). A general theory of pseudo-likelihood functions is given in Severini (1998b).

As is the case with a conditional or marginal likelihood function, a signed likelihood ratio statistic may be based on the modified profile likelihood function. This approach is considered by Satori *et al.* (1999) who show that the properties of such a statistic are often superior to the properties of the usual signed likelihood ratio statistic.

One area of much recent research not discussed in this book is the construction of nonparametric versions of a pseudo-likelihood function. These methods include empirical likelihood (Owen 1988, 1990; Qin and Lawless 1994) and bootstrap likelihood (Davison, Hinkley, and Worton 1992; Davison and Hinkley, 1997).

Example 9.1 is due to Berger, Liseo, and Wolpert (1999).

9.7 Exercises

9.1 Consider a full-rank exponential family model in which the parameter of interest ψ is a canonical parameter. Calculate $L_M(\psi)$ and compare it to the approximate conditional likelihood function derived in Section 8.2.4.

9.2 Suppose that λ is a scalar parameter and that there exists a version of the log-likelihood function that depends on λ and $\hat{\lambda}$ only through $\hat{\lambda}/\lambda$. Find a simplified expression for $L_M(\psi)$ in this case.

9.3 Let $Y_1, \ldots, Y_n$ are independent random variables each distributed according to a Weibull distribution with rate parameter λ and index ψ. Find $\ell_M(\psi)$.

9.4 Consider p independent samples from gamma distributions with common index ψ and rate parameters $\lambda_1, \ldots, \lambda_p$. Find $\ell_M(\psi)$.

9.5 Show directly that $L_M(\psi)$ is invariant under interest-respecting reparameterizations.

9.6 Calculate $L_M(\psi)$ for the comparison of exponential distributions example discussed in Example 9.4. Compare the result to the marginal likelihood calculated in Example 9.4.

9.7 Calculate $L_M(\psi)$ for the special normal distribution considered in Example 9.1 and compare the result to the marginal likelihood function based on $y_1, \ldots, y_n$.

9.8 Calculate $L_M(\psi)$ for the ratio of normal means example considered in Examples 8.19 and 8.20. Compare the result to the conditional likelihood function. In particular, consider the cases $x = 0$, $y > 0$ and $y = 0$, $x > 0$.

9.9 Consider a model with a scalar parameter of interest ψ and a scalar nuisance parameter λ orthogonal to ψ. Determine an expansion with error $O(n^{-1})$ for $E[\ell_A'(\psi); \theta]$. Compare the result to the expansion for $E[\ell_p'(\psi); \theta]$ used in Section 5.3.2.

9.10 Consider n independent time-dependent Poisson processes each with rate function

$$\psi\lambda \exp\{\psi t\}, \quad t > 0$$

where $\psi > 0$ and $\lambda > 0$ are unknown parameters. Suppose that each process is observed until the first event occurs and let $Y_1, \ldots, Y_n$ denote the event times. Find $\tilde{\ell}_M(\psi)$.

9.11 Consider a normal-theory nonlinear regression model as in Example 9.16. Take the scale parameter as the parameter of interest. Find $\tilde{\ell}_M(\psi)$.

9.12 Calculate $\bar{\ell}_M(\psi)$ for the linear regression model considered in Example 9.5 and compare the result to $\ell_M(\psi)$.

9.13 Calculate $\bar{\ell}_M(\psi)$ for the normal-theory nonlinear regression model considered in Example 9.16.

9.14 Let $Y_1, \ldots, Y_n$ denote independent p-dimensional multivariate normal random variables each with mean vector μ and covariance matrix $\Sigma(\psi)$; here μ is an unknown parameter vector of length p and $\Sigma(\psi)$ is a $p \times p$ positive definite matrix that depends on unknown parameter ψ, which is taken as the parameter of interest. Find $\bar{\ell}_M(\psi)$.

Appendix
Data sets used in the examples

Data set 1

The data in Table A.1 represent the intervals between successive failures in the air conditioning equipment of four aircraft of a certain model. The units are operating hours of the aircraft.

Table A.1 Failure times of aircraft air-conditioning equipment

Aircraft			
1	2	4	8
194	359	50	487
15	9	254	18
41	12	5	100
29	270	283	7
33	603	35	98
181	3	12	5
	104		85
	2		91
	438		43
			230
			3
			130

Source: Cox and Lewis (1966), Table 1.2.

Data set 2

The data in Table A.2 represent the length in millimeters of 1000 specimens of the trypanosome protozoon. The observations were grouped in 1 mm intervals; for the analyses in this book however this grouping is ignored.

Table A.2 Trypanosome data

Length	Frequency
15	10
16	21
17	56
18	79
19	114
20	122
21	110
22	85
23	85
24	61
25	47
26	49
27	47
28	44
29	31
30	20
31	11
32	4
33	4

Source: Everitt and Hand (1981, Table 2.6).

Data set 3

The data in Table A.3 represent the failure times, in hours, of certain pressure vessels subjected to constant pressure.

Table A.3 Failure times of pressure vessels

274	1661	1787
28.5	236	970
1.7	828	.75
20.8	458	1278
871	290	776
363	54.9	126
1311	175	

Source: Keating, Glaser, and Ketchum (1990).

Data set 4

The data in Table A.4 represent the velocity of an enzymatic reaction as a function of substrate concentration. Velocity is measured as counts per minute of radioactive product from the reaction; substrate concentration is measured in parts per million.

Table A.4 Velocity of an enzymatic reaction

Concentration	Velocity
0.02	76
0.02	47
0.06	97
0.06	107
0.11	123
0.11	139
0.22	159
0.22	122
0.56	191
0.56	201
1.10	207
1.10	200

Source: Bates and Watts (1988), Data Set A1.3.

Data set 5

The data in Table A.5 represent the number of cans damaged in a boxcar shipment of cans and the speed of the boxcar at impact.

Table A.5 Damaged cans in railway cars

Speed	Cans
4	27
3	54
5	86
8	136
4	65
3	109
3	28
4	75
3	53
5	33
7	168
3	47
8	52

Source: Draper and Smith (1981), Chapter 1, problem I.

Data set 6

The data in Table A.4 represent independent times of failure for failures occurring on two electrical transmission circuits in New Mexico. The units are mile–years.

Table A.6 Failure times of transmission circuits

Circuit	
1	2
3.06	1.08
4.35	4.72
7.88	5.27
20.76	6.05
53.16	6.35
58.98	6.89
96.20	27.66
222.50	30.07
345.86	37.60
41.24	
101.77	
146.15	

Source: Martz and Waller (1982), Example 4.13.

Data set 7

The data in Table A.7 represent the time to death, in weeks after diagnosis, along with the white blood count (WBC), in thousands, of 33 leukemia patients.

Table A.7 Failure times of leukaemia patients

WBC	Failure time
2.3	65
0.8	156
4.3	100
2.6	134
6.0	16
10.5	108
10.0	121
17.0	4
5.4	39
7.0	143
9.4	56
32.0	26
35.0	22
100.0	1
100.0	1
52.0	5
100.0	65

Source: Feigl and Zelen (1965).

Data set 8

The data in Table A.8 represent the results of a small bioequivalence study. Each patient received three treatments, an *approved* treatment, a *new* treatment that is claimed to be identical to the approved treatment, and a placebo. The response is the blood level of a hormone measured after receiving the treatment. X denotes the difference between the *approved* measurement and the *placebo* measurement and Y denotes the difference between the *new* measurement and the *approved* measurement.

Table A.8 Results from a bioequivalence study

Patient	X	Y
1	8406	−1200
2	2342	2601
3	8187	−2705
4	8459	1982
5	4795	−1290
6	3516	351
7	4796	−638
8	10238	−2719

Source: Efron and Tibshirani (1993, Table 25.1).

Data set 9

The data in Table A.9 represent the breaking load in oz. of a certain type of worsted yarn. The tested yarn came from one of 6 bobbins.

Table A.9 Strength of worsted yarn

Bobbin					
1	2	3	4	5	6
15.0	15.7	14.8	14.9	13.0	15.9
17.0	15.6	14.2	14.2	16.2	15.6
13.8	17.6	15.0	15.0	16.4	15.0
15.5	17.1	12.8	12.8	14.8	15.5

Source: Cox and Snell (1981, Example Q).

Data set 10

The data in Table A.10 represent the weight of onions as a function of growing time.

Table A.10 Weight of onions versus growing time

Time	Weight
1	16.08
2	33.83
3	65.80
4	97.20
5	191.55
6	326.20
7	386.87
8	520.53
9	590.03
10	651.92
11	724.93
12	699.56
13	689.96
14	637.56
15	717.41

Source: Ratkowsky (1983, Data Set 2, Appendix 4.A).

References

Aitkin, M. and Stasinopoulos, M. (1989). Likelihood analysis of a binomial sample size problem. In *Contributions to probability and statistics*, eds. L. J. Gleser, M. D. Perlman, S. J. Press, and A. Sampson. New York: Springer.

Amari, S.-I. (1985). *Differential-geometric methods in statistics.* Heidelberg: Springer.

Andersen, E. B. (1970). Asymptotic properties of conditional maximum likelihood estimators. *Journal of the Royal Statistical Society* B **32**, 283–301.

Andersen, E. B. (1971). Asymptotic properties of conditional likelihood ratio tests. *Journal of the American Statistical Association* **66**, 630–3.

Andersen, E. B. (1973). *Conditional inference and models for measuring.* Copenhagen: Mentalhygiejnisk Forlag.

Andersen, E. B. (1980). *Discrete statistical models with social sciences applications.* Amsterdam: North-Holland.

Atkinson, K. E. (1978). *An introduction to numerical analysis.* New York: Wiley

Azzalini, A. (1996). *Statistical inference based on the likelihood.* London: Chapman and Hall.

Babu G. J. and Rao C. R. (1992). Expansions for statistics involving mean absolute deviations. *Annals of the Institute Statistics and Mathematics* **44**, 387–403.

Bahadur, R. R. (1954). Sufficiency and statistical decision functions. *Annals of Mathematical Statistics* **25**, 423–62.

Bahadur, R. R. (1958). Examples of inconsistency of maximum likelihood estimates. *Sankhyā* **20**, 207–10.

Bahadur, R. R. (1964). On Fisher's bound for asymptotic variances. *Annals of Mathematical Statistics* **35**, 1545–52.

Bahadur, R. R. (1971). *Some limit theorems in statistics.* Philadelphia: Society of Industrial Applications of Mathematics (SIAM).

Barnard, G. A. (1949). Statistical inference (with discussion). *Journal of the Royal Statistical Society* B **11**, 115–39.

Barnard, G. A. (1951). Theory of information (with discussion). *Journal of the Royal Statistical Society* B **13**, 46–64.

Barnard, G. A. (1976). Conditional inference is not inefficient. *Scandinavian Journal of Statistics* **3**, 132–4.

Barnard, G. A. (1982). Conditionality versus similarity in the analysis of 2×2 tables. *Statistics and probability: essays in honor of C. R. Rao.* Amsterdam: North-Holland, 59–65.

Barndard, G. A., Jenkins, G. M., and Winsten, C. B. (1962). Likelihood inference and time series (with discussion). *Journal of the Royal Statistical Society* A **125**, 321–75.

Barnard, G. A. and Sprott, D. A. (1971). A note on Basu's examples of anomalous ancillary statistics (with discussion). In *Foundations of statistical inference*, eds. V. P. Godambe and D. A. Sprott. Toronto: Holt, Rinehart, and Winston, 163–76.

Barndorff-Nielsen, O. E. (1973). On M-ancillarity. *Biometrika* **60**, 447–55.

Barndorff-Nielsen, O. E. (1976). Nonformation. *Biometrika* **63**, 567–71.

Barndorff-Nielsen, O. E. (1978). *Information and exponential families.* Chichester: Wiley.

Barndorff-Nielsen, O. E. (1980). Conditionality resolutions. *Biometrika* **67**, 293–310.

Barndorff-Nielsen, O. E. (1983). On a formula for the distribution of the maximum likelihood estimator. *Biometrika* **70**, 343–65.

Barndorff-Nielsen, O. E. (1984). On conditionality resolution and the likelihood ratio for curved exponential families. *Scandinavian Journal of Statistics* **11**, 157–70.

Barndorff-Nielsen, O. E. (1985). Properties of modified profile likelihood. In *Contributions to probability and statistics in honour of Gunnar Blum*, eds. J. Lanke and G. Lindgren. Dept. Math. Statist., Lund University, 25–38.

Barndorff-Nielsen, O. E. (1986). Inference on full and partial parameters based on the standardized signed log likelihood ratio. *Biometrika* **73**, 307–22.

Barndorff-Nielsen, O. E. (1988). *Parametric statistical models and likelihood.* Heidelberg: Springer-Verlag.

Barndorff-Nielsen, O. E. (1990a). Approximate interval probabilities. *Journal of the Royal Statistical Soceity* B **52**, 485–96.

Barndorff-Nielsen, O. E. (1990b). A note on the standardised signed log likelihood ratio. *Scandinavian Journal of Statistics* **17**, 157–60.

Barndorff-Nielsen, O. E. (1990c). p^* and Laplace's method. REBRAPE **4**, 89–103.

Barndorff-Nielsen, O. E. (1991a). Likelihood theory. In *Statistical theory and modelling*, eds. D.V. Hinkley, N. Reid, and E. J. Snell. London: Chapman and Hall.

Barndorff-Nielsen, O. E. (1991b). Modified signed log likelihood ratio. *Biometrika* **78**, 557–63.

Barndorff-Nielsen, O. E. (1994). Adjusted versions of profile likelihood and directed likelihood, and extended likelihood. *Journal of the Royal Statistical Society* B **56**, 125–40.

Barndorff-Nielsen, O. E. (1995). Stable and invariant adjusted profile likelihood and directed likelihood for curved exponential models. *Biometrika* **82**, 489–500.

Barndorff-Nielsen, O. E. and Blæsild, P. (1975). S-ancillarity in exponential families. *Sankhyā* **37**, 354–85.

Barndorff-Nielsen, O. E. and Blæsild, P. (1993). Likelihood yokes and stable combinations of cumulants. *Scandinavian Journal of Statistics* **20**, 83–8.

Barndorff-Nielsen, O. E., Blæsild, P., and Eriksen, P. S. (1989). *Decomposition and invariance of measures, and statistical transformation Models.* Heidelberg: Springer-Verlag.

Barndorff-Nielsen, O. E. and Chamberlin, S. R. (1991). An ancillary invariant modification of the signed log likelihood ratio. *Scandinavian Journal of Statistics* **18**, 341–52.

Barndorff-Nielsen, O. E. and Chamberlin, S. R. (1994). Stable and invariant adjusted directed likelihood. *Biometrika* **81**, 485–99.

Barndorff-Nielsen, O. E. and Cox, D. R. (1979). Edgeworth and saddle-point approximations with statistical applications (with discussion). *Journal of the Royal Statistical Society* B **41**, 279–312.

Barndorff-Nielsen, O. E. and Cox, D. R. (1984). Bartlett adjustments to the likelihood ratio statistic and the distribution of the maximum likelihood estimator. *Journal of the Royal Statistical Society* B **46**, 483–95.

Barndorff-Nielsen, O. E. and Cox, D. R. (1989). *Asymptotics techniques for use in statistics.* London: Chapman and Hall.

Barndorff-Nielsen, O. E. and Cox, D. R. (1994). *Inference and asymptotics.* London: Chapman and Hall.

Barndorff-Nielsen, O. E. and Hall, P. (1988). On the level-error after Bartlett adjustment of the likelihood ratio statistic. *Biometrika* **75**, 374–8.

Barndorff-Nielsen, O. E. and McCullagh, P. (1993). A note on the relation between modified profile likelihood and the Cox-Reid adjusted profile likelihood. *Biometrika* **80**, 321–8.

Barndorff-Nielsen, O. E. and Pedersen, K. (1968). Sufficient data reduction and exponential families. *Math. Scand.* **22**, 197–202.

Barndorff-Nielsen, O. E. and Sørensen, M. (1993). A review of some aspects of likelihood theory for stochastic processes. *International Statististical Review* **62**, .

Barndorff-Nielsen, O. E. and Wood, A. T. A. (1998). On large deviations and choice of ancillary for p^* and r^*. *Bernoulli* **4**, 35–63.

Bartlett, M. S. (1936). The information available in small samples. *Proceedings of the Cambridge Philosophical Society* **32**, 560–6.

Bartlett, M. S. (1937). Properties of sufficiency and statistical tests. *Proceedings of the Royal Society London Series A* **160**, 268–82.

Bartlett, M. S. (1938). Further aspects of the theory of multiple regression. *Proceedings of the Cambridge Philosophical Society* **34**, 33–40.

Bartlett, M. S. (1953a). Approximate confidence intervals. *Biometrika* **40**, 12–19.

Bartlett, M. S. (1953b). Approximate confidence intervals II: More than one unknown parameter. *Biometrika* **40**, 306–317.

Basu, D. (1955). On statistics independent of a complete sufficient statistic. *Sankhyā* **15**, 377–80.

Basu, D. (1958). On statistics independent of sufficient statistics. *Sankhyā* **20**, 223–6.

Basu, D. (1964). Recovery of ancillary information. *Sankhyā A* **26**, 3–16.

Basu, D. (1975). Statistical information and likelihood (with discussion). *Sankhyā A* **37**, 1–71.

Basu, D. (1977). On the elimination of nuisance parameters. *Journal of the American Statistical Association*, **72**, 355–66.

Basu, D. (1978). On partial sufficiency: a review. *J. Statist. Plann. Inf.* **2**, 1–13.

Bates, D. M. and Watts, D. G. (1989). *Nonlinear regression analysis and its applications.* New York: John Wiley and Sons.

Berger, J. O. (1985). *Statistical decision theory and Bayesian analysis*, 2nd edn. New York: Springer.

Berger, J. O., Liseo, B., and Wolpert, R. (1999). Integrated likelihood functions for eliminating nuisance parameters (with discussion). *Statistical Science* **14**, 1–28.

Berger, J. O. and Wolpert, R. L. (1988). *The likelihood principle*, 2nd edn. Hayward: IMS.

Bernardo, J. M. and Smith, A. F. M. (1994). *Bayesian theory.* Chichester: Wiley.

Bhapkar, V. P. (1989). Conditioning on ancillary statistics and loss of information in the presence of nuisance parameters. *J. Statist. Plann. Inference* **21**, 139–60.

Bhapkar, V. P. (1991). Loss of information in the presence of nuisance parameters and partial sufficiency. *J. Statist. Plann. Inference* **28**, 185–203.

Bhattacharya, R. N. and Ghosh, J. K. (1978). On the validity of the formal Edgeworth expansion. *Annals of Statistics* **6**, 434–51.

Bhattacharya, R. N. and Rao, R. R. (1976). *Normal approximation and asymptotic expansions.* New York: Wiley.

Bickel, P. and Doksum, K. (1977). *Mathematical statistics: basic ideas and selected topics.* Englewood Cliffs: Prentice-Hall.

Bickel, P. J. and Ghosh, J. K. (1990). A decomposition for the likelihood ratio statistic and the Bartlett correction—a Bayesian argument. *Annals of Statistics* **18**, 1070–90.

Birnbaum, A. (1962). On the foundations of statistical inference (with discussion). *Journal of the American Statistical Association* **57**, 269–306.

Birnbaum, A. (1970). On Durbin's modified principle of conditionality. *Journal of the American Statistical Association* **65**, 402–3.

Birnbaum, A. (1977). The Neyman–Pearson theory as decision theory and as inference theory: with a criticism of the Lindley-Savage argument for Bayesian theory. *Synthese* **36**, 19–49.

Blackwell, D. (1947). Conditional expectation and unbiased sequential estimation. *Annals of Mathematical Statistics* **18**, 105–10.

Blyth, C. (1990). Letter to the editor. *American Statistician* **44**, 329.

Box, G. E. P. (1949). A general distribution theory for a class of likelihood criteria. *Biometrika* **36**, 317–46.

Brown, L. (1964). Sufficient statistics in the case of independent random variables. *Annals of Mathematical Statistics* **35**, 1456–74.

Brown, L. D. (1988). *Fundamentals of statistical exponential families.* IMS lecture notes—monograph series 9. Hayward: IMS.

Buehler, R. J. (1982). Some ancillary statistics and their properties (with discussion). *Journal of the American Statistical Association* **77**, 581–94.

Cakmak, S. and Fraser, D. A. S. (1994). Multivariate asymptotic model: exponential and location approximations. *Utilitas Mathematica* **14**, 21–31.

Casella, G. and Berger, R. L. (1990). *Statistical inference.* Pacific Grove: Wadsworth.

Chernoff, H. (1954). On the distribution of the likelihood ratio. *Annals of Mathematical Statistics* **25**, 573–8.

Cordeiro, G. M. (1987). On the corrections to the likelihood ratio statistis. *Biometrika* **74**, 265–74.

Cox, D. R. (1958). Some problems connected with statistical inference. *Annals of Mathematical Statistics*, **29**, 357–72.

Cox, D. R. (1971). The choice between alternative ancillary statistics. *Journal of the Royal Statistical Society* B **33**, 251–5.

Cox, D. R. (1975). Partial likelihood. *Biometrika* **62**, 269–76.

Cox, D. R. (1980). Local ancillarity. *Biometrika* **67**, 279–86.

Cox, D. R. (1988). Some aspects of conditional and asymptotic inference: A review. *Sankhyā A* **50**, 314–37.

Cox, D. R. and Hinkley, D. V. (1974). *Theoretical statistics.* London: Chapman and Hall.

Cox, D. R. and Lewis, P. A. W. (1966). *The statistical analysis of series of events.* London: Methuen.

Cox, D. R. and Reid, N. (1987). Parameter orthogonality and approximate conditional inference. *Journal of the Royal Statistical Society* B **49**, 1–39.

Cox, D. R. and Reid, N. (1992). A note on the difference between profile and modified profile likelihood. *Biometrika* **79**, 408–11.

Cox, D. R. and Reid, N. (1993). A note on the calculation of adjusted profile likelihood. *Journal of the Royal Statistical Society* B **55**, 467–71.

Cox, D. R. and Snell, E. J. (1981). *Applied statistics: principles and examples.* London: Chapman and Hall.

Cramér, H. (1946). *Mathematical methods of statistics.* Princeton: Princeton University Press.

Daniels, H. E. (1954). Saddlepoint approximations in statistics. *Annals of Mathematical Statistics* **25**, 631–50.

Daniels, H. E. (1980). Exact saddlepoint approximations. *Biometrika* **67**, 53–8.

Daniels, H. E. (1983). Saddlepoint approximations for estimating equations. *Biometrika* **70**, 89–96.

Daniels, H. E. (1987). Tail probability approximations. *International Statististical Review* **55**, 37–48.

Davison, A. C. (1988). Approximate conditional inference in generalized linear models. *Journal of the Royal Statistical Society* B **50** 445–61.

Davison, A. C. and Hinkley, D. V. (1997). *Bootstrap methods and their applications.* Cambridge: Cambridge University Press.

Davison, A. C., Hinkley, D. V., and Worton, B. J. (1992). Bootstrap likelihoods. *Biometrika* **79**, 113–30.

Dawid, A. P. (1975). On the concepts of sufficiency and ancillarity in the presence of nuisance parameters. *Journal of the Royal Statistical Society* B **37**, 248–58.

Dawid, A. P. (1980). A Bayesian look at nuisance parameters. *Bayesian Statistics*, eds. J. M. Bernardo, M. H. DeGroot, D. V. Lindley, and A. F. M. Smith. Valencia: University Press, 167–203.

DiCiccio, T. (1984). On parameter transformations and interval estimation. *Biometrika* **71**, 477–85.

DiCiccio, T. J. (1986). Approximate conditional inference for location families. *Canadian Journal of Statistics* **14**, 5–18.

DiCiccio, T. J. (1988). Likelihood inference for linear regression models. *Biometrika* **75**, 29–34.

DiCiccio, T. J. and Efron, B. (1992). More accurate confidence intervals in exponential families. *Biometrika* **79**, 231–45.

DiCiccio, T. J., Field, C. A., and Fraser, D. A. S. (1990). Approximations of marginal tail probabilities and inference for scalar parameters. *Biometrika* **77**, 77–95.

DiCiccio, T. J., Hall, P., and Romano, J. P. (1989). Comparison of parametric and empirical likelihood functions. *Biometrika* **76**, 465–76.

DiCiccio, T. J. and Martin, M. A. (1991). Approximations of marginal tail probabilities for a class of smooth functions with applications to Bayesian and conditional inference. *Biometrika* **78**, 891–902.

DiCiccio, T. J. and Martin, M. A. (1993). Simple modifications for signed roots of likelihood ratio statistics. *Journal of the Royal Statistical Society* B **55**, 305–16.

DiCiccio, T. J., Martin, M. A., Stern, S. E., and Young, G. A. (1996). Information bias and adjusted profile likelihoods. *Journal of the Royal Statistical Society* B **58**, 189–203.

DiCiccio, T. J. and Stern, S. E. (1993). An adjustment to the profile likelihood based on observed information. *Technical report*, Department of Statistics, Stanford University.

DiCiccio, T. J. and Stern, S. E. (1994). Constructing approximating standard normal pivots from signed roots of adjusted likelihood ratio statistics. *Scandinavian Journal of Statistics* **21**, 447–60.

Draper, N. R. and Smith, H. (1981). *Applied regression analysis*, 2nd edn. New York: Wiley.

Durbin, J. (1970). On Birnbaum's theorem on the relation between sufficiency, conditionality and likelihood. *Journal of the American Statistical Association* **65**, 395–8.

Durbin, J. (1980). Approximations for densities of sufficient estimators. *Biometrika* **67**, 311–33.

Eaton, M. L. (1988). *Group invariance applications in statistics.* Hayward: IMS.

Edwards, A. W. F. (1972). *Likelihood.* Cambridge: Cambridge University Press.

Edwards, A. W. F. (1974). The history of likelihood. *International Statististical Review* **42**, 9–15.

Efron, B. (1975). Defining the curvature of a statistical problem (with applications to second order efficiency. *Annals of Statistics* **3**, 1189–242.

Efron, B. (1978). The geometry of exponential families. *Annals of Statistics* **6**, 362–76.

Efron, B. (1993). Bayes and likelihood calculations from confidence intervals. *Biometrika* **80**, 3–26.

Efron, B. and Hinkley, D. V. (1978). Assessing the accuracy of the maximum likelihood estimator: observed versus expected Fisher information (with discussion). *Biometrika* **65**, 457–87.

Efron, B. and Tibshirani, R. J. (1993). *An introduction to the bootstrap.* London: Chapman and Hall.

Evans, M., Fraser, D. A. S., and Monette, G. (1986). On principles and arguments to likelihood. *Canadian Journal of Statistics* **14**, 181–99.

Everitt, B. S. and Hand, D. J. (1981). *Finite mixture distributions.* London: Chapman and Hall.

Feigl, P. and Zelen, M. (1965). Estimation of exponential survival probabilities with concomitant information. *Biometrics* **21**, 826–38.

Ferguson, H., Cox, D. R., and Reid, N. R. (1991). Estimating equations from modified profile likelihood. In *Estimating functions*, ed. V. P. Godambe. Oxford: Clarendon.

Ferguson, T. (1996). *A course in large sample theory.* London: Chapman and Hall.

Field, C. A. (1982). Small sample asymptotic expansions for multivariate M-estimates. *Annals of Statistics* **10**, 672–89.

Field, C. and Ronchetti, E. (1990). *Small sample asymptotics.* IMS lecture notes—monograph series 13. Hayward: IMS.

Fisher, R. A. (1912). On an absolute criterion for fitting frequency curves. *Mess. Math.* **41**, 155–60.

Fisher, R. A. (1922). On the mathematical foundations of theoretical statistics. *Phil. Trans. Roy. Soc. A* **222**, 309–68.

Fisher, R. A. (1925). Theory of statistical estimation. *Proceedings of the Cambridge Philosophical Society* **22**, 700–25.

Fisher, R. A. (1934). Two properties of mathematical likelihood. *Proceedings of the Royal Society Series A* **144**, 285–307.

Fraser, D. A. S. (1956). Sufficient statistics with nuisance parameters. *Annals of Mathematical Statistics* **27**, 838–42.

Fraser, D. A. S. (1964). Local conditional sufficiency. *Journal of the Royal Statistical Society* B **26**, 52–62.

Fraser, D. A. S. (1968). *The structure of inference.* New York: Wiley.

Fraser, D. A. S. (1979). *Inference and linear models.* New York: McGraw-Hill.

Fraser, D. A. S. (1990). Tail probabilities from observed likelihoods. *Biometrika* **77**, 65–76.

Fraser, D. A. S. (1991). Statistical inference: likelihood to significance. *Journal of the American Statistical Association* **86**, 258–65.

Fraser, D. A. S., Monette, G., and Ng, K. W. (1984). Marginalization, likelihood, and structural models. In *Multivariate analysis VI*, ed. P. R. Krishnaiah. Amsterdam: North-Holland.

Fraser, D. A. S. and Reid, N. (1988). On conditional inference for a real parameter: a differential approach on the sample space. *Biometrika* **75**, 251–64.

Fraser, D. A. S. and Reid, N. (1989). Adjustments to profile likelihood. *Biometrika* **76**, 477–88.

Fraser, D. A. S. and Reid, N. (1993). Simple asymptotic connections between densities and cumulant functions leading to accurate approximations for distribution functions. *Statistica Sinica* **3**, 67–82.

Fraser, D. A. S. and Reid, N. (1995). Ancillaries and third-order significance. *Utilitas Mathematica* **47**, 33–53.

Fraser, D. A. S. and Reid, N. (1999). Ancillary information for statistical inference. In *Proceedings of a CRM Symposium on Empirical Bayes and Likelihood Inference*, eds. E. Ahmad and N. Reid. New York: Springer-Verlag.

Fraser, D. A. S., Reid, N., and Wu, J. (1999). A simple formula for tail probabilities for frequentist and Bayesian inference. *Biometrika* **86**, 249–64.

Frydenberg, M. and Jensen, J. L. (1989). Is the 'improved likelihood ratio statistic' really improved in the discrete case? *Biometrika* **76**, 655–61.

Gelman, A., Carlin, J. B., Stern, H. S., and Rubin, D. B. (1995). *Bayesian data analysis.* London: Chapman and Hall.

Ghosh, J. K. (ed.) (1988). *Statistical information and likelihood.* Lecture Notes in Statistics **45**. Heidelberg: Springer-Verlag.

Ghosh, J. K. (1994). *Higher order asymptotics.* Hayward: IMS.

Godambe, V. P. (1976). Conditional likelihood and optimum estimating equations. *Biometrika* **63**, 277–84.

Godambe, V. P. (1980), On sufficiency and ancillarity in the presence of a nuisance parameter, *Biometrika*, **67**, 155–62.

Goldstein, M. and Howard, J. V. (1991). A likelihood paradox (with comments). *Journal of the Royal Statistical Society* B **53**, 619–28.

Haberman, S. J. (1977). Maximum likelihood estimation in exponential response models. *Annals of Statistics* **5**, 815–41.

Haldane , J. B. S. and Smith, S. M. (1956). The sampling distribution of a maximum likelihood estimate. *Biometrika* **43**, 96–103.

Hayakawa, T. (1977). The likelihood ratio criterion and the asymptotic expansion of its distribution. *Annals of the Institute of Statistics and Mathematics* **29**, 359–78.

Hill, B. M. (1963). The three-parameter lognormal distribution and Bayesian analysis of a point-source epidemic. *Journal of the American Statistical Association* **58**, 72–84.

Hinkley, D. V. (1970). Inference about the change-point in a sequence of random variables. *Biometrika* **57**, 1–17.

Hinkley, D. V. (1977). Conditional inference about a normal mean with known coefficient of variation. *Biometrika* **64**, 105–8.

Hinkley, D. V. (1978). Likelihood inference about location and scale parameters. *Biometrika* **65**, 253–61.

Hinkley, D. V. (1980a). Likelihood. *Canadian Journal of Statistics* **8**, 151–63.

Hinkley, D. V. (1980b). Likelihood as approximate pivotal distribution. *Biometrika* **67**, 287–92.

Hinkley, D. V. (1982). Can frequentist inference be very wrong? A conditional 'yes'. In *Scientifice inference, data analysis and robustness*, eds. G. E. P. Box, T. Leonard, C. F. Wu. New York: Academic Press.

Hougaard, P. (1985). Saddlepoint approximations for curved exponential families. *Statistics and Probability Letters* **3**, 161–6.

Huber, P. J. (1967). The behavior of maximum likelihood estimates under non-standard conditions. *Proceedings of the Fifth Berkeley Symposium on Mathematical Statistics and Probability* **1**, 221–34, University California Press.

Huzurbazar, V. S. (1950). Probability distributions and orthogonal parameters. *Proceedings of the Cambridge Philosophical Society* **46**, 281–4.

Huzurbazar, V. S. (1956). Sufficient statistics and orthogonal parameters. *Sankhyā A* **17**, 217–20.

Ibragimov, I. A. and Has'minskii, R. Z. (1981). *Statistical estimation: asymptotic theory.* New York: Springer-Verlag.

Jensen, J. L. (1986). Similar tests and the standardized log-likelihood ratio statistic. *Biometrika* **73**, 567–72.

Jensen, J. L. (1987). Standardized log-likelihood ratio statistics for mixtures of discrete and continuous observations. *Annals of Statistics* **15**, 314–24.

Jensen, J. L. (1992). The modified signed likelihood ratio and saddlepoint approximations. *Biometrika* **79**, 693–703.

Jensen, J. L. (1993). A historical sketch and some new results on the improved log likelihood ratio statistic. *Scandinavian Journal of Statistics* **1993**, 1–15.

Jensen, J. L. (1995). *Saddlepoint approximations.* Oxford: Oxford University Press.

Jensen, J. L. (1997). A simple derivation of r^* for curved exponential families. *Scandinavian Journal of Statistics* **24**, 33–46.

Johnson, N. L. and Kotz, S. (1970a). *Continuous univariate distributions—I.* New York: Wiley.

Johnson, N. L. and Kotz, S. (1970b). *Continuous univariate distributions—II.* New York: Wiley.

Johnson, N. L. and Kotz, S. (1972). *Distributions in statistics: continuous multivariate distributions.* New York: Wiley.

Jørgensen, B. (1993). A review of conditional inference: is there a universal definition of nonformation? *Bulletin of the International Statistics Institute* **55**, 323–40.

Joshi, V. M. (1989). A counter-example against the likelihood principle. *Journal of the Royal Statistical Society* B **51**, 215–16.

Kabaila, P. (1998). A note on sufficiency and information loss. *Statistics and Probability Letters* **37**, 111–14.

Kalbfleisch, J. D. (1975). Sufficiency and conditionality. *Biometrika* **62**, 251–68.

Kalbfleisch, J. D. (1985). *Probability and statistical inference, Volume 2: statistical inference.* New York: Springer-Verlag.

Kalbfleisch, J. D. and Sprott, D. A. (1970). Application of likelihood methods to models involving large numbers of parameters (with discussion). *Journal of the Royal Statistical Society* B **32**, 175–208.

Kalbfleisch, J. D. and Sprott, D. A. (1973). Marginal and conditional likelihoods. *Sankhya* A **35**, 311–28.

Kalianpur, G. and Rao, C. R. (1955). On Fisher's lower bound to the asymptotic variance of a consistent estimate. *Sankhyā* **15**, 331–42.

Kass, R. E., Tierney, L., and Kadane, J. B. (1990). The validity of posterior expansions based on Laplace's method. In *Bayesian and likelihood methods in statistics and econometrics: essays in honor of George A. Barnard*, eds. S. Geisser, J. S. Hodges, S. J. Press, and A. Zellner. Amsterdam: North-Holland, 473–88.

Kass, R. E. and Vos, P. W. (1997). *Geometrical foundations of asymptotic inference.* New York: Wiley.

Kass, R. E. and Wasserman, L. (1996). Formal rules for selecting prior distributions. *Journal of the American Statistical Association* **91**, 1343–70.

Keating, J. P., Glaser, R. E., and Ketchum, N. S. (1990). Inference about the shape parameter of a gamma distribution. *Technometrics* **32**, 67–82.

Kendall, M. G. and Stuart, A. (1977). *The advanced theory of statistics*, 4th edn. New York: Macmillan.

Kiefer, J. and Wolfowitz, J. (1956). Consistency of the maximum likelihood estimator in the presence of infinitely many nuisance parameters. *Annals of Mathematical Statistics* **27**, 887–906.

Kolassa, J. E. (1994). *Series approximation methods in statistics.* Lecture notes in statistics **88**. New York: Springer-Verlag.

Kullback, S. (1959). *Information theory and statistics.* New York: Wiley.

Kumon, M. and Amari, S. (1983). Geometrical theory of higher-order asymptotics of test, interval estimator, and conditional inference. *Proceedings of the Royal Society London Series A* **387**, 429–58.

Lawley, D. N. (1956). A general method for approximating the distribution of likelihood ratio criteria. *Biometrika* **43**, 295–303.

LeCam, L. (1953). On some asymptotic properties of maximum likelihood estimates and related Bayes estimates. *University California Publications in Statistics* **1**, 277–330.

LeCam, L. (1970). On the assumptions used to prove asymptotic normality of maximum likelihood estimates. *Annals of Mathematical Statistics* **41**, 802–28.

LeCam, L. (1986). *Asymptotic methods in statistical decision theory.* Berlin: Springer-Verlag.
Lehmann, E. L. (1981). An interpretation of completeness and Basu's theorem. *Journal of the American Statistical Association* **76**, 335–40.
Lehmann, E. L. (1983). *Theory of point estimation.* New York: Wiley.
Lehmann, E. L. (1986). *Testing statistical hypotheses*, 2nd edn. New York: Wiley.
Liang, K. Y. (1983). On information and ancillarity in the presence of a nuisance parameter. *Biometrika* **70**, 607–12.
Liang, K. Y. (1984). The asymptotic efficiency of conditional likelihood methods. *Biometrika* **71**, 305–13.
Lindsay, B. G. (1980). Nuisance parameters, mixture models, and the efficiency of partial likelihood estimators. *Phil. Trans. Roy. Soc.* **296**, 639–65.
Lindsay, B. G. (1983a). The geometry of mixture likelihoods: a general theory. *Annals of Statistics* **11**, 86–94.
Lindsay, B. G. (1983b). Efficiency of the conditional score in a mixture setting. *Annals of Statistics* **11**, 486–97.
Lindsay, B. G. (1995). *Mixture models: theory, geometry and applications.* Hayward: IMS.
Lindsey, J. K. (1996). *Parametric statistical inference.* Oxford: Oxford University Press.
Liseo, B. (1992). A note on a counterexample against the likelihood principle. *Communications in Statistical Theory and Methods* **21**, 547–56.
Liseo, B. (1993). Elimination of nuisance parameters with reference priors. *Biometrika* **80**, 295–304.
Lloyd, C. J. (1985). Ancillaries sufficient for the sample size. *Australian Journal of Statistics* **3**, 264–72.
Lloyd, C. J. (1992). Effective conditioning. *Australian Journal of Statistics* **34**, 241–60.
Lugannini, R. and Rice, S. (1980). Saddlepoint approximation for the distribution of the sum of independent random variables. *Advances in Applied Probability* **12**, 475–90.
Mann, H. B. and Wald, A. (1943). On stochastic limit and order relationships. *Annals of Mathematical Statistics* **14**, 217–26.
Martz, H. F. and Waller, R. A. (1983). *Bayesian reliability analysis.* New York: Wiley.
McCullagh, P. (1984). Local sufficiency. *Biometrika* **71**, 233–44.
McCullagh, P. (1987). *Tensor methods in statistics*. London: Chapman and Hall.
McCullagh, P. and Cox, D. R. (1986). Invariants and likelihood ratio statistics. *Annals of Statistics* **14**, 1419–30.
McCullagh, P. and Nelder, J. (1989). *Generalized linear models*, 2nd edn. London: Chapman and Hall.
McCullagh, P. and Tibshirani, R. (1990). A simple method for the adjustment of profile likelihoods. *Journal of the Royal Statistical Society* B **52**, 325–44.
Michel, R. (1979). Asymptotic expansions for conditional distributions. *Journal of Multivariate Analysis* **9**, 393–400.
Mukerjee, R. (1992). Comparison between the conditional likelihood ratio test and the usual likelihood ratio test. *Journal of the Royal Statistical Society* B **54**, 184–96.
Neyman, J. and Scott, E. L. (1948). Consistent estimates based on partially consistent observations. *Econometrica* **16**, 1–32.
Osborne, J. A. and Severini, T. A. (1999). Inference for exponential order statistic models based on an integrated likelihood. To appear in *Journal of The American Statistical Association.*
Owen, A. B. (1988). Empirical likelihood ratio confidence intervals for a single functional. *Biometrika* **75**, 237–49.
Owen, A. B. (1990). Empirical likelihood ratio confidence intervals. *Annals of Statistics* **18**, 90–120.

Pace, L. and Salvan, A. (1992). A note on conditional cumulants in canonical exponential families. *Scandinavian Journal of Statistics* **19**, 185–91.

Pace, L. and Salvan, A. (1997). *Principles of statistical inference.* Singapore: World Scientific.

Pedersen, B. V. (1981). A comparison of the Efron-Hinkley ancillary and the likelihood ratio ancillary in a particular example. *Annals of Statistics* **9**, 1328–33.

Peers, H. W. (1978). Second-order sufficiency and statistical invariants. *Biometrika* **65**, 489–96.

Peers, H. W. and Iqbal, M. (1985). Asymptotic expansions for confidence limits in the presence of nuisance parameters, with applications. *Journal of the Royal Statistical Society* B **47**, 547–54.

Perlman, M. (1970). On the strong consistency of approximate maximum likelihood estimators. *Proceedings of the Sixth Berkeley Symposium on Mathematical Statistics and Probability.*

Pfanzagl, J. (1979). Asymptotic expansions in parametric statistical theory. In *Developments in statistics*, ed. P. R. Krishnaiah. New York: Academic, 1–97.

Pfanzagl, J. (1993). Incidental versus random nuisance parameters. *Annals of Statistics* **21**, 1663–91.

Pierce, D. A. and Peters, D. (1992). Practical use of higher order asymptotics for multiparameter exponential families. *Journal of the Royal Statistical Society* B **54**, 701–37.

Pierce, D. A. and Peters, D. (1994). Higher-order asymptotics and the likelihood principle: One-parameter models. *Biometrika* **81**, 1–10.

Proschan, F. (1963), Theoretical explanation of observed decreasing failure rate, *Technometrics*, **5**, 375–83.

Qin, J. and Lawless, J. (1994). Empirical likelihood and general estimating equations. *Annals of Statistics* **22**, 300–25.

Rao, C. R. (1945). Information and accuracy attainable in the estimation of statistical parameters. *Bulletin of the Calcutta Mathematical Society* **37**, 81–91.

Rao, C. R. (1947). Large sample tests of statistical hypotheses concerning several parameters with applications to problems of estimation. *Proceedings of the Cambridge Philosophical Society* **44**, 40–57.

Rao, C. R. (1973). *Linear statistical inference and its applications*, 2nd edn. New York: Wiley.

Ratkowsky, D. A. (1983). *Nonlinear regression modelling: a unified practical approach.* New York: Marcel Dekker.

Reid, N. (1988). Saddlepoint methods and statistical inference. *Statistical Science* **3**, 213–38.

Reid, N. (1995). The roles of conditioning in inference. *Statistical Science* **10**, 138–57.

Reid, N. (1996). Likelihood and higher-order approximations to tail areas: a review and annotated bibliography. *Canadian Journal of Statistics* **24**, 141–66.

Remon, M. (1984). On a concept of partial sufficiency: L-sufficiency. *International Statististical Review* **52**, 127–36.

Royall, R. (1997). *Statistical evidence.* London: Chapman and Hall.

Ryall, T. A. (1981). Extensions of the concept of local ancillarity. *Biometrika* **68**, 677–83.

Sandved, E. (1965). A principle for conditioning on an ancillary statistic. *Skand. Aktuar. Tidskr.* **49**, 39–47.

Sandved, E. (1972). Ancillary statistics in models without and with nuisance parameters. *Skand. Aktuar. Tidskr.* **56**, 81–91.

Satori, N., Bellio, R., Salvan, A. and Pace, L. (1999). The directed modified profile likelihood in models with many parameters. *Biometrika* **86**, 735–42.

Savage, L. J. (1970). *The foundations of statistics.* New York: Dover.

Schervish, M. J. (1997). *Theory of statistics.* New York: Springer-Verlag.

Sen, P. K. and Singer, J. M. (1993). *Large sample methods in statistics.* London: Chapman and Hall.

Severini, T. A. (1989). Conditional properties of likelihood-based significance tests. *Biometrika* **77**, 343–52.

Severini, T. A. (1993a). Local ancillarity in the presence of a nuisance parameter. *Biometrika*, **80**, 305–20.

Severini, T. A. (1993b). A note on sufficiency and conditional inference. *Statistics and Probability Letters* **17**, 303–5.

Severini, T. A. (1994). On the approximate elimination of nuisance parameters by conditioning. *Biometrika* **81**, 649–61.

Severini, T. A. (1995). Comment on 'The roles of conditioning in inference' by N. Reid. *Statistical Science* **10**, 187–9.

Severini, T. A. (1998a). An approximation to the modified profile likelihood function. *Biometrika* **85**, 403–11.

Severini, T. A. (1998b). Likelihood functions for the elimination of nuisance parameters. *Biometrika* **85**, 507–22.

Severini, T. A. (1999a). On the relationship between Bayesian and non-Bayesian elimination of nuisance parameters. *Statistica Sinica*, to appear.

Severini, T. A. (1999b). An empirical adjustment to the likelihood ratio statistic. *Biometrika* **86**, 235–47.

Severini, T. A. (2000a). Modifications to the profile likelihood function. *Technical Report*, Department of Statistics, Northwestern University.

Severini, T. A. (2000b). The likelihood ratio approximation to the conditional distribution of the maximum likelihood estimate in the discrete case. *Technical Report*, Department of Statistics, Northwestern University.

Severini, T. A. (2000c). Approximation of sample space derivatives. *Technical Report*, Department of Statistics, Northwestern University.

Severini, T. A. and Wong, W. H. (1992). Profile likelihood and conditionally parametric models. *Annals of Statistics* **20**, 1768–802.

Sham, P. (1998). *Statistics in human genetics*. London: Arnold.

Shenton, L. R. and Bowman, K. O. (1977). *Maximum likelihood estimation in small samples*. London: Griffin.

Skates, S. J. (1993). On secant approximations to cumulative distribution functions. *Biometrika* **80**, 223–35.

Skovgaard, I. M. (1981a). Transformation of an Edgeworth expansion by a sequence of smooth functions. *Scandinavian Journal of Statistics* **8**, 207–17.

Skovgaard, I. M. (1981b). Edgeworth expansions for the distributions of maximum likelihood estimators in the general (non-i.i.d.) case. *Scandinavian Journal of Statistics* **8**, 227–36.

Skovgaard, I. M. (1985). A second order investigation of asymptotic ancillarity. *Annals of Statistics* **13**, 534–51.

Skovgaard, I. M. (1986). A note on the differentiation of cumulants of log-likelihood derivatives. *International Statististical Review* **54**, 29–32.

Skovgaard, I. M. (1987). Saddlepoint expansions for conditional distributions. *Journal of Applied Probability* **24**, 875–87.

Skovgaard, I. M. (1990). On the density of minimum contrast estimators. *Annals of Statistics* **18**, 779–89.

Skovgaard, I. M. (1996). An explicit large-deviation approximation to one-parameter tests. *Bernoulli* **2**, 145–65.

Skovgaard, I. M. (2000). Likelihood asymptotics. To appear in *Scandinavian Journal of Statistics*.

Sprott, D. A. (1975). Marginal and conditional sufficiency, *Biometrika*, **62**, 599–605.

Stein, C. (1956). Efficient nonparametric testing and estimation. *Proceedings of the Third Berkeley Symposium on Mathematical Statistics and Probability, 1954–1955, Vol. 1*, 197–195, Berkeley: University of California Press.

Sverdrup, E. (1966). The present state of decision theory and Neyman–Pearson theory. *Reveue of the Institute International Statistics* **34**, 309–33.

Sweeting, T. J. (1980). Uniform asymptotic normality of the maximum likelihood estimator. *Annals of Statistics* , 1375–81.

Sweeting, T. J. (1992). Parameter-based asymptotics. *Biometrika* **79**, 219–30.

Sweeting, T. J. (1995). Bayesian and likelihood approximations. *Biometrika* **82**, 1–23.

Temme, N. M. (1982). The uniform asymptotic expansion of a class of integrals related to cumulative distribution functions. *SIAM Journal of Mathematical Analysis* **13**, 239–53.

Thisted, R. A. (1988). *Elements of statistical computing*. London: Chapman and Hall.

Tjur, T. (1980). *Probability based on radon measures*. New York: Wiley.

Tweedie, M. C. K. (1957). Statistical properties of inverse Gaussian distributions I. *Annals of Mathematical Statistics* **28**, 362–77.

van der Vaart, A. W. (1988). Estimating a real parameter in a class of semiparametric models. *Annals of Statistics* **16**, 1450–74.

Viveros, R. and Sprott, D. A. (1986). Conditional inference and maximum likelihood in a capture-recapture model. *Communications in Statistics: Theory and Methods*, **15**, 1035–46.

Wald, A. (1943). Tests of statistical hypotheses concerning several parameters when the number of observations is large. *Transactions of the American Mathematics Soceity* **54**, 426–82.

Wald, A. (1949). Note on the consistency of the maximum likelihood estimator. *Annals of Mathematical Statistics* **20**, 595–601.

Welch, B. L. (1939). On confidence limits and sufficiency with special reference to parameters of location. *Annals of Mathematical Statistics* **10**, 58–69.

Wilks, S. S. (1938). The large-sample distribution of the likelihood ratio for testing composite hypotheses. *Annals of Mathematical Statistics* **9**, 60–2.

Wong, W. H. (1986). Theory of partial likelihood. *Annals of Statistics* **14** 88–123.

Wong, W. H. (1992). On asymptotic efficiency in estimation theory. *Statistica Sinica* **2**, 47–68.

Yao, Y. and Davis R. A. (1986). The asymptotic behavior of the likelihood ratio for testing a shift in mean in a sequence of independent normal variates. *Sankhyā A*, **48**, 339–53.

Zhu, Y. and Reid, N. (1994). Information, ancillarity, and sufficiency in the presence of nuisance parameters. *Canadian Journal of Statistics* **22**, 111–23.

Author index

Aitkin, M. 320
Amari, S.-I. 23, 171, 320
Andersen, E. B. 24, 319
Atkinson, K. E. 68
Azzalini, A. 69

Babu, G. J. 134
Bahadur, R. R. 23, 133, 134
Barnard, G. A. 24, 100
Barndorff-Nielsen, O. E. 23, 68, 69, 101, 134, 171, 234, 235, 249, 275, 319, 320, 352
Bartlett, M. S. 100, 134, 171
Basu, D. 24, 101, 319
Bates, D. M. 356
Bellio, R. 352
Berger, J. O. 23, 101, 102, 320, 352
Berger, R. L. 23
Bernardo, J. M. 102
Bhapkar, J. M. 101, 319
Bhattacharya, R. N. 68
Bickel, P. J. 23, 101, 171
Birnbaum, A. 101
Blackwell, D. 24
Blæsild, P. 23, 235, 319, 320
Blyth, C. 102
Bowman, K. O. 171
Box, G. E. P. 171
Brown, L. D. 23, 24
Buehler, R. J. 24

Cakmak, S. 275
Carlin, J. B. 102
Casella, G. 23
Chamberlin, S. R. 275
Chernoff, H. 134
Cordeiro, G.M. 171
Cox, D. R. 24, 68, 69, 101, 102, 134, 171, 234, 235, 249, 275, 319, 320, 352, 354, 360

Daniels, H. E. 68, 234
Davis, R. A. 134
Davison, A. C. 234, 352
Dawid, A. P. 319, 320
DiCiccio, T. J. 134, 234, 274, 275, 320, 352
Doksum, K. 23, 101
Draper, N. R. 357
Durbin, J. 68, 101, 234

Eaton, M. L. 23
Edwards, A. W. F. 100, 102
Efron, B. 23, 24, 101, 134, 212, 234, 235, 352, 360
Eriksen, P. S. 23, 320
Evans, M. 101
Everitt, B. S. 355

Feigl, P. 359
Ferguson, H. 352
Ferguson, T. 68, 101, 133, 134
Field, C. A. 68, 171, 234
Fisher, R. A. 100, 234
Fraser, D. A. S. 23, 234, 235, 275, 319, 320, 352
Frydenberg, M. 171

Gelman, A. 102
Ghosh, J. K. 101, 171
Glaser, R. E. 356
Godambe, V. P. 319
Goldstein, M. 101

Haberman, S. J. 320
Haldane, J. B. S. 171
Hall, P. 134, 171
Hand, D. J. 355
Has'minskii, R. Z. 134
Hayakawa, T. 171
Hill, B. M. 134
Hinkley, D. V. 23, 24, 101, 102, 133, 134, 234, 235, 352
Hougaard, P. 171
Howard, J. V. 101
Huber, P. J. 133
Huzurbazar, V. S. 101

Ibragimov, I. A. 133
Iqbal, M. 171

Jenkins, G. M. 100
Jensen, J. L. 68, 69, 171, 234, 235, 274, 275
Johnson, N. L. 24
Jørgensen, B. 319

Joshi, V. M. 101

Kabaila, P. 101
Kadane, J. B. 69
Kalbfleisch, J. D. 24, 101, 102, 319, 320
Kalianpur, G. 134
Kass, R. E. 23, 69, 171, 320
Keating, J. P. 356
Kendall, M. G. 23
Ketchum, N. S. 356
Kiefer, J. 319
Kolassa, J. E. 68
Kotz, S. 24
Kullback, S. 101
Kumon, M. 320

Lawless, J. 352
Lawley, D. N. 171, 274
LeCam, L. 133, 134
Lehmann, E. L. 24, 101, 133
Lewis, P. A. W. 354
Liang, K. Y. 319
Lindsay, B. G. 320
Lindsey, J. K. 23
Liseo, B. 101, 320, 352
Lloyd, C. J. 24
Lugannini, R. 68, 234

Mann, H. B. 69
Martin, M. A. 234, 275, 352
Martz, H. F. 358
McCullagh, P. 23, 24, 68, 101, 171, 234, 235, 274, 352
Michel, R. 69
Monette, G. 101
Mukerjee, R. 352

Nelder, J. 23
Neyman, J. 133, 319
Ng, K. W. 101

Osborne, J. A. 320
Owen, A. B. 352

Pace, L. 23, 69, 319, 352
Pedersen B. V. 235
Pedersen, K. 24
Peers, H. W. 171, 234
Perlman, M. 133
Peters, D. 101, 275
Pfanzagl, J. 171, 320
Pierce, D. A. 101, 275

Qin, J. 352

Rao, C. R. 24, 101, 134
Rao, R. R. 68
Ratkowsky, D. A. 361
Reid, N. 24, 68, 101, 234, 235, 275, 319, 320, 352
Remon, M. 319
Rice, S. 68, 234
Romano, J. P. 134
Ronchetti, E. 68, 171
Royall, R. 102
Rubin, D. B. 102
Ryall, T. A. 235

Salvan, A. 23, 69, 319, 352
Sandved, E. 319
Satori, N. 352
Savage, L. J. 101
Schervish, M. J. 133
Scott, E. L. 133, 319
Severini, T. A. 101, 234, 235, 275, 319, 320, 352
Shenton, L. R. 171
Singer, J. M. 68, 133
Skates, S. J. 68
Skovgaard, I. M. 69, 171, 234, 235, 275
Smith, A. F. M. 102
Smith, H. 357
Snell, E. J. 360
Sørensen, M. 101
Sprott, D. A. 24, 319, 320
Stasinopoulos, M. 320
Stein, C. 134
Stern, H. S. 102
Stern, S. E. 320, 352
Stuart, A. 23
Sverdrup, E. 319
Sweeting, T. J. 134

Temme, N. M. 68
Thisted, R. A. 102
Tibshirani, R. J. 352, 360
Tierney, L. 69
Tjur, T. 320
Tweedie, M. C. K. 320

van der Vaart, A. W. 319
Viveros, R. 320
Vos, P. W. 23, 171

Wald, A. 69, 133, 134
Waller, R. A. 358
Wasserman, L. 320
Watts, D. G. 356
Welch, B. L. 24
Wilks, S. S. 134
Winsten, C. B. 100
Wolfowitz, J. 320

Wolpert, R. L. 101, 320, 352
Wong, W. H. 134, 320
Wood, A. T. A. 275
Worton, B. J. 352
Wu, J. 235, 275

Yao, Y. 134
Young, G. A. 352

Zelen, M. 359
Zhu, Y. 101, 319

Subject index

adjusted profile likelihood 341
ancillarity
  approximate
    Efron-Hinkley 210–12
    likelihood ratio 217
    location 212–15
    pivotal 215–17
  general 19–22, 175
  local 208, 211–12, 214–15, 216, 217
  in the presence of a nuisance parameter
    approximate 285
    P-ancillarity 283–5
    S-ancillarity 282–4, 326–7
asymptotic distribution 27

Barndorff-Nielsen's formula *see* likelihood ratio approximation
Bartlett identities 86–8, 101, 171
Basu's theorem 22, 24
Bayesian inference 102, 320
Bernoulli random variables 4–5, 7, 78
binomial distribution 13–14, 247–8, 316–17
bootstrap 232, 352

capture–recapture model 281, 307–8, 335–7
central limit theorem 28–30
chi-square distribution
  approximation of 35, 48
  for quadratic forms 23
conditional distribution
  approximation of 58–66
  of log-likelihood derivatives 176–83
  of maximum likelihood estimates 183–96
  in multivariate normal 23
conditionality principle 20
confidence region 123–5
convergence
  in distribution 27
  in probability 28
  in quadratic mean 69
covariate 3
cumulants
  definition 11
  of expansions 140–1
  formal 57
  generalized 16
  generating function 11, 14
  joint 14
  relationship to moments 11–12
  stable combinations 196–201, 235
  standardized 12
curvature, statistical 200, 237

differential element 294, 304
direct approximation 38
dispersion model, exponential 25

Edgeworth series approximations
  of densities 31–5, 36–7
  of distribution functions 32–5
  to the distribution of the likelihood ratio statistic 156, 161
  to the distribution of the maximum likelihood estimate 143
  lattice case 33–5
  multivariate case 36–7
empirical likelihood 352
equivalent statistics 17
equivariance 9
Euler–MacLaurin formula 34, 68
exponential distribution
  comparison of 162–3, 191–2, 253–4, 260–1, 314, 327
  hyperbola 209–10, 215, 216, 217, 224–5, 229, 233–4
  matched samples 301–2
  one-parameter 57–8, 61, 109, 113, 119, 122, 124, 143–4, 145–6, 157, 159–60,166, 170–1, 244–5
  regression 11, 84–5, 302–3, 333–5, 343, 345–7, 351
  several 116–17
  two-parameter 131–2
exponential model
  approximation of conditional likelihood 289–92
  canonical statistic of 6, 22
  conditional likelihood in 289–92
  consistency of maximum likelihood estimates 106–7
  cumulant generating function of 13, 15
  curved 4
  full 4
  general 4–7

exponential model (*cont.*)
mean parameter of 6
natural parameter of 6, 24
orthogonal parameter in 94
regression 7
sufficient statistic of 18

Fisher's exact test 317
Fisher information *see* information, expected

gamma distribution 11, 62–3, 67, 148–9, 153–4, 280, 282, 283, 285, 290–1
geometric distribution 195, 196
group 8–9

Hermite polynomials 32, 37

incidental nuisance parameters 286–7
index parameter 10
indirect approximation 38
information
expected 89–91, 101, 103, 134, 201
inequality 91
limiting average 109
observed 89–91, 101, 103, 134, 201
partial 91–2, 104, 283
inverse Gaussian distribution 299, 300, 310, 335, 342

Kullback–Leibler divergence 101, 105

Laplace approximation 66–8, 69
Laplace distribution 12–13, 35, 39, 48, 97, 107, 129–30
large deviation 33, 218, 328,
latent variable 103
lattice distribution 33–5, 49–53, 166–7, 194–6, 240
least favorable direction 134
likelihood
conditional 279–98, 307–8, 313–17
equation 96
general 73–5, 100
integrated 306–9, 318–19
marginal 298–305, 308–9, 309–12
principle
counterexample to 79–80
strong 77–80
weak 80
properties of 75–7
relationship between integrated and marginal and conditional 307–8
likelihood ratio approximation
marginalization of 191–3
to the distribution of the maximum likelihood estimate 165, 193, 194–6, 234
to the distribution of the signed likelihood ratio statistic 241, 249–51
likelihood ratio statistic
as an ancillary statistic 217
asymptotic distribution 113–15
Bartlett correction 156–60, 161–3, 170, 171
cumulants of 154–63
distribution under local alternative 117–19
higher-order asymptotic distribution 156, 161
saddlepoint approximation 169–71
stability of 203–4
linear model 189–91, 192–3
local alternative 117
local model 209, 210–12, 212–15
location model 176, 177, 181, 184, 212–15
location-scale model 182–3
logistic
distribution 158
growth model 267–8, 273
regression 74–5
log-likelihood derivatives
asymptotic properties 88
Bartlett identities for 86–7
conditional cumulants 176–83
cumulants 141–2
relationship between conditional and unconditional cumulants 197–200
log odds ratio 54, 55–6, 57
log-normal distribution 132–3
Lugannani–Rice formula 46–8, 51–2, 242
Lyapounov's condition 29, 85

maximal invariant statistic
ancillarity of 21
definition 7
maximum likelihood estimate
asymptotic normality of 108–12, 133–4
bias correction of 142
conditional cumulants of 178–83
conditional distribution of 183–96
consistency of 105–8, 133
cumulants of 141–54
general 96–7
saddlepoint approximation 163–8, 171
sandwich estimate of variance 135
stability of 201–3
measurement model 2, 5
model
conditional
approximation of 204–9
definition 175
function 1
nonregular 129–33

with many nuisance parameters 286–9
parametric 2
regular 80–5
see also exponential model, regression model, transformation model
moderate deviation 33, 218, 328
modified profile likelihood
approximations to 340–51
calculation without an explicit nuisance parameter 337–40
definition 328
derivation 329–31
modified signed likelihood ratio statistic
approximations to 261–74
based on a conditional likelihood 316–17
based on a marginal likelihood 310–11
calculation without an explicit nuisance parameter 257–61
derivation of
for one parameter models 241–2
for a parameter of interest 248–51
general 238–9
large-deviation properties 242–4
for lattice distributions 246–8
multinomial distribution 15–16, 21, 30, 299–300

negative binomial distribution 306, 318
Newton's method 134
Neyman–Pearson lemma 101
normal distribution
change point in 130–1
dispersion parameter 349–50
distribution theory 22–3
effect size 339–40
inference for mean 98–9, 100, 127, 306–7
inference for standard deviation 147, 153, 299, 309–12
with known coefficient of variation 186–7, 188, 246
mixture of 274
ratio of means 256–7, 293–4, 295–6, 303–5, 312, 317
regression
linear 92, 251–3, 259–60, 264–5, 281–2, 284, 285, 297–8, 305, 324–6, 331–2, 338–9, 341–2
nonlinear 75, 110–11, 111–12, 210, 215, 217, 371–2, 344
several
with common mean 5–6, 268–9, 270, 347–8
with common standard deviation 107–8, 286, 288–9, 332–3
special 324
two measuring instruments 20, 180–1, 200–1, 203, 204, 245–6

O_p, o_p notation 55, 69

p^* formula *see* likelihood ratio approximation
parameterization
identifiable 3
interest-respecting 93
orthogonal 93–4, 94–5, 137
partial likelihood 320
partially tilted approximation 63
pivotal quantity 209, 215–16
Poisson distribution
comparison of 315–16
one-sample 17–18, 29–30, 36, 50, 52–3, 65–6, 99, 100, 158–9, 167, 168, 183
regression 116, 122–3, 144–5, 280–1, 282, 283, 290, 308, 326–7, 348–9
Poisson process
stationary 2–3, 10, 74
time-dependent 82–4, 286–7
profile likelihood 126–9, 323–7
pseudo-likelihood 279

R^* *see* modified signed likelihood ratio statistic
random effects model 287
Rao–Blackwell theorem 19, 24
Rao statistic *see* score statistic
regression model 3–4, 10
repeated sampling principle 79
response variable 3

saddlepoint approximation
continuity correction 52
of densities 37–9, 49–50, 53–4
of distribution functions 46–8, 50–2
to the distribution of the maximum likelihood estimate 163–5
to the distribution of the likelihood ratio statistic 169–70
for lattice variables 49–53
for multivariate distributions 53–4
renormalization 38–9
sample space derivatives
approximation of
based on an approximate ancillary statistic 219–25
based on covariances 225–9
based on empirical covariances 230–4
general 176–7, 218
score statistic
asymptotic distribution 120–3
confidence region based on 124–6
conditional cumulants 178–9, 181–3
distribution under a local alternative 120
test statistic 98

signed likelihood ratio statistic
asymptotic distribution' of 117, 121
cumulants of 154–6
definition 100
higher-order asymptotic distribution of 156–9
modified *see* modified signed likelihood ratio statistic
Slutsky's theorem 55
stochastic asymptotic expansion 54–7
sufficiency
approximate 207–9
complete 22
definition 16
factorization theorem 16
in the presence of nuisance parameters
approximate 300
G-sufficiency 302
P-sufficiency 300–1
S-sufficiency 300–1, 326–7
principle 18, 23

Temme's method 40–4, 68
transformation model
composite 10, 191, 302–3
definition 9–10
transformations
normalizing 239–40
skewness-reducing 70
triangular array 28–9

uniform distribution 21, 103

Wald statistic
asymptotic distribution 120–3
confidence region based on 124–5
distribution under a local alternative 120
test based on 98
see also maximum likelihood estimate
Weibull distribution 90, 95, 125–6, 254–6, 266–7, 273, 296–7